COMPRESSOR HANDBOOK

COMPRESSOR HANDBOOK

Paul C. Hanlon Editor

McGRAW-HILL
New York San Francisco Washington, D.C. Auckland Bogotá
Caracas Lisbon London Madrid Mexico City Milan
Montreal New Delhi San Juan Singapore
Sydney Tokyo Toronto

Library of Congress Cataloging-in-Publication Data

Compressor handbook / Paul C. Hanlon, editor.
    p.   cm.
   Includes index.
   ISBN 0-07-026005-2
   1. Compressors—Handbooks, manuals, etc. I. Hanlon, Paul C.

TJ990.C623   2001
621.5′1—dc21
                                                                                       00-051129

McGraw-Hill

A Division of The **McGraw·Hill** Companies

Copyright © 2001 by The McGraw-Hill Companies, Inc. All rights reserved. Printed in the United States of America. Except as permitted under the United States Copyright Act of 1976, no part of this publication may be reproduced or distributed in any form or by any means, or stored in a data base or retrieval system, without the prior written permission of the publisher.

6 7 8 9 10 11 12 IBT/IBT 1 9 8 7 6 5 4 3 2 1 0

ISBN-13 978-0-07-026005-4
ISBN 0-07-026005-2

The sponsoring editor for this book was Linda Ludewig and the production supervisor was Sherri Souffrance. It was set in Times Roman by Pro-Image Corporation.

McGraw-Hill books are available at special quantity discounts to use as premiums and sales promotions, or for use in corporate training programs. For more information, please write to the Director of Special Sales, Professional Publishing, McGraw-Hill, Two Penn Plaza, New York, NY 10121-2298. Or contact your local bookstore.

 This book is printed on recycled, acid-free paper containing a minimum of 50% recycled, de-inked fiber.

Information contained in this book has been obtained by The McGraw-Hill Companies, Inc., ("McGraw-Hill") from sources believed to be reliable. However, neither McGraw-Hill nor its authors guarantee the accuracy or completeness of any information published herein and neither McGraw-Hill nor its authors shall be responsible for any errors, omissions, or damages arising out of use of this information. This work is published with the understanding that McGraw-Hill and its authors are supplying information, but are not attempting to render engineering or other professional services. If such services are required, the assistance of an appropriate professional should be sought.

CONTENTS

Contributors   vii
Preface   ix

Chapter 1. Compressor Theory   *Derek Woollatt*   1.1

Chapter 2. Compressor Performance—Positive Displacement
Derek Woolatt and Fred Heidrich   2.1

Chapter 3. Compressor Performance—Dynamic   *Paolo Bendinelli,*
Massimo Camatti, Marco Giachi, Eugenio Rossi, and Nuovo Pignone   3.1

Chapter 4. Centrifugal Compressors—Construction and Testing   *Ted Gresh*   4.1

Chapter 5. Compressor Analysis   *Harvey Nix*   5.1

Chapter 6. Compressor and Piping System Simulation   *Larry E. Blodgett*   6.1

Chapter 7. Very High Pressure Compressors (over 100 MPa [14500 psi])
Enzo Giacomelli, Alessandro Traversari, and Nuovo Pignone   7.1

Chapter 8. CNG Compressors   *Mark Epp*   8.1

Chapter 9. Liquid Transfer/Vapor Recovery   *William A. Kennedy Jr.*   9.1

Chapter 10. Compressed Natural Gas for Vehicle Fueling
Adam Weisz-Margulescu, P. Eng. 10.1

Chapter 11. Gas Boosters *Karl-Heinz Bark* 11.1

Chapter 12. Scroll Compressors *Robert W. Shaffer* 12.1

Chapter 13. Straight Lobe Compressors *A.G. Patel, PE* 13.1

Chapter 14. The Oil-Flooded Rotary Screw Compressor *Hasu Gajjar* ... 14.1

Chapter 15. Diaphragm Compressors *G. Reighard* 15.1

Chapter 16. Rotary Compressor Seals *James Netzel* 16.1

Chapter 17. Reciprocating Compressor Sealing *Paul Hanlon* 17.1

Chapter 18. Compressor Lubrication *Glen Majors, P.E.* 18.1

Chapter 19. Principles of Bearing Design *Hooshang Heshmat, Ph.D. and H. Ming Chen, Ph.D., P.E.* 19.1

Chapter 20. Compressor Valves *Walter J. Tuymer and Dr. Erich H. Machu* ... 20.1

Chapter 21. Compressor Control Systems *Robert J. Lowe* 21.1

Chapter 22. Compressor Foundations *Robert L. Rowan, Jr.* 22.1

Chapter 23. Packaging Compressors *Judith E. Vera* 23.1

Appendix   A.1
Index   I.1

CONTRIBUTORS

Bark, Karl-Heinz *MaxPro Technologies* (CHAPTER 11 GAS BOOSTERS)

Bendinelli, Paolo *Turbocompressors Chief Engineer, Nuovo Pignone* (CHAPTER 3 COMPRESSOR PERFORMANCE—DYNAMIC)

Blodgett, Larry E. *Southwest Research Institute* (CHAPTER 6 COMPRESSOR AND PIPING SYSTEM SIMULATION)

Camatti, Massimo *Turbocompressors Design Manager, Nuovo Pignone* (CHAPTER 3 COMPRESSOR PERFORMANCE—DYNAMIC)

Chen, H. Ming, Ph.D., P.E. *Mohawk Innovative Technology, Inc.* (CHAPTER 19 PRINCIPLES OF BEARING DESIGN)

Epp, Mark *Jenmar Concepts* (CHAPTER 8 CNG COMPRESSORS)

Gajjar, Hasu *Weatherford Compression* (CHAPTER 14 THE OIL-FLOODED ROTARY SCREW COMPRESSOR)

Giachi, Marco *Turbocompressors R&D Manager, Nuovo Pignone* (CHAPTER 3 COMPRESSOR PERFORMANCE—DYNAMIC)

Giacomelli, Enzo *General Manager Reciprocating Compressors, Nuovo Pignone* (CHAPTER 7 VERY HIGH PRESSURE COMPRESSORS)

Gresh, Ted *Elliott Company* (CHAPTER 4 CENTRIFUGAL COMPRESSORS—CONSTRUCTION AND TESTING)

Hanlon, Paul *C. Lee Cook, A Dover Resources Company* (CHAPTER 17 RECIPROCATING COMPRESSOR SEALING)

Heidrich, Fred *Dresser-Rand Company* (CHAPTER 2 COMPRESSOR PERFORMANCE—POSITIVE DISPLACEMENT)

Heshmat, Hooshang, Ph.D. *Mohawk Innovative Technology, Inc.* (CHAPTER 19 PRINCIPLES OF BEARING DESIGN)

Kennedy, William A., Jr. *Blackmer/A Dover Resource Company* (CHAPTER 9 LIQUID TRANSFER/VAPOR RECOVERY)

Lowe, Robert J. *T. F. Hudgins, Inc.* (CHAPTER 21 COMPRESSOR CONTROL SYSTEMS)

Machu, Erich H. *Consulting Mechanical Engineer, Hoerbiger Corporation of America, Inc.* (CHAPTER 20 COMPRESSOR VALVES)

Majors, Glen, P.E. *C.E.S. Associates, Inc.* (CHAPTER 18 COMPRESSOR LUBRICATION)

Netzel, James *Chief Engineer, John Crane Inc.* (CHAPTER 16 ROTARY COMPRESSOR SEALS)

Nix, Harvey *Training-n-Technologies* (CHAPTER 5 COMPRESSOR ANALYSIS)

Patel, A.G., PE *Roots Division, Division of Dresser Industries Inc.* (CHAPTER 13 STRAIGHT LOBE COMPRESSORS)

Reighard, G. *Howden Process Compressors, Inc.* (CHAPTER 15 DIAPHRAGM COMPRESSORS)

Rossi, Eugenio *Turbocompressors Researcher, Nuovo Pignone* (CHAPTER 3 COMPRESSOR PERFORMANCE—DYNAMIC)

Rowan, Robert L., Jr. *Robert L. Rowan & Associates, Inc.* (CHAPTER 22 COMPRESSOR FOUNDATIONS)

Shaffer, Robert W. *President, Air Squared, Inc.* (CHAPTER 12 SCROLL COMPRESSORS)

Tuymer, Walter J. *Hoerbiger Corporation of America, Inc.* (CHAPTER 20 COMPRESSOR VALVES)

Traversari, Alessandro *General Manager Rotating Machinery, Nuovo Pignone* (CHAPTER 7 VERY HIGH PRESSURE COMPRESSORS)

Vera, Judith E. *Project Engineer, Energy Industries, Inc.* (CHAPTER 23 PACKAGING COMPRESSORS)

Weisz-Margulescu, Adam, P. Eng. *FuelMaker Corporation* (CHAPTER 10 COMPRESSED NATURAL GAS FOR VEHICLE FUELING)

Woollatt, Derek *Manager, Valve and Regulator Engineering, Dresser-Rand Company & (Screw Compressor Section)* (CHAPTER 1 COMPRESSOR THEORY; CHAPTER 2 COMPRESSOR PERFORMANCE—POSITIVE DISPLACEMENT)

PREFACE

Compressors fall into that category of machinery that is "all around us" but of which we are little aware. We find them in our homes and workplaces, and in almost any form of transportation we might use. Compressors serve in refrigeration, engines, chemical processes, gas transmission, manufacturing, and in just about every place where there is a need to move or compress gas.

The many engineering disciplines (e.g. fluid dynamics, thermodynamics, tribology, and stress analysis) involved in designing and manufacturing compressors make it impossible to do much more than just "hit the high spots," at least in this first edition.

This is such a truly broad field, encompassing so many types and sizes of units, that it is difficult to cover it all in one small volume, representing the work of relatively few authors. Possibly, more than anything else, it will open the door to what must follow—a larger second edition.

In compressors, the areas of greatest concern are those parts with a finite life, such as bearings, seals and valves, or parts that are highly stressed. Treatment of these components takes up a large portion of the handbook, but at the same time space has been given to theory, applications and to some of the different types of compressors.

Much in this handbook is based on empirical principals, so this should serve as a practical guide for designers and manufacturers. There are also test and analysis procedures that all readers will find helpful. There should be something here for anyone who has an interest in compressors.

Paul C. Hanlon

CHAPTER 1
COMPRESSOR THEORY

Derek Woollatt
Manager, Valve and Regulator Engineering
Dresser-Rand Company

1.1 NOMENCLATURE

		Units (See note below)
a	Speed of Sound in Gas	ft/sec
a, b	Constants in Equation of State (Pressure Form)	
A, B	Constants in Equation of State (Compressibility Form)	
B	Bore	Inch
CL	Fixed Clearance as a Fraction of Swept Volume	
c_P	Specific Heat at Constant Pressure	ft.lb$_f$/lb$_m$.R
c_V	Specific Heat at Constant Volume	ft.lb$_f$/lb$_m$.R
e	Specific Internal Energy (i.e. Internal Energy per unit mass)	ft.lb$_f$/lb$_m$
E	Internal Energy	ft.lb$_f$
F	Flow Area	Inch^2
h	Specific Enthalpy (i.e. Enthalpy per unit mass)	ft.lb$_f$/lb$_m$
H	Enthalpy	ft.lb$_f$
HP	Horsepower	
J	Joule's Equivalent	ft.lb$_f$/BTU
k	Ratio of Specific Heats ($= c_P/c_V$)	
m	Mass Flow Rate	lb$_m$/sec
M	Mass	lb$_m$
n_T	Isentropic Temperature Exponent	
n_V	Isentropic Volume Exponent	
N	Compressor Speed	rpm
P	Pressure	lb$_f$/in^2 Abs
PW	Power	ft.lb$_f$/min

q	Heat Transfer Rate	BTU/sec
Q	Heat Transfer	BTU
R	Gas Constant	ft.lb$_f$/lb$_m$.R
s	Specific Entropy	ft.lb$_f$/lb$_m$.R
S	Stroke	Inch
T	Temperature	Rankine
u	Gas Velocity	ft/sec
U_P	Piston Velocity	ft/min
v	Specific Volume (Volume per unit mass)	ft^3/lb$_m$
V	Volume	Inch^3
W	Work Done on Gas during Process	ft.lb$_f$
WD	Work Done on Gas During one compressor cycle	ft.lb$_f$
Z	Compressibility (sometimes called Supercompressibility)	
ΔP	Pressure Drop	lb$_f$/in^2
ρ	Density	lb$_m$/ft^3
θ	Crank Angle	Degree
λ	Integration Constant in expression for Average Pressure Drop	

Suffixes

C	Critical Pressure or Temperature
D	Discharge from the Compressor Cylinder
eq	Equivalent (Area)
in	At Entry to a Control Volume
o	Stagnation Value
out	At Exit from a Control Volume
R	Reduced (Pressure or Temperature)
SW	Swept (Swept Volume is Maximum minus Minimum Cylinder Volume)
1, 2, 3, 4	At Corresponding Points in the Cycle (Fig. 1.2)
1, 2	Before and after process

NOTE:. The basic equations given in this section can be used with any consistent system of units. The units given above are not consistent and the numerical factors required to use the equations with the above units are given at the end of each equation in square brackets. If the above units are used, the equations can be used as written. If an alternate, consistent, system of units is used, the numerical factors at the ends of the equations should be ignored.

1.2 THEORY

1.2.1 Gas Laws

By definition, compressors are intended to compress a substance in a gaseous state. In predicting compressor performance and calculating the loads on the various components, we need methods to predict the properties of the gas. Process compressors are used to compress a wide range of gases over a wide range of conditions. There is no single **equation of state** (an equation that allows the density of a gas to be calculated if the pressure and temperature are known) that will be accurate for all gases under all conditions. Some of the commonly used ones, starting with the most simple, are discussed below.

The simplest equation of state is the perfect gas law:

$$Pv = \frac{P}{\rho} = RT \left[\frac{1}{144}\right]$$

This equation applies accurately only to gases when the temperature is much higher than the critical temperature or the pressure much lower than the critical pressure. Air at atmospheric conditions obeys this law well.

To predict the properties of real gases more accurately, the perfect gas law is often modified by the addition of an empirical value "Z", called the **compressibility**, or sometimes the **supercompressibility**, of the gas. The value of Z is a function of the gas composition and the pressure and temperature of the gas. The modified equation is:

$$\frac{P}{\rho} = ZRT \left[\frac{1}{144}\right]$$

This equation is accurate if, and only if, Z is known accurately. Z can be estimated with reasonable accuracy in many cases using the **Law of Corresponding States** which states that the value of Z as a function of the reduced pressure and temperature is **approximately** the same for all gases. That is:

$$Z = fn\,(P_R, T_R) = fn\left(\frac{P}{P_C}, \frac{T}{T_c}\right)$$

A curve of Z as a function of reduced pressure and temperature is shown as Fig. 1.1. This gives reasonable results for most gases when the gas state is not close to the critical point or the two phase region.

It is frequently useful to have an equation to predict Z. This allows calculation of other properties such as entropy, enthalpy and isentropic exponents that are needed to predict compressor performance. The use of an equation rather than charts is also convenient when a computer is used to perform the calculations. Many equations are available: one of the most simple, the **Redlich-Kwong Equation of State** is given below. Other equations are more accurate over a wider range of gases and conditions, but are more complex. Some of these are discussed in Refs. 2 and 3. The Redlich-Kwong equation of state is:

$$P = \left(\frac{RT}{v - b} - \frac{a}{v^2 + bv}\right)\left[\frac{1}{144}\right]$$

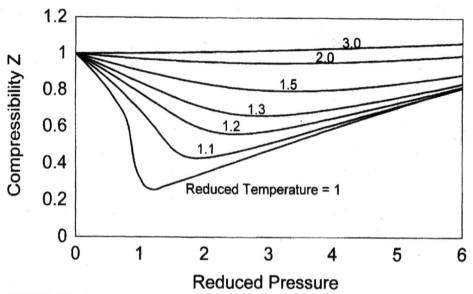

FIGURE 1.1 Compressibility chart (based on the Redlich Kwong equation of state).

where $a = 0.42748 \dfrac{R^2 T_C^{2.5}}{P_C T^{0.5}} \left[\dfrac{1}{144}\right]$

$b = 0.08664 \dfrac{R T_C}{P_C} \left[\dfrac{1}{144}\right]$

or

$$Z^3 - Z^2 + (A - B - B^2)Z - AB = 0$$

where $A = 0.42748 \dfrac{P_R}{T_R^{2.5}}$

$B = 0.08664 \dfrac{P_R}{T_R}$

Solving the above cubic equation for Z once P_R and T_R are known is equivalent to looking up the value of Z on Fig. 1.1.

Other equations of state commonly used in predicting compressor performance include the Soave Redlich Kwong, Peng Robinson, Benedict Webb Rubin, Han Starling, Lee-Kesler, and API Method equations. Details of these methods can be found in the literature (e.g. Refs. 2 and 3).

1.2.2 Thermodynamic Properties

To predict compressor performance ways to calculate the enthalpy, internal energy and entropy of the gas are needed. It is also often convenient to use the **isentropic volume exponent** n_V and the **isentropic temperature exponent** n_T.

The isentropic exponents are defined such as to make the following equations true for an isentropic change of state.

$$PV^{n_V} = \text{Constant}$$

$$\frac{P^{\frac{n_T - 1}{n_T}}}{T} = \text{Constant}$$

For a **perfect gas**, the above properties are easily calculated. The following is for a gas that obeys the ideal gas laws and has constant specific heats. Specific properties are those per unit mass of gas.

$$\text{Specific Internal Energy} = e = c_v T$$

$$\text{Specific Enthalpy} = h = c_p T$$

$$n_V = n_T = c_p/c_V = k$$

$$\text{Change of Specific Entropy} = s_2 - s_1 = c_P \ln\left(\frac{T_2}{T_1}\right) - R \ln\left(\frac{P_2}{P_1}\right)$$

For a **real gas**, the above properties can be obtained from a Mollier chart for the gas or from the equation of state and a knowledge of how the specific heats at low pressure vary with temperature. Methods for this are given in Refs. 2 and 3.

An **approximation** that allows isentropic processes to be calculated easily for a real gas if the Z values are known is often useful. Consider an isentropic change of state from 1 to 2.

$$\frac{\rho_2}{\rho_1} = \frac{Z_1}{Z_2} \frac{P_2}{P_1} \frac{T_1}{T_2} = \left(\frac{P_2}{P_1}\right)^{1/n_V}$$

$$\frac{T_1}{T_2} = \left(\frac{P_1}{P_2}\right)^{n_T - 1/n_T}$$

$$\therefore \left(\frac{P_2}{P_1}\right)^{1/n_V} = \frac{Z_1}{Z_2} \left(\frac{P_2}{P_1}\right)^{1/n_T}$$

It is found that if the gas state is not too near the critical or two phase region, and is therefore acting somewhat like an ideal gas, then n_T is approximately equal to $k = c_p/c_V$. Then

$$\left(\frac{P_2}{P_1}\right)^{1/n_V} \cong \frac{Z_1}{Z_2} \left(\frac{P_2}{P_1}\right)^{1/k}$$

1.2.3 Thermodynamic Laws

For calculating compressor cycles, the energy equation, relationships applying to an isentropic change of state, and the law for fluid flow through a restriction are needed.

The **Energy equation** for a **fixed mass of gas** states simply that the increase of energy of the gas equals the work done on the gas minus the heat transferred from the gas to the surroundings. For the conditions in a compressor, we can ignore changes in potential and chemical energy. In applications where the energy equation for a fixed mass of gas is used, we can usually also ignore changes in kinetic energy. The energy equation then reduces to:

$$E_2 - E_1 = M(e_2 - e_1) = W - Q [J]$$

If we consider a **control volume,** that is a volume fixed in space that fluid can flow into or out of, we must consider the work done by the gas entering and leaving the control volume, and in many cases where this equation is used, we must consider the kinetic energy of the gas entering and leaving the control volume. The **energy equation** then becomes:

$$E_2 - E_1 = M_{in} h_{o\ in} - M_{out} h_{o\ out} + W - Q [J]$$

where $h_o = h + \dfrac{1}{2} u^2 \left[\dfrac{1}{32.18} \right]$

$h = e + Pv [144]$

For a steady process, there is no change of conditions in the control volume and $E_2 = E_1$

$$\text{Then } M_{out} h_{o\ out} - M_{in} h_{o\ in} = H_{o\ out} - H_{o\ in} = W = Q [J]$$

The equations for **isentropic** change of state were given above. They apply to any change during which there are no losses and no heat transfer to the gas. The change of properties can be obtained from a Mollier chart for the gas, or if the gas behaves approximately as a perfect gas, by the equations given above.

$$PV^{n_v} = \text{Constant}$$

$$\dfrac{P^{\frac{n_T - 1}{n_T}}}{T} = \text{Constant}$$

The law for **incompressible fluid flow through a restriction** is:

$$m = F \sqrt{(2\rho\, \Delta P)} \left[\sqrt{\dfrac{32.18}{144}} \right]$$

F = Effective Flow Area = Geometric Flow Area × Flow Coefficient

For a **perfect gas**, if the pressure drop is low enough that the flow is **subsonic**, as should always be the case in reciprocating compressors, the pressure drop is given by:

$$m = k \frac{p_1}{a_1} \left(\frac{p_2}{p_1}\right)^{k+1/2k} F \sqrt{\left[\frac{2}{k-1}\left(\left(\frac{p_1}{p_2}\right)^{k-1/k} - 1\right)\right]} \quad [32.18]$$

if $\frac{p_2}{p_1} < \left(\frac{2}{k+1}\right)^{k/k-1}$ the flow is sonic and $m = k \frac{p_1}{a_1}\left(\frac{2}{k+1}\right)^{k+1/2(k-1)} F$ [32.18]

1.2.4 Compression Cycles

The work supplied to a compressor goes to increasing the pressure of the gas, to increasing the temperature of the gas and to any heat transferred out of the compressor. In most cases, the requirement is to increase the pressure of the gas using the least possible power. If the compression process is adiabatic, that is, there is no heat transfer between the compressor and the outside, then the least work will be done if the process is isentropic. This implies that there are no losses in the compressor and which is an unachievable goal, but one that can be used as a base for the compression efficiency. The **isentropic efficiency** of a compressor is defined as the **work required** to compress the gas in an **isentropic process** divided by the **actual work** used to compress the gas. The efficiency of a compressor is most often given as the isentropic efficiency.

However, it is possible to construct a compressor with an isentropic efficiency greater than 100%. The work done in a **reversible isothermal** process is less than that done in an isentropic process. In a reversible isothermal process, the temperature of the gas is maintained at the suction temperature by reversible heat transfer as the compression proceeds. There must, of course, be no losses in this process. Many compressors have a final discharge temperature that is much lower than the isentropic discharge temperature, and the power required is reduced by this. However, the power required is almost always still greater than the isentropic power and so the isentropic efficiency is universally used to rank compressors.

1.2.5 Ideal Positive Displacement Compressor Cycle

As an example of a positive displacement compressor, consider a **reciprocating compressor** cylinder compressing gas from a suction pressure P_S to a discharge pressure P_D. In compressor terminology, the ratio P_D/P_S is known as the **compression ratio**. This can be contrasted to reciprocating engine terminology where the compression ratio is a ratio of volumes.

For a reciprocating compressor, the **ideal** compression cycle is as shown on Fig. 1.2. The cycle is shown on pressure against crank angle and pressure against cylinder volume coordinates. The cycle can be explained starting at point 1. This represents the point when the piston is at the dead center position that gives the

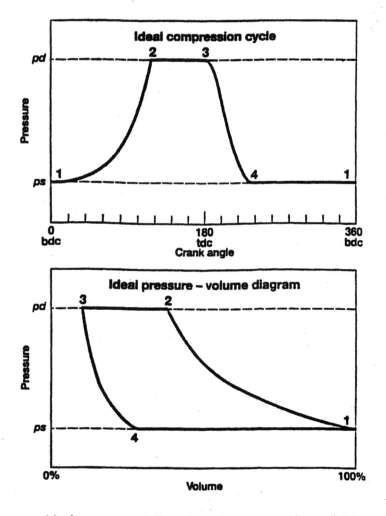

Ideal pressure-crank angle and pressure-volume diagrams.
FIGURE 1.2 Ideal compressor cycle.

maximum cylinder volume. The gas in the cylinder is at the suction pressure P_S. As the piston moves to decrease the cylinder volume, the mass of gas trapped in the cylinder is compressed and its pressure and temperature rise. In the ideal case, there is no friction and no heat transfer and so the change is isentropic and the change of pressure and temperature can be calculated from the known change of volume using the above equations for isentropic change of state.

At point 2, the pressure has increased to equal the discharge pressure. In the ideal compressor, the discharge valve will open at this point and there will be no pressure loss across the valve. As the piston moves to further decrease the cylinder volume, the gas in the cylinder is displaced into the discharge line and the pressure in the cylinder remains constant.

At point 3, the piston has reached the end of its travel, the cylinder is at its minimum volume and the discharge valve closes. As the piston reverses and moves to increase the cylinder volume, the gas that was trapped in the **clearance volume** (sometimes called the **fixed clearance**) at point 3, expands and its pressure and temperature decrease. Again there are no losses or heat transfer and the change of pressure and temperature can be calculated using the expressions for isentropic change of state.

At point 4, the pressure has decreased to again equal the suction pressure. The suction valve opens at this point. As the piston moves to further increase the cylinder volume, gas is drawn into the cylinder through the suction valve. When the piston again reaches the dead center, point 1, the cylinder volume is at its maximum, the suction valve closes, and the cycle repeats.

The **work required per cycle** and hence the **horsepower** required to drive the compressor can easily be calculated from the pressure against volume diagram or from the temperature rise across the compressor.

The work done on the gas during a small time interval during which the cylinder volume changes by dV is equal to $P\,dV$ and the work done during one compressor cycle is the integral of this for the cycle. That is, the work done equals the area of the cycle diagram on pressure against volume axes (Fig. 1.2). Note that the equivalence of work done per cycle and diagram area holds for real as well as ideal cycles. That is, the magnitude of losses that cause a horsepower requirement increase can be measured off the **indicator card** as the pressure vs. volume plot is often called. (If the pressure on the indicator card is in psi and the volume in cubic inches, the work done as given by the card area will be in inch lb. and must be divided by 12 to give the work done in ft. lb.)

Once the work done per cycle is known, the horsepower can be calculated. If the work done is in ft. lb., and the speed in rpm:

$$HP = WD\,N/33{,}000$$

If the heat transfer from the gas in the cylinder can be measured or estimated, the work done per unit time can be calculated from the energy equation.

$$\text{Work Done per Unit Time} = m(h_2 - h_1) + q\,[J]$$

For a cycle with **no heat transfer** with a **perfect gas,** Q is zero and $h = c_p\,T$, then

$$\text{Power},\ PW = mc_p(T_2 - T_1)\,[60]$$

Now for an **ideal cycle** and a **perfect gas**, the compression is isentropic and the discharge temperature T_2 can be calculated from the pressure ratio and the suction temperature T_1 using the isentropic relationship.

$$\frac{T_2}{T_1} = \left(P_D u P_S\right)^{k-1/k}$$

$$\therefore \text{Power, } PW = mc_p T_1 \left[\left(\frac{P_D}{P_S}\right)^{k-1/k} - 1\right] \quad [60]$$

For an **ideal cycle** with a gas for which the **compressibility at suction** and the **average isentropic volume exponent are known**, the power can be derived as follows.

For **unit mass** of gas compressed:

$$\text{Work Done by Gas Flowing Into Cylinder} = P_1 V_1 \quad \left[\frac{1}{12}\right]$$

$$\text{Work Done to Compress Gas In Cylinder} = -\int_1^2 P\, dV \quad \left[\frac{1}{12}\right]$$

$$\text{Work Done on Gas Flowing Out Of Cylinder} = P_2 V_2 \quad \left[\frac{1}{12}\right]$$

$$\therefore \text{Work Done by Shaft per Unit Mass of Gas}$$

$$= \left[-P_1 v_1 - \int_1^2 P\, dv + P_2 v_2\right] \quad [144]$$

Noting that $Pv^{n_v} =$ Constant, and integrating

$$\text{Work Done per Unit Mass} = \left[\frac{1}{n_v - 1}[P_2 v_2 - P_1 v_1] + P_2 v_2 - P_1 v_1\right] \quad [144]$$

$$= \frac{n_v}{n_v - 1}(P_2 v_2 - P_1 v_1)[144]$$

$$= \frac{n_v}{n_v - 1} P_1 v_1 \left[\left(\frac{P_2}{P_1}\right)^{n_v - 1/n_v} - 1\right] \quad [144]$$

$$\therefore \text{Work Done per Unit Time} = \text{Power},$$

$$PW = \frac{n_v}{n_v - 1} mP_1 v_1 \left[\left(\frac{P_2}{P_1}\right)^{n_v - 1/n_v} - 1\right] \quad [(144).(60)]$$

(Or using the modified perfect gas equation of state ($P_v = ZRT$))

$$= \frac{n_v}{n_v - 1} Z_1 RT_1 m \left[\left(\frac{P_2}{P_1}\right)^{n_v - 1/n_v} - 1\right] \quad [60]$$

The **capacity** of the **ideal** compressor end, that is the flow rate through the end, can also be calculated from the pressure vs volume diagram. The amount of gas drawn into the cylinder, which, in the ideal compressor, equals the amount of gas discharged from the cylinder, is equal to $m_1 - m_4$ where points 1 and 4 are defined

COMPRESSOR THEORY

on Fig 1.2. This is often given in terms of the **volumetric efficiency** as defined below. Note that a cylinder with no losses will have a volumetric efficiency less than 100%. The volumetric efficiency only relates the actual capacity to the capacity of a cylinder with no fixed clearance, and gives no information on the efficiency of the cylinder.

$$\text{The Capacity per Cycle} = M_1 - M_4 = \rho_1(V_1 - V_4)\left[\frac{1}{1728}\right]$$

$$\text{The Volumetric Efficiency is Defined as } VE = \frac{V_1 - V_4}{V_1 - V_3}$$

$$\therefore \text{The Capacity per Cycle} = \rho_1 \, VE(V_1 - V_3)\left[\frac{1}{1728}\right]$$

The Flow Rate (Capacity per Unit Time),

$$m = \rho_1 \, VE \, N(V_1 - V_3)\left[\frac{1}{(60)(1728)}\right]$$

Now $V_1 - V_3$ is the Swept Volume (V_{SW})

If the average isentropic volume exponent is known, the volumetric efficiency can be calculated as follows.

$$VE = \frac{V_1 - V_4}{V_1 - V_3}$$

$$= 1 - \frac{V_3}{V_1 - V_3}\left(\frac{V_4}{V_3} - 1\right)$$

Now $V_3/(V_1 - V_3)$ is the fixed clearance expressed as a fraction of the swept volume. This is often called the **clearance** (CL) and is expressed as a fraction or a percent. The term V_4/V_3 can be expressed in terms of the pressure ratio using the definition of the isentropic volume exponent n_V. Then

$$VE = 1 - CL\left[\left(\frac{p_3}{p_4}\right)^{1/n_V} - 1\right]$$

$$\text{i.e. } VE = 1 - CL\left[\left(\frac{p_D}{p_S}\right)^{1/n_V} - 1\right]$$

1.2.6 Approximate Valve Losses

For a compressor with **real valves**, there will be a pressure drop across the valves during the suction and discharge processes. This will increase the power required to drive the compressor and decrease the capacity of the compressor. These losses can be estimated as follows.

To estimate the power loss caused by the valves, it is often **assumed** that the **gas is incompressible during the valve event.** This is reasonable as the gas pressure remains relatively constant during the suction and discharge processes. We will also assume for the moment that the connecting rod is long so that the **piston motion is sinusoidal.** Then, the piston velocity is given by:

$$U_p = \pi NS \, \text{Sin}(\theta) \left[\frac{1}{12}\right]$$

Now, if the gas is incompressible, the mass of gas passing through the valve equals the mass displaced by the piston.

$$m = \rho \frac{\pi B^2}{4} U_p \left[\frac{1}{(60)(144)}\right] = \frac{\pi^2}{4} \rho NSB^2 \, \text{Sin}(\theta) \left[\frac{1}{(1728)(60)}\right]$$

Then, assuming incompressible flow through the orifice representing the valve,

$$\Delta P = \frac{m^2}{2\rho F_{eq}^2}\left[\frac{144}{32.18}\right] = \frac{\left(\frac{\pi^2}{4}\rho NSB^2 \, \text{Sin}(\theta)\right)^2}{2\rho F_{eq}^2}\left[\frac{1}{(32.18)(60^2)(144^2)}\right]$$

Note that in this equation, F_{eq} is the **equivalent area** of all the valves for the corner being considered. That is the area of an ideal orifice that will give the same pressure drop as the valves for the same flow rate of the same gas. If it is assumed that the valves are fully open for the full suction or discharge event, then F_{eq} is a constant that is known from the valve design. Adding the effects of the valve pressure drops modifies the cycle pressure diagrams as shown on Fig 1.3.

To calculate the work or power loss caused by the valve loss, we need to know the area of the valve loss on the pressure vs volume diagram (Fig. 1.3). This is most easily obtained by calculating the average valve pressure drop and then multiplying by the volume change. The average pressure drop on a cylinder volume basis is obtained by integrating the above expression. For the suction valves with reference to Fig. 1.3.

$$\overline{\Delta P} = \frac{\int_{\frac{1}{4}}^{1} \Delta P \, dv}{V_1 - V_4}$$

Substituting for ΔP, integrating and simplifying gives:

$$\overline{\Delta P} = \frac{\rho}{2}\left(\frac{\pi^2 NSB^2}{4F_{eq}}\right)^2 \lambda \left[\frac{1}{(32.18)(60^2)(144^2)}\right]$$

where $\lambda = \dfrac{6VE - 4VE^2}{3}$ is an Integration Factor defined by this equation

COMPRESSOR THEORY   1.13

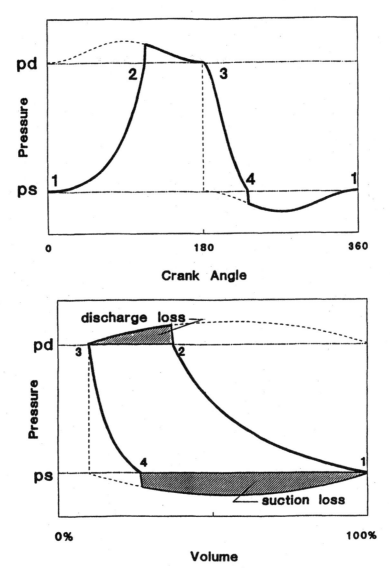

FIGURE 1.3 Cycle with approximate valve losses.

An identical expression is obtained for the discharge valves.

Note that a compressor compressing a heavier (higher molecular weight) gas or running at a higher speed will require larger valve equivalent area for a given size cylinder to give the same efficiency.

As stated earlier, the above only applies if the connecting rod is long compared to the stroke. For a realistic connecting rod length, the value of 8 is as given on Fig. 1.4.

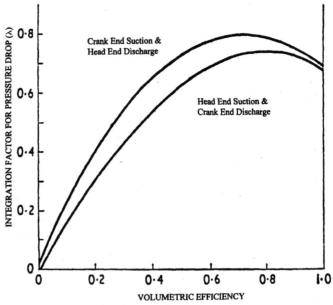

FIGURE 1.4 Integration factor used to calculate valve pressure drop.

The power loss caused by the valves in the corner is then easily obtained.

$$\text{Power Loss, } PW = N\, \Delta V\, \overline{\Delta P} \left[\frac{1}{12}\right]$$

$$= N\, VE\, V_{SW}\, \overline{\Delta P} \left[\frac{1}{12}\right]$$

$$\text{i.e. Horsepower} = N\, VE\, V_{SW}\, \overline{\Delta P} \left[\frac{1}{(12)(33,000)}\right]$$

Note that for discharge valves, the discharge volumetric efficiency must be used in the above. The **Discharge Volumetric Efficiency** is defined as the actual volume of gas discharged from the cylinder each stroke ($V_2 - V_3$ with points 2 and 3 as defined by Fig. 1.2) divided by the swept volume (V_{SW}).

$$\text{That is: Discharge } VE = \frac{V_2 - V_3}{V_{SW}}$$

1.2.7 Ideal Dynamic Compressor Cycle

In a dynamic compressor, the moving part increases the velocity of the gas and the resulting kinetic energy is converted into pressure energy. Typically, both processes occur simultaneously in the rotating element and the gas leaves the rotor at higher pressure and with a higher velocity than it entered. Some of the kinetic energy is then converted into pressure energy in the stator by means of a diffusion process, that is, flow through a diverging channel.

If we ignore the effects of heat transfer, the steady flow energy equation states that the increase in stagnation enthalpy for flow in the rotor equals the work done. As there is no work done on the gas in the stator, the stagnation enthalpy remains constant. These relationships are true regardless of the efficiency of the process. In a completely inefficient process, the temperature of the gas will be increased, but the pressure will not. In an efficient process, the pressure of the gas will be increased as well as the temperature.

For a compressor with no losses and no heat transfer, the process will be isentropic. The increase in enthalpy for compression from a given initial pressure and temperature to a given final pressure can be obtained from a Mollier chart, or from an equation of state. For an ideal gas, it can be calculated as follows.

$$T_{2\,is} = T_1 \left(\frac{P_2}{P_1}\right)^{k-1/k}$$

$$h_{2\,is} - h_1 = c_P(T_{2\,is} - T_1)$$

The isentropic efficiency which is the work required for an isentropic compression divided by the actual work can be calculated as:

$$\text{Isentropic Efficiency} = \frac{h_{2\,is} - h_1}{h_2 - h_1}$$

It is sometimes considered that any excess kinetic energy in the discharge gas over that of the inlet gas is also a useful output of the compressor. It can, after all, be recovered in a diffuser. In this case, the actual stagnation enthalpies should be used and:

$$\text{Isentropic Efficiency} = \frac{h_{2\,is} - h_1}{h_{o2} - h_{o1}}$$

1.3 REFERENCES

1. *Gas Properties and Compressor Data,* Ingersoll-Rand Company Form 3519D.
2. Edmister, Wayne C., *Applied Hydrocarbon Thermodynamics,* Gulf Publishing, 1961, L. of C. 61-17939.
3. Reid, Robert C., John M., Prausnitz, and Bruce E. Poling, *The Properties of Gases and Liquids,* 4th Ed., McGraw Hill, ISBN 0-07-051799-1.

CHAPTER 2
COMPRESSOR PERFORMANCE—POSITIVE DISPLACEMENT

Derek Woolatt
Manager, Valve and Regulator Engineering
Dreser-Rand Company & (Screw Compressor Section)

Fred Heidrich
Dresser-Rand Company

2.1 COMPRESSOR PERFORMANCE

2.1.1 Positive Displacement Compressors

Positive displacement compressors all work on the same principle and have the same loss mechanisms. However, the relative magnitude of the different losses will be different in each type. For example, leakage losses will be low in a lubricated reciprocating compressor with good piston rings, but may be significant in a dry screw unit, especially if the speed is low and the pressure increase, high. Cooling of the gas, which is beneficial, will be small in a reciprocating compressor, but may be almost complete in a liquid flooded screw compressor.

All compressor types have a clearance volume that contains gas at the discharge pressure at the end of the discharge process. This volume may be small in some designs and significant in others. Some types, for example reciprocating compressors may have a large clearance volume, but recover the work done on this gas by expanding it back to suction pressure in the cylinder; other types, for example screw compressors, let the gas in the clearance space expand back to suction pressure without recovering the work.

Some compressor types, specifically those that use fixed ports for the discharge, are designed to operate at a fixed volume ratio. (For a given gas, this is equivalent to a fixed pressure ratio.) As the ratio varies from this value, the compressor efficiency will be less than the optimum. Other compressor types use either ports that can be varied with slides or they use pressure actuated valves. These types are optimized at any pressure ratio.

2.1.2 Reciprocating Compressor Rating

Each component in a compressor frame and cylinder has design limits. To ensure that these are not exceeded in operation, each frame and each cylinder has a design rating above which it may not be used. The loads used to rate compressors are discussed below.

Every **cylinder** has a **maximum allowable discharge pressure**. All compressor components are subjected to alternating loads and the rated pressure of a cylinder will be based on fatigue considerations.

Every **cylinder** has a **minimum clearance** it can be built with. This controls the volumetric efficiency of the cylinder and hence the capacity for a given pressure ratio and gas composition. The clearance of a cylinder can usually be increased if the maximum capacity is not needed for a given application.

Every **cylinder** has a fixed **number of valves and valve size**. A cylinder with a few or small valves for its size will have high losses and will give poor efficiency if used at its normal piston speed when compressing a high molecular weight gas, especially if the pressure ratio is small.

Each cylinder exerts a **rod load** on the running gear components, and a **frame load** on the stationary components. These can be evaluated by considering the forces acting on the various components (Fig. 2.1).

$$Frame\ Load = P_{HE} A_P - P_{CE} (A_P - A_{ROD})$$

Where

P_{HE} and P_{CE} = Pressure in the Head End and Crank End of the Cylinder
A_P and A_{ROD} = The Area of the Piston and Piston Rod

The frame load will vary through the cycle as the pressures in the head end and crank end of the cylinder vary. The maximum tensile and the maximum compres-

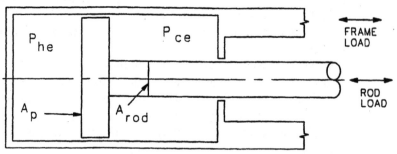

FIGURE 2.1 Frame and rod loads.

sive stresses are calculated. These are the loads the stationary components and bolting must be designed to resist.

The rod load, the force exerted on the piston rod, crosshead, crosshead pin, connecting rod and crankshaft is different for each component. It is the frame load plus the inertia of all the parts outboard of the component that is of interest. For example, the rod load at the crosshead pin, the value that is usually calculated, is the frame load plus the inertia of the piston with rings, the piston rod and the crosshead. The inertia is the mass times the piston acceleration, and varies through the cycle. The rod load quoted is usually the maximum value, compression or tension, at the crosshead pin.

The crosshead pin bearings do not see full rotary motion. Rather the connecting rod oscillates through a fairly small arc. This makes lubrication of these bearings difficult as a hydrodynamic film is never generated. The bearing relies on a squeeze film being formed. This requires that the load change direction from compressive to tensile and back every revolution. Once the rod load diagram has been calculated, the **degrees of reversal**, that is the lesser of the number of degrees of crankshaft rotation that the rod load is compressive and the number of degrees it is tensile, is known. The minimum acceptable number of degrees of reversal depends on the details of the design and will be available for each frame.

Each **frame** will also be limited by the power that can be transmitted through the crankshaft at a given speed. There will be a limit on the **power of each throw** and a higher limit on the **total power of the compressor**. Note that the total compressor power is all transmitted through the crankshaft web closest to the driver.

2.1.3 Reciprocating Compressor Sizing

Once the suction and discharge pressures, the suction gas temperature, the required flow rate and the gas composition are determined, a compressor can be selected to do the job. The selection will depend on the relative importance of efficiency, reliability and cost, but certain principles will always apply.

Compressors for a wide range of applications tend to run with about the same piston speed. That is compressors with a long stroke tend to run slower than those with a short stroke. Further, short stroke compressors tend to be of lighter construction with lower allowable loads. For the best efficiency and reliability at the expense of increased cost, a piston speed at the low end of the normal range will be used. The **compressor speed** and the **stroke** will then be determined by the horsepower requirement. A low horse power application will require a light, low stroke, high speed compressor. A high horse power application will require a heavy, long stroke, low speed compressor. If possible, larger compressors are directly coupled to the driver. Thus the speed range of available drivers may influence the selection of the compressor.

The **number of stages** must then be selected. One consideration here is the allowable discharge temperature; another is the pressure ratio capability of the available cylinders as determined by their fixed clearance; another is efficiency. If the calculated discharge temperature using one stage is too high, obviously more stages are needed. During preliminary sizing, the isentropic discharge temperature can be used, but if a certain number of stages creates a marginal situation, the discharge temperature should be estimated more accurately. As a first estimate, it can be assumed that equal pressure ratios are used for all stages. In practice it is often good to take a higher pressure ratio in the low pressure ratio stages and unload the more critical higher pressure stages a little.

In almost all multi-stage applications the gas will be cooled between stages. In this case, increasing the number of stages, up to a limit, will increase the efficiency of the compressor. This is because with intercooling, the compression more closely approximates an isothermal compression with resulting lower power requirement. An alternative way of looking at this is on a pressure volume diagram. The work required to compress the gas is given by the area of the pressure vs volume diagram. Fig. 2.2 shows a single- and a two-stage compression for a given application. The diagram for single stage compression is 1-2-3-4-1. For two-stage compression, it is 1-5-6-7-3-8-4-1. As the interstage gas is cooled (5-6), its volume decreases. The

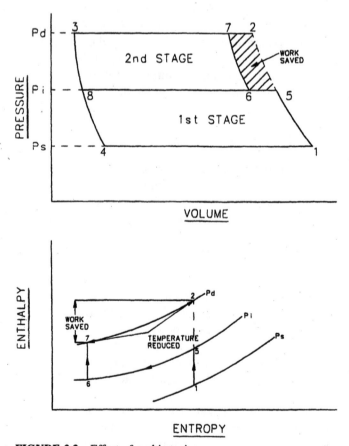

FIGURE 2.2 Effect of multi staging.

work done as given by the areas of the diagrams is obviously less in the two-stage case than in the single-stage case. Further, if any liquids are condensed out of the gas in the intercoolers, the liquids must be separated from the gas and the mass of gas compressed from that interstage to the final discharge is reduced with a further resulting power reduction. However as stages are added, the number of compressor valves the gas must flow through in series, and the amount of interstage piping and coolers increase. If too many stages are used, the pressure losses in the valves and piping will offset the gains from intercooling and the efficiency will be reduced.

The cost of a compressor to do a given task usually increases as the number of stages is increased because of the additional compressor cylinders, coolers and piping.

In a few applications, there will be side streams where gas either enters or leaves the process at fixed pressures. These requirements may determine the interstage pressures used.

Once the number of stages is selected, the **cylinders for each stage** can be selected. Usually a selection will be made from cylinder designs available. Knowing the inlet conditions and the required capacity, and with the speed and stroke already selected, the required cylinder bore can be estimated. The available cylinders can then be checked to see which, if any, meet the requirements. The following must be checked. First, the pressure rating of the cylinder must be adequate to be safe at the design and any upset conditions. The cylinder rating should be higher than the relief valve setting. Second, the frame load, rod load and degrees of reversal must be within the rating for the frame components. Third, the capacity calculated with the minimum cylinder clearance allowing for all losses must meet the requirements. Fourth, the power requirement of this cylinder must not exceed the power rating per throw of the frame components. If all these requirements are met, a suitable cylinder has been chosen. Additional optimization may be needed to determine the best possible cylinder for this application. If no cylinder can be found to meet the requirements, then either a new cylinder must be designed, a frame rated for a higher frame load or horsepower per throw must be selected, or two or more, smaller cylinders must be chosen to run in parallel to meet the required flow. Note that if smaller cylinders are used, the frame load and the power per throw will be reduced. It is usual for smaller cylinders to be available in higher pressure ratio versions, so all the requirements can usually be met by using multiple cylinders per stage.

The basic compressor sizing is then complete, but must be checked at alternate design or upset conditions. Additional factors such as the out-of-balance force transmitted from the compressor to the foundation, the potential for harmful torsional vibrations in the crankshaft and drive train, optimization of the compressor layout, efficiency, and cost will be considered before the design is finalized.

2.1.4 Capacity Control

In many applications, it is necessary to be able to reduce the capacity of the compressor to meet changing process needs. There are several ways to accomplish this.

A very simple control system used mainly on air compressors is **stop/start** control. In a compressed air system with a large receiver, the compressor can be run to fill the receiver to greater than the required pressure and the compressor can then be stopped. When the receiver pressure falls to the lowest acceptable value, the compressor is started again. The system is very simple and requires no additional equipment on the compressor, but large pressure swings must be accepted and the frequent stops and starts can be hard on the compressor.

If a **variable speed driver** is available, varying the speed of the compressor is an excellent way to control capacity. It will give close, infinite step control, without additional equipment on the compressor. Reduced speed operation is usually easy for the compressor and maintenance intervals may be increased. This method is normally used with compressors driven by an engine and is increasingly frequent with compressors driven by an electric motor. In many cases, the speed range is not sufficient to give the full capacity range needed and speed control is used in conjunction with other methods. Many unloading methods give step changes in capacity and speed control can be used to trim the flow rate between these steps.

The output of the compressor can also be adjusted by use of a **bypass**. This allows some of the compressed gas to be leaked back to the suction. This obviously is very inefficient and the bypassed gas may have to be cooled. It is the only unloading method discussed here that significantly decreases the efficiency of the process. It is, however, simple, reliable and inexpensive and is very suitable for unloading a compressor for a short period during start up or shut down. A similarly energy inefficient method sometimes used to adjust the flow of small compressors is to throttle the suction. This is effective, but care must be taken not to overload the compressor. It may be noted here that most dynamic compressors cannot be unloaded without loss of efficiency or surge problems. Positive displacement compressors usually can be unloaded with either little reduction or increase of efficiency.

Some screw compressors have **slides** that change the inlet port timing and allow the capacity to be adjusted to any value between the full capacity and some minimum capacity.

Various schemes are available to **unload an end** of a reciprocating compressor cylinder. These reduce the capacity of that end to zero. With a single cylinder per stage, double acting compressor, this gives three-step control. The compressor can be run at approximately 0, 50 or 100% capacity. If there are more than one cylinder per stage, additional steps can be arranged. For example, if a stage has two identical double acting cylinders and each end of each cylinder has the same clearance, then five-step control can be achieved. The compressor will run at 0, 25, 50, 75 or 100% capacity. If the various ends have different clearances or swept volumes, then additional steps are available. In multi-stage units, the first stage usually controls the capacity of the complete machine, but if only the first stage is unloaded, the interstage pressures will be greatly changed and the design limits of one of the higher stage cylinders will probably be exceeded. It is usually necessary to unload all stages. It is also essential that the degrees of reversal be checked on any cylinder that is unloaded as reversal can easily be lost by unloading a cylinder end. There may also be limitations on how long a cylinder can be run completely unloaded without excessive heat build-up, or problems associated with a build-up of lubricating oil or process liquids in the cylinder. During normal operation, liquids that get into the cylinder in small quantities are passed through the discharge valves with the process gas.

Three methods are in common use to unload a cylinder end. They all work by connecting the cylinder to the suction passage so gas flows back and forth between the cylinder and suction passage rather than getting compressed. It is essential that sufficient area for the flow is provided. If it is not, the losses will be high and as well as consuming unnecessary power, the cylinder will overheat. If a compressor is to be run at a constant capacity less than full load for an extended time, the compressor can be shut down and the **suction valves removed**. This gives very low loss and hence very low heat build up. It requires no additional equipment, but the compressor must be shut down and worked on whenever the load must be changed. If the load step must be changed while the compressor is running, either **finger unloaders (valve depressors)** or **plug or port unloaders** can be used. A typical finger unloader arrangement is shown as Fig. 2.3. The fingers are usually operated pneumatically. To unload the cylinder, the fingers push the moving elements in the valve away from their seat, thus opening the valve and holding it open throughout the cycle. To get sufficient flow area, it is usually necessary to provide an unloader for every suction valve. If it is arranged that the fingers move in and out for each compressor cycle, and if these movements are timed to delay the closing of the valve by a varying amount, then the capacity of the cylinder end can be varied between full and zero capacity. Typical plug or port unloaders are shown on Fig. 2.4. With these, a hole between the cylinder and the suction passage is opened when the end is to be unloaded. With a port unloader, one of the suction valve ports is used for an unloading plug rather than a valve. In some cases, removing a suction valve causes an unacceptable decrease in efficiency, and a plug unloader is used. In this, a special suction valve with a hole in its center is used. When the cylinder is loaded, this hole is sealed, and when it is unloaded, the hole is opened to allow flow between the cylinder and suction passage. The plug or port unloaders are usually operated pneumatically.

If it is necessary to reduce the flow in a cylinder end to some value greater than zero, a **clearance pocket** can be used, Fig. 2.5. This is an additional volume that can be connected to the cylinder, or isolated from it, by a pneumatically controlled valve very similar to that used for a port or plug unloader. When the cylinder clearance volume is increased by opening the clearance pocket valve, the cylinder will compress a reduced amount of gas. In some cases, the volume of the clearance pocket is varied with a sliding piston. This allows any capacity within the range of the unloader to be selected. The required volume of the clearance pocket will depend on the amount of capacity reduction required, the size of the cylinder and the pressure ratio. If the pressure ratio is low, a very large pocket will be required to give a small reduction in capacity. In a larger cylinder, it is possible to fit several fixed volume clearance pockets in one end of the cylinder. This allows a number of different capacity steps to be used. If pockets are used on both ends of the

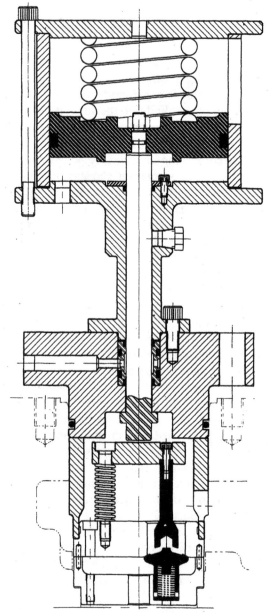

FIGURE 2.3 Finger unloader.

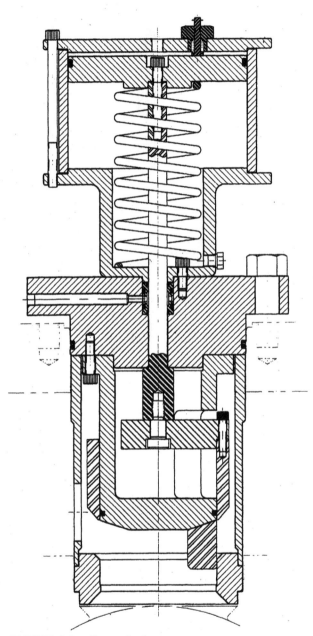

FIGURE 2.4a Port unloader.

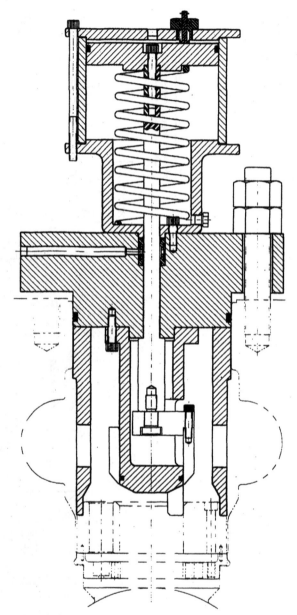

FIGURE 2.4b Plug unloader.

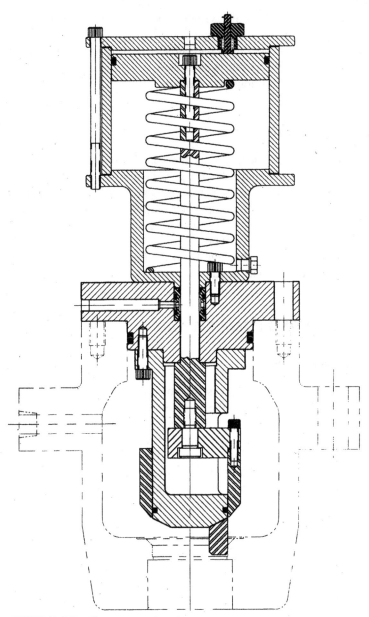

FIGURE 2.5 Clearance pocket.

cylinder, and on all cylinders if there are multiple cylinders on the stage, a large number of different capacity steps can be provided. It is important to check the compressor operation carefully at every possible unloaded condition to ensure that all cylinders operate within their design limits with acceptable discharge pressures, rod loads and degrees of reversal.

2.1.5 Compressor Performance

The performance, that is the capacity (mass of gas compressed) and the power required to compress the gas, is affected by many details of the compressor's design. Several of these are discussed below. They are discussed first with reference to a reciprocating compressor, and then with reference to a screw compressor. The losses in other types of positive displacement compressors will be similar to those discussed here. All types of compressors have losses caused by flow losses, by heat transfer, and by leakage from the high pressure to the low pressure zone and some types have losses associated with the valves.

2.2 RECIPROCATING COMPRESSORS

2.2.1 Compressor Valves

The compressor valves are the most critical component in a reciprocating compressor because of their effect on the efficiency (horsepower and capacity) and reliability of the compressor. Compressor valves are nothing more than check valves, but they are required to operate reliably for about a billion cycles, with opening and closing times measured in milliseconds, with no leakage in the reverse flow direction and with low pressure loss in the forward flow direction. To make matters worse, they are frequently expected to operate in highly corrosive, dirty gas, while covered in sticky deposits.

Compressor valves affect performance due to the pressure drop caused by flow through the valve; the leakage through the valve in the reverse direction; and the fact that the valves do not close exactly when an ideal valve would. Typical valve dynamics are shown in Fig. 2.6. Note that: a) due to its inertia, the valve does not open instantaneously; b) due to the springing, the valve does not stay at full lift for the full time it is open; and c) the valve does not close exactly at the dead center. All of these factors affect both the capacity and the power of the compressor.

A simple method of calculating the power loss due to the pressure drop across the valve was given in the section on theory (Chapter 1). However, this assumed that the valve was at full lift for the entire time gas was flowing through it. For a more accurate estimate of the power loss, the weighted average valve lift should be used. This can be calculated from the valve lift diagram, Fig. 2.6. Obviously, the average lift is less than the full lift and so the average valve flow area is less than the full lift flow area. Thus the actual power loss is greater than that calculated

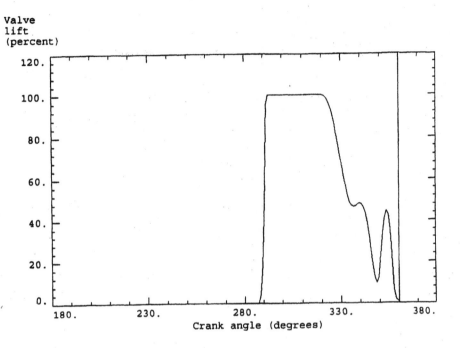

TYPICAL VALVE DYNAMICS DIAGRAM
(Lift vs. Crank Angle)

FIGURE 2.6 Typical valve dynamics diagram.

by the method given under "Theory." Fortuitously, that method also contains an error that makes it overestimate the power loss and in many cases it gives a good estimate of the true loss. One assumption of that method is that the gas is incompressible. That is, it is assumed that at valve opening, the pressure loss increases instantaneously to the value calculated from the piston velocity. In fact, due to the compressible nature of the gas and as shown in Fig. 2.7, the pressure drop rises gradually from zero at the instant the valve opens. As the valve takes a finite time to open because of its inertia, the pressure drop, after initially being less than that estimated by the simple theory, then overshoots. These effects, taken with the fact that the valve starts to close well before the end of the stroke, cause offsetting errors.

The power losses caused by the valves are well known. The effects of the valves on the capacity of the compressor are less obvious, but equally important. The valves affect the capacity in three ways.

1. As the valves never close exactly at the dead center, the amount of gas trapped in the cylinder is never that predicted from simple theory. The springs in a compressor valve should be designed to close the valve at about the dead center. In practice, the exact closing angle will vary as the conditions of service vary

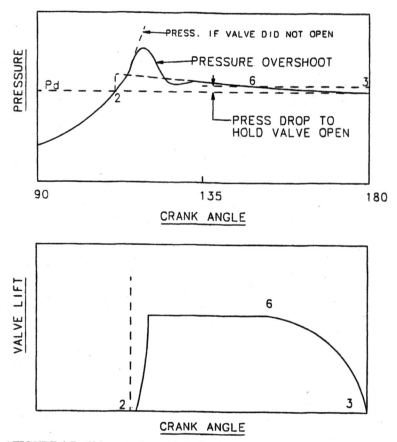

FIGURE 2.7 Valve opening and closing.

and will depend on how strongly the moving parts of the valve adhere to their stops. This will depend on the amount and nature of liquids and deposits on the valve. For the suction process, the cylinder volume when the valves close is smaller than the maximum cylinder volume so less gas is trapped in the cylinder and compressed. Note that either too heavy a spring, which causes the valve to close early, or too light a spring which causes the valve to close late, will reduce the capacity. For the discharge process, the cylinder volume when the valve closes is larger than the minimum cylinder volume, and the mass of gas is larger than the ideal. This extra gas is re-expanded to suction conditions instead of being discharged at high pressure.

2. The gas is heated by the loss associated with flow through the suction valve. This causes the gas trapped in the cylinder when the valve closes to be at a temperature higher than the suction gas temperature. Thus the density is reduced and less gas is trapped in the cylinder to be compressed. The temperature rise can be discussed with reference to Fig. 2.8. Consider a particle of suction gas at condition "s" throttled through the suction valve to condition "5", the pres-

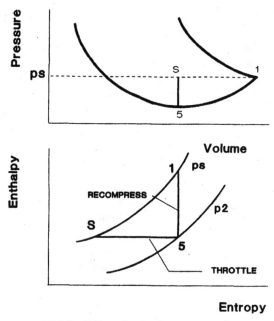

FIGURE 2.8 Effect of valve loss on capacity.

sure in the cylinder at that time. From the steady flow energy equation, with no work done and any heat transfer neglected, this is a process at constant enthalpy. For an ideal gas, it will be at constant temperature. As the piston slows down towards the end of the stroke, the valve pressure drop declines and the pressure in the cylinder increases. The particle of gas we are considering is compressed isentropically from condition "5" to condition "1" with consequent temperature rise.

3. The valve pressure drop, if it is large, can directly affect the capacity loss. This occurs if the valve equivalent area is so small relative to the application that the gas cannot flow in through the suction valves fast enough to fill the cylinder. The pressure at the end of the suction stroke will then be less than the suction pressure and the amount of gas compressed will be reduced.

2.2.2 Passage Losses

The valve losses as discussed above are always considered when predicting compressor performance. The additional loss caused by the pressure drop resulting from the flow through the remainder of the cylinder are often ignored, although they can be comparable in magnitude to the valve losses. Over the years, the equivalent area of the valves has been increased and the effects of the other flow losses have become more important. These losses typically occur in three places. First, to decrease clearance volume, the cylinder is frequently designed so there is only a small passage through which the gas can flow to get from the suction valves into the cylinder, and from the cylinder to the discharge valves. Second, the components used to hold the valves in the cylinder and any unloading devices may restrict the flow. Third, there will be losses in the cylinder passages that conduct the gas between the valves and the cylinder flanges.

These losses affect the compressor performance—power and capacity—in exactly the same way as valve losses.

2.2.3 Pulsation Losses

In the above discussion, we have assumed that the pressure in the cylinder varies, but that the pressure on the line side of the valves is constant. In practice, due to the unsteady nature of the flow entering or leaving the compressor cylinder, there are pulsations in the piping to and from the cylinder. The form and amplitude of these pulsations depends on the cylinder, the valves and the piping. Methods for calculating the pulsations are given in Chapter 1 and a typical result is shown as Fig. 2.9. The details of the pulsations depend on the complete piping system, but as a first approximation it can be assumed that the cylinder is connected by a nozzle to a reservoir at constant pressure. Considering the suction process first. The pressure in the suction passage of the cylinder will usually fall as the valve opens and gas starts flowing into the cylinder. The duration of this reduced pressure will depend on the length of the suction nozzle. If the nozzle is short, the pressure will rapidly rise back to and will then oscillate about the suction pressure. If the nozzle is long, the pressure may stay lower than the suction pressure for the complete suction process. The reduced pressure has a similar effect to valve losses as

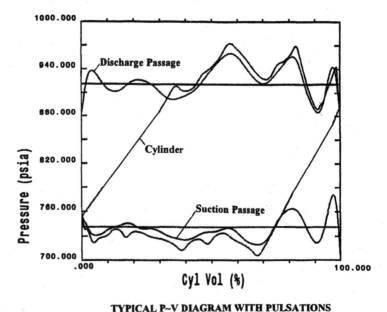

TYPICAL P-V DIAGRAM WITH PULSATIONS
FIGURE 2.9 Typical pressure–volume diagram with pulsations.

far as the cylinder is concerned. That is more work required to draw the gas into the cylinder. Similarly, for the discharge process, the pressure on the line side of the valve during the discharge process will be higher than the average discharge pressure and the power required to drive the compressor will be increased. The pulsations also affect the capacity of the compressor by changing the pressure on the line side of the valve at the instant the valve is closed. If the pressure in the suction nozzle outside the valve is higher than the average suction pressure at the instant the valve closes, the cylinder will contain more gas than expected and the capacity will be increased. The power required will, of course, be increased proportionately. Conversely, if the pressure is lower the capacity will be decreased. Similar effects apply to the discharge process. In a real life system, the pulsations can only be predicted by a detailed analysis and the compressor power and capacity may each be either increased or decreased. However, as the energy to sustain the pulsations is provided by the compressor, it is not possible to use pulsations to increase the efficiency. Any increase in capacity will require a corresponding increase in power.

2.2.4 Heat Transfer

Some compressor cylinders are cooled by liquid, usually water, some smaller cylinders are actively air cooled, and others are essentially uncooled with a small amount of heat lost to the atmosphere. If the cooling liquid—water or oil—is mixed with the compressed gas as it is in many screw compressors, the cooling can be sufficient to make the compression nearly isothermal with resulting beneficial effect on the compression efficiency. While many reciprocating compressors run with a discharge temperature far below the isentropic discharge temperature, much of the cooling occurs in the discharge passages and the efficiency improvement is usually small. In most cases the cooling is used to reduce part temperatures to decrease wear, especially of plastic parts; prevent distortion caused by uneven component temperatures; and reduce lubricating oil degradation.

Calculating the effects of heat transfer is difficult and imprecise. The magnitude of the effects must usually be determined by testing.

The most important effect of heat transfer on performance in most compressors is the heating of the suction gas as it flows through the suction passage, the suction port and the valves. This is equivalent to increasing the suction temperature. It decreases the mass flow compressed, because the gas density is reduced, without changing the required power significantly. The compression efficiency is therefore reduced.

Dynamic heat transfer in the cylinder of a positive displacement compressor must also be considered. The cylinder walls and the piston will run at a temperature between the suction and discharge temperatures and this temperature will be close to steady even though the gas temperature varies through the cycle. The temperature of the cylinder wall will not be uniform. It will be higher close to the discharge valves and lower close to the suction valves. Details of the temperature distribution will depend on the cylinder design and the cooling. During the suction process,

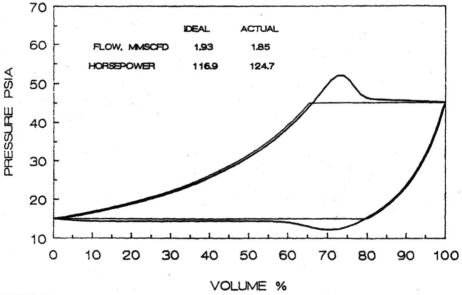

FIGURE 2.10 Effect of valve and passage flow losses.

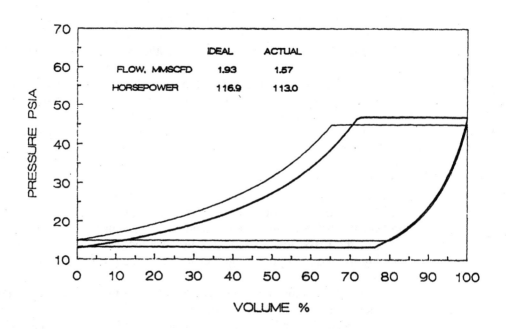

EFFECT OF VALVE SPRING PRELOAD
(Low Suction Pressure)

FIGURE 2.11 Effect of valve spring preload.

COMPRESSOR PERFORMANCE—POSITIVE DISPLACEMENT 2.19

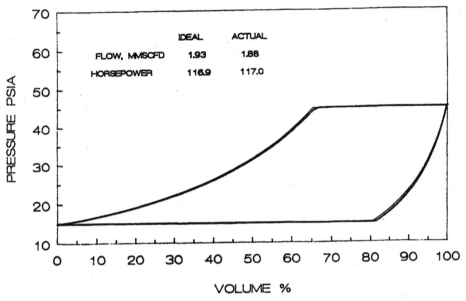

FIGURE 2.12 Effect of packing leakage.

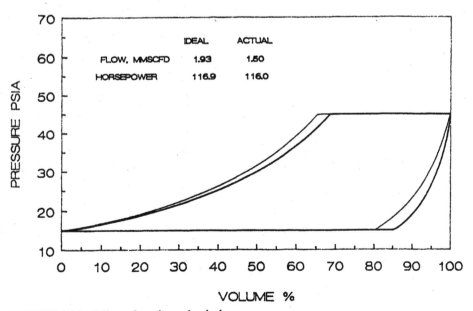

FIGURE 2.13 Effect of suction valve leakage.

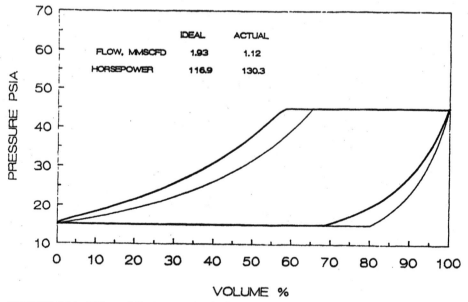
FIGURE 2.14 Effect of discharge valve leakage.

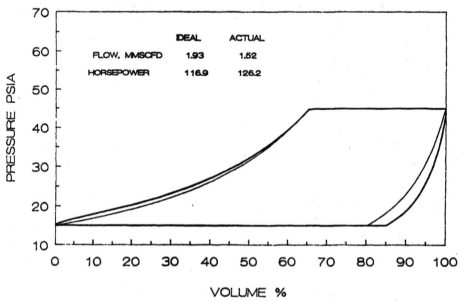
FIGURE 2.15 Effect of piston ring leakage.

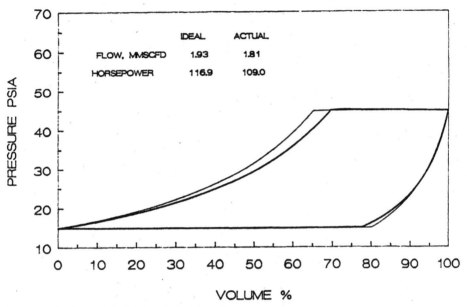

FIGURE 2.16 Effect of internal heat transfer.

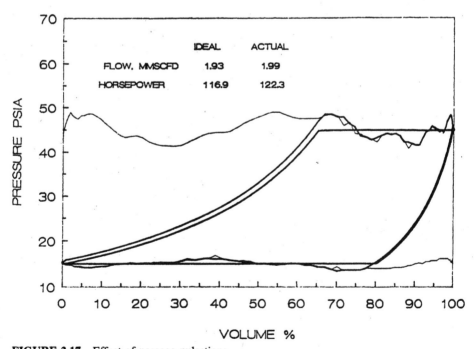

FIGURE 2.17 Effect of passage pulsations.

heat will be transferred from the cylinder wall to the gas, increasing the gas temperature. This will decrease the mass of gas trapped in the cylinder at the end of the stroke and hence will reduce the capacity. Note that the power will be approximately unchanged by this and so the efficiency will be reduced.

Heat transfer during the compression stroke also affects the compressor performance. During the first part of the compression, the cylinder wall is hotter than the gas so heat is transferred to the gas, increasing its temperature and pressure. During the second part of the compression, heat is transferred from the gas to the cylinder wall. This decreases the temperature and the pressure. The relative magnitude of these two effects depends on the effectiveness of the cylinder cooling. However as the gas in the cylinder spends more time at or near suction temperature than it does at or near discharge temperature, the average cylinder wall is usually closer to suction than discharge temperature, and the net effect of heat transfer during compression is to reduce the temperature and the power requirement. The capacity is slightly reduced by the heat transfer to the cylinder wall as the amount of gas remaining in the cylinder at the end of the discharge process is slightly increased by the lower temperature resulting from the cooling.

2.2.5 Leakage

All compressors have sliding seals between high and low pressure zones. These always leak to some extent which always has a negative effect on compression efficiency. In reciprocating compressors, the usual leakage paths are through the piston rings, the rod packing of double acting compressors, and the valves, which do not seal perfectly against reverse flow.

The effects of leakage through the rod packing on double acting compressors and through the piston rings of single acting compressors are easily understood. Some gas that has been at least partially compressed leaks to the atmosphere or a flare line. The power used to compress this gas is wasted and the capacity at discharge is reduced by the amount of the leakage.

Leakage through a suction valve has a double effect. In addition to the loss described for packing leakage, suction valve leakage causes hot gas to enter the suction passage. This is equivalent to heat transfer to the suction gas and has the same negative effect on capacity and efficiency. The change in power is usually small, but may act to reduce the power due to the reduced pressure during the compression process. In most cases, the heating effect causes greater losses than the direct effect.

Discharge valve leakage has the direct effect of wasting gas that has already been compressed, thus decreasing the flow without any decrease in the power. It also increases the pressure in the cylinder during the compression process, thus increasing the power requirement. In addition, the gas that leaks during the expansion and suction processes will heat the gas in the cylinder and reduce the capacity by decreasing the trapped gas density.

Piston ring leakage in a double acting cylinder is a little more complex. Gas leaks into each end of the cylinder during the low pressure part of the cycle and out of the end during the high pressure part of its cycle. In all cases, some of the work that has been done on the gas is wasted as it must be recompressed. In addition, the gas always heats the gas in the end it leaks into. A large amount of the leakage into an end occurs during the suction process. Thus the heating decreases the trapped gas density and decreases the capacity.

2.2.6 Examples

The results of some calculations in which the compression is ideal except for a single loss mechanism are given as Figs 2.10 to 2.17.* Each diagram shows the effect of the loss mechanism on the pressure volume cards and the effect on the power requirement and the capacity. The magnitude of the losses has been fixed at a high value so the effects can be clearly seen. The reader should be able to predict the shape of the pressure volume cards with the losses from the above discussion and an understanding of the processes involved.

It has been common for people to measure the volumetric efficiency of the pressure volume card. This is only adequate as a way to measure capacity in a compressor with no leakage or heating effects. As an illustration of this consider the example with discharge valve leakage. From the pressure volume card it would appear that the volumetric efficiency is increased by the leakage, whereas in fact the capacity is decreased by 40%. In contrast, the power required can be calculated accurately from the pressure volume card.

2.3 SCREW COMPRESSORS

2.3.1 Port and Passage Losses

Screw compressors do not rely on suction and discharge valves to regulate the flow of gas through the compressor, therefore the valve loss equations typically used for reciprocating compressors do not apply. However, a series of alternate factors need to be examined. Typically, gas is moved through a screw compressor via ports machined in the compressor housing. The design and location of these ports is crucial to the overall efficiency of the machine. The inlet port must be sized so that entrance flow losses are minimized. The same can be said of exit losses at the discharge port. However, with regards to the discharge port, the most critical factor is its location. As with any positive displacement compressor, pressure is increased by steadily decreasing the volume of the gas trapped in the compression chamber. Since there are no discharge valves, the compression process continues until the discharge port is uncovered. Therefore, given a fixed port location, the compressor always compresses to the same volume ratio. If the discharge port is not properly

---

* Woollatt, D. "Factors affecting reciprocating compressor performance". Hydrocarbon Processing Magazine (June 1993). Copyright (1993) by Gulf Publishing Co. All rights reserved.

located, inefficiencies can occur. The time to reach pressure is a function of the gas properties, namely the isentropic volume exponent. If the discharge port is located early in the compression phase, the port uncovers before the gas has reached the proper discharge pressure. This means the pressure downstream of the compressor will leak back into the groove and reduce the overall volumetric efficiency. Alternately, if the discharge port is located late in the compression phase, discharge pressure is reached too soon, thus causing an overcompression of the gas which wastes power. This loss is evident in the isentropic efficiency. Compressor manufacturers utilize various means to locate the discharge port properly and maintain peak compressor efficiency.

2.3.2 Heat Transfer Effects

Unlike a reciprocating compressor, screw compressors have the ability to compress up to 20 compression ratios in a single stage. This feature is achievable because a significant amount of coolant is injected into the compression chamber during the cycle. The heat transfer effects between the gas and the coolant allow much higher pressure ratios without the penalty of extremely high discharge temperatures. The compression horsepower relationship is represented as:

$$\text{Power Input} = m_{gas} h_{gas} + m_{coolant} h_{coolant}$$

where

m_{gas} = Mass Flow Rate of Gas
$h_{coolant}$ = Specific Enthalpy Rise of Coolant
$m_{coolant}$ = Mass Flow Rate of Coolant
h_{gas} = Specific Enthalpy Increase of Gas

Screw performance is balanced by a two-fold effect. If there is an increase in the volume of coolant injected into the compression chamber, the effective volume left for the gas is reduced. This, in theory, reduces the capacity. However, the increase in the amount of coolant helps lower the discharge temperature, thereby producing near isothermal compression which improves the compressor isentropic efficiency. Both factors are modeled in screw performance prediction.

2.3.3 Pulsation Effects

Screw compressors are subjected to the same effects of piping pulsation as reciprocating compressors yet at much higher frequencies. As with reciprocating compressors, piping leading to and from the compressor must be sized properly, i.e. proper lengths and diameters. Typical screw machines operate at 3600 rpm. Depending on the type of screw, compression occurs six to 12 times per revolution. Therefore, higher frequencies are important in their effect on performance.

2.3.4 Leakage Effects

The leakage paths for a screw compressor differ from those for a reciprocating compressor in that instead of valves, rings and packing, a screw compressor relies solely on tight running clearances to establish sealing. In a screw compressor, the most significant leakage occurs at

1. The interaction between the meshing rotors
2. The clearance between the rotors and the compressor housing

The impact of these leak paths must be determined to accurately predict screw compressor performance. The leakage is reduced in an oil- or water-flooded compressor by the sealing effect of the liquid.

In a screw machine, compression is the result of two rotating rotors meshing, thereby reducing a volume in the groove of one rotor as part of the other rotor moves in the groove. Tight clearances must be used to maintain the increased pressure in the groove. If the clearance between the two rotors is large, pressurized gas will flow back to a low pressure zone. In a typical screw design, this usually means back to suction. This loss will cause a decrease in volumetric efficiency or capacity.

The potential leakage between the rotors and the housing is harder to quantify in terms of performance loss. The main rotor lobe has two edges, the leading and trailing edges. The leading edge faces the discharge pressure zone, while the trailing edge faces the suction pressure zone. Leakage occurs from the leading edge to the trailing edge. This means that during the compression cycle, the gas in the groove can leak back to the next groove in the screw since it is at lower pressure, and receive higher pressure gas from the groove that precedes it. These two effects do not cancel each other out. They are the direct effect of the running clearance between the rotor and the housing. The performance effects are two fold. Since gas is constantly being transferred from one groove to another, this becomes a fixed flow loss. Similarly, since the machine needs to recompress already worked gas, this becomes a fixed power loss. It is important to determine the effect of these leakages to determine the overall efficiency of the unit.

2.4 ALL COMPRESSORS

2.4.1 Friction

All compressors have sliding parts in the various bearings and seals. Additional power is required to overcome the friction. Any friction that occurs in components exposed to the gas will tend to heat the gas. Depending on the point in the cycle at which the heating occurs, this may, or may not, have a significant effect on the capacity.

CHAPTER 3
COMPRESSOR PERFORMANCE—DYNAMIC

Paolo Bendinelli
Turbocompressors Chief Engineer
Nuovo Pignone

Massimo Camatti
Turbocompressors Design Manager
Nuovo Pignone

Marco Giachi
Turbocompressors R&D Manager
Nuovo Pignone

Eugenio Rossi
Turbocompressors Researcher
Nuovo Pignone

LIST OF SYMBOLS

A = Area
α = Absolute flow angle
b = Blade height
β = Relative flow angle
β_b = Blade angle
B = Blockage factor
C = Absolute velocity module
C_θ = Absolute velocity tangential component
C_m = Absolute velocity meridian component
C_P = Specific heat at constant pressure or pressure recovery coefficient
D = Diameter or diffusion factor
δ = Deviation angle
E = Kinetic energy
ϕ = Flow coefficient
h = Enthalpy
h_0 = Total enthalpy

H = Load
K = Loss coefficient based on total pressure
i = Incidence angle
μ_0 = Viscosity at reference condition
λ = Total pressure recovery coefficient
m = Mass flow rate
M = Mach number
M_U = Peripheral Mach number
n = Polytropic exponent
N = Rotational speed (rpm)
η = Efficiency
Q = Heat exchange or volumetric flow rate
γ = Ratio between heat values of the gas
ρ = Density
ρ_0 = Total density
p = Pressure
ψ = Load coefficient
p_0 = Total pressure
r = radius
R = Gas constant
Re = Reynolds number
σ = Slip factor
τ = Torque or working factor
T = Static temperature
θ = Blade deflection angle
T_0 = Total temperature
U = Tip speed
V_S = Absolute tangential velocity effect
ω = Rotational speed (rad/s)
Z = Compressibility factor or blade number
W = Relative velocity
W_A = Friction losses

3.1 GENERAL DESCRIPTION OF A CENTRIFUGAL COMPRESSOR

A centrifugal compressor is a "dynamic" machine. It has a continuous flow of fluid which receives energy from integral shaft impellers. This energy is transformed into pressure—partly across the impellers and partly in the stator section, i.e., in the diffusers. This type of machine is composed (see Fig. 3.1) of an outer casing (A) which contains a stator part, called a diaphragm bundle (B), and of a rotor formed by a shaft (C), one or more impellers (D), a balance drum (E), and thrust collar (F).

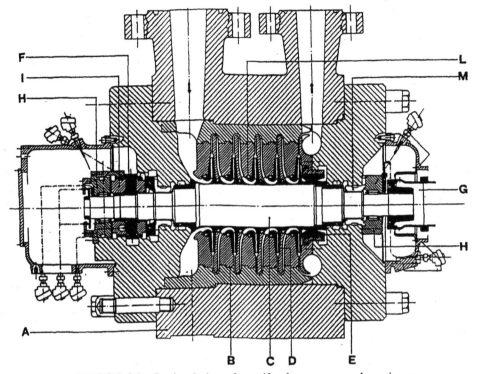

FIGURE 3.1 Sectional view of centrifugal compressor schematic.

The rotor is driven by means of a hub (G) and is held in position axially by a thrust bearing (I), while rotating on journal bearings (H). The rotor is fitted with labyrinth seals (L) and, if necessary, oil film end seals (M).

Gas is drawn into the compressor through a suction nozzle and enters an annular chamber (inlet volute), flowing from it towards the center from all directions in a uniform radial pattern (see Fig. 3.2). At the opposite side of the chamber from the suction nozzle is a fin to prevent gas vortices.

The gas flows into the suction diaphragm and is then picked up by the first impeller (see Fig. 3.3).

The impellers consist of two discs, referred to as the disc and shroud, connected by blades which are shrunk onto the shaft and held by either one or two keys. The impeller pushes the gas outwards raising its velocity and pressure; the outlet velocity will have a radial and a tangential component (see section 3.7 for further details). On the disc side, the impeller is exposed to discharge pressure (see Fig. 3.4) and on the other side partly to the same pressure and partly to suction pressure. Thus a thrust force is created towards suction.

The gas next flows through a circular chamber (diffuser), following a spiral path where it loses velocity and increases pressure (similar to fluid flow through conduits). The gas then flows along the return channel; this is a circular chamber

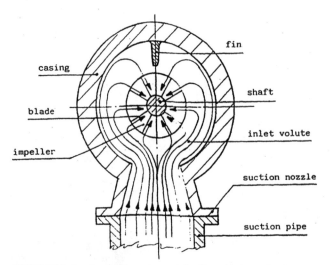

FIGURE 3.2 Qualitative view of the flow in the volute.

bounded by two rings that form the intermediate diaphragm, which is fitted with blades (see Fig. 3.5) to direct the gas toward the inlet of the next impeller. The blades are arranged to straighten the spiral gas flow in order to obtain a radial outlet and axial inlet to the following impeller. The gas path is the same for each impeller.

Labyrinth seals are installed on the diaphragms to minimize internal gas leaks (see Fig. 3.5). These seals are formed by rings made in two or more parts. The last impeller of a stage (the term stage refers to the area of compression between two consecutive nozzles) sends the gas into a diffuser which leads to an annular

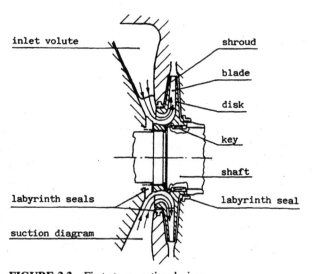

FIGURE 3.3 First stage sectional view.

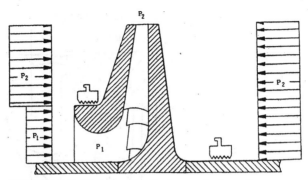

FIGURE 3.4 Pressure distribution on the impeller.

chamber called a discharge volute (see Fig. 3.6). The discharge volute is a circular chamber which collects the gas from the external boundary of the diffuser and conveys it to the discharge nozzle. Near the discharge nozzle there is another fin which prevents the gas from continuing to flow around the volute and directs it to the discharge nozzle (see Fig. 3.7).

The balance drum (E) is mounted on the shaft after the end impeller (see Fig. 3.1). It serves to balance the total thrust produced by the impellers. Having end impeller delivery pressure on one side of the drum, compressor inlet pressure is applied to the other by an external connection (balancing line, see Fig. 3.8). In this way, gas pressures at both ends of the rotor are roughly balanced. To get even closer pressure levels and, therefore, the same operating conditions for the shaft-

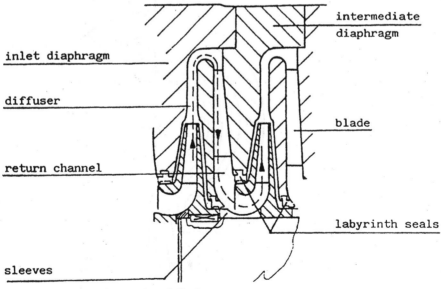

FIGURE 3.5 Labyrinth seals and diaphragms.

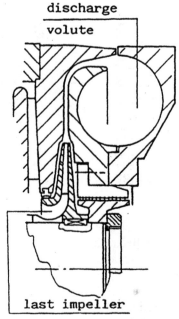

FIGURE 3.6 Last impeller of a stage.

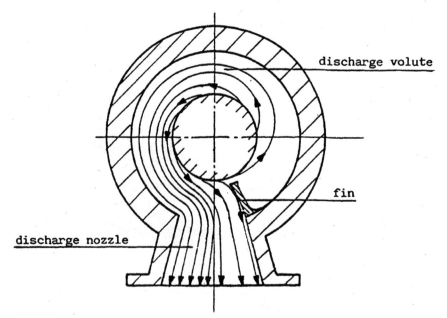

FIGURE 3.7 Discharge volute: qualitative view of the flow.

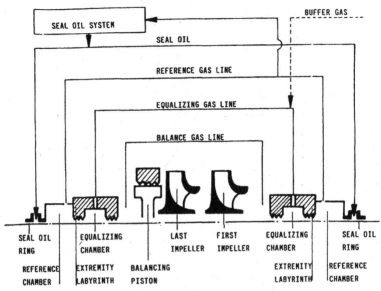

FIGURE 3.8 External connection of the oil system.

end oil seals, another external connection is made between the balancing chambers (balancing line, see Fig. 3.8).

The gas chambers are positioned outside the shaft-end labyrinths. They are connected to achieve the same pressure as that used as reference for the oil seal system (see Fig. 3.8 for a block diagram). In special cases, when the seal oil and process gas have to be kept separate, inert gas is injected into the balancing chamber (buffer gas system) at a pressure that allows it to leak both inwards and outwards forming a seal.

3.2 CENTRIFUGAL COMPRESSORS TYPES

Centrifugal compressors may have different configurations to suit specific services and pressure ratings. They may be classified as follows:

3.2.1 Compressors with Horizontally-split Casings

Horizontally-split casings consisting of half casings joined along the horizontal center-line are employed for operating pressures below 60 bars.

The suction and delivery nozzles as well as any side stream nozzles, lube oil pipes and all other compressor-plant connections are located in the lower casing. With this arrangement all that is necessary to raise the upper casing and gain access to all internal components, such as the rotor, diaphragms and labyrinth seals is to remove the cover bolts along the horizontal center-line.

Horizontally-split casing compressors may be further identified according to the number of stages.

- Multistage compressors with one compression stage only (Fig 3.9).
- Multistage compressors with two compression stages. The two compression stages are set in series in the same machine. Between the two stages, cooling of the fluid is performed in order to increase the efficiency of compression.
- Multistage compressors with more than two compression stages in a single casing. As a rule they are used in services where different gas flows have to be compressed to various pressure levels, i.e., by injecting and/or extracting gas during compression.
- Sometimes compression stages are arranged in parallel in a single casing. The fact that both stages are identical and the delivery nozzle is positioned in the center of the casing makes this solution the most balanced possible. Moreover, a double flow is created by a common central impeller (see Fig. 3.12).

3.2.2 Compressors with Vertically-split Casings

Vertically-split casings are formed by a cylinder closed by two end covers: hence the denotation "barrel," used to refer to compressors with these casings. These machines, which are generally multistage, are used for high pressure services (up

FIGURE 3.9 Horizontally-split casing.

COMPRESSOR PERFORMANCE—DYNAMIC **3.9**

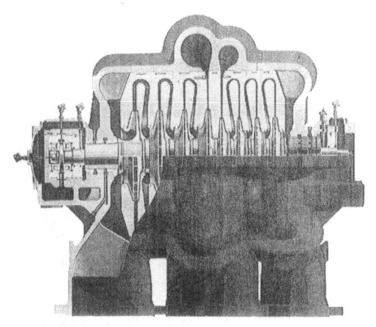

FIGURE 3.10 Multistage two phase compressor.

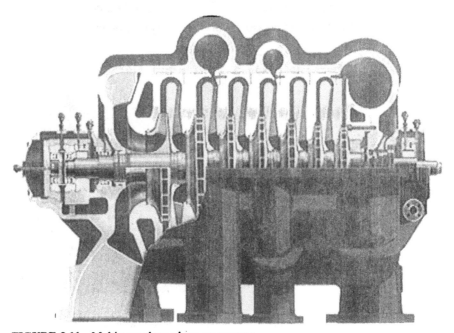

FIGURE 3.11 Multistage three phase compressor.

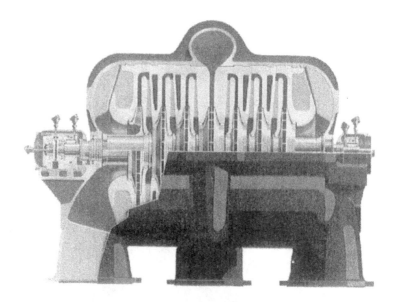

FIGURE 3.12 Two phase compressor with a central double flow impeller.

to 700 kg/cm²). Inside the casing, the rotor and diaphragms are essentially the same as those for compressors with horizontally-split casings.

- Barrel type compressors which have a single compression stage
- Barrel type compressors with two compression stages in series in a single casing

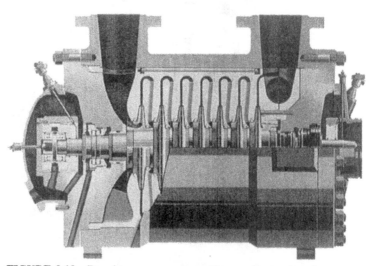

FIGURE 3.13 Barrel type compressor with one compression phase.

FIGURE 3.14 Barrel type compressor with two compression phases.

- Compressors which incorporate two compression stages in parallel in a single casing

3.2.3 Compressors with Bell Casings

Barrel compressors for high pressures have bell-shaped casings and are closed with shear rings instead of bolts (see Fig. 3.15).

3.2.4 Pipeline Compressors

These have bell-shaped casings with a single vertical end cover. They are generally used for natural gas transportation (see Fig. 3.16). They normally have side suction and delivery nozzles positioned opposite each other to facilitate installation on gas pipelines.

3.2.5 SR Compressors

These compressors are suitable for relatively low pressure services. They have the feature of having several shafts with overhung impellers. The impellers are normally open type, i.e., shroudless, to achieve high tip speeds with low stress levels and high pressure ratios per stage. Each impeller inlet is coaxial whereas the outlet is tangential. These compressors are generally employed for air or steam compression, geothermal applications etc. (see Fig. 3.17.).

FIGURE 3.15 High pressure barrel type compressor.

3.3 BASIC THEORETICAL ASPECTS

3.3.1 Preliminary Definitions

The term turbomachinery is used to indicate systems in which energy is exchanged between a fluid, evolving continuously and in a clearly determined quantity, and a machine equipped with rotary blading.

FIGURE 3.16 Pipeline compressor.

FIGURE 3.17 SR type compressor.

Turbomachines can be classified as:

- Process machines, in which the machine transfers energy to the fluid
- Drivers, in which the machine receives energy from the fluid

An initial classification of turbomachines may be made on the basis of the predominant direction of the flow within the machine:

- Axial machines, in which the predominant direction is parallel to the axis of rotation
- Radial machines, in which the predominant direction is orthogonal to the axis, although portions of the flow may have an axial direction
- Mixed machines, where the situation is intermediate between the described above

Turbocompressors (more briefly, compressors) constitute a special category of process machines. They operate with compressible fluids and are characterized by an appreciable increase in the density of the fluid between the first and the last compression stages. The compression process is frequently distributed among several stages, a term used to indicate an elementary system composed of mobile blading, in which the fluid acquires energy, and fixed blading, in which the energy is converted from one form to another.

3.3.2 The Compression Process

Consider Fig. 3.18, which represents a generic compression process in the Mollier plane (enthalpy-entropy) taking place in a single compressor stage. The fluid, taken in determined conditions p_{00} and T_{00}, is subsequently accelerated up to the inlet to the stage where it reaches the conditions defined by thermodynamic state 1. The acceleration process is accompanied by dissipation phenomena linked to the increase in speed of the fluid. In flowing along the rotor the fluid undergoes a transformation that brings it to the conditions p_2 and T_2. During this phase there is an increment in potential energy per mass unit of fluid given by:

$$\Delta E_{P,1-2} = h_2 - h_1 \quad (3.1)$$

and an increment in kinetic energy per mass unit of fluid given by:

$$\Delta E_{K,1-2} = \frac{C_2^2}{2} - \frac{C_1^2}{2} \quad (3.2)$$

The entropy of the fluid, as it flows through the stage, increases as a consequence of the dissipation processes involved in compression. In the stator part the kinetic energy of the fluid is converted into potential energy. The total enthalpy for state 4 can thus be evaluated as:

$$h_{0,4} = h_4 + \frac{C_4^2}{2} \quad (3.3)$$

The fluid then leaves the stage in the conditions defined by state 4, with residual velocity C_4.

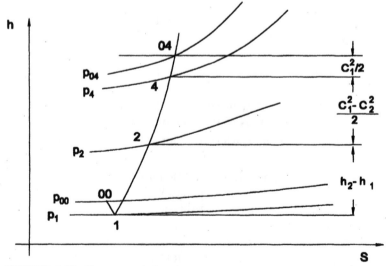

FIGURE 3.18 Entropy-enthalpy diagram of a compression process.

3.3.3 Basic Quantities of Compression Process

The basic quantities utilized to quantify the exchanges of energy in compressors are given below. Note that the quantities indicated apply both to complete compressors and to individual stages. It is also assumed that the thermodynamic characteristics of the fluid are represented by the perfect gas model.

Effective Head. The effective head H_R is defined as the effective work exchanged between blading and fluid per mass unit of fluid processed:

$$H_R = \int_{p_1}^{p_4} dp/\rho \tag{3.4}$$

We also have

$$H_R = (h_{04} - h_{01}) + Q_{\text{EXT}} \tag{3.5}$$

In the hypothesis of adiabatic conditions $Q_{\text{EXT}} = 0$ and we further obtain:

$$H_R = (h_{04} - h_{01}) \tag{3.6}$$

Polytropic Head. The polytropic head H_P is defined as the energy per mass unit accumulated by the fluid under the form of increment in potential energy; it is expressed by:

$$H_P = \int_{p_1}^{p_4} dp/\rho \tag{3.7}$$

in which the relationship between pressure and density is expressed in the form

$$p\rho^{-n} = \cos\tan te \tag{3.8}$$

where n represents the mean exponent of polytropic transformation between the two states 1 and 2. Polytropic head can thus be expressed by the following formula:

$$H_P = \frac{n}{n-1} Z_0 R T_{00} \left[\left(\frac{p_{04}}{p_{00}}\right)^{(n-1/n)} - 1 \right] \tag{3.9}$$

Isentropic Head. Isentropic head is defined as the energy per mass unit accumulated by the fluid subsequent to a reversible (and thus isentropic) adiabatic transformation between states 1 and 2. This gives the following equation:

$$H_S = \int_{p_0}^{p_4} dp/\rho \tag{3.10}$$

with

$$p\rho^{-\gamma} = \cos \tan te \tag{3.11}$$

in which γ constitutes the ratio between the specific heat values of the gas.

$$H_S = \frac{\gamma}{\gamma - 1} Z_0 R T_{00} \left[\left(\frac{p_{04}}{p_{00}}\right)^{(\gamma-1/\gamma)} - 1 \right] \tag{3.12}$$

Polytropic Efficiency. The polytropic efficiency is defined as the ratio between polytropic head H_P and effective head H_R necessary to effect compression between states 0 and 4. Applying the preceding definition we obtain:

$$\eta_P = \frac{H_P}{H_R} = \frac{\dfrac{n}{n-1} Z_0 R T_{00} \left[\left(\dfrac{p_{04}}{p_{00}}\right)^{(n-1/n)} - 1 \right]}{(h_{04} - h_{00})} \tag{3.13}$$

by developing the above equation we obtain:

$$\eta_P = \frac{n\gamma - 1}{(n-1)\gamma} \tag{3.14}$$

The polytropic head can be further rewritten in the form

$$\eta_P = \frac{(\gamma - 1)\ln(p_{04}/p_{00})}{\gamma \ln(T_{04}/T_{00})} \tag{3.15}$$

Polytropic efficiency possesses the important property of being dependent only on the properties of the gas, the pressure, and temperature ratios. It is independent of the absolute pressure level from which the compression process starts.

Isentropic Efficiency. The isentropic efficiency is defined as the ratio between isentropic head H_S and effective head H_R associated with compression between states 0 and 4. From this definition we obtain:

$$\eta_S = \frac{H_S}{H_R} = \frac{\dfrac{\gamma}{\gamma - 1} Z_0 R T_{00} \left[\left(\dfrac{p_{04}}{p_{00}}\right)^{(\gamma-1/\gamma)} - 1 \right]}{(h_{04} - h_{00})} \tag{3.16}$$

The isentropic efficiency can be rewritten as:

$$\eta_S = \frac{\left(\dfrac{p_{04}}{p_{00}}\right)^{(\gamma-1/\gamma)} - 1}{\left(\dfrac{T_{04}}{T_{00}}\right) - 1} \quad (3.17)$$

It may be stated that, for a compressor, the polytropic efficiency is always greater than the isentropic efficiency relevant to the same transformation.

3.3.4 Euler Equation for Turbomachines

With reference to Fig. 3.19, we may consider a rotor belonging to a generic turbomachine, taking into examination the conditions existing in section 1 (inlet) and section 2 (discharge). Utilizing the equation of balance of momentum for the stationary flow between two sections, it is possible to obtain:

$$\tau = m(r_2 C_{\theta 2} - r_1 C_{\theta 1}) \quad (3.18)$$

The work transferred through the blading per mass unit of fluid processed is thus given by:

$$W_x = \tau\omega/m = \omega(r_2 C_{\theta 2} - r_1 C_{\theta 1}) \quad (3.19)$$

The first principle of thermodynamics establishes that the work per mass unit is equal to, for an adiabatic flow, the variation in total enthalpy. We thus obtain:

$$\Delta h_{0,1-2} = h_2 - h_1 = \omega(r_2 C_{\theta 2} - r_1 C_{\theta 1}) = U_2 C_{\theta 2} - U_1 C_{\theta 1} \quad (3.20)$$

The above equation, known as the Euler equation, is one of the fundamentally important equations for the study of turbomachines. Application of the Eq. 3.20 shows that in a generic stage composed of a rotor and a stator, there is no transfer of mechanical work outside of the rotary parts; in particular, then, the enthalpy of

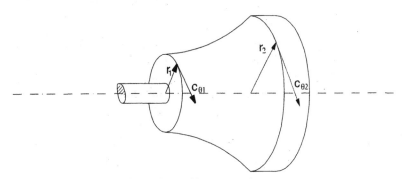

FIGURE 3.19 The Euler turbomachinery equation.

the fluid does not change in traversing the stationary components, but only in traversing the rotary ones. Consequently it may be stated:

$$\Delta h_{0,1-4} = h_{04} - h_{00} = h_{02} - h_{00} = U_2 C_{\theta 2} - U_1 C_{\theta 1} \qquad (3.21)$$

and in the case of perfect gas:

$$\Delta h_{0,1-4} = c_P(T_{04} - T_{02}) = \Delta h_{0,1-2} = c_P(T_{02} - T_{01}) \qquad (3.22)$$

The total temperature is thus constant throughout the stationary components.

3.3.5 Dimensionless Parameters

The behavior of a generic stage can be characterized in terms of dimensionless quantities which specify its operating conditions as well as its performance. The dimensionless representation makes it possible to disregard the actual dimensions of the machine and its real operating conditions (flow rate and speed of rotation) and is thus more general as compared to the use of dimensional quantities.

The number of dimensionless parameters necessary and sufficient to describe the characteristics of a stage is specified by *Buckingham's theorem*, which is also used to *determine their general form*. The dimensionless parameters used to describe the performance of axial and centrifugal compressors are given below.

Flow Coefficient. The flow coefficient for an axial machine is defined as the ratio between the axial velocity at the rotor inlet section and the tip speed of the blade

$$\phi_1 = \frac{C_{1A}}{U_2} \qquad (3.23)$$

For centrifugal machines the flow coefficient is defined as follows:

$$\phi_1 = \frac{4Q}{\pi D_2^2 U_2} \qquad (3.24)$$

Both of these definitions can be interpreted as dimensionless volume flow rate of the fluid processed by the machine.

Machine Mach Number. The Mach number, M_U, is defined as the ratio between the machine tip speed and the velocity of sound in the reference conditions:

$$M_U = \frac{U_2}{a_{00}} = \frac{U_2}{\sqrt{\gamma R T_{00}}} \qquad (3.25)$$

The Mach number can be interpreted as a dimensionless speed of rotation of the machine.

Reynolds Number. The Reynolds number, *Re*, is generally defined as the ratio between inertial forces and viscous forces, evaluated in relation to assigned reference conditions, acting on a fluid particle for the particular fluid-dynamic problem in question.

In the axial machine field, a frequently used formulation for the Reynolds number is the following:

$$Re = \frac{UD}{\mu_{00}/\rho_{00}} \tag{3.26}$$

For centrifugal machines the following formulation is frequently used:

$$Re = \frac{U_2 D_2}{\mu_{00}/\rho_{00}} \tag{3.27}$$

Specific Heat Ratio. The specific heat ratio is simply defined as the ratio between specific heat at constant pressure and at constant volume for the gas in question:

$$\gamma = \frac{C_P}{C_V} \tag{3.28}$$

The specific heat ratio is used to take account of the thermodynamic properties of the fluid.

Coefficients of Work and of Head. The coefficient of work for an axial machine is defined as the ratio between the work per mass unit transferred by the blading to the fluid and the square of the tip speed.

$$\psi = \frac{h_{02} - h_{00}}{U_2^2} = \frac{C_{\vartheta 2} - C_{\vartheta 1}}{U_2} \tag{3.29}$$

In the above formula, the Euler equation and the kinematic equality $U_1 = U_2$, valid in first approximation for an axial machine, have been introduced.

For a centrifugal machine an identical parameter is defined, which is however called head coefficient τ. It is expressed by the following equation:

$$\tau = \frac{h_{02} - h_{00}}{U_2^2} = \frac{U_2 C_{\vartheta 2} - U_1 C_{\vartheta 1}}{U_2^2} \tag{3.30}$$

The two quantities defined above can be interpreted as dimensionless work per mass unit transferred by the blading to the fluid.

Polytropic Efficiency. The same definition given in 3.3.3 is applied.

$$\eta_P = \frac{H_P}{H_R} = \frac{\dfrac{n}{n-1} Z_0 R T_{00} \left[\left(\dfrac{p_{04}}{p_{00}}\right)^{(n-1/n)} - 1\right]}{(h_{04} - h_{00})} \tag{3.31}$$

This formula applies to both centrifugal and axial machines.

3.4 PERFORMANCE OF COMPRESSOR STAGES

3.4.1 General Information

In any one of the compressor stages, work is transferred by the rotary blading to the fluid in modes depending on the geometry, the fluid-dynamic conditions and the properties of the gas processed. The study of these energy interactions, governed by the Euler Eq. (3.20), calls for analysis of the speed of the fluid in suitable sections of the stage. This analysis is usually carried out utilizing speed triangles determined in suitable sections of the stage.

The quantity of energy absorbed by the compressor cannot be entirely converted into a pressure increment in the fluid due to dissipation phenomena of various kinds involving the machine as a whole. Among these, the pressure drops directly attributable to effects of aerodynamic type will be examined here.

Knowledge of the energy transfer and dissipation mechanisms in a stage provides the necessary tools for understanding the factors that determine and influence its performance. These aspects are examined in the following paragraphs, along with the representations normally utilized to describe performance.

3.4.2 Speed Triangles

In studying turbomachines the concept of speed triangles is frequently used to represent the kinematic conditions, for both fluid and blade, existing at the inlet and discharge sections of a generic fixed or rotary blading.

The speed triangles for an axial compressor stage are shown in Fig. 3.20. Note that the absolute velocity C of the fluid in a given section of the stage is obtained

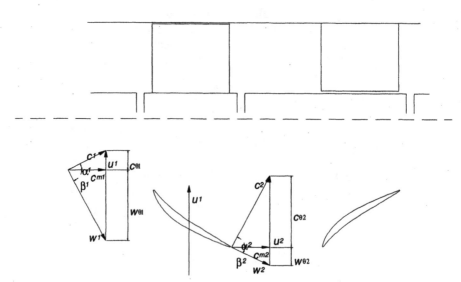

FIGURE 3.20 Velocity (speed) triangles for an axial compressor stage.

by combining a relative velocity W with a velocity U determined by the rotation of the blade. The absolute velocity C can be further broken down into an axial velocity C_A and a tangential velocity C_q.

The speed triangle on discharge from the rotor is characterized by the fact that the direction of the absolute velocity vector does not exactly coincide with the direction indicated by the trailing edge of the blade. This phenomenon, termed deviation, determines a reduction in the value of C_q in respect to the value that could be theoretically obtained in the case of null deviation.

Recalling the Euler equation and the definition of work coefficient, we may write:

$$\psi = \frac{h_{02} - h_{00}}{U_2^2} = \frac{C_{\theta 2} - C_{\theta 1}}{U_2} \tag{3.32}$$

With reference to the speed triangle in Fig. 3.20 and also considering the simplifying assumption that the flow can be considered incompressible between sections 1 and 2, the formula for the work coefficient can be rewritten as follows:

$$\psi = U\left[1 - \frac{C_x}{U}(tg\alpha_1 + tg\beta_2)\right] = U[1 - \phi_1(tg\alpha_1 + tg\beta_2)] \tag{3.33}$$

The above equation shows that, in the further hypothesis that the direction of flow does not change from blade inlet to outlet, the relation between flow coefficient and work coefficient depends in linear manner on $(tg\alpha_1 + tg\beta_2)$ in the mode shown in Fig. 3.21. In the hypothesis of compressible flow and non-constant angles the relation is no longer linear but the qualitative description is still valid.

The speed triangles for a centrifugal stage are shown in Fig. 3.22 The physical interpretation of the quantities is the same as that of the axial machine, although the meridian rather than the axial components of the quantities represented should be taken into consideration. In centrifugal machines too the phenomenon of devi-

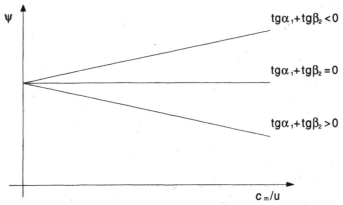

FIGURE 3.21 Loading coefficient vs. flow coefficient for an axial stage.

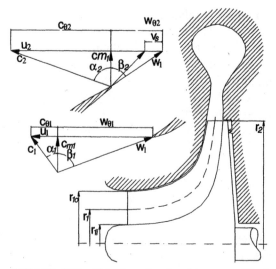

FIGURE 3.22 Velocity triangles for a centrifugal stage.

ation, conventionally termed slip, can be observed, so that the relative velocity on discharge from the impeller is not aligned with the direction of the blade.

The head coefficient for a centrifugal compressor can be expressed as follows:

$$\tau = \frac{h_{02} - h_{00}}{U_2^2} = \frac{U_2 C_{\vartheta 2} - U_1 C_{\vartheta 1}}{U_2^2} = \frac{U_2 C_2 \cos \alpha_a - U_1 C_1 \cos \alpha_1}{U_2^2} \quad (3.34)$$

The dependency between the structural angle b_2 and t can be expressed in explicit form by introducing the quantity:

$$\phi_2 = \frac{C_{2m}}{U_2} = \frac{Q_2}{\pi b_2 D_2 U_2} \quad (3.35)$$

called flow coefficient at the impeller discharge section. A quantity σ, termed slip factor, which takes account of the imperfect guiding action of the impeller, is also introduced; it may be defined as:

$$\sigma = 1 - \frac{V_S}{U_2} \quad (3.36)$$

where the term V_S represents the tangential velocity defect associated with the slip effect.

Utilizing these definitions and hypothesizing inlet guide vane conditions null ($C_{q1} = 0$), Eq. (3.32) is rewritten as:

$$\tau = \frac{C_{\theta 2}}{U_2} = \sigma - \phi_2 tg\beta_{b2} \tag{3.37}$$

The above equation is illustrated in Fig. 3.23, which is the equivalent of the one already given for axial machines. For centrifugal compressors, geometries with structural angles b_{b2} greater than zero (i.e., blades turned in the same direction as that of rotation) are not utilized insofar as they generate high pressure drops. Radial blades or those turned in the direction opposite that of rotation up to b_{b2} values of about −60 degrees are normally used in common applications.

3.4.3 Conventional Representation of Pressure Drop in Compressors

Pressure drops in compressors are conventionally divided into two main categories.

1. Pressure drop due to friction
2. Pressure drop due to incidence

These two phenomena are discussed in the following paragraphs.

Pressure Drop Due to Friction. These are dissipation terms associated with friction phenomena between the walls of the ports of the machine (both rotor and stationary) and fluid flowing through it. In general, the flow in compressors is characterized by turbulence, so it can be considered that the energy dissipated is proportional, in first approximation, to the square of the fluid velocity and thus to the square of the volume flow in inlet conditions. This energy is not transferred to the fluid under the form of potential energy, but only under the form of heat.

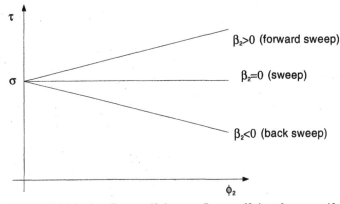

FIGURE 3.23 Loading coefficient vs. flow coefficient for a centrifugal stage.

Accordingly, indicating by W_A the work per mass unit associated with dissipation due to friction, we may write:

$$W_A = k_A Q_{00}^2 \tag{3.38}$$

where k_A represents a suitable constant that takes account of the specific fluid-dynamic characteristics of the stage in question. In dimensionless terms this can be expressed as:

$$W_A = k_A \phi_1^2 \tag{3.39}$$

Pressure Drop Due to Incidence. With reference to the classic studies on bi-dimensional wing contours, it should be recalled that the pressure drop of a generic contour, stated in relation to the incidence, follows a trend of the type shown in Fig. 3.24. This distribution can be approximated with a parabolic law that presents a minimum point at a certain incidence i^*.

Although the behavior described here applies, strictly speaking, to wing contours alone, it may be extended with reasonable accuracy to the blading of centrifugal machines as well. It is thus possible to define, for a generic turbomachine blading, whether stationary or rotary, an optimum incidence condition, at which the pressure loss phenomena deriving from incidence are minimum.

This optimum incidence value depends on the geometry of the blade and on the speed triangle immediately upstream of the blade leading edge. When the speed of rotation and the geometry have been assigned, the speed triangle and the incidences depend only on the volume flow of the processed fluid.

Pressure drop due to incidence may therefore be expressed in the form:

$$W_I = k_U (Q_{00} - Q_{00}^*)^2 + k_0 \tag{3.40}$$

This equation can be expressed in dimensionless form:

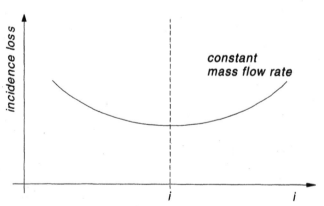

FIGURE 3.24 Typical airfoil losses distribution as a function of incidence.

$$WyI = k_U(\phi_1 - \phi_1^*)^2 + k_0 \qquad (3.41)$$

The constants k_0 and k_1 are once again associated to the particular problem considered.

Overall Pressure Drop. In going on to consider the overall pressure drop of a compressor stage, note that in general both of the above-mentioned contributions will be present. The overall pressure loss can thus be represented as:

$$W_T = W_I + W_A = k_A Q_{00}^2 + k_U(Q_{00} - Q_{00}^*)^2 + k_0 \qquad (3.42)$$

or in dimensionless form:

$$W_T = k_A \phi_1^2 + k_U(\phi_1 - \phi_1^*)^2 + k_0 \qquad (3.43)$$

The above equation can be given significant graphic interpretation, as shown in Fig. 3.25.

Curve A represents the evaluated relationship between flow coefficient and head, evaluated taking account of the effective deviation phenomena that occur in a real blading. Curve A thus represents all of the energy per mass unit that is transferred to the fluid and that is thus theoretically available to be converted under the form of pressure. This quantity is diminished by the dissipation associated with pressure drop due to friction (curve B) and pressure drop due to incidence (curve C), both expressed by parabolic equations. Point C1 thus expresses the work which is effectively contained in the fluid under the form of potential energy and kinetic energy for an assigned flow coefficient f_1.

From this analysis, it can be stated that, due to the shapes of the various curves considered, the quantity defined above tends to present a maximum in coincidence with a clearly determined value of f_1.

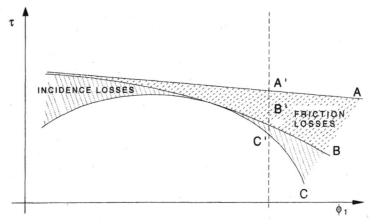

FIGURE 3.25 Losses breakdown as a function of the flow coefficient.

Other Pressure Drop Contributions. The description given in the preceding paragraphs of subsection 3.4.3 provides a substantially correct illustration of the main effects linked to pressure drop in a generic compressor and their influence on overall performance. However, real compressors present dissipation effects, which do not fall within the simplified scheme of pressure drops due to incidence and pressure drops due to friction. In these cases, it is often necessary to classify the pressure drop contributions through detailed reference to the physical modes in which dissipation takes place.

For centrifugal machines the main effects of additional pressure drop are linked to the presence of the blade tip and casing recess (in open machines), to ventilation phenomena between the rotating and stationary surfaces in the spaces between hubs and diaphragms, and to the presence of end seals and interstage seals. Further pressure drops can be attributed to the presence of separation areas in the impeller.

For axial machines, the representation of pressure drop is slightly different to take account of the different aerodynamic phenomena involved. One possible distinction could be the following:

- Contour pressure drop. This is pressure drop deriving from the presence of boundary layers which develop along the blade surfaces. It can be estimated through the methods used for calculating turbulent boundary layers.
- Endwall pressure drop. Pressure drop of this kind depends on the presence of localized limit states on the casing surface or the compressor rotor. These effects are usually evaluated through experimental correlations.
- Pressure drop due to impact. This term indicates phenomena of the dissipation type linked to the generation of impact waves and to consequent production of entropy. In general these consist of leading edge impact waves and port impact waves, depending on the place where these effects occur. These phenomena tend to involve all types of compressors, both axial and centrifugal, with the exception of the totally subsonic ones.
- Pressure drop due to mixing. This consists of irreversibility associated with transition between a non-uniform fluid-dynamic state, linked for example to local separation effects, and a uniform condition. These phenomena take place in the regions downstream of the stator or rotor blade arrays and are estimated through experimental correlations.

In spite of the physical diversity of the pressure drop contributions involved, the qualitative considerations on the overall pressure drop curves, presented in subsection 3.4.3 in the paragraph entitled Overall Pressure Drop, remain valid in a general sense for both axial and centrifugal compressors.

3.4.4 Operating Curve Limits: Surge and Choking

The operating curves of the stages, both centrifugal and axial, present limits to the flow ranges that can be processed by the stage itself or by the machine of which it is a part. These limits are established by two separate phenomena, called *surge* and *choking*, described below.

Surge. The term "surge" indicates a phenomenon of instability which takes place at low flow values and which involves an entire system including not only the compressor, but also the group of components traversed by the fluid upstream and downstream of it. The term "separation" indicates a condition in which the boundary layer in proximity to a solid wall presents areas of inversion of the direction of velocity and in which the streamlines tend to detach from the wall. Separation is in general a phenomenon linked to the presence of "adverse" pressure gradients in respect to the main direction of motion, which means that the pressure to which a fluid particle is subjected becomes increasingly higher as the particle proceeds along a streamline.

The term "stall," referring to a turbomachine stage, describes a situation in which, due to low flow values, the stage pressure ratio or the head do not vary in a stable manner with the flow rate. Stall in a stage is generally caused by important separation phenomena in one or more of its components.

Surge is characterized by intense and rapid flow and pressure fluctuation throughout the system and is generally associated with stall involving one or more compressor stages. This phenomenon is generally accompanied by strong noise and violent vibrations which can severely damage the machines involved.

Experience has shown that surge is particularly likely to occur in compressors operating in conditions where the Q-H curve of the machine has a positive slope. Less severe instability can moreover take place also in proximity to areas of null slope. This depends on the presence of rotary stall, defined as the condition in which multiple separation cells are generated which rotate at a fraction of the angular velocity of the compressor.

Surge prevention is effected through experimental tests in which pressure pulsation at low flow rates is measured on the individual stages. On this basis, it is possible to identify the flow values at which stable operation of the stage is guaranteed. A knowledge of the operating limits of each stage can then be used to evaluate the corresponding operating limits of the machine as a whole.

Choking. Assume that a stage of assigned geometry is operating at a fixed speed of rotation and the flow rate of the processed fluid is increasing. A condition will ultimately be reached at which, in coincidence with a port, the fluid reaches sonic conditions. In this situation, termed "choking," no further increase in flow rate will be possible and there will be a rapid, abrupt decrease in the performance of the stage.

The occurrence of choking depends not only on the geometry and operating conditions of the stage, but also on the thermodynamic properties of the fluid. In this regard, choking can be particularly limiting for machines operating with fluids of high molecular weight, such as coolants.

Many types of compressors, including industrial process compressors, normally operate in conditions quite far from those of choking. For these machines, the maximum flow limit is frequently defined as the flow corresponding to a prescribed reduction in efficiency in respect to the peak value.

3.4.5 Performance of Stages

The discussion contained in the previous paragraphs provides the necessary elements for understanding and interpreting the global performance of a generic stage and the manner in which it is usually represented through suitable diagrams. This subject is further discussed in the next two paragraphs.

Dimensionless Representation of Performance. A possible dimensionless presentation of stage performance can be effected as shown in Fig. 3.26. The interpretation of the various parameters utilized is the one given by the definitions provided above.

The dimensionless representation is such that once the design values for the flow coefficients and the Mach number have been established, the behavior expressed by the curves is independent of the actual size of the stage.

Dimensional Representation of Performance. The dimensionless performance of the stage being known, it is possible to obtain a representation in dimensional form with the use of equations given in (3.4) to (3.17). One possible description of this type is given in Fig. 3.27.

The conditions of the gas on discharge from the stage can be evaluated once the gas properties and the stage inlet conditions, defined by the pressure p_{00} and the temperature T_{00}, have been specified. In cases where the behavior of the gas can be diagrammed through the perfect gas model, we will have for instance:

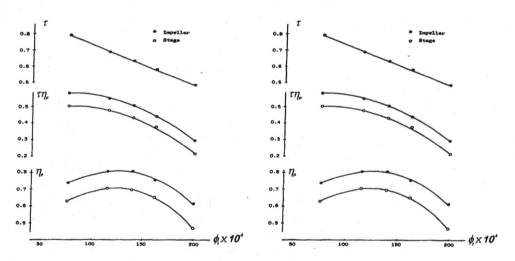

FIGURE 3.26 Non-dimensional performance curves for a stage.

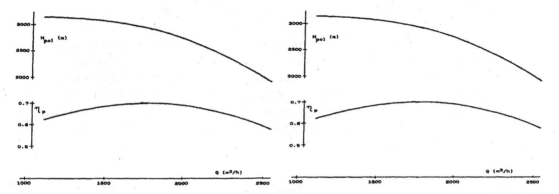

FIGURE 3.27 Dimensional performance curves for a stage.

$$\lambda = \frac{p_{04}}{P_{00}} = \left[1 + \frac{H_P}{\frac{\gamma}{\gamma-1}\eta_P Z_0 RT_0}\right]^{(\gamma/\gamma-1)\eta_P}$$

$$T_{04} = T_{04}\lambda^{\gamma-1/\gamma\eta_P}$$

$$\rho_{04} = \frac{P_{04}}{Z_4 RT_{04}} \tag{3.44}$$

In cases where the perfect gas model is not applicable, it becomes necessary to apply an equation of state for real gases, for example, of the type defined by the *Benedict-Webb-Rubin-Starling model*.

3.5 MULTISTAGE COMPRESSORS

3.5.1 General Information

The pressure ratio obtainable with a simple single-stage compressor is normally limited by constraints of both aerodynamic and structural type.

In the field of centrifugal compressors for aeronautic applications, unitary pressure ratios of about 12 have been obtained. In industrial applications the values are much lower, usually not exceeding the limit of three. The unitary pressure ratios of the centrifugal stages are limited mainly by the maximum tip speed allowable in relation to the structural integrity requirements of the rotor and thus of the material of which it is built.

For axial compressors, the maximum unitary pressure ratio obtained in advanced compressors for aeronautic applications is about 2.5. In this case, the unitary pressure ratio is constrained essentially by limitations of the aerodynamic type linked to the need to keep the work transferred to the fluid within acceptable limits so as to avoid stall.

In all situations where the pressure ratio exceeds the maximum unitary value for the particular type of compressor in question it becomes necessary to recur to a multistage arrangement with two or more stages arranged in series in a repetitive configuration. The methods employed are analyzed here, with determination of the operating curves of a generic multistage compressor, taking into consideration the problems involved in the coupling of the various stages in both design and off-design conditions.

3.5.2 Multistage Compressor Operating Curves

In selecting the stages that make up the complete machine, an obvious consideration is that each of them should be utilized in conditions of maximum efficiency. The efficiency of a stage is maximum in the design condition identified by a given value of the flow coefficient f_1, a value which decreases progressively in moving away from this condition.

In designing a multistage compressor, each individual stage must be utilized around the design condition, accepting a performance slightly lower than that of design, since it is impossible, in practice, to size the individual stage for each specific design condition relevant to the complete compressor. It thus becomes necessary to establish suitable operating conditions, different from those of design, at which the efficiency of each individual stage is satisfactory while margins are provided as regards stall and choking.

Determination of the global compressor curves requires knowledge of the performance curves of each of its individual stages. In the case of a multistage centrifugal compressor which will be examined below, the performance of the individual stage can be represented by the following parameters:

$(f_1)_i^*$ = design flow coefficient of nth stage

$\left(\phi_1 \dfrac{\rho_{01}}{\rho_{06}}\right)_i^*$ = design flow coefficient of nth stage corrected for variation in density between inlet and discharge

$t_{M_u'}^*$ = head coefficient corresponding to $\left(\phi_1 \dfrac{\rho_{01}}{\rho_{06}}\right)_i^*$ and to $M_u = M_u^*$

$h_{PM_u^*,D2}^*$ = polytropic efficiency for $(f_1)_i = (f_1)_i^*$ and for $M_u = M_u^*$ corresponding to a given reference diameter

The mode in which the performance of a stage varies around design conditions must also be specified. This can be done utilizing curves that describe the behavior of the head coefficient and the efficiency in relation to independent parameters. A possible general form of this representation is:

$$\left(\frac{\eta_P}{\eta_P^*}\right)_i = f_1 \left(\frac{\phi_1}{\phi_1^*}\right)_i \bigg/ \left[1 + f_2 \left(\frac{\phi_1}{\phi_1^*}\right)_i\right] \qquad (3.45)\text{a}$$

$$\left(\frac{\tau_P}{\tau_P^*}\right)_i = f_3 \left[\left(\phi_1 \frac{\rho_{01}}{\rho_{06}}\right)_i \bigg/ \left(\phi_1 \frac{\rho_{01}}{\rho_{06}}\right)_i^*\right] \qquad (3.45)\text{b}$$

in which the functions f_1, f_2 and f_3 express dependencies that can be made explicit through experimental tests where the performance of each individual stage is measured for a particular set of operating conditions. The above-mentioned curves are also associated to suitable constraints that represent the operating limits for the stage in question relevant to choking and surge, also determined through testing.

Calculation proceeds from the first stage, first evaluating the dimensionless parameters f_1 and M_u relevant to a generic operating condition defined by volume flow rate, Q and speed of rotation N (e.g., for the design condition). The conditions on outlet from the first stage are then calculated utilizing equations of the type (3.44) and introducing various corrections to take account of the effects of the Reynolds number. The subsequent stages are then calculated in sequence, ultimately determining the compressor discharge conditions.

For off-design conditions, the volume flow rate and speed of rotation are varied in parametric manner to obtain the performance levels relevant to a prescribed set of operating conditions. Through calculation it is also possible to verify the conditions corresponding to the operating limits of the compressor and to identify the stages responsible for any surge or choking. If the working gases cannot be represented through the perfect gas diagram it will be necessary to use a real gas model to calculate the thermodynamic state on inlet to and discharge from each stage.

A typical complete compressor map, evaluated for different speeds of rotation, is shown in Fig. 3.28.

3.5.3 Effect of Variation in Flow Rate on Stage Coupling

In evaluating the behavior of a multistage compressor, changes in the operating conditions of the individual stages consequent to variations in flow rate should be examined.

For this purpose we may consider Fig. 3.29, which shows the Q curve for all of the stages of a multistage compressor at design speed of rotation. Assume that the first stage operates at its own design flow rate Q_1. In this condition, the density of the fluid on discharge from the stage is known and it is possible to evaluate the volume flow rate Q_2 for the second stage, which is hypothesized as being that of design.

If the flow rate Q_1 is decreased by a quantity DQ_1, the first stage will then operate at a pressure ratio higher than in the preceding situation. In this case, it can be seen that the density of the fluid on inlet to the second stage is increased,

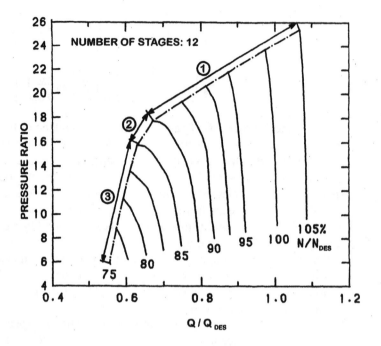

① STALL LIMIT IN END STAGES
② STALL LIMIT IN INTERMEDIATE STAGES
③ STALL LIMIT IN FRONT STAGES

FIGURE 3.28 Performance map for a multistage compressor.

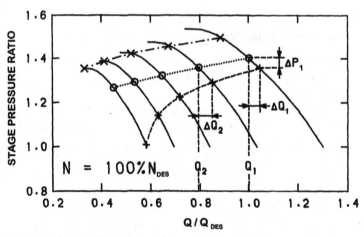

FIGURE 3.29 Effect of mass flow rate in a multistage compressor.

so that the volume flow rate of the second stage is decreased by a quantity $DQ_2 > DQ_1$ in respect to the value Q_2. All this shows that flow perturbation tends to "amplify" in proceeding from the first to the last stage, increasingly so in proportion to the number of stages. From the figure it can be seen that a slight variation in flow rate on the front stage ultimately produces stall in the last stage.

When the flow rate is increased by a quantity DQ_1 the pressure ratio in the front stage decreases, so that the density of the fluid on inlet to the second stage decreases in respect to the design value. In this case too there is a change in flow rate $DQ_2 > DQ_1$, resulting in an "amplification" effect capable of determining final choking in the last stage.

These considerations show that in a multistage compressor where the stages have been correctly coupled, compressor stall and possible surge are always determined by stall in the final stage due to diminution in its volume flow rate. In the same way, compressor choking is determined by choking in the final stage, operating at increased volume flow rate values.

3.5.4 Effect of Variation in Speed on Stage Coupling

In similar manner, the behavior of a multistage compressor is influenced by variations in speed regardless of the characteristics of the individual stages.

Consider Fig. 3.30 which shows the Q curves of the individual stages for a speed of rotation lower than that of design. In this case, the reduction in speed of rotation determines an increment in fluid density from one stage to the next that is lower than at design speed. Since the stable operating range of the compressor is determined by the range of the last stage, it will be the latter to determine the volume flow rate of stall and of choking. Moreover, since the increment in density is lower

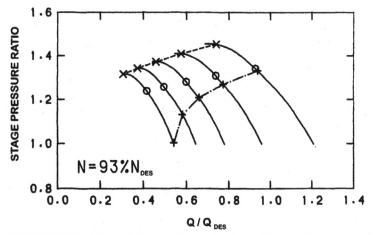

FIGURE 3.30 Effect of decrease in rotational speed in a multistage compressor.

than at design speed, it follows that the front stages will move toward low flow rates and high pressure ratios as compared to the design values.

These considerations show that at very low speeds of rotation, the front stage may operate in conditions of pronounced stall, while the final stage is working in conditions approaching those of choking. In this situation, the operating range is practically reduced to a single point and the compressor entirely loses its flexibility.

Let us now consider an increment in the speed of rotation in respect to that of design (Fig. 3.31). In this case, there is a greater increase in fluid density along the compressor, so that the front stages are operating in fields of high flow rates and low pressure ratios.

3.5.5 Families of Centrifugal Stages

The concept of families of stages is frequently utilized in the multistage centrifugal compressor field. This term indicates a group of stages having the same basic geometry and the same design parameters f_2^* and M_U^*, studied to cover a certain range of flow coefficients f_1. The individual stages belonging to a certain family are designed for a given value of f_1^* and have an assigned range of operating flow rates. The values of the flow coefficients and the flow ranges of the individual stages are defined in such a way as to continuously cover a range of flow coefficients, thus defining the characteristic range of the family. For this purpose, the following are considered:

f_{1i}^*	design flow coefficient of nth stage
f_{1+1}^*	design flow coefficient of nth stage
$f_{1,MAX}^*$	maximum design flow coefficient for the family
$f_{1,MIN}^*$	maximum design flow coefficient for the family
$e_S = f_{1i,S}/f_{1i}^*$	left limit of nth stage selection range

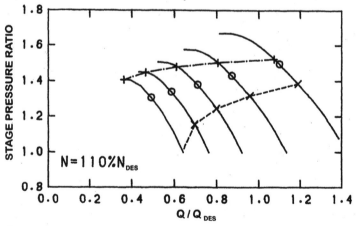

FIGURE 3.31 Effect of increase in rotational speed in a multistage compressor.

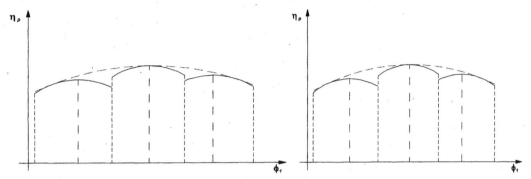

FIGURE 3.32 Ranges of design flow coefficients for a family of stages.

$e_D = f_{1i,D}/f_{1i}^*$ left limit of nth stage selection range
$f_{1i,S}$ left limit of nth stage flow range
$f_{1i,D}$ left limit of nth stage flow range

Assuming contiguity of the flow ranges between the nth stage and the $(n + 1)$th stage, we obtain:

$$e_S f_{1i}^* = e_S f_{1i+1}^* \tag{3.46}$$

Assuming that the operating range of the family is covered by n stages and that e_S = cos t, e_D = cos t, we will have:

$$\frac{\phi_{1,i+1}^*}{\phi_{1,i}^*} = \frac{\varepsilon_D}{\varepsilon_S} = C = \cos t \tag{3.47}$$

and thus:

$$C = \left(\frac{\phi_{1,\text{MAX}}^*}{\phi_{1,\text{MIN}}^*}\right)^{n/n-1} \tag{3.48}$$

so that ultimately:

$$n = 1 + \ln\left(\frac{\phi_{1,\text{MAX}}^*}{\phi_{1,\text{MIN}}^*}\right) \Big/ \ln C \tag{3.49}$$

In practical applications, however, selection must take into account the various effects that contribute to determining impeller performance, so that it is more complex than the simplified diagram shown here.

3.5.6 Standardization of Centrifugal Stages

The vast and highly diversified nature of applications for industrial centrifugal compressors calls for stages capable of working in extremely variable operating conditions. From the engineering viewpoint, this means designing and testing stages have very different geometries and highly variable values of f_1, f_2 and M_U. Considering the limitations of a single stage in terms of operating range, this would call for the realization of an enormous number of stages, with consequent high costs and uncertainty in predicting performance.

The "family of stages" concept, by extending the operating range of a generic stage, provides a tool for simplifying these problems inasmuch as it reduces variation in the design parameters involved and thus reduces, in the final analysis, the number of stages to be designed and tested. In this case, a suitable group of families is defined to cover ample variations in the design parameters, particularly those of the quantities f_1 and M_U. Within a single family, the individual stages are typically designed to cover a much narrower range. This procedure makes it possible to reduce the number of stages to be tested, thus cutting down on time and costs of engineering and development.

In summary, the availability of an effectively standardized group of stages, accompanied by suitable procedures for coupling them, is an element of primary importance in the realization of multistage compressors.

3.6 THERMODYNAMIC AND FLUID-DYNAMIC ANALYSIS OF STAGES

3.6.1 General Information

As mentioned in the introduction, thermodynamic and fluid-dynamic analysis of compressor stages is at present conducted through methodologies based on a number of highly diversified physical models and assumptions. A convenient classification of these methods may be made on the basis of the type of hypothesis formulated to analyze the machine flow rate. On the most general level we may distinguish between:

- Monodimensional methods. This term indicates a group of models deriving from application of the hypothesis of monodimensional flow in the stage.
- Non-viscous methods. This refers to numerical techniques based on flow analysis in the individual components of the stage in the approximation of non-viscous flow.
- Viscous methods. These methods are based on flow analysis conducted through numerical integration of the flow viscous equations.

3.6.2 Monodimensional Methods

The monodimensional approach may be considered the most elementary level of representing the fluid-dynamic characteristics of a centrifugal stage. It is based on the assumption that the fluid-dynamic and thermodynamic states in a given section of the machine can be described in terms of a single condition, which represents a mean value of the actual conditions present in the section.

The basic aspects of the single-area monodimensional approach are outlined below. A specific operating condition is assumed, defined by the following parameters, assumed to be known:

p_{00} = total inlet pressure
T_{00} = total inlet temperature
m = mass flow rate
N = impeller speed of rotation

It is also considered that the fluid thermodynamic properties γ, C_p, R are known. It is assumed that the flow is uniform in the inlet section, adiabatic and stationary in respect to a datum point integral with the rotating components.

Analysis of Impeller Inlet Section. The flow between sections 0 (stage inlet section) and 1 (impeller inlet section) can usually be considered isentropic. In accordance with the hypotheses formulated above it can be stated that:

$$p_{01} = p_{00} \tag{3.50}$$

$$T_{01} = T_{00}$$

The conditions in section 1 can be evaluated by applying the continuity energy and momentum equations. Determination of the quantities relevant to the streamline passing in proximity to the blade tip ($r = r_{1o}$) is particularly important since it is here that the highest relative Mach numbers are found.

The meridian component of the absolute velocity C_{m1} can be determined through the continuity equation:

$$C_{m1} = m/r_1 \ A_1 \tag{3.51}$$

where $A_1 = c_D \ (r_{1o}^2 - r_{1i}^2)$
c_D = blockage factor due to presence of the blades.

The tangential component of the absolute velocity C_{q1} depends on whether or not inlet guide vanes are utilized. In the absence of vanes, we will have $C_{q1} = 0$.

Consequently, it is possible to resolve the rotor inlet speed triangle, illustrated in Fig. 3.33, through the following equations:

$$C_1 = (C_{m1}^2 + C_{t1}^2)^{1/2} \tag{3.52}$$

and also:

$$W_1 = ((U_{1o} - C_{q1})^2 + C_{m1}^2)^{1/2}$$

$$U_{1o} = 2 \ p \ r_{1o} \ N \tag{3.53}$$

The local absolute Mach number is given by:

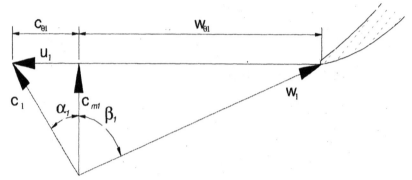

FIGURE 3.33 Impeller inlet velocity triangle.

$$M_{1o} = C_1/(g R T_2)^{1/2} \tag{3.54}$$

Pressure and temperature are linked to the Mach number through the following equations:

$$p_{01}/p_1 = (1 + (g - 1) M_{1o}^2/2)^{(\gamma/(\gamma-1))}$$

$$T_{01}/T_1 = (1 + (g - 1) M_{1o}^2/2) \tag{3.55}$$

A further relation is provided by the perfect gas equation of state:

$$r_1 = p_1/R T_1 \tag{3.56}$$

Equations (3.51) through (3.56), applied, if necessary, to determination of the speed triangle in coincidence with an arbitrary radius r, fully characterize the conditions present in the rotor inlet section.

Analysis of Impeller Discharge Section. As the next step the basic equations of fluid mechanics can be utilized to evaluate the conditions existing in the rotor discharge section.

In this regard, consider the speed triangle relevant to section 2, shown in Fig. 3.34.

The velocity U_2 can be obtained through the simple kinematic equation:

$$U_2 = 2 p r_2 N \tag{3.57}$$

The meridian component of the absolute velocity of the fluid is calculated here too through the flow continuity equation:

$$C_{m2} = m/r_{2A2} \tag{3.58}$$

with $A_2 = 2 p r_2 b_2$.

The tangential component C_{q2} is given by:

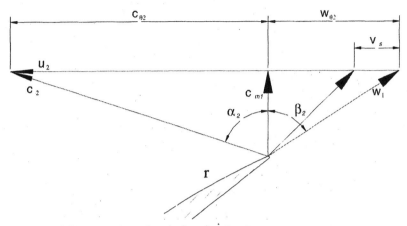

FIGURE 3.34 Impeller exit velocity triangle.

$$C_{q2} = U_2 + C_{m2}\tan(\beta_{b2}) - V_s \qquad (3.59)$$

where β_{b2} is the structural angle of the blade at the discharge section and V_s represents tangential speed defect associated with the slip factor s:

$$V_S = U_2(1 - s) \qquad (3.60)$$

Numerous correlations between slip factor and rotor geometry, obtained both theoretically and experimentally, are available for an estimation of C_{q2} to be used in design problems. A frequently used correlation is the following, proposed by Wiesner:

$$\sigma = 1 - \frac{\sqrt{\cos(\beta_{b2})}}{Z^{0.7}} \qquad (3.61)$$

valid for $R_{1i}/R_2 < e^{-8.16}(\cos(\beta_{b2}))/Z$.

Application of the Euler equation for turbomachines produces:

$$Dh_0 = U_2 C_{q2} - U_1 C_{q1} \qquad (3.62)$$

and thus the increment in total temperature, assuming that secondary energy contributions deriving from the effects of recirculation, friction, etc., can be ignored, is given by:

$$Dh_0 = (U_2 C_{q2} - U_1 C_{q1})/C_P \qquad (3.63)$$

The equations for pressure and temperature are again of the type:

$$p_{02}/p_2 = (1 + (g - 1)M_2^2/2)^{(\gamma/(\gamma-1))}$$
$$T_{02}/T_2 = (1 + (g - 1)M_2^2/2) \qquad (3.64)$$

in which:

$$C_2 = (C_{m2}^2 + C_{q2}^2)^{1/2} \tag{3.65}$$

$$M_2 = C_2/(\gamma R T_2)^{1/2}$$

The isentropic efficiency of the impeller can be defined as follows:

$$\eta_{S,ROT} = \frac{\left(\dfrac{p_{02}}{p_{00}}\right)^{(\gamma-1/\gamma)} - 1}{\left(\dfrac{T_{02}}{T_{00}}\right) - 1} \tag{3.66}$$

The utilization of a value correlated for thermodynamic efficiency of the type defined in (3.66), normally assumed by the design engineer on the basis of experimentation on real machines, makes it possible to close the system (3.58) to (3.65) and to formulate an estimate of the conditions existing in the impeller discharge section.

In conclusion, it should be mentioned that the single-area monodimensional model is dealt with comprehensively in the majority of reference texts on radial turbomachinery.

3.6.3 Monodimensional Analysis of Diffusers

Analysis methods based on monodimensional flow approximation are frequently utilized in the field of diffusors. The main function of these methods is that of predicting the performance of a given configuration, in relation to determined flow conditions existing at impeller discharge.

The most important diffusor performance parameter is the pressure recovery coefficient C_p defined by the equation:

$$C_p = \frac{p_4 - p_2}{p_{02} - p_2} \tag{3.67}$$

This parameter is utilized to quantify the capacity for converting into pressure the kinetic energy transferred to the fluid by the impeller.

The diffusors most frequently utilized in centrifugal stages can be classified under two headings: free vortex and bladed. The approach most frequently used in analyzing free vortex diffusors hypothesizes a succession of monodimensional condition sections with $r =$ constant lying between impeller discharge section and diffusor discharge section. The fluid-dynamic balance equations relevant to this representation, inclusive of the friction terms deriving from the presence of side walls, can be integrated numerically starting from known conditions in the discharge section. This procedure can be used to evaluate the fluid-dynamic state on discharge from the diffusor and the consequent performance of the component.

With the bladed diffusor, the substantial complexity of the conditions precludes the use of monodimensional methods based on the application of theoretical principles alone. Consequently, the approach most commonly employed for evaluating the performance of this component consists of experimental correlations.

The best-known of these correlations refers to experiments conducted by Runstadler on diffusors of bidimensional geometry with straight walls diverging on a single plane. It shows that the recovery coefficient depends on a number of geometric and aerodynamic parameters, such as the length/width ratio, throat section L/w, and divergence angle $2q$. A typical performance map, obtained from Runstadler's work, is shown in Fig. 3.35, where the recovery coefficient is represented in relation to the previously introduced geometric parameters.

3.6.4 Non-viscous Numerical Methods

The monodimensional methods described above present some disadvantages which can be summarized as follows: impossibility of obtaining an accurate representation of the fluid-dynamic field at all machine points; impossibility of diagramming the detailed geometry of the components and its influence on the fluid-dynamic characteristics; and need to introduce empirical data in the form of various experimental correlations.

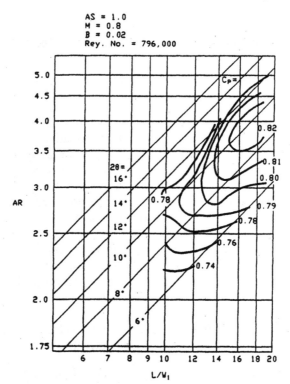

FIGURE 3.35 Diffuser data for compressor diffuser design.

The attempt to overcome at least some of these limitations has revealed the need for analysis methods capable of resolving, through numerical calculation procedures, the fluid-dynamic field within the components of the stage.

In view of the complexity and expense of using viscous models, attention was initially focused on models based on the hypothesis of non-viscous, stationary flow. These methods frequently incorporate further hypotheses, e.g., assuming that the surfaces along the fluid trajectories can be represented by suitable bidimensional surfaces, termed streamline surfaces.

Note that the hypothesis of non-viscous flow does not correspond to the conditions observable in experimentation, particularly as regards centrifugal machines. On the contrary, the latter show a vast range of phenomena in which viscous effects have significant importance and extent. Accordingly, the representation obtainable through the non-viscous approach should be considered at most an approximation of the conditions encountered in reality.

In spite of this considerable limitation, non-viscous methods can be utilized for diagramming that is satisfactory from the engineering viewpoint. It can in fact be assumed that the behavior of the regions subject to viscous effects, and of the boundary layers in particular, can be reconstructed from a knowledge of the velocity and pressure distributions obtained from the non-viscous model.

The non-viscous methods can be divided into four categories:

- Bidimensional solutions relevant to streamline surfaces lying in the hub-to-shroud direction
- Bidimensional solutions relevant to streamline surfaces lying in the blade-to-blade direction
- Quasi-three-dimensional solutions
- Three-dimensional solutions

In each of these categories the methods can be classified still further as streamline curvature methods and partial derivative methods. The streamline curvature methods are based on the integration of ordinary differential equations of the first order: these describe the momentum balance along directions defined by the so-called "quasi-normals" to the streamlines. The partial derivative methods are based on the integration of differential equations with the partial derivatives which describe the balance of mass, that of quantity of motion and that of energy at a point in the calculation domain.

Most of the partial derivative methods consist of developments of the formulation proposed by Wu in 1952. Through these it is possible, thanks to the introduction of particular derivatives, to divide the original three-dimensional problem into two bidimensional problems relevant to hub-to-shroud surfaces and blade-to-blade surfaces respectively.

Having briefly introduced the main categories of methods, we will go on to describe the salient characteristics of each of them and the results obtainable.

Bidimensional Solutions Relevant to Streamline Surfaces in the Hub-to-shroud Direction. These methods are based on representation of the conditions existing on a hypothetical mean streamline surface, extending in the hub-to-shroud direction within the area lying between two adjacent blades. The geometry of this surface is usually established in relation to the position and orientation of the blades.

A typical calculation code for this category, utilizing the streamline curvature approach, is based on integration of the momentum balance equations, evaluated in reference to a grid, defined on the hub-to-shroud surface, formed of streamlines and quasi-normals. These equations are placed in a system with further mass balance equations, and evaluated in coincidence with the quasi-normals. The position of streamlines and quasi-normals is modified through an iterative procedure up to convergence with the desired flow rate value. The codes based on the partial derivations approach frequently utilize Wu's formulation, mentioned above.

As regards application of the results, note firstly that the assumptions made concerning the geometry of the hypothesized streamlines do not coincide with what has been found in experimentation, where the movement of the streamlines is often highly distorted. Furthermore, the methods described above presume conditions of the axial-symmetric type, which differ from the situations observed, especially in impellers with high pressure ratios. Greater accuracy can however be obtained by associating these procedures with methods for blade-to-blade flow analysis, discussed in the following paragraph.

Bidimensional Solutions Relevant to Streamlines in the Blade-to-blade Direction. These methods are based on the representation of conditions in hypothetical streamlines consisting of surfaces of revolution between two contiguous blades. Many of these methods employ the streamline curvature formulation. These procedures are based on solving the equations along quasi-normals oriented in the blade-to-blade direction, according to a scheme similar to the one described in the preceding paragraph. The surfaces of revolution are obtained by rotation around the axes of streamlines calculated through a bidimensional method in the hub-to-shroud direction.

The most widely used approach consists however of utilizing finite differences methods, frequently based on the formulation proposed by Stanitz. As regards application of the results, the remarks concerning the arbitrary nature of the presumed streamlines, which do not usually coincide with the real streamlines, should apply.

The most useful aspect of the methods described here is their capacity for evaluating the conditions existing on the blade surfaces. This makes it possible, as will be demonstrated, to evaluate the pressure and velocity distribution, and consequently to predict the behavior of the boundary layers in the real machine. Moreover, the methods described here can be utilized as constituent elements of quasi-three-dimensional or three-dimensional procedures, as will be shown in the following paragraph.

Quasi-three-dimensional and Three-dimensional Solutions. The methods discussed above refer, in all cases, to bidimensional representations of the flow. As previously mentioned, these methods do not take account of the actual conditions existing in a centrifugal compressor, where there are important three-dimensional effects. It is thus necessary to find models which go beyond the bidimensional hypothesis.

A frequently used technique consists of producing quasi-three-dimensional representations, obtained by combining two bidimensional solutions of the types described above.

In the model developed by NREC for instance, a bidimensional solution relevant to a hub-to-shroud surface is superimposed on another bidimensional solution relevant to a blade-to-blade surface. The first solution is obtained through a conventional streamline curvature method; the second is evaluated through an approximate method developed by Stanitz, based on imposing a condition of absolutely null circulation along a closed line lying between two adjacent blades.

The study of this distribution yields a first approximation of the behavior of the boundary layers relevant to a given compressor geometry. The information obtained in this way is of fundamental importance to design. For this reason, calculation methods of the type described here are by now a well-consolidated procedure in designing impellers.

The next level of approximation consists of utilizing methods which make use of an actual three-dimensional representation. Among these is the model developed by Hirsch, Lacor, and Warzee which utilizes a finite-element procedure.

3.6.5 Viscous Methods

The term "viscous methods" indicates a family of calculation codes based on procedures of numerical integration of the viscous, compressible, and three-dimensional equations of motion.

Generally speaking, the system formed of the complete Navier-Stokes equations in non-stationary form, the constitutive laws of fluid, and the equations that specify the dependency of viscosity and thermal conductivity on other variables, provide the most general representation of a generic fluid-dynamic phenomenon.

In the case of laminar flow, a numerical simulation based on such an approach provides a strict description of the problem in question and, moreover, does not require the introduction of further information based on empirical data. Most of the applications, however, and almost all of the cases significant for the analysis of centrifugal compressors, concern situations in which the flow is to be considered turbulent.

In principle, it is possible to simulate a turbulent flow through integration of the Navier-Stokes equations in non-stationary form. This approach does not require the introduction of additional information on the structure and properties of the turbulence. However, it implies the availability of calculation resources exceeding those available at the moment or in the near future.

This makes it necessary to recur to formulations where the effect of turbulence is represented by an appropriate model based on empirical data. The procedures based on this approach utilize Reynolds's formulation of equations of motion, in which the non-stationary variables are brought to a mean value calculated in respect to an appropriate time interval. A turbulence model is utilized to diagram the Reynolds stress tensor terms appearing in the equations of motion.

A general formulation of the fluid-dynamic problem defined in this way is the following:

$$\frac{\partial Q}{\partial t} + \frac{\partial E}{\partial x} + \frac{\partial F}{\partial y} = \frac{\partial E_V}{\partial x} + \frac{\partial F_V}{\partial y} \qquad (3.68)$$

where

$$Q = \begin{pmatrix} \rho \\ \rho u \\ \rho v \\ e \end{pmatrix} \quad E = \begin{pmatrix} \rho u \\ \rho u^2 + p \\ \rho u v \\ u(e + p) \end{pmatrix} \quad F = \begin{pmatrix} \rho v \\ \rho u v \\ \rho v^2 + p \\ u(e + p) \end{pmatrix} \quad E_v = \begin{pmatrix} 0 \\ \tau_{xx} \\ \tau_{xy} \end{pmatrix} \quad F_v = \begin{pmatrix} 0 \\ \tau_{yx} \\ \tau_{yy} \end{pmatrix} \qquad (3.69)$$

in which terms of the type τ_{xx}, τ_{xy} represent friction effects depending on the viscosity μ and the velocity derivations. This system of equations should be associated with a suitable representation of the thermodynamic properties of the fluid, generally represented by the perfect gas model. Suitable equations, based in general on Sutherland's law, are also included to make explicit the dependency of the viscosity on temperature.

The simplest representation of turbulence is that which considers an effective viscosity determined by the sum total of a molecular contribution and a turbulent one (Boussinesq's hypothesis).

$$\mu = \mu_L + \mu_T \qquad (3.70)$$

The molecular viscosity term μ_T can be calculated in different ways: in the class of methods termed algebraic models, an approach based on Prandtl's hypothesis of mixing length is used. Among these methods, the model developed by Baldwin and Lomax is especially well known.

3.7 THERMODYNAMIC PERFORMANCES TEST OF CENTRIFUGAL COMPRESSORS STAGES

3.7.1 General

Even if computing resources are developing faster, the only solid indicator of a compressor's performances still lies in extensive testing. Tests can be carried out on single stage scale models for basic understandings of stage behaviour and on the whole machine for performances measurements.

Normally the gas available to the manufacturer has different properties (mainly different molecular weight) with respect to the real one, hence an extensive use of similitude laws is done.

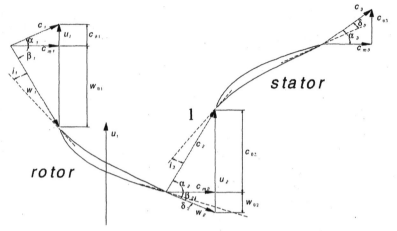

FIGURE 3.36 Velocity triangles for an axial stage.

Single Stage Testing. This type of test is normally performed using scale models and closed or open loop facilities handling air or some heavy gases available on the market. Peripheral Mach number and Volume Flow coefficient are reproduced in the lab.

Tests are normally carried out at lower Reynolds number with respect to full scale conditions. The higher the Reynolds number, the better the compressor efficiency and the higher the head. ASME standards governing thermodynamic tests state the relevant corrections.

A facility for single stage test is shown in the Fig. 3.39.

Because of the fact that the tests are carried out not on the real machine components, many probes can be installed inside the model and temperature and pressure are measured at several positions for a full description of both impeller and statoric parts performances.

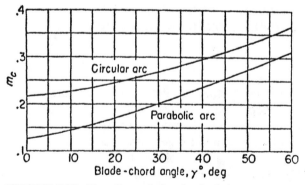

FIGURE 3.37 Howell correlation for deviation.

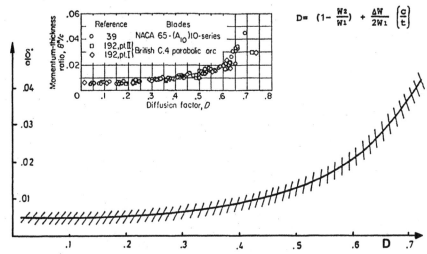

FIGURE 3.38 Correlation between diffusion factor and momentum thickness.

3.8 MECHANICAL TESTS

Mechanical tests can be of various types and complexities depending on the information that it is required to obtain. The test conditions ought to be as close as possible to the actual contractual conditions. During the tests, the item of greatest interest is the location of the lateral critical speed.

In general, a running test at maximum continuous speed is carried out (100% of design speed for motor driven units, 105% for steam or gas turbine units). During this test, shaft vibration measurements are taken at various speeds, close to the bearings. An overspeed test is carried out up to the overspeed trip setting of the turbine to check the safety of the compressor in the event of control failure of turbine. In this case, the speed can reach 10% over the trip after which the machine shuts down.

It is known that the coupling has a notable influence above all on the second critical speed; for this reason, it is advisable to carry out the tests with the job coupling. If a different coupling is used, care must be taken to ensure that the overhang (consider weight and axial dimensions) is the same. In particular, the same sleeve weight and position of center of gravity, the same distance of the teeth from the bearing, and the same weight and flexibility as the job coupling are required.

The connection between compressor shaft and coupling (carried out by means of matching teeth) is considered as a hinge: in fact the bending moment is thus not transmitted to the compressor shaft. In particular, the weight of the spacer is considered divided by two identical forces acting on the end toothings.

Further, the test driver ought to be that used for the project, in fact, the driver-coupling compressor group is a whole, the interaction of the component parts of

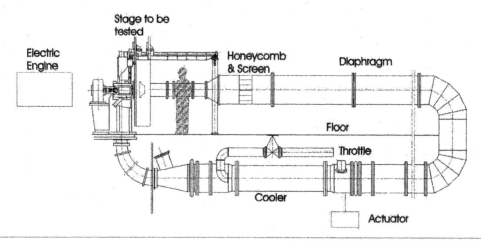

FIGURE 3.39 Single stage model test facility.

which is not easy to be reproduced. In general, however, the drivers of the test facilities are used; only with particularly critical machines are tests made using the project driver. This is generally what concerns the elastic response of the compressor. One must take into account also the bearings and the seals, since the critical speeds are much influenced by these (type of bearings and seals, their clearances, oil viscosity etc.). It has already been seen when considering the lateral critical speed and instability problems, that the type of bearing and seal have a fundamental importance in reducing the destabilizing forces that act on the rotor-support system. It is useful to remember that the case in which asynchronous vibrations occur at speeds which are multiples of the rotation speed, indicates misalignment, bearing failure, or other causes of this kind. Asynchronous vibrations at speeds lower than that of rotation are to be attributed to instability of the oil film in the bearings or in the seals. The oil used in the test must have the same viscosity as that used at the operating site; this can be arranged by adjusting the oil temperature to get the viscosity required at the bearing inlet.

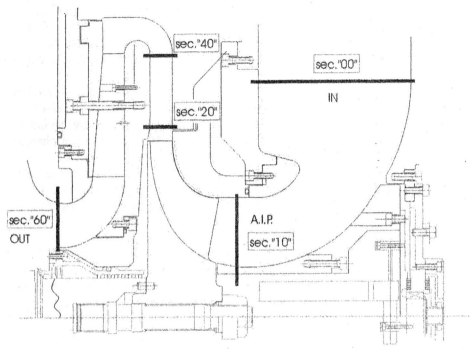

FIGURE 3.40 Single stage test model.

To be able to reproduce the same working conditions in the high pressure seals, the oil or gas should circulate in closed high pressure loop. This may be complicated and costly.

The old edition of API standards stated to carry out tests without seals installed. The new edition, on the contrary, states to carry out the test with the seals installed. This, in practice, means carrying out the test at a pressure equal to at least a quarter of that required in operation. In fact, the capacity across the low pressure ring is proportional to the pressure: at reduced pressure the cooling is less which leads to an increase in temperature.

During the tests, oil temperature and pressure are measured at inlet and the temperature at the bearing discharge. Sometimes the bearing temperature is measured by embedding a thermocouple into the white metal. Measurements on the lubricating and seal oil capacities are not frequently made. The measurement of vibrations is carried out both on the shaft and on the case. The case measurements are made close to the bearing positions in the vertical, horizontal and axial planes. The measuring instrument consists of an external part joined to the case and one free, practically fixed in space, which has a very low frequency of oscillation and is not affected by the high frequency case vibration. The variation of the magnetic field in the air gap due to the relative movement between the two parts generates electromagnetic forces which are suitably amplified and presented on a monitor, and give information on the vibrations of the case. Filters are used to select the various harmonics for monitoring on an oscilloscope. The vibrations on the shaft are measured in different positions with probes at 90° so that on the oscilloscope it is possible to see the orbits of the points of the shaft axis corresponding to the particular section.

It is also interesting to see the phase variations, that is, to see how the amplitude of the vibration moves with regard to a fixed point on the shaft at various rotation speeds (the fixed point on the shaft can be arranged by having a reference mark monitored by a photo-electric cell). By observation of the phase, it is possible to find between which critical speeds the operating point exists, since in passing across one of these there is a phase change in the vibrations. For example, before the first critical speed, unbalance and vibrations are in phase, beyond the first critical speed they are in phase opposition; in reality, these phase changes are never instantaneous but are distributed within a speed range. Naturally, it is necessary that the shaft is perfectly cylindrical and concentric with respect to the supports, otherwise considerable vibrations will be detected even if the shaft does not vibrate. During vibration measurements, it is necessary to take into account the electric and mechanical run-outs. The electric run-out is a phenomenon due to the fact that during the forging operations magnetic fields are created which subsequently disturb the measurements. It is necessary to avoid these difficulties before the tests by de-magnetizing the rotor with a solenoid. The mechanical run-out, due to unavoidable eccentricity and ovality of the mechanical parts, can be examined by means of other instruments.

At one time the trend was to limit vibration amplitude, now instead, the trend is to limit vibration speed or vibration acceleration. The vibration speed is proportional to the product of the amplitude by the frequency and to the dissipated energy; that's why it is an important reference for evaluating vibrations. In general the vibration amplitudes acceptable on the case are half of those on the shaft; to give an idea of the order of these amplitudes, for a shaft running at about 5000 rpm, the vibration amplitude acceptable is up to about 40 microns. For the casing, the vibration speed limits are acceptable in the order of 10 to 20 mm/sec.

3.9 ROTOR DYNAMICS AND DESIGN CRITERIA

3.9.1 Introduction

The greatest effort in the design of centrifugal compressors mainly for high pressure applications is at present devoted to problems connected with the lateral stability of the rotor. Stability problem concerns the compressor in all its components, since, all basic parts of the machine contribute to stability: rotor, bearings, oil seals, coupling and all flowdynamic parts such as impellers, diffusers, return channels.

Theoretical prediction and experimental investigation methods made available in recent years have contributed much to the progress in this field. Referred to here is the availability in industry of large computers capable of carrying out very elaborate calculation programs, and of electronic equipment for detection of vibrations and pressure pulsations (non-contact probes, key-phasors, pressure transducers, real time analysers etc.) which have allowed more accurate diagnosis.

The measure of mechanical behaviour of a compressor is given by the amplitude and frequency of the rotor vibrations.

Rotor vibration amplitude must not cause: contact between rotor and small clearance stator parts (labyrinths), overloading of oil seals, or fatigue stress in the bearings. The frequency of the vibrations is a very important element in evaluating stability of the system.

Vibration may have a frequency corresponding to the machine rotation (synchronous vibration) or a different frequency (asynchronous vibration). Usually in rotating machines both types of vibration can be present.

3.9.2 Synchronous Vibration

Synchronous vibrations are usually attributable to one or a combination of the two following causes:

a) Accidental defects of the rotors (as for example unbalance)

b) Design defects; that is to say operating speed too close to resonance and/or insufficient damping of the system

As regards point a), machine manufacturers now have equipment which permits achievement of a very accurate balancing. This considerable accuracy in balancing is however sometimes upset by accidental causes so that point b) assumes great importance; correct design of the rotor-bearing system must assure acceptable vibration levels even when accidental causes destroy the original state of perfect balance.

Two approaches are usually used to predict the synchronous dynamic behaviour of a rotor.

The first approach is the Myhlestad-Prohl numerical calculation that considers the rotor as a dynamic system consisting of a number of concentrated masses attached to a zero mass shaft supported by bearings. The computer program solves the system for a variety of constant support values over the entire possible range. A diagram can be made in which the lateral critical speeds are a function of the equivalent stiffness of the supports. The actual values of lateral critical speeds can be established on the basis of the knowledge one has of the bearing stiffness (Fig. 3.41).

The original speed program also calculates the rotor mode shapes at the critical speeds for each specified value of the bearing stiffness (Fig. 3.42). The mode shapes are important because they indicate the relative vibration amplitude at each station along the rotor. If relative amplitudes at the bearings are low a high unbalance producing considerable deflection in some sections of the shaft will cause very

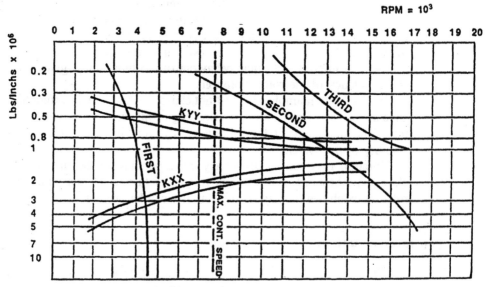

FIGURE 3.41 Map of lateral critical speeds.

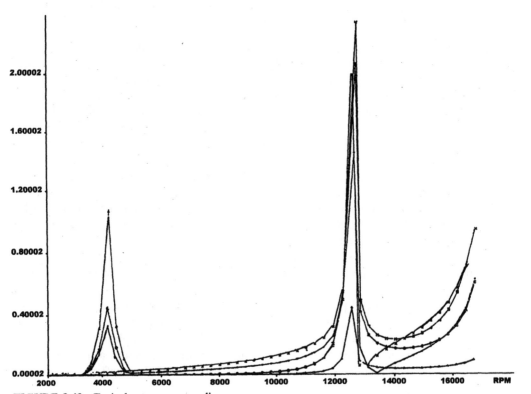

FIGURE 3.42 Typical rotor response diagram.

small relative motion in the bearings. Without relative motion, damping of the bearings cannot be effective. Thus the bearings are not placed in the most efficaceous position, and their position must be corrected.

The second approach is to carry out the shaft response calculation in which the rotor motion throughout its operating speed range is studied as a damped system response to an unbalancing excitation. The unbalances are generally placed where they may be expected to occur, i.e., at impellers, couplings etc. The amplitude of rotor motion is calculated at selected stations along the rotor.

Coefficients simulating the dynamic stiffness and damping of the bearing are included in the calculation. The calculated whirl orbits are generally elliptical due to the difference between the vertical and horizontal stiffness and damping. A response diagram represents the variation with speed of the semi-major axis of the elliptical whirl orbit at selected stations along the rotor (Fig. 3.43).

Various tests carried out directly in actual operating conditions have shown that the frequencies and amplitudes measured are close to the expected values.

The design parameters available to act upon damping capacities and resonance values are: bearing positions, especially with respect to the shaft overhangs, bearing

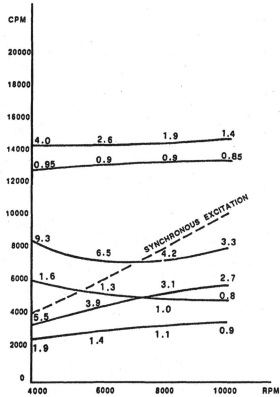

FIGURE 3.43 Damped lateral frequencies and decrement diagram.

type, type of lubricating fluid, coupling type and obviously elastic characteristics of the rotor.

3.9.3 Asynchronous Vibration

In the asynchronous vibration field it is necessary to make a further distinction between vibration frequencies that are multiples of the rotating speed and vibration frequencies lower or higher than rotating speed but not multiples.

To the first type belong vibrations usually caused by local factors such as: misalignment, rubbing between rotating and static parts, excessive stresses in the piping, foundations etc.

To the second type belong vibrations that have been the cause of more serious problems especially in the field of high pressure compressors. They may be caused by external phenomena (forced vibrations: for example, the effect of aerodynamic forces) or by phenomena intrinsic to the movement of the rotor itself (self-exciting vibrations), which impair stability at its base.

Stability is a function of a balance of several factors. The main ones are:

A—Rotor-support system with its elastic characteristics
B—Aerodynamic effects
C—Oil seals
D—Labyrinth seals

Each factor plays its part in the balance of stability and may be either positive or negative. The system is more or less stable or unstable according to the result of this balance.

A theoretical approach for predicting the stability of a rotating system is the log decrement calculation.

The program calculates the natural damped frequencies of the rotor-support system at selected speeds and supplies, for each frequency, the value of the log decrement that is a sound indication of the stability of the system itself.

A—As far as the rotor is concerned, we have already seen how the natural frequencies are determined and how the bearing effectiveness can be evaluated on the basis of the bending shapes.

To avoid or minimize internal hysteresis, shrunk assembled elements (such as sleeves, spacers, impellers etc.) must be as axially limited as possible.

The keyways may cause differentiated elastic response in the various planes. For this reason they are reduced to the minimum size, staggered at 90 degrees between one impeller and the next, and in some cases they are eliminated.

As regards bearings, in order to avoid oil whip problems the tilting pad type is generally used. In some cases damper type bearings are also used (Fig. 3.44.). These offer the advantage of allowing independent adjustment of damping and stiffness coefficients.

B—The occurrence of rotating stall in one or more impellers may explain the presence of pulsations indicating vibrations at the same frequency (forced vibrations).

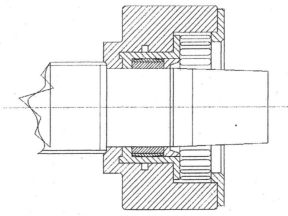

FIGURE 3.44 Damper bearing.

All centrifugal compressors, whatever the pressure, are affected by aerodynamic excitation. Other conditions being equal, these effects increase in intensity in proportion to the actual density of the gas. The determinant parameter is not only pressure, but also temperature, molecular weight and compressibility together. This is the reason why the problems of vibrations excited by aerodynamic effects occur more frequently in reinjection or urea synthesis plants than in ammonia synthesis or refinery compressors, even when running at the same pressure levels.

The "unsteady flow phenomena" has been studied in its standard stage configuration. The conclusions were that the aerodynamic disturbance and the consequent pressure pulsations were coming from stator blades of the return channel well before coming from the impeller itself. In this case, the relevant shaft vibration had the following characteristics:

- Stability in amplitude
- Very low frequency (order of magnitude about 10% of the running speed)
- Amplitude function of the tip speed and the density of the gas

C—The shaft end oil seals are still one of the most critical parts in the manufacture of high pressure centrifugal compressors.

An important requirement that the oil seals must satisfy is to contribute to the stability of the system or at least not to disturb it too much. It is easy to understand that seals, owing to their nature, would be very negative components in the stability balance of the system if they were "locked", because they would act as lightly loaded, perfectly circular bearings. This negative tendency is generally countered by making the rings floating as much as possible in operating conditions.

This can be obtained by distributing the oil pressure drop on the atmospheric side amongst several rings, and reducing the surface of each ring where the pressure acts by lapping the surfaces. When these techniques are insufficient to avoid "locking" (i.e., a high detachment limit force value), circumferential or axial grooves on floating rings may make a positive contribution to the stability influencing damping and stiffness characteristics of the system.

D—Another important possible cause of instability and sub-synchronous vibration can arise from the labyrinth seals.

In the annular surfaces the gas circumferential motions, because of the rotor displacement, can become uneven; therefore, they can cause a non-symmetrical distribution of the pressure, with a resultant force perpendicular to the displacement itself (so called cross-coupling effect). This is a typical self-exciting phenomenon causing instability.

The importance of the phenomenon grows with gas density (therefore with the pressure) and with the location of the seal. In fact, the vibration which always initiates above the first critical speed, has a characteristic frequency just equal to the first critical speed with the same mode shape.

Therefore particularly important from this point of view are the back-to-back compressors in which the biggest labyrinth is in the middle (as in the highest pressure) where the shaft motions are greater.

The sealing system in Fig. 3.45b represents a first attempt to decrease or try to interrupt the circumferential motions by means of many septums placed axially on the labyrinth.

The honey comb seal in Fig. 3.46, is derived from the previous one by putting the annular surface between two consecutive teeth in communication with an inner toroidal chamber in order to equalize pressure inside as much as possible.

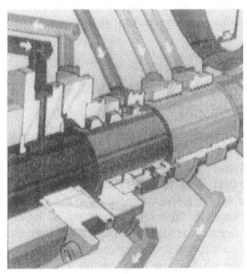

a) Flow in The Sealing System

b) Special Labyrinth Seal

FIGURE 3.45 Sealing system.

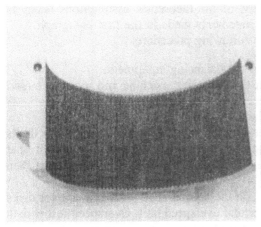
a) Honey Comb Seal

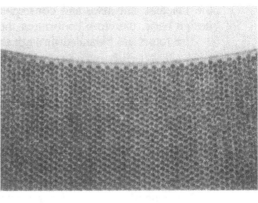
b) Honey Comb Seal Detail

c) Honey Comb Seal on Casing

FIGURE 3.46 Honey comb seal.

3.9.4 Balancing and Overspeed

The most important causes of either asynchronous or synchronous vibrations can be very well simulated during calculation so that a good forecast of the rotor dynamic behaviour is available.

Moreover, the parallel growth of instrumentation technology provides the possibility of thorough verification not only of the mechanical running conditions of the machine but also, and consequently, of the theoretical assumptions taken as design basis, therefore confirming the statements made in the first paragraph.

The rotors are balanced through the following procedure:

Impeller. The impeller is mounted on the balancing equipment.

The whole unit is then mounted on the balancing machine and must be turned by hand to check correct mounting; eccentricity is measured on external diameter of impeller seal (max. permissible value: 0.02 mm).

Next impeller is balanced at a higher speed, compatible with the machine's limits in accordance with its weight, by removing material on hub and shroud until final unbalance is within permissible range given by API 617. The impeller must then be subjected to overspeed test.

Subsequently the impeller, mounted on the special shaft with two adaptor disks used for balancing, is fitted onto the vertical overspeed unit. Overspeed test is then carried out maintaining the same level for about 10 minutes, in accordance with the values given in the specification after the trip device of the unit has been set at a speed 2% greater than per specification.

Vacuum should remain at an absolute pressure of less than 1 Torr and vibrations measured on the driving turbine should be less than 6 mm/sec.

Test values must be recorded; the impeller must then be rechecked by penetrant dye liquids and then assembled onto the shaft.

Rotor. Mount shaft on balancing machine resting it on its journal bearings; fit false half-keys and begin balancing process, temporarily adding filler on surfaces for end locking rings, using adhesive tape for this purpose. Balancing speed is selected with reference to the characteristics of the machine, in compliance with the degree of accuracy required by API 617 (Oct. '73). Next, mount one impeller at a time and after each mounting, balance by removing material from hub and shroud. The end seal rings must be fitted on and the temporary weights, added previously, removed. Balance by removing material, by drilling, from the rings themselves. Mount thrust bearing block and correct its unbalance by machining the outer diameter. Mount specified joints and check balance. For final checking, turn connection joint on balancing machine by 180° and check balance again.

Note: The above is based on the basic criterion for a flexible rotor: to prevent internal moments arising during assembly of rotor, the rotor is balanced at different intervals, i.e., after mounting each individual part (impeller, spacer etc.) the rotor undergoes balancing.

3.10 STRUCTURAL AND MANUFACTURING CHARACTERISTICS OF CENTRIFUGAL COMPRESSORS

3.10.1 Casings

Horizontally-split Casings. Both half-casings are obtained from conventional castings. The material is chosen depending on operating pressure and temperature, size, gas handled, and regulations provided by API stds. Generally used is material similar to Meehanite GD cast iron with 25-30 Kg/mm tensile strength and 70 Kg/mm compressive strength. When steel has to be used to cast these casings, ASTM A 216 WCA steel is employed; should the compressor operate at low temperatures ASTM A 352 steel is used in one of its four grades depending on the operating temperature; lastly, ASTM 351 Gr. CA15 steel (13% Cr) or Gr. CF8 is used in case of corrosive media.

The usual test these castings undergo is the magnetic particle inspection. In particular cases, when a check through the section is required, the ultrasonic test is carried out. Sometimes radiographic inspection is required; it is useful as stresses affecting these elements are limited and the flaws existing in castings, yet acceptable and not detrimental to such castings, can be displayed in this way. The latest tendency is to use welded casings, (Fig. 3.47) this has advantages over casting in that, it reduces rejections, repairs etc.

Vertically-split Casings. Both casings and end covers should be obtained from forgings so that material might be as homogeneous as possible, hence more resistant to failure, considering the high pressures these compressors have to contend with.

ASTM A 105 Gr. II carbon steel is generally used for the barrel, supports and end covers: the carbon content applied (0.2 to 0.25% instead of 0.35%) is enough to get good mechanical characteristics, at the same time granting characteristics of weldability. Alloy steel with higher mechanical characteristics is used for compressors running under very high pressure.

Suction and discharge nozzles are welded to the casing, generally forged in the same material; as to pipeline compresor, owing to their complicated structure hence not suited to be forged (see Fig. 3.47), are often made of castings.

FIGURE 3.47 Welded casing.

Diaphragms. The diaphragms form the flow path of the gas inside the compressor's stator. They are divided into four types: suction, intermediate, interstage and discharge diaphragms.

The suction diaphragm conveys the gas to the first impeller inlet. It is supplied with adjustable vanes when the compressor control is performed by changing the inlet angle of gas to the impeller, operating outside the compressor.

The intermediate diaphragms perform the double task of forming the diffuser, where kinetic energy is converted to pressure energy, and the return channel to lead gas to the next impeller inlet. The diffusers can be free-vortex type or vaned: these vanes while improving the conversion efficiency, reduce the flexibility of the machine.

The discharge diaphragm forms the diffuser of the last impeller, as well as the discharge scroll.

The interstage diaphragms separate the discharge sides of the two stages in the compressors with back-to-back impellers.

Each diaphragm has labyrinth rings to make the impeller shroud tight (to prevent gas at impeller outlet from returning to suction side) and on the spacer rings to cut out interstage leaks. The seal rings, of split construction, can be easily removed.

For reasons of rotor installation, the diaphragms are halved; whether they are mounted on barrels or on horizontally split casings, the difference is not as great; they only differ in their being housed in the casing.

In barrel-type compressors, the diaphragm halves are kept together by tierods thus making up two separate bundles; after installing the rotor they are bolted to each other: the resulting assembly (see Fig. 3.48) is placed in the casing axially.

In the horizontally-split compressors, each diaphragm half is singly installed in the two casing halves; the outer surface on each diaphragm has a groove to combine

FIGURE 3.48 Barrel type compressor assembly.

with the corresponding relief on the casing. Each diaphragm is lowered onto the half-casing. (See Fig. 3.49)

Concerning the design criteria, a distinction must be made between sizing of the gas path, on the ground of thermodynamic requirements to guarantee the speed and gas angle requested, and the sizing of thickness which is based on the Δp established on the two faces of each diaphragm.

In normal installations (in-line, low or medium pressure compressors) this Δp value is that produced by each impeller: hence it does not reach very high values. In a barrel compressor, e.g., with 8 impellers and 30 ata suction and 80 ata discharge, we have approximately:

$$\text{single impeller}^\rho = \sqrt[8]{\frac{80}{30}} = 1.13$$

therefore the max. Δp produced by the last impellers amounts to about 9.2 kg/cm².

For these installations, the diaphragms are nearly always cast owing to their complicated structure. Generally Meehanite GD type or spheroidal cast iron is used, sometimes adding nickel percentages to improve its characteristic impact resistance at low temperature (1 to 1.5% Ni). If the operating temperature is below $-100°C$,

FIGURE 3.49 Diaphragm assembly with rotor assembly on lower casing.

ASTM A 352 steel is used in the four available grades or ASTM A 351, grade CF8.

Under a certain size of gas channel, casting is somewhat difficult, therefore diaphragms are manufactured in two parts, usually one cast and the other of sheet-metal, bolted to each other.

Under severe conditions such as in high-pressure compressor diaphragms or interstage diaphragms in compressor with back-to-back impellers (undergoing the Δp of one whole stage), the design Δp of a diaphragm can reach very high values. In this instance, it is necessary to use materials such as forged carbon steels (ASTM A 182 F22).

Should very high pressures be involved, it is necessary to stiffen the diaphragm bundle structure. The solution consists in manufacturing the countercasing in forged steel ASTM A 182 F22) made up of two half-casings where diaphragms are mounted like in horizontally split compressors. This has the advantage of smaller diameter diaphragms in which deflection is reduced (see Fig. 3.50.)

3.10.2 Rotor

The rotor of a centrifugal compressor is made up of shaft, impellers, balancing drum, thrust bearing collar, the coupling hub and sleeves and spacer rings.

Shaft. The shaft consists of a central section, usually with constant diameter, on which impellers and spacers are mounted, and two ends with diameters suitably tapered to house bearings and seals. The shaft is sized to be as stiff as possible (reducing the distance between bearing centers and increasing the diameter according to the flow-dynamic design) to reach the best flexural behaviour. The material used to manufacture shafts for many compressors is steel 40 NiCrMo7 UNI. The mechanical characteristics of this steel are better than common carbon steel, which is sometimes used in compressor shafts.

Steel 40NiCrMo7 is very suitable for hardening and tempering; the normal-size shafts for centrifugal compressors made up of this steel, benefit from hardening through the cross section while plain carbon steel will be affected only superficially.

Since the aim is to reach good toughness and ductility, rather than very high yield point and ultimate tensile strength, tempering is carried out at temperature which allows the material to reach an ultimate tensile strength over 100 Kg/mm^2 and yield point over 65–75 Kg/mm^2.

Impellers. Impellers are shrink-fitted on the shaft (see Fig. 3.51).

Under the impellers, splines are provided to transmit torque.

Impellers are interference-fitted not only because of torque transmission, but also to avoid loosening under high speed of rotation. Stresses due to centrifugal forces, can cause impeller to loosen, become eccentric and thus create unbalance.

Impellers may be, structurally, of closed or open type. The closed impellers are made up of one hub, a number of blades and one shroud. Blading is generally

FIGURE 3.50 Rotor assembly on lower casing.

slanted backwards. These parts are joined in different ways, but typically by welding.

Blades are generally milled (see Fig. 3.52.) on the hub (or shroud), then the shroud (or the hub) is internally welded. The blades are milled on the hub or shroud depending on the impeller shape and, hence, on the possibility for the electrode to get into the channel.

FIGURE 3.51 Impeller assembly on the rotor.

If, owing to the narrower width of the impeller, it is difficult to weld internally, external welding is carried out: on the shroud (or hub) near blading and according to its shape, grooves are carried out superficially. Hub and shroud are connected to each other by temporary butt-welding. By filling these grooves with weld material, the facing surfaces between blade and shroud are melted thus resulting in a weld.

The impeller manufacturing cycle is the following: welding carried out as described before; followed by stress relieving heat treatment; inspection of welded parts; hardening and tempering; and final machining.

Open impellers are different from the closed impellers in that the open lack a shroud. Usually this kind of impeller has tridimensional blades produced by milling. Blading can be radial or slanted backwards.

As to the mechanical design, it has to be taken into account that the impellers are the most stressed elements in a compressor. Reducing the number of stages leads to higher and higher tip speeds and, hence, stresses. The stress trend in the various impeller parts varies, of course, according to the impeller type. Use of Finite Element Method (F.E.M.) allows for a fine analysis of the stress distribution.

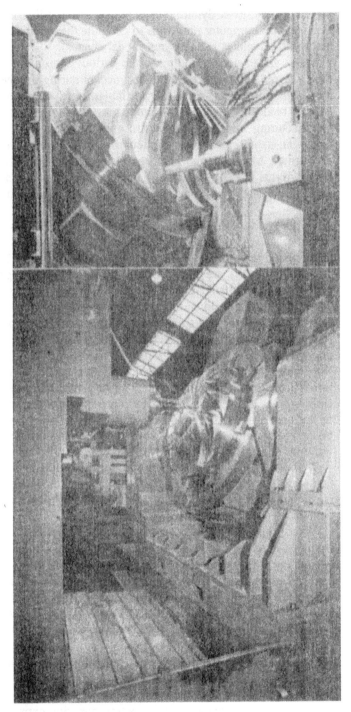

FIGURE 3.52 Impeller automatic milling.

More recently, even dynamic calculations are possible in a reasonable time using average level computer platforms to compute the dynamic behaviour of the impeller (natural frequencies and relevant modal shapes).

Stress values corresponding to the various speeds are proportional to the speed ratio squared. The severest condition occurs during the overspeed test (at 115% of the continuous max. speed). In particular there are stressed areas on the leading edge of blades. Stress concentrations must be avoided, and as a general rule, when manufacturing impellers much care must be taken in finishing their surfaces and designing them, considering particularly the thickness, the key slots and rounding off corners. Materials and heat treatments are chosen taking account of stress due to centrifugal force (as a function of the tip speed at which the impeller has to run) and the working conditions, such as corrosion, stress corrosion, low temperature etc.

To get good welds in blades, they have to be made of steel with high mechanical characteristics, yet low carbon content. Typical material would be a low-alloy steel containing 2% chromium, 1% molybdenum and 0.13 to 0.17% carbon.

When impellers are manufactured using steel with higher carbon content, thus getting better mechanical characteristics, there would be some doubts about quality of welding, as the weld and the area around are subject to intercrystalline corrosion. This is the reason why some manufacturers call for limits in carbon percentage. Intercrystalline corrosion leads to the relaxation of the metallographic bonds among grains and, hence, to degradation of strength.

A carbon content in steel higher than its austenitic matrix solubility limit determines the potentiality for the steel to be subject to intercrystalline corrosion, as carbon is the main cause for carbide precipitation and chromium impoverishment in the area around the grain boundary. The carbides precipitated along the grain boundary can initiate fracture, while the chromium impoverishment makes the material more liable to corrosion.

If steel remains at the sensitization temperature (from 400 to 900°C) during the quenching process after heat treatment, as well as during heating for welding, the chromium carbides may precipitate to a greater extent depending on the carbon content.

When impellers handle corrosive media, steel with higher chromium content is used, such as X15C13 (13% Cr); in particularly corrosive areas the chromium percentage is further increased such as with KXOA2-FNOX steels (from 15 to 19% Cr); if higher strength along with corrosion resistance is required, then MARAGING steels, series 17% Cr, 4% Ni are used, age-hardened at low temperature. Steel with up to 9% nickel is used for impellers running at low temperature; this content was arrived at to get good impact strength down to $-196°C$.

Balancing Drum. During normal operation, inside the compressor a thrust is generated against the rotor, which has to be taken up by the thrust bearing. Such a thrust is mainly due to the pressures acting on the impeller. The Δp produced by the impeller generates a force in suction direction expressed by the product of Δp multiplied by the area underneath the seal on the shroud.

The sum of these thrusts is generally very high and often beyond the thrust bearing capacity. Consider a medium pressure compressor, with 5 impellers, mean Δp for each impeller = 6 Kg/cm^2, shaft ø = 17 cm, seal ø = 27 cm, the generated thrust is:

$$\frac{\pi}{4}(27^2 - 17^2) \cdot 6 \cdot 5 \simeq 10370 \text{ Kg}$$

Therefore a balancing drum is provided for, after the last impeller; placing its opposite face under suction pressure and sizing its diameter adequately, a thrust is generated from suction to discharge side, such as to balance the thrust coming from the impellers. Some unbalance is provided to obtain a residual thrust capable of being taken up by the bearing, to avoid any axial instability of the rotor. Other thrusts are generated besides those described before, such as the thrust caused by the variation of gas flow entering the impeller axially and leaving it radially, or such as the thrust resulting from the irregularity of pressure acting on the impeller in high-pressure machines. Generally, these thrusts are not so high to change the state of things.

As regards shape, it has to be noted that the width of this drum be such as to support the whole Δp developed by the compressor. Inadequate sizing of the labyrinth seal results in strong gas leakage towards suction, thus impairing compressor performance. Generally, the balancing drum is made of X12C13 steel shrink-fitted with key like the impellers.

Coupling. The coupling transmits power from driver to the compressor. Coupling can be direct or through a speed increasing gear. Usually, toothed couplings are used with force-feed or filling lubrication. The couplings with force-feed lubrication are designed for high speed of rotation and for the most part they only are used in compressors. Another type of couplings is sealed, generally with lubricating grease to be filled every so often; these couplings are used only on slower speed drive shafts.

When transmitting a torque a toothed coupling can originate an axial thrust if the shafts to be coupled vary their relative position, during transient load state or owing to thermal expansion, getting axially closer or farther apart. A relative displacement of the two shafts is not allowed, until an axial force is generated over the friction value on the coupling toothing: up to this moment this force is carried by the thrust bearing. An example of thrust in case of a compressor with power N = 10,000, rpm n = 10,000, and radius R = 100 mm. (radius of the coupling toothing pitch line HP, is:

$$M_t = \frac{716 N}{n} = 716 \text{ Kgm} \quad \text{transmitted torque}$$

$$F_t = M_t/R = 7,160 \text{ Kg} \quad \text{tangential force transmitted}$$

As the toothing friction coefficient range is: $0.15 < f > 0.3$, an axial force is necessary to overcome friction on toothing

$$F_a = F_t f = 1074 \text{ to } 2148 \text{ Kg}$$

To understand the thrust behaviour, it should be taken into account that the fixed point of rotor is the thrust bearing and the thermal expansion of the rotor is lower than that of casing. The use of diaphragm couplings has recently increased. But this type of coupling has some disadvantages. It's main advantage is accomodation of high misalignments, but at same time it is heavier. There's a negative effect on the flexural behaviour of the rotor at the second critical speed, difficulty in balancing, and possibility of fatigue failure in the thin plates.

Thrust Bearing Collar. The thrust bearing collar is made of C40-carbon steel and is generally force-fitted hydraulically.

Spacer Rings. The spacers are sleeves placed between impellers having double function: to protect the shaft from corrosive media (generally they are made of X15C13, a stainless steel with 0.15% carbon and 13% chromium), and to fix the relative position of one impeller versus another.

The spacers are shrink-fitted on the shaft with 0.5 to 1% interference. The stress resulting from such a shrinkage is:

$$\delta = \varepsilon E; \quad \varepsilon = \frac{\Delta 1}{1} = \frac{1}{1000} \quad \delta = 21{,}000 \frac{1}{1000} = 21 \text{ Kg/mm}^2$$

The tangential stresses δ_t caused by the centrifugal force could eliminate this interference, should they overcome those due to shrinking. If it's considered that a spacer is stressed at most by $\delta_t = 8$ Kg/mm^2 at a tip speed = 100 m/sec (rarely this speed is exceeded in usual cases because of the small radii), then the spacer should not detach from the shaft.

Sleeves Under Oil Seals. Sleeves under oil seals are of carbon steel, coated with very hard material (600 Brinell hardness). Colmonoy is typical of coatings used for this purpose. These sleeves are applied to protect shaft from corrosion and any scoring; in addition they can be easily replaced. In case of high pressure, Colmonoy coated sleeves are not used because they are limited in the amount of shrink that can be applied; in this case sleeves of 40 NiCrMo7 hardened and tempered steel are used (300/350 Brinell hardness).

3.10.3 Seals

Seals at the two shaft ends, at the points where shaft comes out of the casing, are used to avoid or minimize leakage of compressed gas or air getting into the compressor casing. This seal can be of three types: labyrinth, oil or mechanical seal.

Labyrinth Seals. These seals are made of blades of light-alloy or material resistant to corrosion, with hardness lower than the shaft, to avoid damaging it in case of accidental contacts. They can be easily removed. The blade number and clearance value depend on the operating conditions. If no slight gas leakage is allowed (poisonous, explosive gases etc.), the labyrinth seals are combined with extraction or injection systems. Labyrinth seals are sometimes made of annealed aluminum alloy (70–80 Brinell hardness); if aluminum is not compatible with gas corrosivity, stainless steels are used with 18% Cr and 8% Ni content. There is no limitation to use other materials such as bronze, babbitt etc.

Oil Seals.   Oil seals consist of two floating rings (H.P. ring on high pressure side, L.P. ring on low pressure side) babbitt-lined (see Fig. 3.45–3.46 showing a picture of a typical compressor equipped with several L.P. seal rings).

Seal oil is introduced, at a pressure slightly over the gas, into the annular space between the two rings and flows through the gap left between rings and shaft. Oil coming from the low pressure side goes back to the reservoir and is recirculated; oil from h.p. side is drained by automatic traps.

Oil is prevented from flowing into the gas by a large labyrinth seal placed between the oil seals and compressors inside, equipped with an intermediate pressure balancing chamber.

The oil seals are made up of a carbon steel support ring, coated with a thin white metal layer or copperless white metal, if not compatible with the gas handled.

Mechanical Seals.   The mechanical seal consists chiefly of a carbon ring, generally stationary, kept in contact with a steel collar rotating with the compressor shaft. This contact is maintained by the combined action of elastic elements

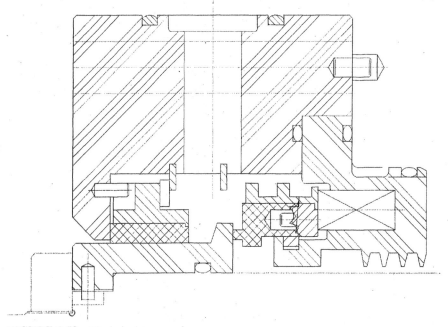

FIGURE 3.53   Mechnical type seal.

(springs or bellows) and the distribution of pressure acting on the ring. Heat is generated by the contact between collar and ring and must be removed by cooling the oil seal. The differential pressure between oil and gas must be high (3 to 5 ata), to lubricate the gap between collar and ring. To keep this Δp constant, a pressure chamber has to be built up, making it necessary to have a low pressure seal between pressurized oil and atmosphere. In most instances the seal is a carbon floating ring.

Mechanical seals are applied where oil contaminating the gas must be kept to minimum; in fact, oil leaking from the H.P. ring is about 10% that from usual oil seals. In case of compressor shutdown for lack of oil, the seal can continue sealing gas with the machine at a standstill (even if not perfectly, depending on condition of the mating surfaces between collar and ring).

3.10.4 Bearings

Journal and thrust bearings are usually plain type with pressure lubrication. They are placed outside the compressor casing and can be inspected without releasing pressure from casing inside. Generally, the thrust bearing is mounted outside the journal bearings and on the side opposite to the coupling. This solution aims at reducing the center distance, thus improving the flexural behavior of the compressor.

For machines constituting a compressor train, having two couplings, the above solution would result in having one shaft end loaded by the coupling and the thrust bearing weights—thus possibly leading to flexural problems, because masses concentrated outside the journal bearings produce the second critical frequency. To overcome this, the thrust bearing is mounted inboard as compared with the journal bearings.

Journal Bearings. At present, many compressors are equipped with tilting pad bearings. They seem to be the most suitable for resisting any unbalancing action of the oil film.

Their use is based on computers studies of the effect on shaft vibration frequencies. Carbon steel is the backing material chosen for most bearings. Working faces of the pads are usually coated with babbitt metal (a tin alloy); the metal application applied by centrifugal casting techniques.

Thrust Bearings. The thrust bearings used on many compressors are also of the tilting pad type, provided with supports to distribute load equally (see Fig. 3.54). The pads act on the collar hydraulically fitted onto the shaft. Although compressors are designed to run generally with a positive thrust, i.e., towards the thrust bearing outside, their bearings are double-acting type. They have, pads also inside to support negative thrusts caused by extraordinary conditions (transient states, start-ups etc.).

Outside the bearing a ring with a gauged hole is provided to control oil flow for lubrication. Typically speed of the collar should not exceed 190 m/s, and load on

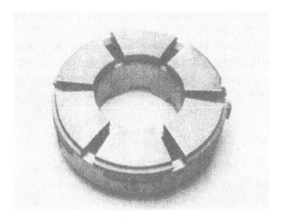

FIGURE 3.54 Thrust bearings.

bearing not over 50% of maximum limit stated by manufacturer. These are typical parameters adopted to choose the bearing to be mounted.

3.11 INDUSTRIAL APPLICATION OF CENTRIFUGAL COMPRESSORS

Centrifugal compressors have a large number of applications in numerous sectors of industry and particularly in many processes that call for very wide performance ranges. The last 30 years in particular have seen continuous expansion in the field of application which has grown to encompass some services traditionally covered by other types of compressors. This is also due to a big rise in individual plant capacities as well as to introduction of new processes. In fact, centrifugal compressors were initially used mainly as atmospheric air blowers for blast furnaces, mines, etc. They had flow rates that varied greatly from case to case, but always low pressure ratios. Later, as they began to become more common in chemical plants these machines had to change considerably. The gases compressed changed from atmospheric air to different gases and then to gas and/or vapor mixtures, all with different characteristics from air. Also the compression flows and pressure ratios increased remarkably to meet process requirements. Some of the main processes requiring centrifugal compressors are outlined below along with the principal features of the machines used for them.

3.11.1 Refineries

Fig. 3.55 shows a schematic diagram of a refinery with the main process lines.

The heart of a refinery is a fractionating column which separates the various crude components that may be used as they are, or further treated

Reforming. In reforming plants hydrocarbon molecular structures are changed from an open-chain to a cyclic structure which translates into an improved octane rating for reformed substances. This reaction which involves evolution of hydrogen takes place in a reactor with a catalyst which is almost always platinum, hence the term platforming generally used to indicate this process.

The compressor's task is to recycle the gas mixture in the reactor with a high percentage of hydrogen which minimizes side reactions that tend to form carbon deposits on the catalyst eliminating its porosity and consequently deactivating it.

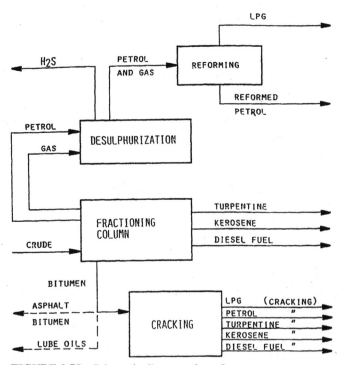

FIGURE 3.55 Schematic diagram of a refinery.

The extra pressure the compressor gives the gas is therefore used essentially to make up for the pressure losses in the system ensuring a constant gas flow.

As a rule, pressure ratios are not very high and operating pressures vary from 20 to 40 bars. It is interesting to note that the molecular weight of compressed gas may vary in time in relation to the condition of the reactor. This is because the catalyst tends to become more and more heavily coated with carbon deposits and therefore the compressor has to recycle gas which has more and more hydrogen in it. Beyond a certain limit it is necessary for the economy of operation to regenerate the catalyst.

Cracking. The heavy distillates produced by the fractionating column can be treated by a cracking process in order to break the heavy molecules into lighter ones to get premium grade fuel.

There are several types of cracking processes, one of which is Fluid Catalytic Cracking (FCC) which requires centrifugal compressors. This process gets its name from the fact that the reactor has a fluid catalytic bed. This feature offers several advantages such as an even distribution of temperature, a very large area of contact and good heat transmission. Air compressors are required with delivery pressures which may vary from 2 to 4 bars depending on the process. Also cracking gas compressors with a relatively high molecular weight (30 to 40) and delivery pressures around 15 to 17 bars are necessary. Axial compressors are sometimes used to compress the air as low pressure ratios and relatively high flows are required.

Lubricant Production. In plants which utilize the heavy distillates produced by the fractionating column, centrifugal compressors are chiefly used in the oil dewaxing process. Since paraffin wax crystallization takes place at low temperatures and separation at low pressures, it follows that two different types of compressors are necessary:

- Refrigerating compressors, which generally handle propane with design temperatures around $-20°C$
- Compressors that maintain a high vacuum level (suction pressure is usually 0.2 bars) inside the large rotary filters where the crystallized paraffin wax is deposited. Both machines have low operating pressures and therefore they normally have horizontally split casings.

Bitumen and Asphalt Production. Bitumen is obtained by oxidating the heavy distillate produced by the fractionating column in a tower with air flowing through it. A low-pressure air compressor is therefore required, generally on-line without intercooling.

3.11.2 Ammonia Synthesis Plants

NH_3 synthesis plants have been extensively applied in industry particularly over the last 20 years. It has been noted that the trend is for bigger and bigger plants while processes have been developed with lower and lower synthesis pressures. At present, the most common size of plant ranges between 800 and 1800 metric tons per day and synthesis pressures vary between 180 and 260 bars depending on the process. An NH_3 synthesis plant is outlined very schematically in Fig. 3.56. The processed materials are natural gas (most common), water or air.

Natural gas compressed to approximately 40 bars by compressor C1 is injected into the desulphurization plant to protect the catalysts used in the process from the sulphur. The gas is then sent to the first reformer where steam is injected; with a catalyst it slowly combusts producing H_2, CO, CO_2 and considerable quantities of methane.

The mixture of hot gases then passes to the second reformer where it is mixed with heated air at 30 to 40 bars from compressor C2. Due to the presence of air combustion in the second reformer is rapid. It may be expressed as follows:

$$CH_4 + O_2 + N_2 \rightleftharpoons 2H_2 + CO_2 + N_2 \quad \text{(exothermic reaction)}$$

The heat developed by this reaction which raises the temperature of the reagents to around 1000°C is used to produce the steam required for the process that is

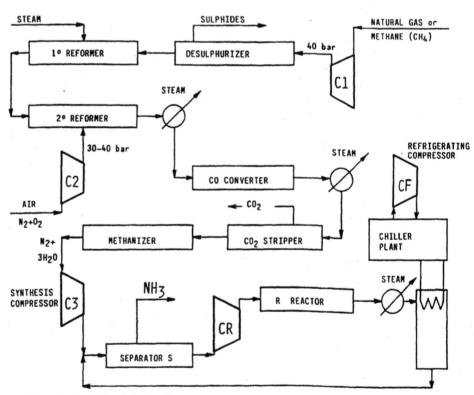

FIGURE 3.56 NH_3 synthesis plant outline.

generated in exchanger S1 downstream from the second reformer. The gas mixture then flows into a separator where the carbon monoxide is removed. It is cooled to eliminate excess water vapor and finally put into a separator to eliminate the carbon dioxide CO_2. To purify the mixture of any remaining CO or CO_2 methanization follows whereby carbon residues are hydrogenated to produce methane and water. At this point of the process, a mixture of H_2 and N_2 is obtained containing a small percentage of methane which forms the synthesis gas mixture. This mixture is then brought to a pressure, which may vary according to the type of process used, in order to make the synthesis reaction shift more easily:

$$N_2 + 3H_2 \rightleftarrows 2NH_3 \quad \text{(exothermic reaction)}$$

Since the reaction is exothermic here too the heat is recovered downstream from the reactor to generate steam. The ammonia is condensed next by cooling with a chiller plant; a refrigerating compressor CF is used for this using ammonia as the refrigerating fluid for obvious reasons. Finally, in separator S the ammonia that has formed is separated and the gas that has not transformed into ammonia is remixed with the fresh gas and recycled through compressor CR. It has been seen how several types of centrifugal compressor are used in an ammonia synthesis plant processing gas differing in nature, pressure and temperature; the characteristics of these machines are briefly outlined below.

Natural Gas Compressors. Compression takes place in one or two casings reaching a final pressure of approximately 40 bars. In units with two casings, there is a low pressure compressor with a horizontally split casing, MCL type, and a high pressure compressor with a barrel type casing open vertically at both ends.

Air Compressors. A final pressure of 30 to 35 bars is normally reached with two horizontally split compressor casings.

Each compressor casing houses two back-to-back compression stages with external cooling. This arrangement is preferred for the lesser number of compressor casings and lower absorbed power.

Synthesis Compressors. A train of synthesis compressors is composed of a minimum of two casings and maximum of four depending on plant dimensions and the processes used, i.e., the pressure ratios to be achieved.

Barrel type compressors are employed incorporating either a single stage (BCL compressor) or two stages (2BCL compressor) in a back-to-back arrangement. The recycled gas is normally compressed by a single impeller positioned in the end compressor casing back-to-back with the impellers that compress the fresh gas. These compressors were the first centrifugals widely applied in industry in the 60's for high pressures.

Refrigerating Compressors. This service, which uses NH_3 as refrigerant fluid, is normally performed with one or two low pressure compressor casings.

3.11.3 Methanol Synthesis Plants

Methanol or methyl alcohol CH_3OH is obtained today by synthesis processes. It is mainly used for the following:

- As solvent in the chemical industry for preparing paints, etc.
- As the starting product for preparing formic aldehyde (formalin)
- Preparing antifreeze aqueous solutions (a solution made up of 40% methyl alcohol in water freezes around $-40°C$)
- As fuel, especially blended with petrol (10 to 20%) in order to considerably increase the fuel grade (the octane rating of methyl alcohol is 120)

As regards the process in recent years there has been a tendency to use low pressure processes while plant capacities have grown considerably. Nowadays 2000 to 2500 MTPD plants are normally required and there are projections for 5000 MTPD plants. A typical low pressure process (ICI) makes it possible to obtain 99.5 to 99.99% methanol starting with natural gas, steam and carbon dioxide.

After the natural gas is desulphurized it is reformed with steam to obtain synthesis gas consisting of CO, CO_2, H_2 and CH_4. The reforming pressures and temperatures depend on the quality of the initial material but range between 10 and 20 bars, 800 and 900°C. The heat developed by the synthesis gas reactions is mainly used to produce steam. The gas is then cooled and sent to the compressors which compress it to 80 to 90 bars and send it to the reactor for conversion into methanol. On leaving the reactor, the gas is cooled and the methanol separated from the gas that has not reacted. As usual there is a recycle impeller to recompress the gas and send it back into the reactor. Machines in methanol synthesis service are normally

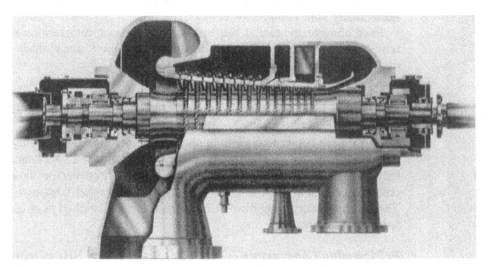

FIGURE 3.57 Axial compressor for process applications.

FIGURE 3.58 Gas turbine axial compressor.

made up of 2 or 3 compression stages in 2 barrel type compressor casings. Compressors that handle the synthesis gas are normally directly coupled to a steam turbine, frequently the type with a double exhaust. The recycle compressor is sometimes driven separate from another steam turbine.

3.11.4 Urea Synthesis Plants

Like the NH_3 synthesis plants, urea synthesis plants have developed in recent years reaching ever greater capacities while the trend has been towards relatively low pressure synthesis processes. Now plants utilizing centrifugal compressors only vary in size from a minimum of 600 to a maximum of 1800 MTPD; up to the early 70's, CO_2 was compressed by a combination of centrifugal and reciprocating compressors. The processes that have become most common can be divided into two categories requiring pressures of 150 to 160 bars and 230 to 250 bars respectively.

Urea is obtained by synthesis of liquid NH_3 and gaseous CO_2 through the known chemical reactions

$$2NH_3 + CO_2 \rightleftarrows NH_4CO_2NH_2 \quad \text{(exothermic reaction)}$$
$$NH_4CXO_2NH_2 \rightleftarrows NH_2CONH_2 + H_2O \quad \text{(exothermic reaction)}$$

In practice, with the first reaction carbamate is formed and then dehydrated to form urea. Centrifugal compressors were used that compress almost completely pure CO_2 from atmospheric pressure to the synthesis pressure. Compression normally takes place in two casings (150 to 160 bar synthesis) or in three casings (230 to 250 bar synthesis).

A train is made up of a low pressure compressor with two compression stages (2MCL compressors) and of one or two high pressure compressors (2BCL and BCL compressors). These machines have to achieve very high pressure ratios consequently with great differences in flow rates between the beginning and end of compression (the pressure ratios are 150 to 160 or 230 to 250 respectively). High gas density even at not particularly high pressures calls for special attention in the mechanical sizing of the rotors. This is peculiar to urea synthesis as compared to as in ammonia synthesis, for instance, the problem does not exist on account of the lower molecular weight of the gas.

Another difference is that CO_2 is corrosive, particularly if wet; its corrosiveness is related to pressure and temperature and therefore also the type of material employed has to vary along the compression path. Although carbon steel is normally used in the first compressor casing all parts in contact with the gas in following casings have to be stainless steel.

3.11.5 Natural Gas Service

Demand for machines to compress natural gas to medium and high pressures has grown considerably in recent years as a result of development of some special services such as reinjection, transportation and gas liquefaction followed by regasification. The first two services basically originate from the fact that in the last ten years or so, it has become more and more economical to recover gas that used to be flared at wellheads or that at any rate failed to be put to profitable use. These services are outlined here below and the machines most commonly used described.

Gas Reinjection. Natural gas reinjection has become widespread over the last ten years. It has several purposes chiefly depending on the nature of a field. In short, one can say that natural gas is reinjected for any of the following reasons:

- To maintain wellhead pressure with the gas that is obtained during extraction of crude oil, thus allowing the production of crude to be kept practically constant. Gas injection has a "piston" effect.
- Gas storage: consisting in storing the gas which is usually obtained along with crude oil in wells until it can be used as a source of energy.
- Gas lift system: consisting in injecting gas into the crude production well in order to lighten the well column. This system is normally adopted in fields where water injection is used for extraction. By injecting water and extracting crude in time the specific gravity of the mixture tends to increase and the percentage of crude drops. The gas lift technique is used to improve well productivity.
- Peak shaving: a system used in highly developed industrial areas.

This is particularly advantageous from a saving point of view when there is a depleted natural reservoir in a highly industrialized area. The natural gas pipeline, in this case, can be sized for the average consumption in the area and during hot periods, excess gas is stored in the reservoir. In cold weather, the stored gas can be used for peak demand.

- Gas is injected in noncondensed gas fields: or into gas fields contain widely varying quantities of condensate. For example, at Groning (Holland) the condensate content is 6 g/m^3 whereas at Hassi R'Mel (Algeria) it is over 220 g/m^3.

The condensates are separated in de-gasolining units and the dry gas can be reinjected into the field. The aim of this injection is to maintain pressure in the field and with the injected dry gas to facilitate absorption of condensates in the gas. The end purpose is therefore that of producing gasoline with the gas acting as vector fluid. Main features of the compressors may be summarized as follows:

- Delivery pressures vary from 100 to 150 bars for gas lift service and some peak shavings, to 250 to 500 bars for the other systems with the exception of some reinjections which requires pressures of 600 to 700 bars (the maximum required today for centrifugal compressors). The problem of designing machines capable of reaching such high pressures has been solved only since the 70's.
- The compressors are therefore barrel type, but train compositions vary greatly from application to application.
- These compressors handle high density gas, therefore as was mentioned for urea synthesis, particular attention has to be given to the mechanical and aerodynamic calculations and tests.
- Finally, since these units are often situated in desert areas or at any rate far from industrialized areas, a characteristic in particular demand for these units is that they are packaged. That is, the machines have to be self-contained, easy to transport and to install (package units), should require minimal maintenance and have remote control.

Gas Transportation. The numerous compressor stations distributed along gas pipelines normally utilize PCL centrifugal compressors. These are machines with modest pressure ratios (r = 1.2 to 1.8) which are required to provide relatively high efficiency as they are usually driven by gas turbines which burn a high quality fuel such as the very gas that is in the gas pipeline.

Particular attention is therefore dedicated to the design and engineering to achieve high efficiencies. Three-dimensional impellers are normally used and the configuration of the machine allows the impellers to be spaced out along the axis with gently curved return channels for good impeller inlet conditions.

Liquefaction and Regasification. The growing demand for natural gas for use in industry and domestic heating makes it necessary to use the huge quantities of gas which are available also from fields such as those in Algeria or Libya from which gas is transported in liquid form.

Among the equipment required for this technology, centrifugal compressors in particular and rotary machines in general have proven to be very versatile by being able to be used for gathering as well as in the later stages of transportation, liquefaction and regasification. The gas that comes out of the wells is compressed by means of centrifugal compressors and put into a gas pipeline which takes it through a series of compression stations, previously described, to the liquefaction station located near the gas loading buoys. The gas is liquified in plants using centrifugal compressors whereas the ships are often driven by gas turbines fueled on part of the substance that changes from liquid to gaseous state during transport. The natural gas is stored at the port of arrival in liquid form in special tanks; it is then taken from these to be heated, vaporized and put into the gas pipeline again by centrifugal compressors.

The mechanical energy required to compress the gas is provided by a gas turbine system and the thermal energy necessary for vaporizing the gas is recovered from the exhaust gases from the same system.

A typical regasification station in Italy is the one at Panigaglia in the gulf of La Spezia.

3.11.6 Oxygen Service in Metallurgical Plants

One of the initial premises for the development made in steel production is the possibility of utilizing great quantities of very cheap oxygen (Fig. 3.59).

Construction of the most advanced low pressure air separation plants alongside the steelworks has been expanding now for years. The average capacity of these plants has risen from approximately. 300-t/d (ton per day) in 1965 to the present 1000 to 1500 t/d for every unit installed and units are being designed for over 5000 t/d.

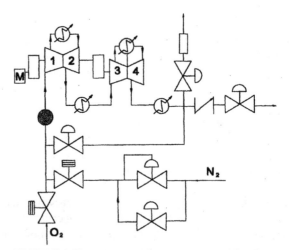

FIGURE 3.59 Diagram of oxygen compression from 1 to 40 bars in four stages.

FIGURE 3.60 Gas turbine centrifugal compressor.

Another determining element in connection with these new production levels is use of the oxygen centrifugal compressors which are now constructed for large flows and high pressures and are therefore capable of satisfying the most exacting requirements of modern metallurgical plants as well as influencing the choice of requirements. The advantages ensuing from introducing centrifugal compressors also in the metallurgical sector soon became apparent as a consequence of the economy due to the simple operation, modest maintenance requirements and relatively low total cost of installation.

Around 1965, centrifugal compressors were installed with capacities ranging from 16,000 to 28,000 Nm^3/hr and final pressures of 25 to 30 bars; in recent years centrifugal compressors have come into operation compressing 40,000 Nm^3/hr at 40 bars and machines are now being designed for chemical service to reach over 60 bars. These achievements have been made possible by overcoming the problems that principally concern the danger of fire as a result of 99% pure oxygen in contact with flammable materials including the metal used to construct the compressors. The risks of fire grow considerably when gas pressures, temperatures and flow rates rise and the causes can only be partially controlled.

Compressors for service in steel works are presently employed at an operating pressure of approximately 40 bars, which is the highest pressure standard of those adopted in metallurgical plants for the oxygen distribution network. The gas is generally compressed in four successive stages in two casings: The cast iron and steel casings are for low and high pressure respectively, they are horizontally-split and each has balanced opposed stages with sidestream suction nozzles and interstage discharges.

The gas is cooled after each discharge in water coolers installed outside the machines.

3.12 ANTISURGE PROTECTION SYSTEM

3.12.1 General

Surge control represent a regulation system to maintain compressors inside their stable working range, assuring a volume flow rate at impeller inlet section, higher than the surging rate. An efficient control method prevents compressors and other turbo-machines from crossing the surge line and avoids rotating stall conditions for compression ratios as wide as possible. These aerodynamic instabilities are intrinsic to almost all kinds of turbo-machines, and often represent a strong limitation to the range of efficient performance. Anti-surge control systems can thus represent a useful instrument to improve the global performance of a compressor.

Fig. 3.61 represents a typical performance map, obtained from compressor test results. In the figure the compression ratio $\rho = (p_2/p_1)$, across the whole machine is plotted against volume flow rate Q_1. One of the most striking features of a typical

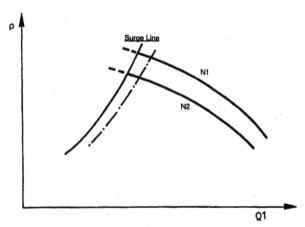

FIGURE 3.61 Typical performance map for a compressor stage.

performance characteristic is the strong dependence shown by the compressor on the rotational speed (N1 and N2 in figure). As previously stated for a centrifugal compressor efficient operation at constant N lies to the right side of a pseudo-parabolic line called "surge" line, approximately falling near maximum point for pressure. Both for a axial or centrifugal compressor the surge line delimits the range of stable working conditions, unstable operation being characterized by severe oscillation of the mass flow rate.

The extreme regulation line (dotted line in the picture), should be parallel and slightly to the right with respect to the actual surge line.

The measure of the volumetric flow rate processed by the turbo-machine is necessarily the key point for any kind of regulation system. In general the surge control system can be chosen according to different specific requirements and relies on other physical variables different from volume flow rate, but some basic features should be guaranteed:

a. The regulation line should be as close as possible to the surge limit line moved parallel to the capacity axis by an established amount.

b. When conditions of the suction fluid vary, the regulation line should not get any closer to the surge line, relating to the design conditions, than may be necessary for proper operation of the antisurge system.

c. The regulation system should protect the machinery in an automatic way during start up, standstill, and all possible off design conditions.

3.12.2 Basic Regulation Theory

The application of general formulation of similarity laws in a turbo-machine states: the working condition for two different rotational speeds are dynamically similar if all fluid particles, in corresponding points within the machine, have the same direction and velocities proportional to blade speed.

If points, belonging to different dimensional characteristic curves, represent dynamically similar states, then the nondimensional groups of variables involved in the physical problem (ignoring Reynolds number effects), are expected to have the same numerical value.

The dotted line in Fig. 3.61 shows that the set of dynamically similar states lies on a parabola, since the nondimensional groups $Q/(ND^3)$ and $gH/(ND)^2$ are in similarity.

If similarity theory is applied to the same turbo-machine, D is constant, so dynamic head becomes proportional to N^2 and volume flow rate to N. In particular, for surge points the following laws can be applied:

$$Q_1 = K'N \qquad (3.71)$$

$$H_p = KN^2 \qquad (3.72)$$

Eliminating N, from (3.71) and (3.72), the equation for the surge curve can be obtained:

$$H_p = K''Q^2 \qquad (3.73)$$

Surge points, in a single stage compressor, for the various speeds lay on a parabola with vertex in the origin and axis coinciding with the axis of ordinates on the polytropic head—volumetric flow rate plane.

The polytropic head can be obtained from (3.8), while volume flow rate can be experimentally derived from measures with a calibrated orifice, through the following relation:

$$Q_1 = C_Q \Psi_2 \varepsilon_c \sqrt{2gh} \qquad (3.74)$$

in which h is the measured pressure loss in the orifice, expressed in mmH$_2$O C_Q is a flow coefficient depending both on the geometrical characteristics of the orifice (Ψ_2/Ψ_1) and, the real flow conditions though C_r (related to viscosity losses and flow contraction):

$$C_Q = \frac{C_r}{\sqrt{1 - \left(\frac{\Psi_2}{\Psi_1}\right)^2}} \qquad (3.75)$$

Fluid compressibility is accounted in equation (3.74) ε_c depending on pressure drops in the orifice.

Once the geometry of the orifice and his fluid-dynamics characteristics are known the following expression for volume flow rate can be used:

$$Q_1 = \beta \sqrt{\frac{h_1}{\gamma_1}} \qquad (3.76)$$

where γ_1 is the specific weight, expressed in the following way for a perfect gas:

$$\gamma_1 = \frac{1}{R} \cdot \frac{P_1}{3Z_1} \cdot \frac{10^4}{T_1} \text{ (Kg/m}^3\text{)} \qquad (3.77)$$

Substituting in equation (3.76) the γ_1 value obtained from equation (3.77), the following final expression for the volume flow rate can be obtained:

$$Q_1 = \beta \sqrt{\frac{h_1 \cdot R \cdot Z_1 T_1}{P_1}} \qquad (3.78)$$

with h_1 expressed in Kg/cm^2

Substituting equations (3.74) and (3.78) in (3.73) the following relation can be obtained between stage compression ratio in incipient stall condition and volumetric flow rate measured in the calibrated orifice:

$$Z_1 RT_1 \frac{n}{n-1} (\rho^{p-1/n} - 1) = K'' \beta^2 \frac{h \cdot RZ_1 T_1}{P_1} \qquad (3.79)$$

Simplifying:

$$\frac{n}{n-1}(\rho^{n-1/n} - 1) = K''\beta^2 \frac{h_1}{P_1} \qquad (3.80)$$

or equivalently:

$$\frac{1}{K''\beta^2} \cdot P_1 \frac{n}{n-1}\left(\rho^{n-1/n} - 1\right) = h_1 \qquad (3.81)$$

For different inlet conditions and speeds surging can be avoided, from formula (3.81) if:

$$h_1 > \frac{1}{K''\beta^2} \cdot P_1 \frac{n}{n-1}(\rho^{n/n-1} - 1)$$

or equivalently if:

$$\frac{h_1}{\frac{n}{n-1}(\rho^{n-1/n} - 1)P_1} > \frac{1}{K''\beta^2} \qquad (3.83)$$

Formula (3.83) is a very useful relation to avoid surge and stall condition for the following reasons:

1) The rotational speed of the compressor is not explicitly involved.
2) The calibrated orifice pressure loss (h), is the only required measurement, regardless pressure, temperature, gas composition.
3) Gas composition can be explicitly accounted through polytropic exponent n, changes of n having however very poor effect on the following value:

$$\frac{n}{n-1}(\rho^{n/n-1} - 1)$$

so they be neglected in first attempt.

From a more physical point of view, in formula (3.83) the compression ratio, depending both from speed and inlet conditions is employed to balance the volume flow rate.

Although extremely simplified, formula (3.83) can be too complex to be realized with simple instrumentation. However, if the compression ratio is small enough, for example in single or two-stage compressors, the quantity $n/n - 1$ ($\rho^{n/n-1} - 1$) can be substituted by ($\rho - 1$) in first approximation. The error is always in the safe direction being:

$$\rho - 1 > \frac{n}{n-1}(\rho^{\frac{n}{n-1}} - 1).$$

With this approximation formula (3.83) becomes:

$$\frac{h_1}{P_1(\rho - 1)} > \frac{1}{K''\beta^2}$$

or better:

$$\frac{h_1}{P_1(\rho - 1)} = K''' \qquad (3.84)$$

where

$$K''' > \frac{1}{K''\beta^2}.$$

A first estimate of percentage error can be obtained from the expression:

$$\varepsilon = 100 \cdot \left[\frac{\rho - 1}{\frac{n}{n-1}(\rho^{n-1/n} - 1)} - 1 \right] \qquad (3.85)$$

Leading to the following expression for the percentage error in the volume flow rate:

$$\varepsilon_{Q1} = 100 \cdot \left(\sqrt{1 + \frac{\varepsilon}{100}} - 1 \right) \qquad (3.86)$$

In general, error in flow rate is about one half of that due to the substitution of $(\rho - 1)$ with $n/n - 1/(\rho^{n/n-1} - 1)$ so that:

$$\varepsilon Q_1 \cong \frac{\varepsilon}{2} \qquad (3.87)$$

Rewriting formula (3.84) in the following way:

$$\frac{h}{P_1 - P_1} = K''' \qquad (3.88)$$

or:

$$\frac{h}{\delta_p} = K''' \qquad (3.89)$$

(3.89) is the equation of a parabola (in the plane ρ, Q_1) with origin in the point $\rho = 1; Q_1 = 0$ (see Fig. 3.62).

This type of anti-surge system is extremely simple. All the necessary information to protect the machine can be collected with only two differential pressure transducers, the first instrument measuring pressure drops through the calibrated orifice and the second the real pressure rise inside the centrifugal compressor stage. The

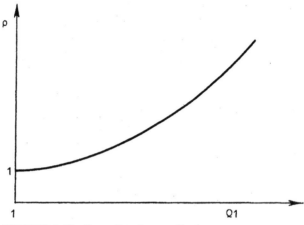

FIGURE 3.62 Surge line in $\rho - Q_1$ plane.

values, previously converted in electric signals are sent to a divider unity, which computes the result of the ratio in (3.89). This value is then transmitted sent to a regulation unit, on which the value of the anti-surge regulation set-point K'' is settled. The regulator acts on the by-pass valve, preventing the ratio from going below the established set-point avoiding unsafe working range to be reached.

3.13 ADAPTATION OF THE ANTISURGE LAW TO MULTI-STAGE COMPRESSORS

When the compression ratios are high as for multistage compressors, surge curves, are no longer parabolic. Surge conditions can be due to different impellers at disparate speeds and thus the previously obtained equation it is no longer safe to protect the machine unless cutting wide operating range areas, otherwise useable (see Fig. 3.64 as a reference).

This kind of situation can be avoided through the transformation of the basic equation with the introduction of a multiplicative factor in the expression of $\delta_p (P_1)$ leading to the following formula:

$$\frac{h_1}{P_2 - aP_1} = K \qquad (3.90)$$

The introduction of a coefficient allows the shifting of the origin of the parabola along the vertical axis, from $\rho > 1$ up to $\rho = 0$ (see Figs. 3.65., 3.66., 3.67.), while maintaining the basic characteristics of the original equation.

If vertex of the parabola is in the origin of the reference system, where $\rho = 0$, equation (3.90) can be simplified in:

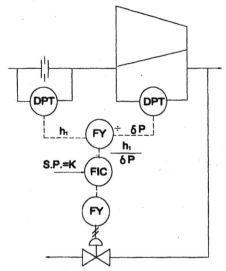

FIGURE 3.63 $h_1/\delta_p = K$ regulation scheme.

$$\frac{h_1}{P_2} = K \qquad (3.91)$$

To make the $h_1/P_2 - aP_1 = K$ equation, one makes use of the diagram in Fig. 3.68.

If $a = 0$ then the simplified schematic diagram in Fig. 3.69 can be used.

If the operating ranges of a multistage compressors are wide, the maximum employment exploitation of the stable compressor field becomes necessary. Nowadays, always more often the preceding equation may not suffice for the purpose and further processing can be required. As a consequence a function generator can be introduced on either the left and right signals of the basic equation (3.90). Thus

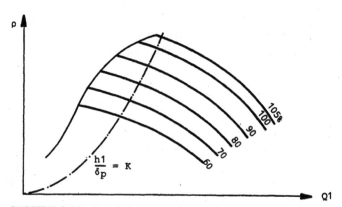

FIGURE 3.64 Regulation curve for multistage compressor.

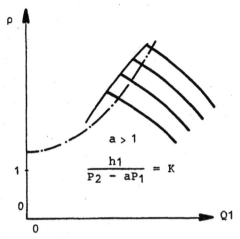

FIGURE 3.65 Regulation curve for multistage compressor (Es. a).

the antisurge curve can be modified in order to fit the surge curve as accurately as possible. A typical form of the basic equation is reported in eq. (3.92) and represented schematically in Fig 3.70:

$$\frac{h_1}{f(\delta p)} = K \tag{3.92}$$

in which function f can have the most general expression (based on the requirements of the function generation unit) as shown in Fig. (3.70).

The use of a general function for the generation of one the two signals can have some drawbacks. In fact, in all the previously stated antisurge laws, h and δp signals

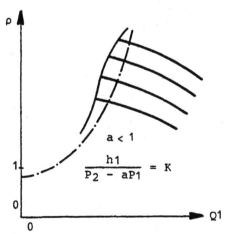

FIGURE 3.66 Flexible regulation curve for multistage compressor (Es. b).

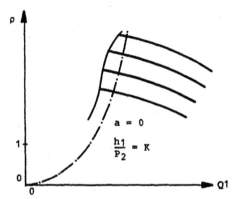

FIGURE 3.67 Flexible regulation curve for multistage compressor (Es. c).

were both linearly dependent with the inlet pressure ($h_1 \equiv Q_1^2 P_1$; $\delta p = P_1(\rho - 1)$) and, as a consequence, the resulting antisurge curves were insensitive to the inlet pressure changes. In this case the generic signal $f(\delta p)$ can depend upon inlet pressure in a more complicated way.

For a generic function $y = f(x)$, the ratio between the relative variation of the dependent variable y is related to the relative variation of the independent variable x in the following way:

$$\frac{\Delta y}{y} = y' \cdot \frac{\Delta x}{y} \cdot = \frac{y'}{y} \Delta x \qquad (3.93)$$

where y' is the first derivative of y in the point $(x;y)$.

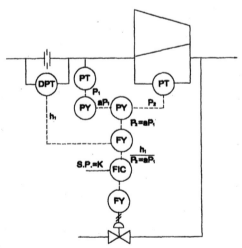

FIGURE 3.68 Regulation scheme type $h_1/(p_2 - aP_1) = K$.

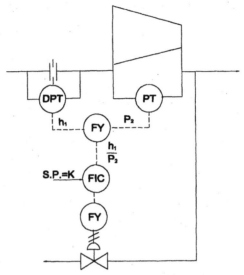

FIGURE 3.69 Regulation scheme type $h_1/p_2 = K$.

Eq. (3.93) can be applied in the surge problem. The change for $f(\delta p)$ expressed as a function of the relative ratio δp, of the inlet pressure P_1, is given by the following expression:

$$\frac{\Delta f(\delta p)}{f(\delta p)} = \frac{f'(\delta p)}{f(\delta p)} \cdot \Delta \delta p \qquad (3.94)$$

where:

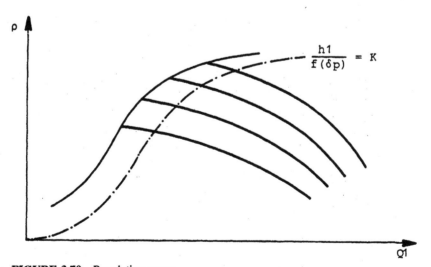

FIGURE 3.70 Regulation curve.

$$\Delta \delta p = (P_1 - \overline{P}_1)(\rho - 1) \tag{3.95}$$

and $f'(\delta p)$ is the derivative of function $f(\delta p)$ in the considered point.

The h_1 relative change is related relative change of P_1 around point $\overline{P}_1$ by the relation:

$$\frac{\Delta h_1}{h_1} = \frac{P_1 - \overline{P}_1}{\overline{P}_1} \tag{3.96}$$

The percentage error on the inlet volume flow rate, expressed as a function of the generic anti-surge curve at the design pressure $\overline{P}_1$ is:

$$\varepsilon Q_1(\%) = 100 \left(\sqrt{\frac{1 + \dfrac{f'(\delta p) \cdot (P_1 - \overline{P}_1)(\rho - 1)}{f(\delta p)}}{1 + \dfrac{P_1 - \overline{P}_1}{\overline{P}_1}}} - 1 \right) \tag{3.97}$$

In a series of machines, where the change in the inlet pressure could be remarkable, the shift in position of the antisurge curve is large enough and only partially balanced by the corresponding movement of the surge curve. In fact, (supposing the inlet pressure is the only modified variable), the latter feels a change in P_1 both in Z_1 factor and n value,

In these cases, the balancing of the two signals (h and δp) with the inlet pressure is an efficient tool to obtain antisurge curves almost insensitive to changes in the inlet pressure.

The general formulation can be therefore written in the following form:

$$\frac{h_1 \cdot \dfrac{\overline{P}_1}{P_1}}{f\left(\delta p \cdot \dfrac{\overline{P}_1}{P_1}\right)} = K \tag{3.98}$$

The practical realization of this equation can be performed through the scheme shown in Fig. 3.71:

3.14 ANTISURGE LAWS FOR SPECIAL APPLICATIONS

The antisurge law (3.97) previously developed is completely general, no assumptions having been made about the particular behavior of the machine. As a consequence, its validity is not compromised even when applied in complex systems. However, sometimes some simplifications can be introduced in plain cases, ac-

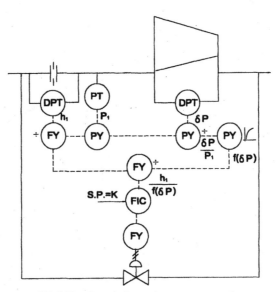

FIGURE 3.71 Regulation scheme.

cording to the special features of the particular systems. Two cases will be briefly examined: Electric Driven Compressors and Variable Speed Compressors with Constant Inlet Conditions.

3.14.1 Electric Driven Compressors (Fixed RPM)

If the inlet pressure conditions remain constant, the following law can be used:

$$\overline{h}_1 = K \tag{3.99}$$

since:

$$\overline{Q}_1 = C \sqrt{\overline{h}_1} \cdot \sqrt{\frac{1}{\gamma}} \tag{3.100}$$

This law can be represented by straight line parallel to the vertical axis crossing horizontal axis at the point $Q_1 = \overline{Q}_1$.

The error in the volume flow rate, expressed as a function of inlet flow characteristics is:

$$\varepsilon_{Q1}\% = 100 \left(\sqrt{\frac{\mu_1 \cdot \overline{P}_1 \cdot T_1 \cdot Z_1}{\mu_1 \cdot P_1 \cdot \overline{T}_1 \cdot \overline{Z}_1}} - 1 \right) \tag{3.101}$$

A schematic practical realization of the equation $h_1 = K$ can be fulfilled with the procedure shown in Fig. (3.74):

3.94 CHAPTER THREE

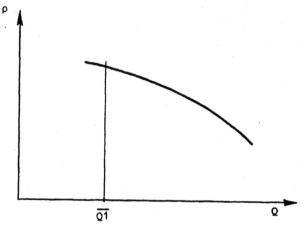

FIGURE 3.72 $\overline{h}_1 = K$ regulation line.

3.14.2 Variable Speed Compressors with Near Constant Inlet Conditions

When inlet pressure and delivery temperature (cooler) are kept constant, or when, for design or economical choices, the continuous measurement of the inlet flow rate is not possible, the following law can be considered in substitution of (3.97) Considering that,

$$Q_1 = C \frac{Z_1 \cdot T_1}{\sqrt{Z_2 \cdot T_1 \cdot \mu_2 \cdot P_1}} \cdot \sqrt{h_2} \cdot \sqrt{\rho} \quad (m^3/h) \tag{3.102}$$

can be represented graphically with a parabola having the origin in the origin of

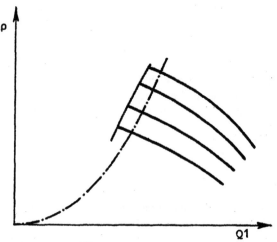

FIGURE 3.73 $\overline{h}_2 = K$ regulation line.

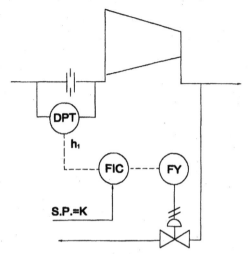

FIGURE 3.74 $h_1 = K$ regulation scheme.

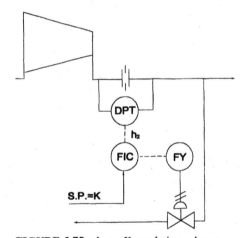

FIGURE 3.75 $h_2 = K$ regulation scheme.

the reference system as shown in Fig. 3.73. The percentage error in the flow rate expressed as a function of suction and delivery conditions is:

$$\varepsilon_{Q1}\% = 100 \left(\frac{Z_1 T_1}{\overline{Z}_1 \overline{T}_1} \sqrt{\frac{\overline{Z}_2 \cdot \overline{T}_2 \cdot \overline{\mu}_2 \cdot \overline{P}_1}{Z_2 \cdot T_2 \cdot \mu_2 \cdot P_2}} - 1 \right) \qquad (3.103)$$

The scheme shown in Fig. 3.75 can be used for the practical realization of (3.99).

CHAPTER 4
CENTRIFUGAL COMPRESSORS—CONSTRUCTION AND TESTING

Ted Gresh
Elliott Company

A centrifugal compressor is a dynamic compressor and thus depends on motion to transfer energy from the compressor rotor to the process gas. Compression of the gas is implemented by means of blades on a rotating impeller. The resulting rotary motion of the gas results in an outward velocity due to centrifugal forces. The tangential component of this outward velocity is then transformed to pressure by means of the diffuser.

4.1 CASING CONFIGURATION

The centrifugal compressor is a very versatile machine that can be readily adapted to a wide range of mechanical and process demands.

For example, depending on the head and flow requirements, the number of impellers used may be varied from one to as many as 10 or more in one casing. Inlet and exhaust nozzles may be located up, down or at some offset angle. Additional nozzles for economizers or other side streams or for cooling between stages, are easily accommodated. This flexibility of configuration is especially augmented by fabricated casing design (Fig. 4.1).

A few of the more common casing configurations are shown in Figs. 4.2 to 4.8.

4.2 CONSTRUCTION FEATURES

The major elements of the centrifugal compressor (Fig. 4.9) consist of: (a) the inlet nozzle; (b) inlet guide vanes; (c) impeller; (d) radial diffuser; (e) return channel; (f) collector volute; and (g) discharge nozzle.[1]

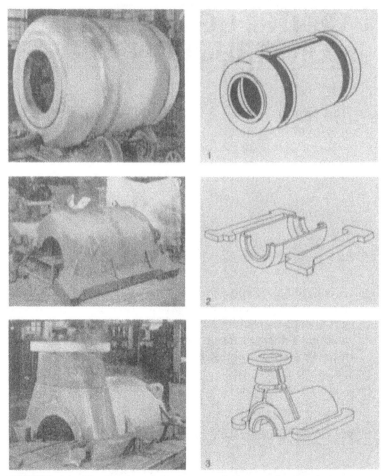

FIGURE 4.1 The Fabricated Compressor Casing is a flexible design for accommodating the specific requirements of the application including special nozzle arrangements. 1) The center section, a cylindrical shell of rolled and welded steel plate is joined by welding to two dished heads formed by hot spinning. Photo shows automatic girth welding of heads to shell. 2) Assembly is cut longitudinally into two equal halves and sturdy flanges are welded to each half of the shell. 3) Holes are burned in the shell and fabricated sections are welded together to form inlet, discharge, and sidestream nozzles.

The inlet nozzle accelerates the gas stream and directs it into the inlet guide vanes, which may be fixed or adjustable.

On a multi-stage compressor, the inlet nozzle is generally radial. In this case, the inlet guide vanes are necessary to properly distribute the flow evenly to the first-stage impeller (Fig. 4.10). Single-stage compressors frequently incorporate an axial inlet. In this case, inlet vanes may not be used.

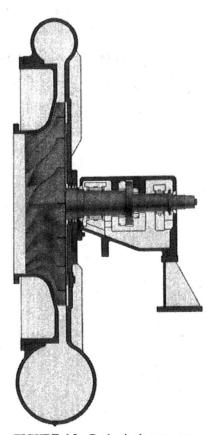

FIGURE 4.2 Basic single-stage compressor. Typical construction is impeller overhung from the bearing housing (*courtesy Elliott Co.*).

Because of the rotational effects of the impeller, the gas travels through the diffuser in a spiral manner. Therefore before entering the next impeller, the flow must be straightened out by the return channel vanes (Fig. 4.11).

4.2.1 Diaphragms

A diaphragm consists of a stationary element which forms half of the diffuser wall of the former stage, part of the return bend, the return channel, and half of the diffuser wall of the later stage. Due to the pressure rise generated, the diaphragm (Fig. 4.12) is a structural as well as an aerodynamic device.

For the last stage or for a single-stage compressor, the flow leaving the diffuser enters the discharge volute. It is common to design these volutes for constant

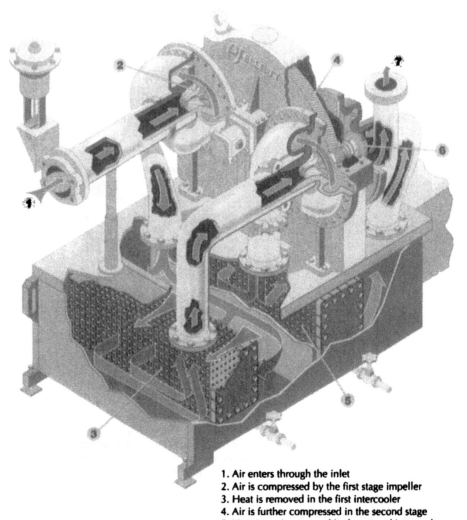

1. Air enters through the inlet
2. Air is compressed by the first stage impeller
3. Heat is removed in the first intercooler
4. Air is further compressed in the second stage
5. Heat is again removed in the second intercooler
6. Air is compressed to final pressure in the third stage
7. Air exits to the aftercooler and plant air system

FIGURE 4.3 For high speed single-stage compressors, maximum efficiency can be realized with intercooling between each stage. Also, cooling is required to keep operating temperatures below material limitations (*courtesy Elliott Co.*).

angular momentum ($R_i V_i$ = constant). This generally results in limited velocity change through the volute (V_2 to V_5). Once the gas leaves the volute, it passes through the discharge nozzle which reduces the velocity somewhat before entering the process piping. Figure 4.13 represents such a constant angular momentum volute.

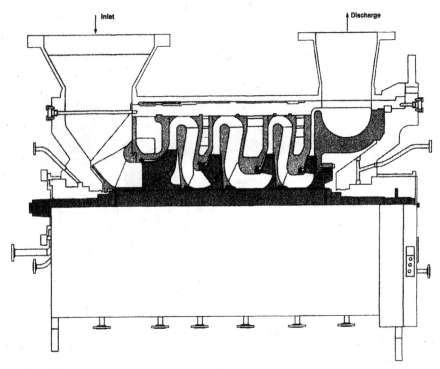

FIGURE 4.4 Basic straight through multistage compressor with a balancing piston. This arrangement may employ 10 or more stages of compression. This arrangement is most often used for low-pressure rise process gas compression. Casing design shown is a barrel construction used for high pressure or low mol weight gases which provides limited leakage areas and thus better contains the process gas (*courtesy Elliott Co.*).

Since velocities are relatively high through the diffuser section (several hundred feet per second), surface finish/friction factor is crucial to overall efficiency of the unit.

In many processes, dirt or polymer buildup on the impeller and diaphragm surfaces will give the aerodynamic surfaces a rough finish (Fig. 4.14). In some cases polymer buildup has been known to severely restrict the diffuser passage. Both conditions cause increased pressure losses and result in reduced overall efficiency of the compressor.[2]

The chemical mechanism that takes place to generate polymerization is not well understood, but experience has shown that under the right conditions polymers do form and bond tenaciously to the component base metal. Factors that have been found to be critical to the fouling process include: 1) temperature—polymerization occurs above 194°F; 2) pressure—the extent of fouling is proportional to pressure level; 3) surface finish—the smoother the surface the less apt the component is to foul; 4) gas composition—fouling is proportional to concentration of reactable hydrocarbons in the process gas.

4.6 CHAPTER FOUR

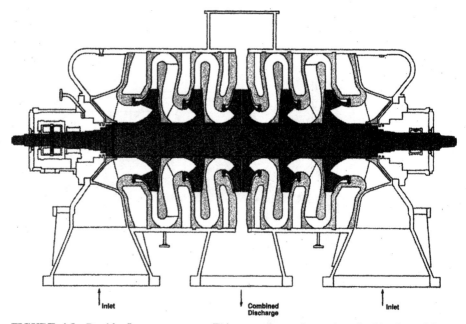

FIGURE 4.5 Double flow compressor. This arrangement is used to double the maximum flow capability for a compressor frame. Since the number of impellers handling each inlet flow is only half of that of an equivalent straight through machine, the maximum head capability is reduced accordingly (*courtesy Elliott Co.*).

Operating temperature can be reduced by water injection at each stage starting after the first wheel. The water should be injected via atomizing spray nozzles, and the amount should bring the gas to just below the saturation level.

Surface finish in these critical areas can be enhanced and preserved by applying a non-stick coating, such as fluorocarbon-based (Teflon) material, or a corrosion-resistant coating, such as electroless nickel. Multi component coating "systems" that provide a barrier coating, an inhibitive coating and a sacrificial coating have provided the best long term service. Compressor performance is best preserved by including a wash system that includes water with detergents or hydrocarbon solvents to wet aerodynamic surfaces preventing attachment of the polymers and to help wash compressor surfaces once bonding of the polymers occurs. Wash liquids introduced should be limited to 3% of the gas mass flow rate to prevent erosion and be injected stage by stage with increasing amounts at the discharge.

The long-term effects that include on-line liquid washing that might utilize hydrocarbon solvents or detergent enhanced water are shown in Fig. 4.15.[2]

4.2.2 Interstage Seals

Due to the pressure rise across successive compression stages, seals are required at the impeller eye and rotor shaft to prevent gas backflow from the discharge to

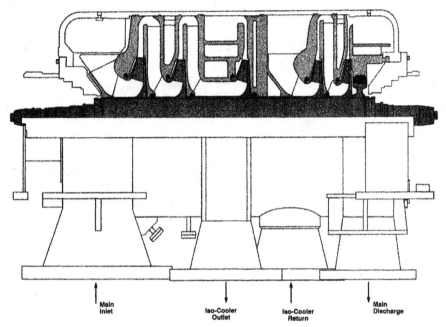

FIGURE 4.6 As in Fig. 4.3, cooling is required to keep operating temperatures below material or process limits as well as to improve operating efficiency. Iso-cooling nozzles permit the hot gas to be extracted from the compressor and to an external heat exchanger, then returned to the following stage at reduced temperature for further compression (*courtesy Elliott Co.*).

inlet end of the casing. The condition of these seals directly affects the compressor performance.

The simplest and most economical of all shaft seals is the straight labyrinth shown in Fig. 4.16. This seal is commonly utilized between compression stages and consists of a series of thin strips or fins, which are normally part of a stationary assembly mounted in the diaphragms. A close clearance is maintained between the rotor and the tip of the fins.

The labyrinth seal is equivalent to a series of orifices. Minimizing the size of the openings is the most effective way of reducing the gas flow. Labyrinths clogged with dirt (Fig. 4.17) and worn or wiped labyrinths with increased clearances (Fig. 4.18) allow larger gas leakage. This can affect compressor operation, and therefore the seals should be replaced.

Labyrinth material has typically been aluminum, because aluminum is compatible with most gases and is ductile enough to prevent rotor damage in the event of rubbing. Where corrosive elements are a concern, plastics such as Arlon CP (PEEK) and Torlon have been successfully used without the need to increase the seal clearances since the material has similar mechanical properties as aluminum. These materials (Arlon and Torlon) have been touted as rub tolerant when a raked tooth design of 15° is used, allowing a 50% reduction is seal clearance relative to aluminum labyrinth seals.

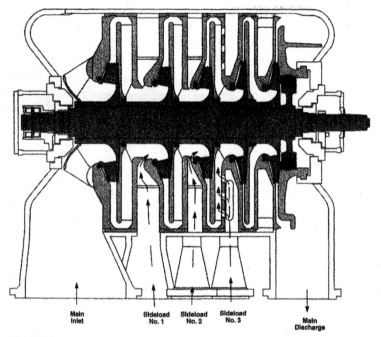

FIGURE 4.7 Side stream nozzles permit introducing or extracting gas at selected pressure levels. These flows may be process gas streams or flows from economizers in refrigeration service. Sideloads may be introduced through the diaphragm between two stages (sideload 3), or if the flow is high as in sideloads 1 and 2, the flow may be introduced into the area provided by omitting one or two impellers (*courtesy Elliott Co.*).

A hard labyrinth material such as stainless steel or cast iron could result in dry whirl and catastrophic failure of the compressor. One such case occurred when aluminum seals were replaced with cast iron seals. Since the clearances were not increased, the rotor touched the cast iron labyrinths while passing through the critical speed and dry whirl occurred. Vibration was so severe that the bearing retainer bolts backed out and the bearing housings fell off the compressor case. Needless to say, the damage was extensive.

Calculations and field performance data indicate that wiped interstage seals can decrease unit efficiency by 7% or more. Operating modes that contribute to labyrinth damage include surging, running in the critical speed, and liquid ingestion. A common problem is a trip on a compressor that has a rotor with buildup (unbalance) and a small, slow opening anti-surge valve. This condition will cause a high response through the first critical and open the seal clearances.

In order to reduce or negate the performance effects common with damaged interstage seals, several improvements have been adopted by compressor manufacturers. Most noteworthy is the use of abradable seals in the impeller eye and shaft seal areas. Advantages include tighter design operating clearances and minimal efficiency effects after a seal rub, as shown in Figs. 4.19 and 4.20.

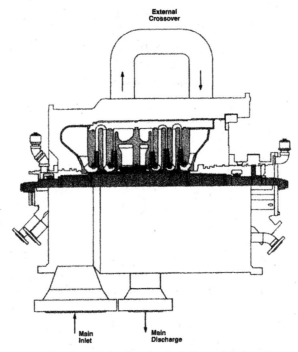

FIGURE 4.8 The back to back design minimizes thrust when a high pressure rise is to be achieved within a single casing. Note that the thrust forces acting across the two sections act in opposing directions, thus neutralizing one another (*courtesy Elliott Co.*).

The efficiency gain of abradable seals is achieved through the reduction of seal clearances, thereby reducing recirculating flow through the impellers. Impeller eye seals, interstage shaft seals, and balance piston seals are effective in improving compressor efficiency when changed to the abradable design. Abradable seals also control impeller thrust, which varies with seal clearance (Fig. 4.21).

Having the fins as the rotating element permits centrifugal force prevents the build up of process deposits. Where conventional static labyrinths are used on a fouling duty, build up of deposits adversely affects the flow characteristic across the labyrinth, with detrimental effect on compressor efficiency. Rotating fins minimize this problem (see Fig. 4.17.)

A rub on an aluminum labyrinth causes the tips of the aluminum fins to mushroom out (Fig. 4.18). This creates undesirable flow characteristics across the labyrinth and increases the radial clearance. These factors are detrimental to compressor efficiency due to the resulting increased leakage and will have an effect on the thrust loading of the machine. With the abradable design, the rotating fins rub into the static element without damage to the fins and without effect on the normal running clearances. No performance deterioration or change in thrust load occurs (Figs. 4.19 and 4.20).

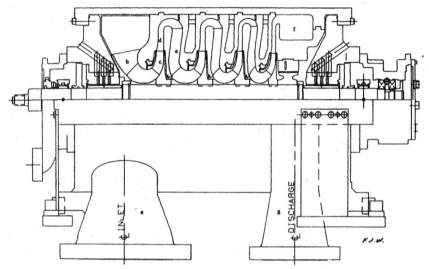

FIGURE 4.9 Major elements of a multi-stage centrifugal compressor: a) inlet nozzle, b) inlet guide vanes, c) impeller, d) radial diffuser, e) return channel, f) collector volute, and g) discharge nozzle.[1]

The overall efficiency improvement attainable by using abradable seals in a compressor varies with several factors, most notably the size of the compressor. Flow capacity increases as the square of the impeller diameter, while seal clearance increases more linearly with impeller size and is also dependent on other factors such as bearing clearances and manufacturing tolerances. Therefore as the compressor size increases, the leakages involved become a smaller portion of the total

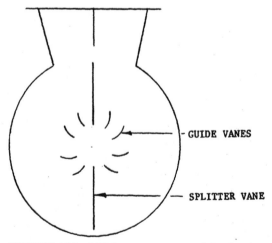

FIGURE 4.10 Multi-stage compressor inlet showing splitter vanes and guide vanes.[1]

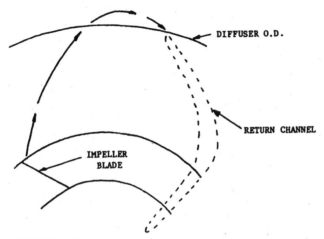

FIGURE 4.11 Flow path of gas from tip to return channel.[1]

flow. As this happens, the improvements gained by reducing these leakages have a diminishing impact on the machine's overall efficiency. Therefore it is the smaller, higher pressure compressors that benefit most from abradable seal.

4.2.3 Balance Piston Seal

A balance piston (or a center seal) is utilized to compensate for aerodynamic thrust forces imposed on the rotor due to the pressure rise through a compressor. The purpose of the balance piston is to utilize the readily available pressure differentials

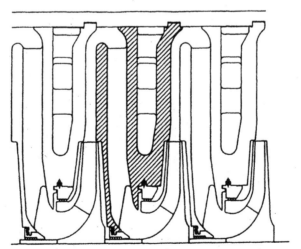

FIGURE 4.12 Multi-stage centrifugal compressor diaphragm.[1]

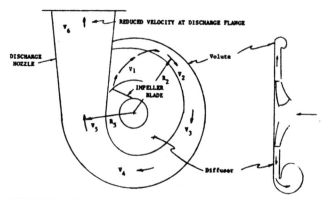

FIGURE 4.13 Discharge volute.[1]

to oppose and balance most of these thrust forces. This enables the selection of a smaller thrust bearing, which results in lower horsepower losses.

A certain amount of leakage occurs across the balance piston since a labyrinth seal is utilized. This parasitic flow is normally routed back to the compressor suction, thus creating a known differential pressure across the balance piston (Fig. 4.22). Occasionally, leakoff may be routed to other sections to gain an efficiency advantage. Air compressors generally route the balance piston leakage to atmosphere.

FIGURE 4.14 Photo of polymer build up in impeller and diffuser passages on an ethylene feed gas compressor.

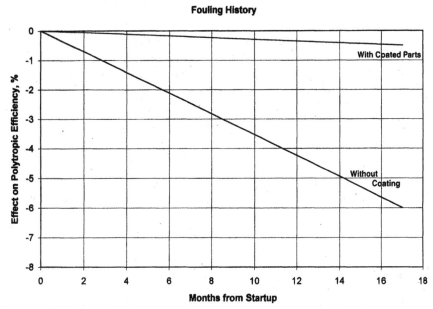

FIGURE 4.15 Effect of coated and non-coated surfaces on an ethylene feed gas compressor.[2]

Since the balance piston seal must seal the full compressor pressure rise, integrity of this seal is crucial to good performance. A damaged seal results in higher leakage rates, higher horsepower consumptions, and greater thrust loads.

One user of a compressor noted the following data before and after a balance piston seal replacement. This machine was in refrigeration service and was required to maintain a constant discharge pressure.

		Before	After
Discharge pressure	(PSIG)	410	410
Discharge temperature	(°F)	142	116
Axial position	(Mils)	24	19
Balance line DP	(PSID)	4.7	1.5
Speed	(RPM)	11440	10770
Thrust metal temperature	(°F)	240+	165

The balance piston damage was a result of surging and vibration excursions. The interstage seals were also extensively damaged, which contributed to the poor compressor efficiency. Note the differences in the various data before and after the seal replacement. The discharge temperature was high, since more work input was

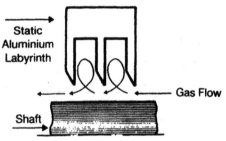

FIGURE 4.16 Aluminum labyrinth in new condition. Tight clearance and flow turbulence creates resistance to leakage flow.[1]

required to achieve the desired discharge pressure. In order to get the higher level of work input, the speed was increased. The wiped seals not only caused increased inefficiencies, but also higher thrust loads. This showed up in the axial position and thrust bearing temperature.

4.2.4 Impeller Thrust

Impeller thrust is generated by the differential force on the cover and hub of the wheel. These forces are the summation of the product of the pressures acting on the cover, hub, and the differential area from the shaft to the tip of the wheel.[3]

The impeller generates thrust between the eye and tip of the wheel, as well as below the eye. These forces (thrust) are caused by several different effects:

1. Rotational inertia field
2. Leakage
3. Friction
4. Diffusion

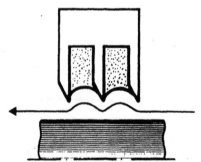

FIGURE 4.17 Fouled labyrinth. Turbulence is reduced and leakage flow is increased.[1]

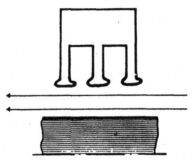

FIGURE 4.18 Rubbed labyrinth. Clearance is increased and turbulence reduced resulting in increased leakage.[1]

The effects of friction and diffusion are insignificant.

As indicated by the flow paths (Fig. 4.21), gas will flow toward the tip of the wheel along the hub and toward the eye of the wheel along the cover. Due to the pressure rise in the diffuser, the return channel pressure is greater than the pressure behind the impeller hub. Leakage therefore occurs from the return channel toward the impeller hub and outward toward the impeller tip. The effect of the pressure established by this leakage, superimposed upon the rotating inertia field, is shown in Fig. 4.21a.

From Fig. 4.21, it can be seen that there is an obvious net pressure differential toward the suction of the machine in addition to the area caused by the eye of the impeller. This is indicated in Fig. 4.21b. Integrating the products of the pressure and area from eye to tip results in the net thrust on a wheel.

For a "perfect" (zero leakage) seal at the impeller eye and shaft areas, the thrust is simply a function of the area inside the eye seal. For "real" seals with large clearance and corresponding high leakage in these areas, the net thrust can be as much as 50% greater.[3] Thus, maintenance of tight seal clearances is crucial for mechanical as well as aerodynamic concerns.

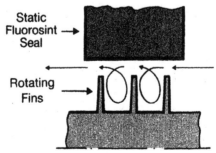

FIGURE 4.19 Abradable seal. Tight clearance and turbulence creates resistance to leakage flow.[1]

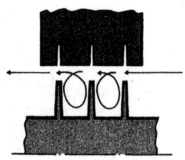

FIGURE 4.20 Rubbed abradable seal. Tight effective running clearance is unaltered and turbulence continues to create resistance to leakage flow.[1]

4.3 PERFORMANCE CHARACTERISTICS

The characteristics of a centrifugal compressor (Fig. 4.23) are determined by the impeller and diffuser geometry. Simply stated, kinetic energy is imparted to the gas via the impeller by centrifugal forces. The diffuser then reduces the velocity and converts the kinetic energy to pressure energy.

There are three important aspects of the compressor curve that will be discussed (Fig. 4.24):

1. Slope of the curve
2. Stonewall (or choke)
3. Surge

4.3.1 Slope

To understand about the slope of the centrifugal compressor head curve, it is necessary to first understand what is going on at the impeller discharge in terms of velocity vector diagrams.

V_{rel} (Fig. 4.25) represents the gas velocity relative to the blade. U_2 represents the absolute tip speed of the blade. The resultant of these two velocity vectors is represented by V, which is the absolute velocity of the gas

$$U_2 + V_{rel} = V \tag{4.1}$$

Knowing the magnitude and direction of this absolute velocity, we can break this vector into its radial (V_R) and tangential (V_T) components (Fig. 4.26).

For a radial inlet impeller (Fig. 4.23), the head output is proportional to the product of U_2 and V_T.

For a typical backward-leaning bladed impeller, as the flow decreases at constant speed, V_{rel} decreases. This causes V_T to increase, which increases head output. This

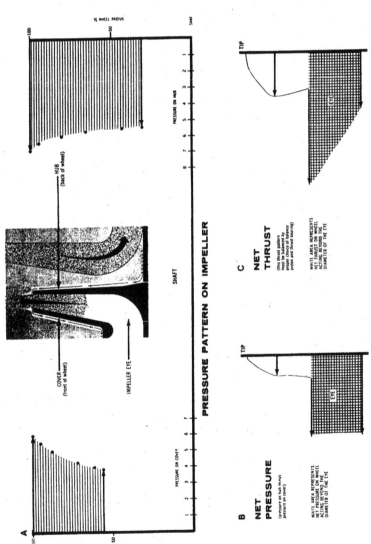

FIGURE 4.21 (*a*) Diagram showing the pressure pattern on the impeller. (*b*) Net Pressure. (*c*) Net thrust.[3]

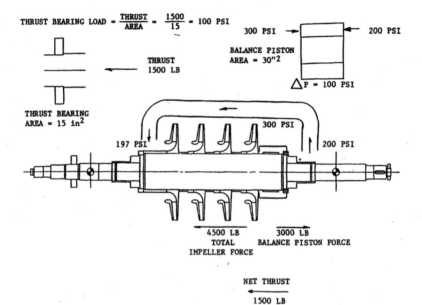

FIGURE 4.22 Schematic of compressor thrust. Pressure drop in the balance line is normally 1 to 3 psi.[3]

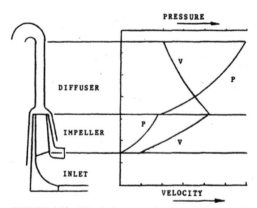

FIGURE 4.23 Velocity/pressure development for a typical radial inlet impeller.[4]

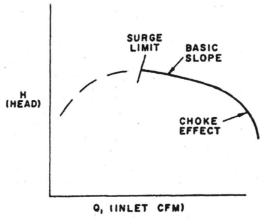

FIGURE 4.24 Head curve for a compressor stage.[5]

head increase with decreasing flow is what causes the basic slope to the centrifugal compressor performance curve (Fig. 4.27).

Figure 4.28 shows characteristic curves for three basic configurations: forward-leaning, radial, and backward-leaning blade profiles. Note that the forward-leaning blades provide a positive sloping head curve and the maximum head output. This is because V_T is increasing with increasing flow.[6]

A radial bladed impeller has a theoretical constant (flat) head curve, since V_T does not change with flow.

Overall stage efficiency is highest for backward-leaning impellers, while efficiency is lowest for forward-leaning blades. For best efficiency, most modern centrifugal compressors use backward-leaning bladed impellers. Directionally speak-

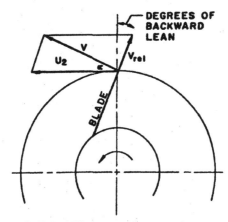

FIGURE 4.25 Vector diagram of the gas velocity relative to the impeller blade. The slope of the characteristic curve is strongly influenced by this relationship.[5]

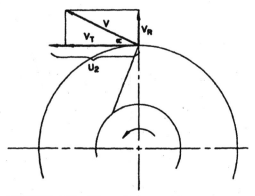

FIGURE 4.26 Vector diagram of the gas velocity shown in Fig. 4.25 in radial and tangential components.[5]

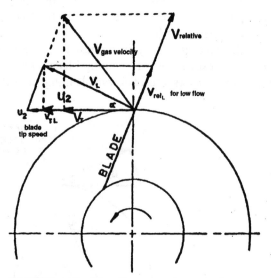

FIGURE 4.27 The effect of a change in flow rate on the vector diagram at the impeller O. D. is shown. Note that V_T decreases as flow increases (V_{rel} increases) for a backward-leaning impeller blade. This gives the backward-leaning impeller the characteristic negative-sloping head curve shown in Fig. 4.24.[5]

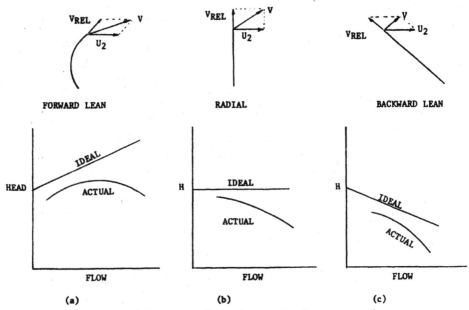

FIGURE 4.28 Three basic head curve shapes for centrifugal compressors.[6]

ing, the greater the backward lean, the better the efficiency. However, as the angle increases, the head is reduced (see Figs. 4.28c and 4.29). A designer can select a blade angle and tip width to best fit the desired head and efficiency characteristics of the particular application.

4.3.2 Stonewall

Stonewall, or choke, is a condition at which increased capacity (flow) results in a rapid decrease in head as flow is increased (see Fig. 4.30). This occurs because the Mach number is approaching 1.0.

Operating at a very high flow rate has very negative effects on the performance of a centrifugal compressor, and can sometimes be damaging. The stonewall effect of the centrifugal compressor stage with vaneless diffuser is controlled by impeller inlet geometry.

U_1 in Fig. 4.31 represents the tangential velocity of the leading edge of the blade. V represents the absolute velocity of the inlet gas, which, having made a 90° turn, is now moving essentially radially (in the absence of prewhirl vanes)—hence the name radial inlet. By vector analysis V_{rel}, which is gas velocity relative to the blade, is of the magnitude and direction shown.

$$V = U_1 + V_{rel} \tag{4.2}$$

At design flow, V_{rel} lines up with the blade angles. As flow increases beyond

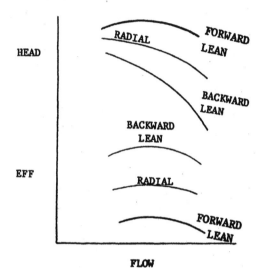

FIGURE 4.29 The effect of impeller blade angle on head.

design, V increases. As V increases, so does V_{rel}. V_{rel} now impinges at a negative angle to the blade, a condition known as negative angle of attack. High negative angles of attack contribute to the stonewall phenomenon because of boundary-layer separation and a reduction of effective area in the blade pack. This area reduction, in addition to the already high V_{rel}, brings on Mach 1 and a corresponding shock wave, as shown in Fig. 4.30.

4.3.3 Surge

Surge flow has been defined as peak head. Below the surge point, head decreases with a decrease in flow (Figs. 4.24 and 4.30).

Surge is especially damaging to a compressor and must be avoided. During surge, flow reversal occurs resulting in reverse bending on nearly all compressor

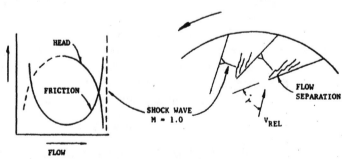

FIGURE 4.30 Stonewall. Flow is limited in impeller throat due to flow separation and developing shock wave.

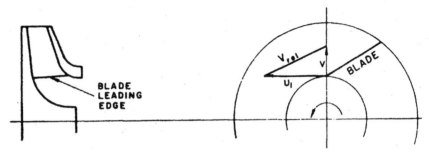

FIGURE 4.31 Flow vectors for impeller design condition.[5]

components. The higher the pressure or energy level, the more damaging the surge forces will be.

As flow is reduced at constant speed, the magnitude of V_{rel} decreases proportionally, causing the flow angle to decrease (see Figs. 4.27 and 4.32). Additionally, the incidence angle i is increased (Fig. 4.33).

The smaller the flow angle α, the longer the flow path of a given gas particle from the impeller tip to the diffuser outside diameter (Fig. 4.32). When angle becomes small enough and the diffuser flow path long enough, the flow momentum of the gas is dissipated by the diffuser walls by friction to the point where the frictional forces are increasing (versus flow reduction) faster than the head is increasing (versus decreasing flow).

The high losses associated with low flow (see Fig. 4.33) are partly caused by a poor incidence angle i, which can result in flow separation at the low pressure side of the blade leading edge. This flow separation frequently starts at one or more blades and continuously shifts around the impeller blades. This occurs at relatively low speeds just before full surge occurs. At higher speeds, the compressor generally

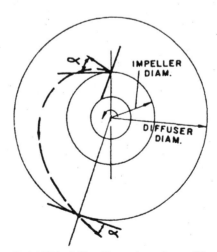

FIGURE 4.32 Flow through a diffuser.[5]

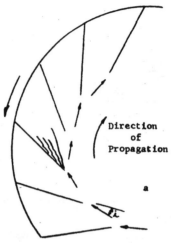

FIGURE 4.33 Rotating stall.

goes directly from stable operation to flow separation on all blades and full reverse flow. Rotating stall can also originate in the diffuser.

The flow separation plus the higher frictional losses result in a positively sloped curve. Since system resistance curves are also positively sloped, the system is unstable.

The point at which a compressor surges can be controlled somewhat by the designer adjusting the diffuser area to increase V_r and flow angle a. Of course higher velocities result in higher frictional losses, so a designer must balance between desired surge point and stage efficiency during the design process.

The surge point is reduced by the addition of vanes in the diffuser (Fig. 4.34). The vanes shorten the flow path through the diffuser, reducing frictional losses and controlling the radial velocity component of the gas. Due to lower friction, head and efficiency are enhanced, but the operating range is reduced. Off-design operation rapidly changes the incidence angle to the vanes and flow separation occurs, resulting in the reduced operating range.

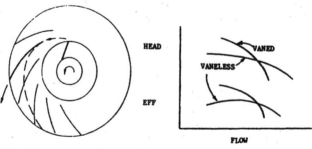

FIGURE 4.34 Diffuser vanes.[4,5]

To better understand what is occurring during surge, visualize the simple system shown in Fig. 4.35. The system consists of a small motor-driven compressor delivering air to a relatively large tank. While in an idle state, the entire system is at ambient conditions. The instant the unit reaches design speed, the pressure in the tank is still zero (Point 1). As time passes, the pressure builds in the tank and flow is reduced due to increased resistance. Eventually Point 2 is reached, where the pressure of the tank causes such a high backpressure on the compressor that flow through the impeller is significantly reduced. Much of the energy input is going to friction instead of building head. This is due to both the mismatch of inlet angle and the longer diffuser passage described earlier. Since this effect continues to build as flow is reduced, the slope of the head curve is reversed. As flow is reduced to Point 3, the head output of the compressor is also reduced. Since the pressure in the tank is still at Point 2, flow occurs from the tank to the compressor. Once the pressure in the tank is reduced (by reverse flow) to a level less than the head capability of the compressor, the process will then recover and the gas will flow from the compressor to the tank. This process will continue to repeat itself indefinitely.

4.4 OFF-DESIGN OPERATION

Off-design operation of a compressor can dramatically affect the performance characteristic curve shape. Any change in inlet conditions can change the discharge pressure and gas horsepower as shown in Fig. 4.36.[5]

In addition to changing the characteristic pressure and horsepower curves, the characteristic head curve also changes. This is due to volume ratio effects and equivalent speed effects.

If a constant discharge pressure is desired and gas conditions are changed (inlet pressure or temperature, mole weight change), a speed change is required. Since the curve shape changes with speed (higher losses at higher speeds), the head curve shape then changes (Fig. 4.37). This effect is further compounded by volume ratio effects (Fig. 4.38).

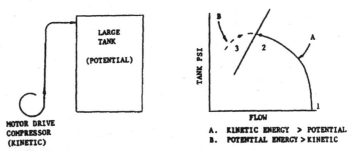

FIGURE 4.35 Surge. Once the pressure in the tank exceeds the capability of the compressor to produce head, reverse flow occurs.

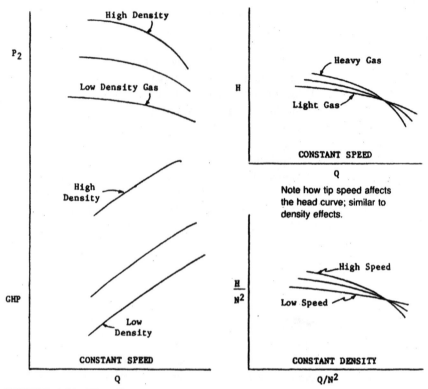

FIGURE 4.36 The effect of varying inlet conditions at constant speed for a single-stage compressor. For a multi-stage compressor, the curve shape and operating range is further compounded by volume ratio effects. See Fig. 4.38.[3,5]

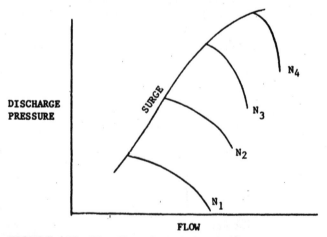

FIGURE 4.37 The effect of speed change on compressor performance curve.

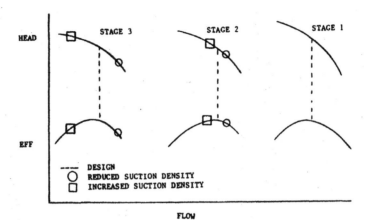

FIGURE 4.38 Volume ratio effects.[3]

The head characteristics are a function of the acoustic velocity of the gas. Knowing this, it is most convenient to refer to some "constant" gas and obtain an "equivalent tip speed." This reference constant is typically air at 80°F, since this is what most "developmental" testing uses as a test medium.

As an example, we know the sonic velocity of air at 80°F is 1140 fps and that of propylene at −40°F is 740 fps. If a compressor stage is operating at a mechanical tip speed of 780 fps on propylene at −40°F, the equivalent tip speed is:

$$U_{eq} = 780 \times 1140/740 = 1200 \text{ fps} \tag{4.3}$$

The stage characteristic head curve shape at 780 fps on propylene is therefore the same as 80°F air at 1200 fps. (Note: This is not exact. There is also an impeller tip volume ratio effect based on gas density that causes head and work input to change somewhat. An air and propylene test will not result in exactly the same head curve at constant equivalent speed, but for all practical purposes the results are close enough to be considered the same.)

In a multi-stage compressor the "equivalent speed" effect is compounded by volume ratio effect (Fig. 4.38). If the gas density varies, the pressure rise and volume ratio will also vary. This will feed a different flow rate to the second stage. The effect on following stages will be compounded. The end result is a premature choke and surge.

4.5 ROTOR DYNAMICS

The primary factor to assure long-term reliability of any rotating machinery is a good understanding of the rotor dynamics of that equipment. While there are many facets to rotor dynamics, the first and primary concern is good balance.

4.6 ROTOR BALANCING

The purpose of balancing rotors is to improve the mass distribution of the rotor and its components (caused by machining tolerances and non-uniform structure) so the mass centerline of the rotating parts will be in line with the centerline of the journals. To accomplish this, it is necessary to reduce the unbalanced forces in the rotor by altering the mass distribution. The process of adding or subtracting weight to obtain proper distribution is called balancing.

Correction of unbalance in axial planes along a rotor, other than those in which the unbalance occurs, may induce vibration at speeds other than the speed at which the rotor was originally balanced. For this reason, balancing at the design operating speed is most desirable in high speed turbomachinery.

To help understand the effects of altering the mass distribution of a compressor rotor, it is important to categorize rotors into three basic groups:

1. Stiff shaft rotors—Rotors that operate at speeds well below the first lateral critical speed. These rotors can be balanced at low speeds in two (2) correction planes and will retain the quality of balance when operating at service speeds.

2. Quasi-flexible rotors—Rotors that operate at speeds above the first lateral critical speed, but below the higher lateral critical speeds. In these cases, modal shapes or modal components of unbalance must be taken into consideration, as well as the static and couple unbalance (Fig. 4.39). Low speed balancing is still possible due to balancing techniques which correct the static and couple unbalance and sufficiently reduce the residual modal unbalance to retain the quality of balance when run at service speeds. Most multi-stage compressor rotors fit in this category.

3. Very flexible rotors—Rotors that operate at speeds above two or more major lateral critical speeds. Due to their flexibility, several changes in modal shape occur as speed is increased to the operating range. These rotors require high speed balancing techniques utilizing numerous balance planes to make the necessary weight distribution correction.

4.6.1 Two-Plane Balancing

The completed rotor is placed in a balancing machine on bearing pedestals. The amount of unbalance is determined and corrections made by adding or removing

FIGURE 4.39 Rotor lateral critical speed mode shapes: left) 1st critical mode shape, right) 2nd critical mode shape.

weight from two predesignated balancing planes. These two planes are usually near the 1/4 span of the rotor.

Following the final corrections, a residual unbalance check is performed to verify that the residual unbalance is within the allowable tolerance.

Maximum allowable residual unbalance guidelines:

$$\text{oz.} - \text{in.} = \frac{4 \times (\text{rotor weight})}{\text{rotor speed}} \qquad (4.4)$$

or

$$\text{oz.} - \text{in.} = \frac{56{,}347 \times (\text{journal static loading})}{N^2 \text{ (max. continuous rpm)}} \qquad (4.5)$$

4.6.2 Three-Plane Balancing

For low speed balancing of quasi-flexible rotors, major components (shafts, wheels, etc.) are individually balanced using two-plane balancing. For clarity, static (or force) and dynamic (or moment) balancing will be referred to as single-plane and two-plane balancing, respectively.

The rotor is completely assembled using pre-balanced components. Using the two-plane technique, the amount of unbalance in the two outer planes (wheels) is determined. This is resolved into a force component and a moment. The single-plane (force) correction is made as near the center of gravity of the rotor as possible. The residual unbalance moment is then corrected in two planes through the end wheels, which are usually situated near the one-quarter point of the rotor span between bearings. Following the final unbalance corrections, a residual unbalance check is performed to verify that the remaining unbalance is within tolerance.

The following items will help to ensure a satisfactory balance:

1. All components (wheels, impellers, balance piston) are individually balanced prior to assembly on the balanced shafts.
2. Mechanical runout is checked and recorded prior to balancing.
3. Combined electrical and mechanical runout checks are performed and recorded.

4.7 HIGH SPEED BALANCE

Axial unbalance distribution along any rotor is likely to be random in nature. Local unbalances can result from shrink fitting discs, impellers, sleeves, etc., along with residual unbalances present in all component parts. The vector sum of the unbalance distribution is compensated for in completely assembled rotors by balancing the assembly in a low speed balancing machine. The low speed machine measures either bearing displacement or bearing forces and provides information for correction in two or three transverse radial planes. The low speed machine can measure only the sum of the unbalances; therefore, the individual unbalances can excite the various flexural modes when the rotor is accelerated to operating speed. Additionally, when a rotor is operated at its maximum continuous speed or trip speed, forces acting on the various component parts, along with temperature changes, will alter the distributed unbalances. Rotors are processed through the high speed facility to measure vibration amplitudes and display mode shapes with the intention of minimizing these deflections.

4.7.1 Setup and Operation

Prior to operation, the rotor is placed in isotropic supports which are contained in a vacuum chamber (Fig. 4.40). The vacuum chamber is sealed and evacuated to 5 to 7 Torr. The rotor is accelerated slowly to maximum continuous speed while monitoring vibration levels and critical speeds.

Vibration tolerances are formulated to limit alternating forces at the bearings to 10% of static load (0.1g).

During the first run, if vibration levels do not exceed limits, the rotor is accelerated to trip speed or rated overspeed and held there for a minimum of 15 minutes. This process "seasons" the rotor by stabilizing the rotor temperature and at the same time seats component parts. After overspeed, the rotor is brought to rest and

FIGURE 4.40 Compressor rotor in at-speed balancing machine. The rotor is balanced at operating speed in a vacuum chamber.

again accelerated to operating speed while vibration levels are recorded. The procedure is repeated until repetitive vibration levels are observed, and the balancing process is started. (If the rotor is within tolerance at this point, a plot of vibration levels through the entire test range is made and the test is terminated.)

When repetitive indications are available from the measuring instrumentation, they are stored in the computer memory and considered to be the reference condition or initial unbalance. Test weights are now fabricated and one test weight is added in one correction plane. The rotor is accelerated to maximum continuous speed and unbalance response at various speeds is recorded. These test runs are repeated with one test weight in each correction plane.

When the test runs are completed, a correction weight set is determined. Calculations are conducted by the computer via the influence coefficient method. The calculated weight set is then applied to the rotor and additional measuring runs are made to check the results of the correction process. The sequence is repeated until the tolerance level is reached.

4.8 ROTOR STABILITY

One of the most important problems affecting the operation of high speed turbomachinery is stability of rotor motion. The susceptibility of a rotor to self excite can mean the difference between a smooth running piece of equipment and virtual self destruction.

4.8.1. Stability

The stability of a vibratory system is determined by observing the motion of the system after giving it a small perturbation about an equilibrium position. If this motion dies out with time and the system returns to its original position, the system is said to be stable; on the other hand if this motion grows with time, it is said to be unstable (see Fig. 4.41).

4.8.2 Idealized Damped System

In order to understand rotor instability, it is first necessary to understand a simple idealized damped system. This is normally represented by a mass (M) supported by a spring (K), with a dash pot (b) to damp the system motion (see Fig. 4.42).

In this example, the system is stable. Both the spring force and the damping force tend to return the system to the original position once the system is disturbed by an external force. The slope of the damping curve is positive. Increasing velocity generates an increasing force which opposes motion. Likewise, the spring force increases and opposes increased displacement from the equilibrium point.[7]

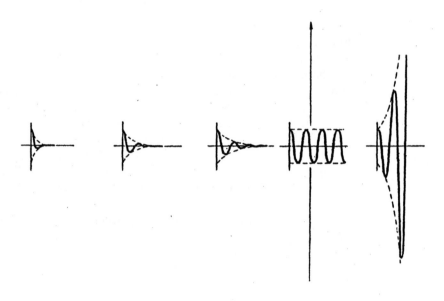

VERY STABLE STABLE NEUTRAL UNSTABLE

FIGURE 4.41 The natural response of stable and unstable systems.

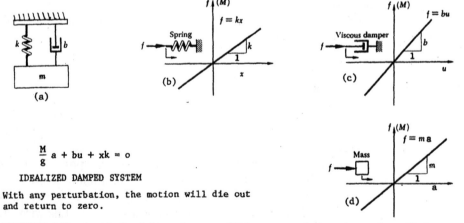

$$\frac{M}{g} a + bu + xk = 0$$

IDEALIZED DAMPED SYSTEM

With any perturbation, the motion will die out and return to zero.

FIGURE 4.42 An idealized damped system. With any perturbation, the motion will die out and return to zero.[7]

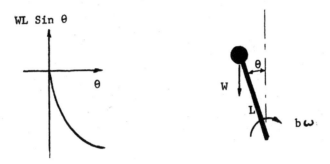

INVERTED PENDULUM

FIGURE 4.43 An inverted pendulum. The "negative spring" force increases with increased angle ϕ. Time behavior is divergent and unstable.[7]

In an unstable system, just the opposite occurs. Negative spring forces and/or negative damping tend to increase velocity and displacement with time.

4.8.3 Negative Spring

A good example of a "negative spring" is an inverted pendulum or a valve in the near closed position (Figs. 4.43 and 4.44). It is easy to visualize what happens to

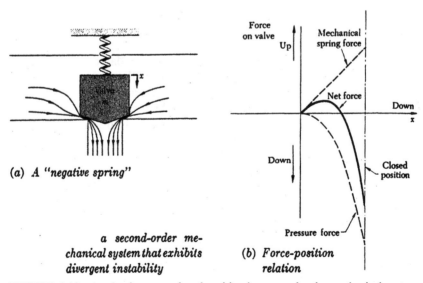

(a) A "negative spring"

a second-order mechanical system that exhibits divergent instability

(b) Force-position relation

FIGURE 4.44 A valve in a near closed position is a second-order mechanical system that is unstable.[7]

the balanced inverted pendulum once disturbed. A small initial disturbance will "push it over the hill." Due to the configuration of the system, the force of gravity will overcome any damping forces in the system and the pendulum will go to and remain at a fully extended position.

Similar to this are the problems of valve chatter. The example in Fig. 4.44 shows a model of a fuel valve. When near the fully open position, the net spring constant is positive, and the system is stable. However, when near closed, the pressure forces exceed the spring forces and the net spring force is negative, causing the valve to become unstable.

4.8.4 Negative Damping

Negative damping is much more common in familiar physical systems than many may realize. A good way to understand what negative damping is, is to simply look at it as the opposite of positive damping. Damping, as usually thought of in a positive value, is represented as a dash pot or shock absorber in an automotive suspension system. The dash pot or shock absorber generate a force that is a function of velocity (see Fig. 4.42c).

$$F = bV \tag{4.6}$$

Since the force generated is increasing with velocity and opposes the velocity, the dash pot tends to reduce oscillation of the system. This is "positive damping."

Now consider a system where the damping force generated decreases with an increase in velocity.

$$F = -bV \tag{4.7}$$

In this situation, the force will tend to increase the oscillation of the system. This is called "Negative Damping."

Examples of instability due to negative damping include the dry-friction vibration produced while playing a violin or the tool chatter produced in a machining operation (Fig. 4.45b). In rotating machinery, a labyrinth seal rub, or friction between elements on a rotor, can produce similar results.

These dry-friction systems are inherently unstable because the effective system damping constant is negative. An increase in velocity leads to a decrease in the friction force opposing that velocity.

The "galloping transmission line" problem is also a good representation of negative damping (Fig. 4.45a). Aerodynamic lift forces in the direction of velocity lead to unstable oscillations.

Vibrations which are sustained by the energy from a steady or non-oscillating force are known as "self-excited oscillations." The unstable, growing motion of each system described above can occur only because energy is being supplied to the system from an external source. The energy to sustain vibration of the violin string is supplied through the bow by the musician's arm.

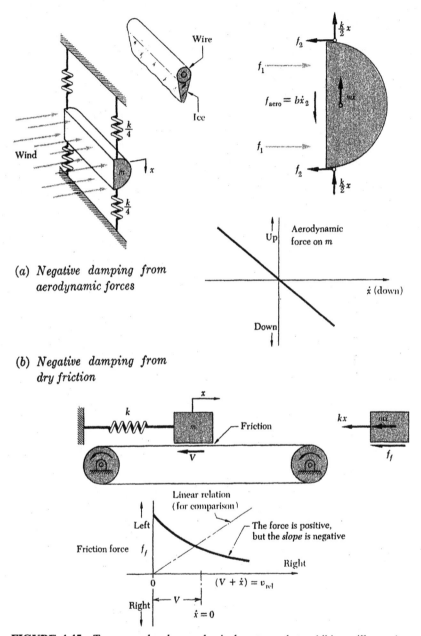

FIGURE 4.45 Two second-order mechanical systems that exhibit oscillatory instability. *a*) The "galloping transmission line" problem is a result of aerodynamic lift forces in the direction of velocity that lead to unstable conditions. *b*) Dry-friction vibration produced while playing a violin is an example of negative damping.[7]

4.8.5 Rotor Stability

During stable motion, a rotor assumes a deflected shape dependent upon the rotor elastic properties and the forces exerted upon it by the bearings, seals, aerodynamics, and unbalance. The deflection is primarily induced by a mass unbalance distribution along the rotor. The deflected rotor then whirls about the axis of zero deflection at a speed equal in direction and magnitude to the rotational or operating speed of the machine, and the whirl motion is therefore termed synchronous. Although the actual path described by each point in the rotor need not be circular, the size of the orbit remains constant in time (assuming the speed and unbalance distribution remain constant). When the motion is unstable, however, the size of the orbits described by each rotor point increases with time and the whirl motion of the rotor, while generally in the direction of the operating rotation, is at a lower frequency. Such whirl motion is generally termed subsynchronous. With very few exceptions, the whirl frequency is half of the operating speed when the rotor speed is less than two times the first critical speed. At and above two times the first critical, the whirl frequency locks onto the first critical frequency.

The operating speed at which the motion becomes unstable is dependent on the type of instability mechanism and its magnitude, but is generally above the first critical speed of the rotor. This speed is generally termed the *instability onset* or *threshold speed*. Although the rotor whirl orbits generally increase only to a certain size because of non-linearities in the system, they can become quite large; and in many instances it is not possible to operate the rotor much faster than the instability threshold speed, and in some instances actual rotor failure may occur.

4.8.6 Instability Mechanisms

The significant mechanisms which cause rotor instability can generally be found in one or more of the following categories:

1. Hydrodynamic bearings
2. Seals
3. Aerodynamic effects
4. Rotor internal friction

The first three categories are usually those encountered in practice, although internal friction damping is generally present in all machines to some degree and reduces the overall stability of the machine. In certain instances, such as with unrelieved shrink fits or unlubricated splines, internal rotor friction may become a significant source of rotor instability.

Although the mechanisms causing instability are varied, they all have certain characteristics in common, the most important of which is a cross coupling effect of the forces (see Fig. 4.46). The cross coupling can exist in both velocity (damping) and displacement (stiffness), but displacement cross coupling is the more im-

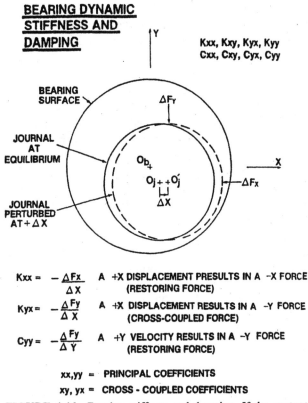

FIGURE 4.46 Bearing stiffness and damping. If the amount of cross coupling is large enough so that the cross coupled forces exceed the damping forces, the rotor will be unstable.[8]

portant of the two, and is characterized by a force due to a displacement of the rotor acting at a right angle to the displacement vector. The cross coupled forces act in a direction such that they oppose the damping forces; and if the amount of cross coupling is large enough so that the cross coupled forces exceed the damping forces, the rotor will become unstable. In effect, the cross coupling forces create negative spring and damping forces.[8]

4.8.7 Susceptible Designs

As discussed in the previous section, mechanisms for instability are: bearings, seals, aerodynamics, and rotor internal friction. Designs most susceptible to instability will have several or all of the above mechanisms at work.

The cross coupling forces in hydrodynamic liner bearings are relatively low on low speed, heavily loaded bearings. On high speed, lightly loaded bearings, the cross coupling forces are relatively high. Similar statements can be made concerning bushing type seals. However, the difference with bushing seals is that they are mounted so as to "float" with the shaft. Actual forces transmitted via the floating bushing seal if properly balanced should be negligible.

Cross coupling forces occur, not only in bearings and seals, but also anywhere along the rotor where there is relatively close proximity between rotating and stationary parts. Cross coupling forces occur in these areas similar to those in a hydrodynamic liner bearing: between impellers and diaphragms, shaft and labyrinth seals—balance piston seal, shaft seal, and impeller eye seal.

Aerodynamic cross coupling forces are most significant for high pressure, high speed (tip speed), close clearances, non concentric, mid span areas. The most pronounced example of these features is the center seal (balance drum) on a back-to-back design compressor, since this design has all of the above features. The center seal has a very close clearance to minimize leakage from one section to the next. The seal is of significant length (large L/D) to further minimize leakage, and concentricity is generally poor due to rotor sag. A regular balance piston seal has the same effect, but to a lesser degree due to it's location at one end of the rotor away from maximum deflection point for the first critical speed (see Fig. 4.39).

Most high speed turbomachinery rotors are made up of several parts that are somehow fitted together. These fits can be slip fits, bolted or riveted arrangements, shrink fitted, or even welded joints. Any of these joints that have any relative movement during operation can generate destabilizing forces or "negative damping."

4.8.8 Analytical Analysis

The analysis of rotor bearing systems for stability reveals a large amount of useful information on the behavior of the system. It answers the question as to whether or not the system is stable at the operating speed for which the analysis was performed.

The rate of decay of a damped vibrating system takes the form of the natural log of the ratio of successive peak amplitudes (Fig. 4.47) and is therefore called the logarithmic decrement (α) and is used as a measure of the rotor damping. If the log dec value for the first forward even mode at the first critical frequency of the rotor is positive, the rotor is stable. If the log dec value is negative, the rotor is unstable and whirling of the rotor at that frequency will occur.[9]

As with any analytical method, some assumptions are necessary; and the model does not quite represent the actual situation. For this reason, a safety factor is applied.

It is very important to realize that there is a significant difference in rotor dynamics programs used.

Some stability programs use damping and stiffness parameters which are frequency depending. That is, the stiffness and damping parameters used are based upon individual pad data and varies with the particular frequency being evaluated and the synchronous speed. Other stability programs use stiffness and damping parameters at synchronous speed or are synchronous dependent.

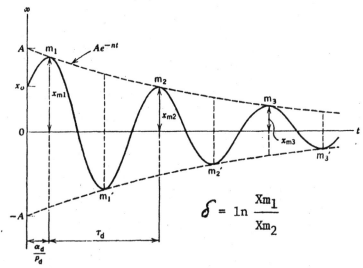

FIGURE 4.47 Logarithmic decrement (log dec.). The time decay of a viscous damped system is a natural log function.[9]

Because of the differences in the various rotor dynamics programs, one must be very careful in comparing the results of one program to another. Log decrements are widely being used as a measure of rotor damping. A particular log decrement value is dependent upon the stability program in use. For example, a log decrement of +.05 may be considered as having adequate stability margin utilizing one program, while another may require a log decrement of +.4.

It is also most important when discussing log decrement values as to whether the value being referenced is a basic log decrement without aero excitation or is an adjusted log decrement value, which includes aero cross coupling effects.

4.8.9 Identifying Instability

As noted previously, rotor instability shows up as just under half running speed frequency when operating below 2x the first critical speed. When operating above 2x the first critical speed, the instability locks onto the first critical speed (see Fig. 4.48). Operating vibration levels during unstable operation can be relatively low or so high that the unit is inoperable. Energy levels can be so high during instability that the unit can self destruct in a few seconds of operation. The energy levels can be so low that the only problem is that bearing life is reduced.

The equipment needed to analyze the rotor is a vibration pickup and a frequency analyzer. Start-up and shutdown response can be recorded to obtain actual NC1 and the point at which instability first begins.

A non-contacting probe is preferred, since it records actual rotor movement. Casing velocity or acceleration measurements can introduce questionable data due to foundations or piping vibration.

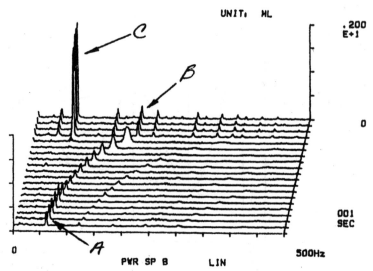

FIGURE 4.48 Waterfall spectra of compressor shutdown showing aerodynamic instability. *a*) High response at running speed indicated rotor first critical speed. *b*) Synchronous amplitude at normal operating speed. *c*) Subsynchronous amplitude at normal operating speed. Note how the subsynchronous frequency (c) aligns with the first critical speed frequency (a).

API limits on subsynchronous vibration is 1/4 of running speed amplitude in the operating speed range. Accordingly, if levels are less than this, operation is satisfactory and corrective procedures are not necessary. If the subsynchronous level is greater than 1/4 of RSV amplitude, then corrective measures should be taken.

4.8.10 Corrective Measures

Whether instability of a rotor system is discovered by operation or by analysis, the best tool for correcting the situation is a rotor stability analysis program. Possible corrective measures can then be analyzed and the best possible solution can be selected. Some methods of improving stability are listed below. Please note these are general comments and are not always true for every case. One change may help one case while of little benefit to another. Each situation must be separately studied.

A. Labyrinth seals
  1. Buffer
     Supplying buffer to a labyrinth seal introduces to the seal a gas flow that is absent of any tangential velocity. This reduces the average tangential velocity of the gas through the seal, thereby reducing the cross coupling forces (which are function of the tangential velocity). See Figs. 4.49 and 4.50.[10]

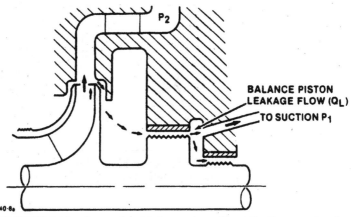

FIGURE 4.49 Conventional balance piston flow. The flow entering the balance piston seal has a circumferential component due to the rotation of the impeller.[10]

2. **Dewhirl vanes**
   Dewhirl vanes (Fig. 4.51) or straightening vanes upstream of a labyrinth seal provide similar effects to buffer by reducing or eliminating tangential velocity before the gas enters the seal.
3. **Honeycomb seals**
   Seals that have a honeycomb pattern (Fig. 4.52) on the stationary portion and smooth rotating drum tend to have a reduced average tangential velocity over standard labyrinth seals. This reduced tangential velocity reduces cross

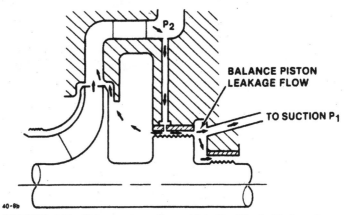

FIGURE 4.50 Balance piston flow with buffer. The buffer introduces the gas without a tangential velocity, thus reducing the average circumferential velocity in the seal and therefore reducing the crosscoupling forces.[10]

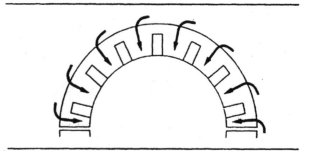

FIGURE 4.51 Swirl brakes eliminate tangential velocity of gas prior to entering labyrinth seals (*courtesy Sultzer*).

coupling forces and therefore delays the onset of instability. Additionally, the cavities of the honeycomb act as dampers contributing to the system stability.

4. TAM seal

   The TAM seal is a labyrinth seal with teeth on the stationary member and a smooth rotating drum. Specially designed barriers in the circumferential cavities between the labyrinth teeth break up the tangential swirl of the gas and generate pressure dams with damping properties.

5. Concentricity

   Cross coupling forces are related to the non-concentricity of the labyrinth seal. Since cross coupling forces are only fully developed in a non-concentric seal, centering of the shaft in the seal will reduce these forces significantly. This is one reason why a compressor which has a balance piston near one journal bearing may be stable, while a similar compressor with a balance piston in the center of the rotor is unstable (eccentricity due to rotor sag).

6. Gas density

   Cross coupling forces are a function of gas density. Higher gas density will give higher cross coupling forces and vice versa. Although it is usually not a practical solution, reducing gas density will improve stability.

B. Journal bearings

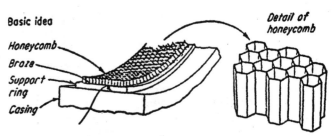

FIGURE 4.52 Honeycomb seals can add stability to the rotor system.

1. Liner bearings
   Liner bearings have inherently high cross coupling forces due to the high density of the oil lubricant, the close clearances, and the usual high offset of the bearing to journal. Several means of altering the liner bearing to reduce or counteract the cross coupling forces are available. These include: dam type bearings, lemon shape, offset bearings, and lobed bearings. Generally the most reliable correction is to change to tilt pad bearings.
2. Tilt pad bearings
   Tilt pad bearings by nature have little or no cross coupling forces. Thus, changing from liner to tilt pads can eliminate the instability if the primary cause is the bearing. If, however, the primary source is elsewhere and tilt pads are already in use, the tilt pad bearing can be enhanced. This can be done by several methods: reduced positive preload, increased L/D, offset pivots, load on pad orientation (for 5 pad bearing), increased clearance. A damped bearing support system can further stabilize the system.
3. Magnetic bearings
   Stability or dampening characteristics of this type of rotor bearing system is dependent on the control system. The interaction of the power and feedback circuits will determine stability of the overall system.
4. Sleeve seals
   Sleeve type seals or bushing seals have the same characteristics as liner bearings, except that the seals, if properly designed, are "free floating." This means that the seals are concentric and negligible forces are transmitted to the shaft. Wear on the seal face can restrict the seal radial movement and lead to subsynchronous vibration.

C. Rotor design
   1. Stiffness
      Increasing rotor stiffness improves rotor stability. Anything, therefore, that increases the critical speed will improve stability. This includes reduction of bearing span, increased shaft diameter, reduced overhang, or reduced rotor weight. A common parameter to consider is the ratio of the running speed to the first critical speed. Lower values are preferred.
   2. Rotor fits
      Any parts that can rub during rotor flexing can cause negative dampening or destabilizing forces. It is therefore prudent to eliminate any source of rubbing between parts such as is caused by loose or slip fit sleeves, marginal shrink fits, or sleeves that touch end to end. Bolted, riveted, or shrunk fits must remain tight at speed so as to eliminate relative motion.

4.9 AVOIDING SURGE

The anti-surge control system should maintain a minimum volume of flow through the machine so that the surge condition is never encountered. This is achieved by bleeding flow from the discharge of the machine to maintain a minimum inlet flow. This flow can either be dumped to atmosphere or recirculated back into the inlet of the compressor. In the latter case, it must be cooled to the normal inlet temperature. For most applications, a simple control based on a flow differential is adequate for this function. However, on machines where the speed or the gas conditions are variable, the control may have to be more sophisticated to insure proper operation under all conditions. This is frequently achieved by modulating the control with a signal for pressure, temperature, speed, or a combination of parameters.[11]

Provisions must be made for start-up and trip-out of the machine with sufficient through flow to prevent surging and excessive heating of the inlet gas.

On any control scheme, a trip-out of the driver should be interlocked to open the anti-surge valve within three seconds and allow the machine to coast to a stop with this line open. Otherwise, the machine could be surging constantly while dropping down in speed, causing mechanical damage to the equipment (Fig. 4.53). This is particularly important for axial compressors and also for high-pressure, high-horsepower centrifugal applications. Basic components of a typical anti-surge control system are shown in Fig. 4.54.

A description of the function of each component is as follows:

FE—The flow element is usually an orifice located in the compressor suction, although it can be a venturi or calibrated inlet such as those used in axial compressors. Its purpose is to cause a temporary pressure drop in the flowing medium in order to determine the flow rate by measuring the difference of static pressures before and after the flow-measuring element.

FIGURE 4.53 Damage from surging.

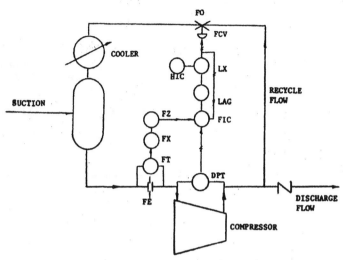

FIGURE 4.54 Typical anti-surge system.[11]

FT—The flow transmitter is a differential pressure transmitter which measures the pressure drop across the flow element and transmits a signal that is proportional to flow squared.

DPT—The differential pressure transmitter measures the differential pressure across the compressor and transmits an output signal that is proportional to the measured pressure differential.

FX—The ratio station receives the signal from the flow transmitter and multiplies the signal by a constant. This constant is the slope of the control line.

FZ—The bias station receives the signal from FX, the ratio station, and biases the surge control line.

The ratio station must have both ratio and bias adjustment to enable the control line to be placed as parallel to the compressor surge line as possible (see Fig. 4.55).

$$\Delta P = Ch + b \qquad (4.8)$$

where ΔP = Calculated compressor differential signal
C = Control line slope (ratio signal)
h = Inlet orifice differential signal measured by FT
b = Control line bias (could be zero)

FIC—The surge controller is a flow control device which compares the calculated output of FZ to the measured ΔP output of the DPT with ΔP as defined above.

When the calculated ΔP is greater than the measured ΔP, the compressor is operating to the right of the control line. When the calculated ΔP is equal to or

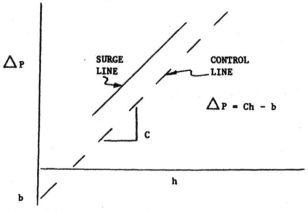

FIGURE 4.55 Surge control line.[11]

less than the measured ΔP, the compressor is on or to the left of the control line, and the surge controller functions as a flow controller and opens the anti-surge valve as necessary to maintain operation of the compressor on this surge control line.

For rapid flow changes, the response of the control system must be rapid to prevent surge.

> LAG—This device functions to enable the surge controller to open the recycle valve quickly, while providing a slow closure rate. This feature provides stability between the anti-surge control system and the process by minimizing the hunting effect between control system and recycle valve.
>
> LX—The low signal selector is set up for two inputs and one output. The inputs are a 100% signal valve and the surge controller output signal. The output of the low selector is sent to the recycle valve as well as back to the surge controller in the form of a feedback signal. This prevents the surge controller from winding up. Windup of the controller penalizes the reaction time of the anti-surge control system.
>
> FCV—The anti-surge recycle valve functions to prevent surge by recycling flow from the compressor discharge back to the compressor inlet. Sizing of the anti-surge valve should be at 1.05% of design flow at design pressure rise.

4.10 SURGE IDENTIFICATION

The following is the preferred procedure for establishing the location of the surge point.

> **1.** Slowly close the recycle or blow off valve, while monitoring the following parameters:

    a. Blow off or recycle valve position, % open.
    b. Audible sound level at the inlet of the compressor. Listen for a pulsing sound.
    c. Audible sound level at the compressor discharge. Listen for a pulsing sound, a low frequency 0 to 25 Hz.
    d. Compressor suction pressure immediately upstream of the compressor inlet flange. Monitor both the local pressure gage (for low pressure, a water manometer works well) and the pressure transmitter. Watch for a bouncing in the pressure level. Note that the pressure transmitter may not show this unless it is rated as a dynamic device, with a rise time below 0.1 sec. When the dynamic amplitude exceeds 20% of the gage static pressure or the compressor pressure rise, consider the unit to be in surge.
    e. Compressor discharge pressure near compressor discharge flange. As the blow off valve is slowly closed, the pressure will rise. Monitor both the pressure gage and the pressure transmitter. Watch for the pressure to bounce (see "d" above). Also watch for any drop in pressure. At the first indication of a drop in pressure (with decreasing flow), consider this to be surge, and record data.
    f. Compressor flow rate. Watch for fluctuations in the flow meter differential pressure. Note that an electronic output on the flowmeter will not indicate surge unless the device is rated for dynamic conditions, with a rise time below 0.1 sec. It is best to locally attach a manometer (for low pressure) or differential pressure gage and monitor this. Any dynamic differential pressure in excess of 20% of the nominal (steady state) differential at the given flow rate is to be considered surge, if no other indications (c, d, e, or f) are observed.
    g. Compressor vibration level. Pay particular attention to subsynchronous amplitudes. Very small increases or bouncing of amplitudes indicate possible onset of surge. An increase of 20% at the given speed of the overall vibration level, or 0.20 mils increase of the subsynchronous, while alone not a sign of surge, indicates the proximity of an instability. Use extra caution when exceeding these values.

2. When any of the above items (except the peak head condition, e above) indicates surge, the position of the blow off or recycle valve should be immediately noted, and then opened to the full open position.

3. Close the valve back to within a few percent of the point where the instability occurred. Example: The suction pressure began to bounce at blow off valve opening of 79%. The valve is immediately opened to 100% open. The blow off valve is then closed back to 81% open.

4. Wait a few minutes to assume data is stable and then record the data.

5. Repeat steps 1 through 4 for the other speed lines, or inlet guide vane positions.

6. Record all data for future reference.

Note that the ideal method of detecting the point of aerodynamic instability, is to monitor dynamic pressure probes near the inlet to the impeller and in the diffuser. Flow instability can develop in either location. In some units it appears in the inlet due to flow separation on the inlet of the impeller blades. The position of this point on the compressor head curve generally lines up with the point of peak head. On other units, stall will occur in the diffuser section. This is caused by the inability of the diffuser to overcome the compressor discharge pressure. This event may not fall in line with peak head.

Sophisticated instrumentation is not required to detect surge. The instability usually can clearly be heard and even felt when standing near the compressor. Sometimes the instability is subtle and you must listen very closely.

If you are standing in the compressor discharge area, you may not hear an inlet stall condition. Likewise, if you are in the control room observing the flow and pressure on the slow responding process monitor equipment, you may only see a deep hard surge condition, when it occurs.

Keep in mind that too much surging will eventually cause equipment failure. Surge the equipment hard enough and long enough and something will eventually break. When setting the surge line, the equipment should experience only one or two surge pulses. Allowing the unit to surge any more is only asking for trouble. In order to accomplish this, the recycle or blow off valve must have a quick opening (1 to 2 sec.) response and a slow (10 to 20 sec.) closing time to keep the system stable.

Be safe and assume that the unit is very sensitive to surge and that the machine could easily wreck if surged very much.

4.11 LIQUIDS

One of the most potentially damaging occurrences for a compressor is the ingestion of liquid with the process gas stream.[3]

Liquids condensing in the recycle line can minimize the effectiveness of any anti-surge system by creating a blockage in the line.

Liquid quench used on refrigeration systems during startup must be minimized to avoid carryover into the compressor suction. This liquid which is in a mist form can easily pass through the compressor inlet knockout drum and its demisters. When large amounts of liquid are ingested, the liquid is vaporized creating an excessive volume of gas in the back of the compressor causing an extreme mismatching (see Fig. 4.38) of the last stages and turbinizing of the last impeller(s). The high velocities associated with this condition have been known to cause mechanical damage to the compressor, particularly to the last stage impeller.

The full range of operation should be studied to avoid having liquids enter the compressor during normal operation and upsets.

1. Trim cooling water or other process conditions to keep the compressor inlet conditions above the liquefaction points for any gas constituent.
2. Heat trace, bleed off, or purge normally stagnant lines when liquids form as the stagnant gas cools down to ambient temperatures.

3. Recycle lines should re-enter the main gas stream either upstream of or at the inlet knockout drum.

4. If any potential for liquid formation exists upstream or downstream of the compressor, drains and level indicators should be provided at all low spots of piping and vessels. This will allow routine checking for liquids and draining as required.

5. After shutdown, be sure all liquids formed by cool down of the stagnant process gas are drained away before the compressor is restarted.

6. Limit liquid quench on startup of refrigeration compressors assuming that the liquid will not vaporize instantly and that the knock out drum will not remove all the remaining liquids.

4.12 FIELD ANALYSIS OF COMPRESSOR PERFORMANCE

Field testing process compressors is not a simple task, but the proper procedures and tools simplify the project. Accurate data and proper analysis is necessary to compare operating data to the manufacturer's guarantee. Proper analysis is required to justify a project whether it be new equipment, a rerate, or an overhaul. Frequent and regular testing is essential to monitor operating expenses and to aid in determining overhaul frequency.[12]

The field test check list:

- A method of obtaining accurate pressure and temperature at each compressor flange. "Snapshot" data is preferable to minimize the effects of transients.
- A means of calculating the mass flow rate to the compressor. This is generally done with an orifice plate or nozzle.
- Vane setting for variable inlet vane units
- Gas analysis via gas chromatograph
- A compressor performance program that utilizes BWR equations of state
- Driver power
- Compressor and driver mechanical losses
- Manufacturers performance curves
- Speed

4.13 GAS SAMPLING

Follow proper precautions when taking a gas sample. Condensate can form on the gas sample container walls and give erroneous results unless proper procedures are followed. Confirm values by comparing the discharge gas analysis to the inlet gas analysis for the same section. The accuracy of test results are no better than the agreement between gas analysis results.

A stainless steel sampling cylinder should be used. It should be at least 300 ml in size and have straight cylinder valves on both ends. The pressure rating of the cylinder should be high enough so that it can withstand full system pressure. To prevent gas condensation (the walls of the cylinder and tubing are relatively cool) during the sampling process, the cylinder and lines leading from the process pipe should be insulated and heat traced. The sampling container should be purged before closing the valves and trapping the gas at process pressure.

The containers are then transported to the laboratory. During the transportation, the samples cool and therefore drop in pressure. The cooling of the sample may result in condensation of some of the gas. This condensation must be gasified before feeding into the gas chromatograph. The only sure way to do this is to heat the sample to or above the sample temperature. It may be wise to then bleed off some pressure before feeding the gas chromatograph. Plot this out on a Mollier diagram of the gas. This brings gas conditions further away from the dew point and provides further assurance of avoiding condensation.

A cross-check on accuracy can be made by checking the weight of a separate sample. This sample container should be at a vacuum and heated before filling with the process gas. Weigh the sample container evacuated and with the sample. Knowing the weight and volume will give the specific volume. Check this against the calculated specific volume for temperature and pressure of the sample point using the composition given by the gas chromatograph.

Of course this procedure is not necessary in all cases, such as low mole weight mixtures. For heavy hydrocarbons, however, it is essential that the above procedure be followed. Errors in gas analysis can give significant errors in performance.

4.14 INSTRUMENTATION

General instrumentation requirements are shown in Fig. 4.56. All temperature and pressure indicators should be dual, i.e., a minimum of two independent instruments per location.[13] If existing single point instrument tapping points must be used, care must be taken that those used are well located. There must be no valves, strainers, silencers, or other sources of significant pressure drop between the pressure tap points and the compressor flanges. Pressure taps near an elbow should be normal to the bend and not in the bend plane.

Ideally, such as under development testing conditions, the static pressure tap hole should be very small (approximately 1/5) compared to the boundary layer thickness.[15] But on a more practical note, the static pressure connection should have a pressure tap hole .25 inch in diameter, but no greater than 0.5 inch, deburred and smooth on its inside edge with a sharp corner. A smaller hole will collect dirt or condensate. Note that the hole must be deburred but have a sharp corner on the measurement side. Take care that the deburring procedure does not round the edges, as this will give erroneous readings (Fig. 4.57).

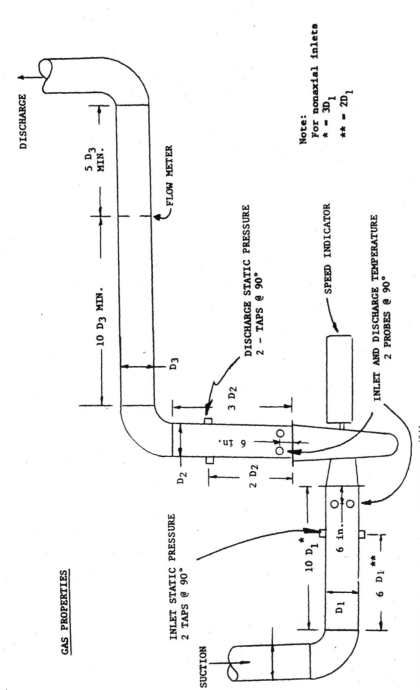

FIGURE 4.56 Typical performance test setup.[13,14]

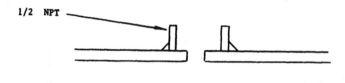

FIGURE 4.57 Static pressure tap. The hole should be 1/4 to 1/2 inch in diameter. It should be deburred but have a sharp edge.[13,15]

All pressures should be measured using quality pressure transducers or pressure gauges, 6-inch or larger diameter, having a 0.25% sensitivity and a maximum error of 0.5% full scale

Pressure readings during testing should be at mid-scale or greater. Mounting should be on a vibration-free local panel, connected with pressure lines of at least 1/2 inch I.D. tubing; lines will continually slope down toward the unit to automatically drain any condensate. Block vent valves should be mounted at the gauges to facilitate their in-place calibration.

Temperatures should be measured using a thermocouple or RTD system having a sensitivity and readability of 0.5°F and an accuracy within 1°F. Care should be taken to avoid intermediate T-C junctions at terminals and switch boxes with a thermocouple system. Glass-stem thermometers are generally unacceptable for safety reasons. The temperature sensing portion of the probe must be immersed into the flow to a depth of 1/3 to 1/2 the pipe diameter. The temperature sensing elements should be in intimate thermal contact if using wells, utilizing a suitable heat transfer filling media, such as graphite paste. Stem conduction errors can be further minimized by wrapping the stem and well with fiberglass or wool insulation.

To minimize effects of system transients, the data should be collected via a digital monitoring system so all data is collected at the same instant in time.

Speed should be determined utilizing two independent systems, one being the compressor key-phasor with calibrated digital readout with 0.25% or better system accuracy.

Pressure taps should be spaced 90° apart. On horizontal runs of pipe, pressure taps must be in the upper half of the pipe only.

Flow rates derived from the process flow indicator should be checked by direct computation of mass flow rates through the metering device. For this reason metering device upstream temperature, upstream pressure, and differential pressure must also be recorded.

4.15 INSTRUMENT CALIBRATION

In general, all instruments used for the measurement of temperature, pressure, flow, and speed should be calibrated by comparison with appropriate standards before the test. General recommendations for calibration procedures are outlined below.

A comprehensive log book should be maintained for all calibrations. Calibration using a certified dead weight tester is preferred. Suggested arrangements are shown in Figs. 4.58 and 4.59. Pressure transducer calibration should state actual deadweight standards throughout the range. Calibration using both increasing and decreasing pressure signals should be done to check for hysteresis. Transducers not within 0.5% error of full scale should not be used. Needles should not be changed or adjusted. The pressure transducers or gauges must have a readable sensitivity to 0.25%.[13]

There should also be a check on the accuracy of the thermocouple system (lead wires, reference junctions, readout devices) for each thermocouple. One method of accomplishing this is to read voltage output of the thermocouple locally, and then compare this to the remote reading of thermocouple output.

It is also recommended that the accuracy of the thermocouple itself be checked by subjecting it to varying temperatures and comparing its output to a reference standard. The thermocouple should be checked throughout its operating temperature range. The thermocouple system should have a readable sensitivity to 0.5°F and an accuracy within 1°F.

Calibration of the flowmeter differential pressure transmitter can be verified by impressing a known differential pressure across it and measuring its output. Finally,

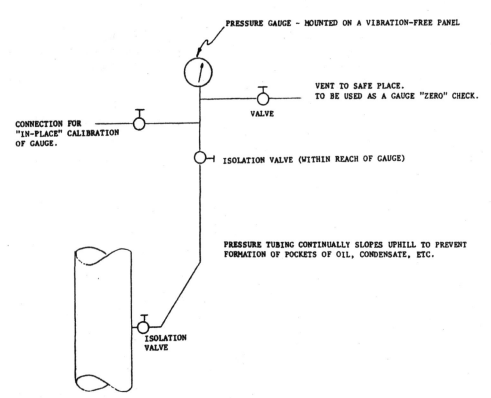

FIGURE 4.58 Typical instrument line with gages mounted above the pressure tap.[13]

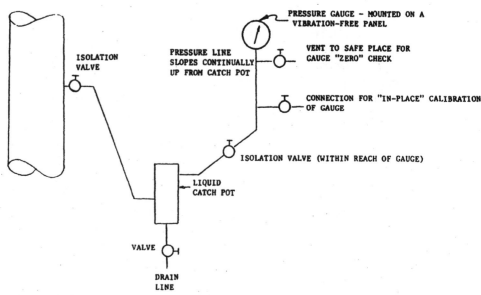

FIGURE 4.59 Typical instrument line where pressure line cannot be sloped continuously upward from the pressure tap to the gage.[13]

an overall system check can be made by impressing a differential pressure across the transmitter and reading the final control room output. The flowmetering device should be removed, and its dimensions should be checked, recorded, and compared to design criteria. Orifice plates should have a sharp edge.

The tolerance for the measurement of compressor speed should not exceed 0.25%. Use of two independent instruments, one to provide a check on the other, is recommended.

4.16 ISO-COOLED COMPRESSORS

Field testing Iso-cooled compressors (Figs. 4.3 and 4.6) can be straight forward, but a few precautions are in order. Each section of the machine should be treated like a separate single section compressor, but with the following items in mind.

Process gas flowing to the second section of the compressor may, due to the gas composition, be different than the first. Liquids may form in the cooler and be drained out prior to continuing on to the next section. Flow to the second section thus will have a lower mass flow rate and the gas will have a lower mole weight. This is especially true for installations like wet (rich) gas compressors.

Measuring the flow of liquid flowing from the cooler and the process gas flow at the compressor main inlet (or discharge) will provide the mass flow rate for the other section by subtracting (or adding).

4.16.1 Heat Transfer

Heat conduction from the discharge of the first section to the next section inlet (Fig. 4.6) can make results confusing. For example, assume that a compressor has a discharge temperature for the first section of 356°F and an inlet to the second section of 90°F. It is easy to understand that there is considerable heat flowing across the wall separating the two sections because of this high temperature differential. This heat is of course flowing from the discharge of the first section to the inlet of the second. The measured temperature at the discharge flange of the first section thus does not accurately represent the true temperature at the discharge of the last wheel of that section. Likewise, the temperature at the inlet flange to the second section does not represent the true temperature at the first impeller in the second section due to the heat transfer effect.

The first stage discharge temperature is artificially low thus a higher than actual efficiency and corresponding low power is calculated. Just the opposite is true for the second section. While this effect should be reflected in the manufacturers predicted performance values, the actual sectional performance can be difficult to accurately predict.

4.16.2 Seal Leakage

To fully understand the performance of Iso-cooled compressors, seal leakage must also be considered. The two areas of importance are the balance piston seal and the seal between the iso-cooled sections.

Normal seal leakage is represented in the compressor design performance. Seal degradation will affect the observed power and efficiency values.

Higher than design flow across the balance piston seal will affect the discharge temperature of the *first* section since the hot balance piston leakage will heat up the compressor first stage inlet gas (assuming the balance line is returned to the main inlet). Calculations with a defective balance piston will thus show a higher than normal power consumption (low efficiency) for the first section. The second section will be affected slightly due to the increased discharge temperature of the first section and resulting increase in heat transfer to the second stage inlet, but usually this effect is insignificant.

If the internal seal between the two sections is damaged, increased flow across this seal will affect test results. Higher leakage from the wiped seal will increase the temperature of the second section inlet since the 356°F gas (discharge of the first section) is flowing to and mixing with the 90°F second stage inlet gas.

4.17 COMPRESSORS WITH ECONOMIZER NOZZLES

Proper analysis of a sideload compressor (Fig. 4.7) requires internal temperature and pressure probes in order to properly calculate the performance of the individual sections of the compressor. This is the normal procedure in any proof test following the purchase of a new refrigeration compressor with economizer nozzles since this is the only way that the compressor performance can be properly analyzed and compared to predicted values.

Preferred instrument locations are as shown in Fig. 4.60. Sideload and extraction lines, if applicable, are to be treated as inlet and discharge lines, respectively.

Pressure and temperature taps can be added to a horizontally split compressor (Fig. 4.60), by drilling and tapping the casing in the return channel crossover area. A minimum of two each pressure and temperature taps should be used.

Once pressures and temperatures are known at the discharge of Section I, a mixing calculation is required to establish suction conditions for the next section.

$$P_8 = P_{10} = P_9 \qquad (4.9)$$

where 8 = Discharge of Section I
        9 = Mixed Suction to Section II
       10 = Sideload Condition

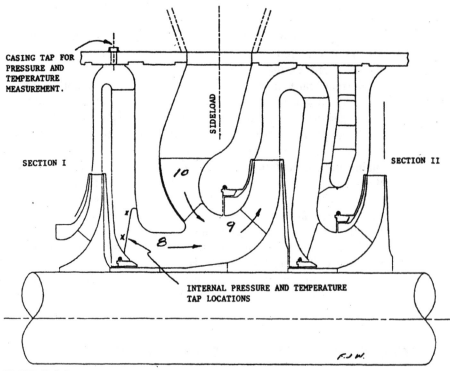

FIGURE 4.60 The internal detail of a sideload compressor where the economizer (sideload) stream meets and mixes with the main refrigeration fluid.

CENTRIFUGAL COMPRESSORS—CONSTRUCTION AND TESTING

$$M_8 h_8 + M_{10} h_{10} = M_9 h_9 \quad (4.10)$$

where $M_8 + M_{10} = M_9$.

T_9 is then found by working back through the gas properties or Mollier diagram knowing h_9 and P_9.

T_9 may be approximated by

$$T_9 = (M_8 T_8 + M_{10} T_{10})/(M_8 + M_{10}) \quad (4.11)$$

A summary of calculation results are shown in Fig. 4.61.

4.18 ESTIMATING INTERNAL TEMPERATURES

Unfortunately, the above type of test may be a little impractical to do for a field location, since the internal instrumentation can be a long-term liability regarding emissions (additional connections that may leak) and maintenance concerns (instruments might break and go through the compressor) as well as additional up front costs. It is generally out of the question to shut down following a field test to remove the internal instrumentation.

Special data reduction techniques can be used on sideload and extraction compressors where internal pressures and temperatures are not available. Internal pressures can be estimated from flange pressures, gas velocity through the compressor nozzle, and standard pressure drop loss coefficients for a given sideload or extraction nozzle design.

Internal gas temperature at the discharge of each section is also required to determine sectional performance. This can be accomplished through an iterative process which makes use of predicted work for each section. The procedure begins for a given test point by establishing the inlet volume flow for Section I. From the predicted work curves, the estimated work input is obtained. These data, along with the internal pressure determined above, are used to establish the estimated discharge temperature for Section I.

A BWR gas properties program or data with the enthalpy as a function of pressure and temperature is used. The sideload flow, as measured on site, will then be mixed with the calculated discharge flow from Section I to establish the inlet flow to Section II. This procedure is then repeated for each following compressor section using its respective work input. The test on the validity of the work input is made by comparing the calculated final discharge temperature with the measured final discharge temperature. If these two temperatures agree, the assumption is made that the correct work input has been used. If, however, the two temperatures do not agree, the work input for each section are varied by the same percentage, and the process is repeated.

Once the sectional inlet and discharge conditions are determined, the sectional heads and efficiencies can be calculated. Note that it is not possible to tell where

Gas Flex Calculation Summary
Sidestream Compressor Estimation Results

Title : Propylene Refrigeration Compressor with Sidestreams
Database name : PROPYLEN.GFE

Gas Analysis   Total Mol Weight 42.08

Gas	Mol Fr
C3H6	1.00000

Flange location/direction		Inlet	SS-1/In	SS-2/In	SS-3/In
Flange data					
Pressure	psia	21.27	41.90	75.40	144.60
Temperature	deg C	-39.7	-17.1	-4.1	18.7
Volume flow	m3/hr	92,322	22,944	1,595	239
Mass flow	lb/min	11,375.3	5,209.9	651.2	186.7
Compressibility		0.9484	0.9242	0.8802	0.8152
Discharge flange data					
Pressure	psia				237.30
Temperature	deg C				84.4
Volume flow	m3/hr				17,464
Mass flow	lb/min				17,423.1
Compressibility					0.8532
Inlet section data					
Compressibility		0.9484	0.9323	0.9136	0.8824
Temperature	deg C	-39.7	-9.0	20.6	54.6
Volume flow	m3/hr	92,322	76,024	47,846	27,172
Mass flow	lb/min	11,375.3	16,585.2	17,236.4	17,423.1
Discharge section data					
Compressibility		0.9356	0.9147	0.8830	0.8532
Temperature	deg C	-5.2	21.6	55.0	84.4
Volume flow	m3/hr	53,061	46,241	26,929	17,464
Overall section data					
Head	kg-m/kg	3,224	3,055	3,666	2,968
Efficiency	%	72.50	74.80	78.80	71.00
Gas power	kW	3,750	5,021	5,945	5,400
Overall data					
Total head	kg-m/kg	12,913			
Total gas power	kW	20,117			

FIGURE 4.61 Calculation summary for a sideload compressor.

the deficiencies if any might be. All sections are treated equally. The efficiency of each section is modified equally up or down until a match is made with the discharge temperature.

4.18.1 Procedure for Calculating Overall Power and Efficiency

Following is an approach that determines *overall* power and efficiency rather than sectional performance. A typical refrigeration cycle with economizers is depicted in Fig. 4.62, while a "modified" cycle is demonstrated in Figs. 4.63 and 4.64. The "modified" cycle uses an approach that eliminates the requirement of hard to obtain data points 8, 11 and 21. Internal mixing is assumed not to occur for the "modified" cycle in order to obtain a manageable solution. Instead, the system is treated as three separate and parallel flow paths with measurable end points.

4.18.2 Overall Power

The gas horsepower for a single stage compressor is:

$$GHP = \frac{778.16}{33000} (h_2 - h_1)M \quad (4.12)$$

where 1 = Inlet conditions
2 = Discharge

For a sideload machine:

$$GHP = \frac{778.16}{33000} \sum_{1,i} (M\Delta h)_i \quad (4.13)$$

where i = the number of sections of the compressor

For a compressor with two sideloads (three sections):

$$GHP = 0.0236[M_a(h_{14a} - h_7) + M_b(h_{14b} - h_{10}) + M_c(h_{14c} - h_{13})] \quad (4.14)$$

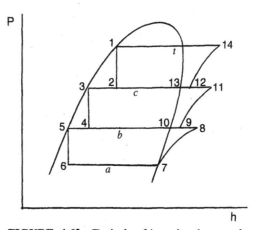

FIGURE 4.62 Typical refrigeration heat cycle with economizers.

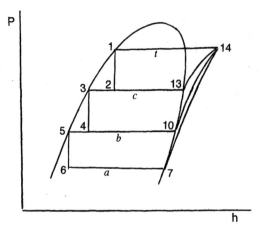

FIGURE 4.63 Figure 4.61 in a simplified "black box" form. The typical economizer cycle is depicted with only external data points.[12]

4.18.3 Overall Efficiency

Efficiency for a single stage compressor is defined as head divided by the work input:

$$\eta = \frac{H}{W} \tag{4.15}$$

For a multi-section compressor, an overall compressor efficiency can be calcu-

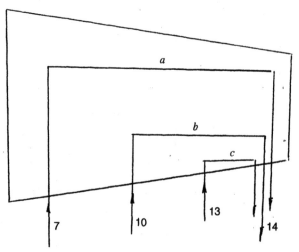

FIGURE 4.64 Equation (4.17) assumes that the sideloads do not mix with the main fluid.[12]

lated by dividing the total of the sum of the head from each section by the sum of the work from each section:

$$\eta = \frac{H_{7-8} + H_{9-11} + H_{12-14}}{W_{7-8} + W_{9-11} + W_{12-14}} \quad (4.16)$$

For the case of the field sideload compressor where points 8 and 11 are not available (Figs. 4.62, 4.63 and 4.64):

$$\eta = \frac{H}{W}$$

$$\eta = \frac{HM}{WM}$$

$$= \frac{H_{7-14a}M_a + H_{10-14b}M_b + H_{13-14c}M_c}{W_{7-14a}M_a + W_{10-14b}M_b + W_{13-14c}M_c}$$

and

$$W = (\Delta h)778.16$$

$$\eta = \frac{H_{7-14a}M_a + H_{10-14b}M_b + H_{13-14c}M_c}{778.16[(h_{14a} - h_7)M_a + (h_{14b} - h_{10})M_b + (h_{14c} - h_{13})M_c]} \quad (4.17)$$

While this procedure may not correctly model the true polytropic process as shown in Fig. 4.62, it does give a very close approximation of the overall compressor performance.

Any error realized from using Eq. (4.17) in place of Eq. (4.16) is reduced as the main inlet flow is increased and the side stream flows are decreased.

Example:

Use Eq. (4.17) only when external data is available:

$$\eta = \frac{H_{7-14a}M_a + H_{10-14b}M_b + H_{13-14c}M_c}{778.16\,[(h_{14a} - h_7)M_a + (h_{14b} - h_{10})M_b + (h_{14c} - h_{13})M_c]}$$

M_a = 3258.5 #/min @ −29.4°F, 18.72 PSIA
M_b = 310.3 #/min @ −6.3°F, 35.03 PSIA
M_c = 1152.7 #/min @ 19.9°F, 58.55 PSIA
M_t = 4721.5 #/min @ 133.8°F, 162.1 PSIA
H_{7-14} = 35,188
H_{10-14} = 25,518
H_{13-14} = 17,126
$h_{14a} - h_7$ = 155.4 − 102.8 = 52.6
$h_{14b} - h_{10}$ = 156.5 − 109.5 = 47

$$h_{14c} - h_{13} = 156.7 - 116.5 = 40.2$$

Note that for this example, each section has a different gas analysis. This is why h_{14a} does not equal h_{14b} or h_{14c}.

$$\eta = \frac{35188 * 3258.5 + 25518 * 310.3 + 17126 * 1152.7}{778.16 \, [52.6 * 3258.5 + 47 * 310.3 + 40.2 * 1152.7]}$$

$$= \frac{142330030}{180781929}$$

$$= 0.7873$$

Use Eq. (4.16) with internal data and sectional head and work can be accurately calculated:

$$\eta = 0.796$$

$$\text{Error} = -1.2\%$$

4.19 FIELD DATA ANALYSIS

Trend the performance data looking for any changes. To confirm accuracy of data, compare the total compressor power to the driver power and monitor the compressor work input, which is a good indication of the collected data accuracy. Work input values remain constant for varying inefficiencies in the compressor. If work input is off design, then there may be instrumentation problems or something affecting the compressors ability to do work, such as flow swirl.

Set up a spread sheet and plot the manufacturers data for the compressor head and efficiency vs. flow as well as work vs. flow (see Eq. (4.15)). Data plotted on these curves should be fan law corrected for speed differences (Figs. 4.65 and 4.66).

Nomenclature:

a = Main inlet

b = 1st sidestream

c = 2nd sidestream

t = Total, discharge flow

7 = Main inlet

8 = Disch first section

9 = Inlet second section

10 = 1st sideload

11 = Disch second section

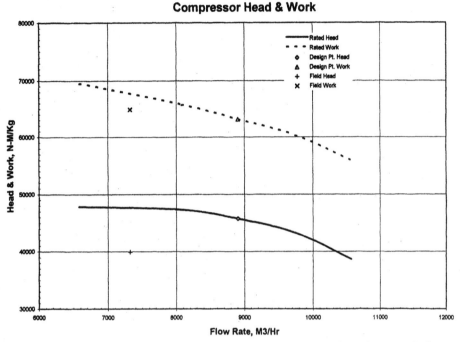

FIGURE 4.65 Plot both head and compressor work input. Work input can be a good indicator that the data used is valid. Fan law correct the data to compensate for speed variations from the reference curve.

12 = Inlet third section
13 = 2nd sideload
14 = Final discharge
h = total enthalpy
M = Mass flow
H = Head
W = Work
η = Efficiency, polytropic
GHP = Gas horsepower

4.20 TROUBLE SHOOTING COMPRESSOR PERFORMANCE

Large dynamic compressors are commonly the heart of various petrochemical and industrial processes. The plant output as well as the power consumption is controlled by the compressor performance. Reduced compressor efficiency will not

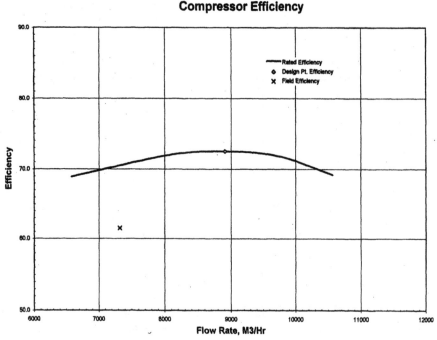

FIGURE 4.66 The efficiency cannot be fan law corrected, but the flow for the data point should be corrected.

only cause increased utility bills, but may also limit plant production rates. Maximization of compressor efficiency is therefore most crucial in assuring maximization of plant profits.

Correction of compressor performance deficiencies can be very complex, and a very methodical procedure is necessary in pinpointing the root cause of the situation. The following outline is a starting point.

1. Define the problem.
   a. What exactly is the problem?
   b. What should the performance be?
   c. What is the performance now?
2. Outline the history of the compressor.
   a. How long has it been operating?
   b. When was the last overhaul?
   c. What changes were made at that time?
   d. When did the problem start?
   e. Was it a quick or gradual change?
   f. Note the trend of various parameters.
   g. What else changed, what other problems occurred at this time
      i. on the compressor?

ii. in the process?
   iii. in operation and control?
3. Verify all data.
   a. Have instruments been calibrated?
   b. Do cross checks agree?

A thorough performance test should be the first step. Follow as closely as possible ASME PTC10 test procedure. If possible get several operating points at one speed so a full curve can be plotted. This can be a big help in determining corrective measures.

Refer to guide in Table 4.1 for help in trouble shooting aerodynamic problems.

TABLE 4.1 Troubleshooting Checklist[1]

- Define problem—what, where, when.
- Outline history of operation—trend data.
- Verify data.

Test data
   Complete power balance.
   Check pressure taps: location, size, and condition.
   Is there liquid in pressure lines?
   Note the temperature probe insertion depth, and heat transfer.
   Calibrate instruments.
   Inspect flow meter: wear and sludge build up.
   Are there condensates in gas analysis?
   Is there a vortex or undeveloped velocity profile upstream of flow meter?
   Conduct a mass flow balance.

Equipment problems
   Vortex or undeveloped velocity profile upstream of compressor suction
   Internal leakage across diaphragm splitline
   Recirculation from rubbed interstage seals or balance piston seals, casing drains, other areas
   Foreign object damage or blockage
   Liquids in process
   Dirt accumulation or polymer buildup
   Erosion of impeller blades and diffuser passages
   Proper direction of rotation
   Balance line sleeve

Economics
   Per diem cost to operate as is
   Associated risks
   Cost for repairs
   Cost for down time to complete repairs
   Safety concerns

4.20.1 Common Sources of Test Error

In order to trouble shoot any problem, it is important to have correct information on the subject. Aerodynamic performance is very involved and data errors can rapidly mushroom, thereby misleading the problem solver. It is therefore essential that accurate data be obtained.

Before trouble shooting the compressor, trouble shoot the testing procedure. The best way to do this is a power balance. If it is not feasible to do a power balance, or if there is a significant error (7%) between the compressor power and the driver power, a thorough analysis of the test procedure is necessary.

4.20.2 Gas Analysis

To have good test results, it is critical to have an accurate gas analysis. This can be a bit complex on high pressure, high mole weight gas. If the sample is taken at high temperatures, some of the heavy gas may condense on the walls of the sample container when it cools. If the sample is taken at the inlet, there may be some liquids in the gas stream that will remain in the sample container. When testing the gas, this condensed liquid will remain in the bottle unless heated.

For best results, take samples at both the inlet and discharge points. Check for condensibles and compensate by heating the sample before testing.

4.20.3 Liquids in the System

If there is liquid anywhere in the system, it is possible that some may carry over into the compressor. Knockout drums and demister pads do not always work the way they should. This liquid carryover will give erroneous results on the performance test.

Another liquid problem is liquid in pressure tap lines. Be sure all lines are properly sloped and drained. If lines are too small (less than 1/2 inch), capillary action will hold liquid in the lines.

Be sure to open drain valves at low spots in process piping and instrument lines before, during, and after test.

4.20.4 Pressure and Temperature Measurement

Be sure that a proper pressure tap is installed (see Fig. 4.57). Inspect the inside edge of the hole to see that it was deburred and that it has not been eroded, corroded, or plugged with dirt.

Check thermocouple installations. Thermocouples should be inserted into the pipe one-third to one-half the pipe diameter. Use graphite paste in the thermowell to assure good heat transfer between the thermowell and thermocouple.

Be sure to only use instruments that have recently been calibrated.

4.20.5 Cleaning Centrifugal Compressors

Sometimes dirt, polymer build up, or other substances can clog the compressor internally and seriously degrade performance. Very small amounts of dirt on axial blades alter the blade profile and degrade performance. Cleaning a compressor may be all that is required to regain "like new" performance.

A centrifugal compressor can be easily cleaned during normal operation (design speed) by using mild abrasives such as cooked rice or walnut shells. More common is the use of liquid cleaning agents sprayed into the process and into the return channel areas of the compressor.

4.20.6 Velocity Profile

A major source of compressor performance problems can be attributed to an incomplete velocity profile or a vortex upstream of the compressor or process flow meter. Either situation will seriously alter the compressor and/or flow meter performance. A flow straightener device is required when flow swirl or a vortex is present. This can occur when there are two or more adjacent elbows in different planes. A flow straightener can be a tube bundle or an "egg crate" as shown in Figs. 4.67 and 4.68.

A flow equalizer is required when the velocity profile is not uniform. This can be caused by flow hugging one side of a pipe due to flow around an elbow or flow through a partly closed butterfly valve. This situation is best corrected by an equalization plate which is essentially a perforated plate. Think of it as parallel orifices in a flow path. At high velocities, the resistance (pressure drop) is greater. The

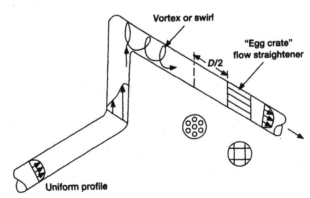

FIGURE 4.67 The vortex or flow swirl shown is caused by two elbows in different planes. The vortex can be corrected by installation of a simple vane flow straightener. Length of vanes should be two times the pipe diameter. A flow straightener may be more effective without a flow equalizer due to the high pressure drop associated with the flow equalizer.

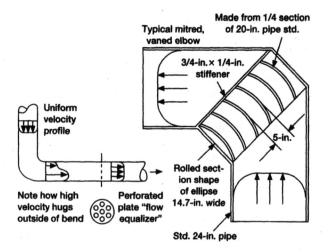

The cost of mitered, vaned elbows is reasonable compared to standard formed elbows, since they can be locally fabricated. A flow equalizer can be used, but at the price of a significant pressure drop. Such a pressure drop in the suction line can seriously limit compressor discharge pressure.

FIGURE 4.68 Mitered, vaned elbow. A flow equalizer can be used, but at the price of a significant pressure drop.[17]

higher velocity side of the velocity profile is restricted more than the lower velocity side, causing a shifting and equalization of the velocity profile (see Fig. 4.67).

When designing a flow equalizer, it is important to realize that pressure drop can be significant, especially if the plate becomes plugged with debris. The best method for calculating pressure drop is to add the area of all the holes in the plate and determine an equivalent single hole orifice while calculating pressure drop accordingly. Be sure to note the effect of the recovery factor.

Make a schematic diagram of the compressor and adjacent piping. Note the length of straight runs of pipe, elbows, flow meters, valves, suction strainers, knock-out drums, silencers, flow straighteners, instrumentation, etc. This will help in resolving system-related problems.

If possible, a flow meter should be installed in each inlet and discharge pipe so a mass flow balance in the system can be carried out. This is done by simply comparing the total mass inflow to the total mass outflow. The difference is the accuracy of the flow meters.

ASME "Fluid Meters" provide comprehensive guidelines on the straight run required upstream of an orifice or flow nozzle.[16]

4.20.7 Inlet Piping

Compressor performance is very dependent on obtaining a uniform flow distribution to the impellers. Great pains are taken by the compressor designer to assure proper flow distribution to intermediate impellers. Although inlet guide vanes may exist on a compressor, this alone does not assure proper flow distribution to the first stage impeller. Compressors are designed assuming a relatively uniform velocity profile at the compressor inlet flange.

Although ASME goes to some detail in describing upstream straight-run requirements for orifice meters, the requirement for compressors is very simply stated. The straight-run requirement for axial inlet compressors is 10 pipe diameters and for non-axial inlets, 3 diameters. Additionally, if the velocity pressure exceeds 1% of the static pressure, a flow equalizer must be used at the exit of the elbow upstream of the compressor inlet flange.

$$\frac{P_V}{P_1} \leq 0.01 \tag{4.18}$$

where

$$P_V = \frac{P_1 V_1^2}{2gZRT_1} \tag{4.19}$$

Values for $P_v/P_1 = .01$ are shown in Fig. 4.69. Note that for 100°F air (MW = 29), the maximum inlet velocity for a three-diameter straight run without an equalizer is 140 fps. For propane (MW = 44) at 0°F, the maximum velocity would be 100 fps.

According to Hackel and King[17] the ASME guideline is adequate but could be modified to call for a reduced straight run for lower P_v/P_1 values, and greater straight run for larger P_v/P_1 values. Also, additional length of straight run should be required for compound piping arrangements.

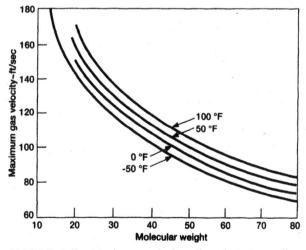

FIGURE 4.69 Maximum velocity allowed at the suction flange based on Eq. (4.18).[17]

For a base case of one elbow in a plane parallel to the compressor axis and for a radial inlet with 50°F gas, Fig. 4.70 gives the minimum straight run required.

To correct for other suction temperatures, use the following equation to find the equivalent velocity for 50°F. First calculate the actual velocity V_1.

$$V_{50°F} = \frac{22.6 V_1}{\sqrt{T_1}} \qquad (4.20)$$

where

$$T_1 = °F$$

For axial inlets and/or other inlet piping arrangements, use Fig. 4.70 along with the multipliers provided in Table 4.2.

Example

Consider a gas with a MW of 25, inlet temperature of 85°F and suction velocity of 100 fps. The piping configuration is a radial inlet multi-stage compressor with

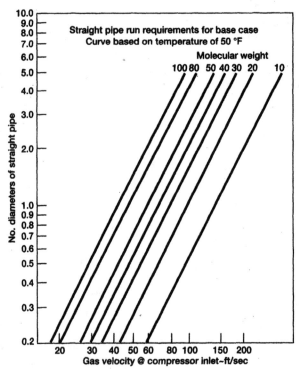

FIGURE 4.70 Straight pipe run requirements for the base case (Fig. 4.71, A and Table 4.2).[17]

TABLE 4.2 Multipliers for Various Inlet Piping Arrangements*

Piping Configuration (Radial Inlet)	Multiplier
One long radius elbow in a plane parallel to compressor shaft (Fig. 4.71)	1.0
One long radius elbow in a plane 90° to compressor shaft	1.5
Two long radius elbows at 90° to each other	2.0
Two elbows in the same plane parallel to compressor shaft	1.15
Two elbows in same plane 90° to rotor	1.75
Butterfly valve	2.0
Gate valve wide open	1.0
Swing Check valve balanced	1.25
Vortex Separator	4.0
Reducer/Expander/Dutchman	1.25
Tee	1.0
Axial inlet	1.25

*Use with Eq. (4.20) and Fig. 4.70 to determine required straight run of piping for a given arrangement.

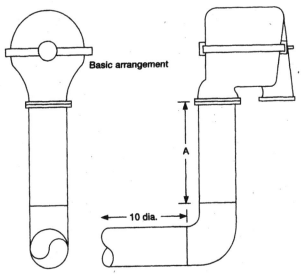

FIGURE 4.71 Base case. Long radius elbow in a plane parallel to the compressor shaft. A minimum of 10 pipe diameters straight run upstream of the elbow is required. Find "A" from Fig. 4.70 and Eq (4.20).[17]

two elbows in different planes upstream of the compressor. The elbow nearest the compressor is in a plane parallel to the rotor. Upstream of the two elbows is a butterfly valve.

Figure 4.70 gives a straight run requirement of 1.45 diameters. The multiplier for the two elbows is 2.0 and butterfly valve factor is 2.0.

Multiplying

$$1.45 \times 2.0 \times 2.0 = 5.8$$

Six pipe diameters of straight run are required for this application.

4.20.8 Double Flow Compressors

A common method of increasing capacity of a system is using two or more compressors in parallel. However it is feasible, since the "identical" units are always somewhat different and system resistance varies, that both units will not be operating at the same point on the performance curve. It is therefore always recommended that each unit have a separate anti-surge system. For a double flow compressor this is not very simple due to the common discharge nozzle. The design of the inlet piping must be such to achieve a well-balanced, distortion-free flow into each inlet of the compressor. Otherwise, as with the parallel compressors, the flow rates to each side may not be balanced and premature surging will occur.

The most reliable inlet piping design for a double flow compressor utilizing a drum to split the flow is shown in Fig. 4.72. While a Y with a proper upstream straight run of pipe may seem like a good design, it should be noted that even the smallest disturbance in the piping upstream of the Y will cause the flow to shift to one leg of the Y or the other.

More often than not, some type of trimming device (orifice plate, butterfly valve, or others) is used in one or both legs of double flow compressors to equalize the flow. For this reason, the most economical method may be to simply install a butterfly valve upstream of a Y connection (Fig. 4.72).

Field Problems. The following are some actual case histories where inlet piping alone was the source of some serious performance problems.

Case 1. During commissioning, a double flow compressor was found to be low in head. Also, the unit surged prematurely.

The inlet piping caused unequal flow distribution to the compressor inlet. This resulted in one section running near surge while the other section was operating near the overload region. The inlet piping to the compressor was modified utilizing a mitered elbow at the tee, a flow equalizer, and a trim valve to improve flow distribution.

Case 2. Two duplicate single-stage air compressors were found to have a significant capacity difference during commissioning.

Both compressors had been performance tested at the factory and were within 1.0% of each other. The suction piping for each unit was identical to the other

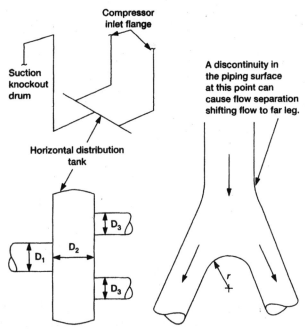

FIGURE 4.72 Suggested piping for double flow compressor (left). $D3$ and $D1$ sized according to Fig. 4.69. Size D_2 to achieve a velocity 1/4 of that in Fig. 4.69. Note that antisurge line should be fitted to the knockout drum further upstream and not to this distribution drum. Piping legs from the drum to each inlet must be identical mirror image of each other. For a Y type splitter (right) note the large radius at the dividing point. A mitered type joint with a sharp, pointed dividing geometry could cause flow separation and uneven distribution. A minimum of 10 pipe diameters is required upstream of the Y joint. Low velocity (relative to Fig. 4.69) will help assure equal flow distribution.

except that they were mirror images. The axial inlet compressors had two elbows at different planes and a suction throttle valve. This piping arrangement caused flow swirl which caused prewhirl at the impeller and effected the head output. The inlet piping was modified to include mitered elbows which minimized the problem.

Case 3. During commissioning, a high pressure multistage compressor was found to be low in head and efficiency. Near surge, control became unstable. The compressor would rumble and continue to surge even with the recycle valve open. Eventually, it would trip on high vibration.

It was found that a vortex separator was being used upstream of the compressor to assure liquids did not enter the compressor. The residue vortex affected both the orifice and compressor performance.

Flow straightening vanes (egg crates) were utilized downstream of the separator to reduce the vortex.

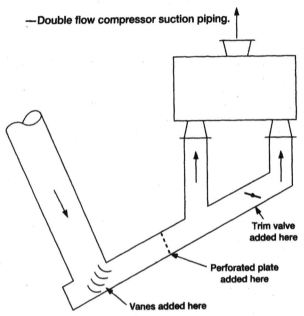

FIGURE 4.73 Double flow compressor suction piping. Flow distortion caused by piping caused the compressor to surge prematurely.[1]

TABLE 4.3 Maintenance Checklist[1]

Check for the following items:
- Preshutdown performance
- Interstage labyrinth seal clearance
- Balance piston seal clearance
- Blade tip clearance on open centrifugal impellers
- Internal splitline leakage
- Impeller to diffuser alignment
- Cleanliness of internal parts
- Surface finish of internal parts (pitting, corrosion, buildup of foreign material, etc.)
- Proper installation of stationary guide vanes, splitter vanes, etc.
- Performance at start-up

4.21 REFERENCE

1. Gresh, M. T., *Compressor Performance: Selection, Operation, and Testing of Axial and Centrifugal Compressors* (Stoneham Mass.: Butterworth-Heinemann, 1991).
2. Chow, R. C. (Novacor Chemicals), R. McMordie (Sermatech International) and R. Wiegand (Elliott Co.), *Performance Maintenance of Centrifugal Compressors Through the Use of Coatings to Reduce Hydrocarbon Fouling*, 1994.

3. *Compressor Refresher* (Jeannette, PA.: Elliott Co., 1975).
4. Paluselli, D. A., *Basic Aerodynamics of Centrifugal Compressors* (Jeannette, PA.: Eliott Co.).
5. Hallock, D. C., *Centrifugal Compressors...The Cause of the Curve* (Jeannette, PA.: Elliott Co., 1968).
6. Sheperd, D. G., (Cornell University), *Principles of Turbomachinery* (New York, NY.: Macmillan, 1967).
7. Cannon, R. H. (Stanford University) *Dynamics of Physical Systems* (New York, NY.: McGraw-Hill, 1967).
8. Nicholas, J., *Fundamental Bearing Design Concepts for Fixed Lobe and Tilting Pad Bearings* (Dresser IN.: 1986).
9. DeChoudhury, P., *Fundamentals of Rotor Stability* (Jeannette, PA.: Elliott Co.).
10. Fox, Aziz A., (Solar Turbines Inc.) *An Examination of Gas Compressor Stability and Rotating Stall,* San Diego, CA.
11. Salisbury, R., *Compressor Performance* (Jeannette, PA.: Elliott Co., 1985).
12. Gresh, K. K. (Flexware Inc.) *Field Analysis of Mulit-Section Compressors* (Jeannette, PA.: Hydrocarbon Processing, Jan. 1998).
13. Bensema, D., *Field Performance Testing* Elliott Co. (Jeannette, PA.: Elliott Co., 1986).
14. ASME PTC 10, *Compressors and Exhausters,* New York, 1965.
15. Leipman, H. W., A. Roshko (California Institute of Technology), *Elements of Gas Dynamics,* (New York, NY.: John Wiley & Sons, 1957).
16. ASME PTC 19.5, Fluid Meters, American Society of Mechanical Engineers, NY., 1971.
17. Hackel, R., and R. King, *Centrifugal Compressor Inlet Piping—A Practical Guide* (Jeannette, PA.: Elliott Co.).

CHAPTER 5
COMPRESSOR ANALYSIS

Harvey Nix
Training-n-Technologies

The practice of utilizing compressor analysis has areas of extreme diversity. This has historically been dependent on the industries or facilities reliance on compressors in the production of product. While production remains a viable rationale for analyzing compressors, maintenance and energy costs are also a concern. Whether the compressor is engine-driven or electric-driven, the costs of energy consumption alone can be very high.

By reducing the losses created by inefficient compressor operation, load can be reduced on the driver (engine or motor). This makes horsepower (hp) available for increased production or reduction in energy costs. Improvements in compressor component life and decreased maintenance costs are within the realm of a compressor machinery analysis program. It has been estimated that up to 50% of horsepower load to the driver from a compressor is not required when compressor components go unchecked. The reasons for increased load are relatively few, but overlooked.

5.1 COMPRESSOR VALVE FAILURES AND LEAKING VALVES

In multiple valve cylinders, one valve may fail and not much difference will be seen in the temperature of the cylinder. However, throughput and efficiency will be down. Unless there is a means of measurement, determination of hp losses are impossible to achieve. Once a cylinder is identified with a valve problem, identification of the particular valve(s) can save maintenance time. Another benefit of analysis is the determination of the severity of capacity loss. This information can help determine when it is economically justified to shut down a unit to perform maintenance.

5.2 COMPRESSOR PISTON RING FAILURES

An analyzer can trend ring degradation over time, thus *scheduled* maintenance can be performed. Many rely on increased cylinder temperatures to determine ring failure and usually a significant cylinder temperature increase warning is correct, but there is no guarantee that it is the rings and not the valves causing the temperature increase. Another way to identify ring wear is to perform periodic maintenance by pulling the piston out after a certain time period. If this is done, the piston may be pulled out too early or too late shutting the unit down, and possibly creating an unnecessary loss of production. Lastly, because the unit is now apart, it may be that life of other components is jeopardized.

5.3 RESTRICTION LOSSES

Valve inefficiencies and gas flow restrictions can be created by a number of faults, typically: design, plugged screens, pulsation of gasses, build up of solids, re-machining of valves, incorrect lift of plates, and improper springing of valves. This is an area where savings in dollars, for both maintenance and production, is realized.

5.4 IMPROPER CYLINDER LOADING

Cylinder loading is an area of compressor analysis that sometimes has been left to flow charts and curves using ideal and theoretical assumptions. There are programs on the market today that create operating curves based upon *actual* operating conditions that can save hundreds of thousands of dollars for corporations; lost dollars of which the corporations are not even aware. The measurement of volumetric efficiencies and the determination of true rod loads can be greatly enhanced to save machines, production and money.

Increased emphasis will be placed upon reciprocating machines in the future. It seems to be the general consensus that many are inefficient and need to be replaced. Replacement of compressors with rotary compressors is expensive. There are means to improve existing units so they deliver what they should, cost less, and are more dependable than what existed in the past.

5.5 REQUIRED INFORMATION

Here are some components needed to perform compressor analysis:

- Rated speed
- Rated load
- Bore
- Stroke
- l/r ratio (some analyzers)
- Connecting rod length
- Rod size/loads
- Suction/discharge pressure
- Cylinder clearance information
- Unloader information
- Crank angle positions between cylinders
- Valve design information
- Gas composition

The manufacturer will provide this information when requested. The design clearance volume of the cylinder is required to perform calculated theoretical throughput. The Crank angles for piston to piston and crankshaft relationships are necessary.

Valve design information may be left out, but has great value when determining why components are failing. Pertinent information is required on the design lift, throughput, temperature, gas composition, and required valve springs. Many times, incorrect spring sizes lead to early valve failure.

5.6 ANALYSIS OF THE COMPRESSOR USING A PRESSURE-VOLUME (PV) DIAGRAM

5.6.1 PV Sequence of Events

The primary tool for the determination of reciprocating compressor performance is the pressure-volume (PV) diagram. The PV diagram describes the relationship of internal pressure relative to the volume (% of stroke) of a particular end of a compressor cylinder.

Figure 5.1 represents a PV diagram and its significant features with a discussion following. This discussion will consider the head-end (HE) of a compressor only, while all also applies to the crank-end (CE) of a cylinder.

Line 4-1: The suction valve opens at point 4. As the piston travels toward BDC, the volume in the cylinder increases and gas flows into the cylinder. The pressure inside the cylinder is slightly less than suction line pressure. This small differential allows the valve to open and holds it open during the suction stroke.

Line 1-2: The suction valve closes as pressure across the valve equalizes as the piston has reached BDC and changes direction at point 1. The cylinder volume

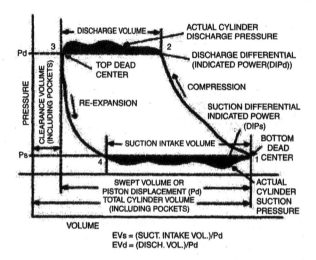

FIGURE 5.1 Near ideal PV diagram.

decreases as the piston moves towards TDC, raising the pressure inside the cylinder. The shape of the compression line (Line 1-2) is determined by the molecular weight of the gas or compression exponent. For an ideal gas (adiabatic-process—no flow of heat to or from the gas being compressed), the compression exponent is the isentropic (constant entropy) exponent that is equal to the ratio of specific heat of the gas being compressed.

Line 2-3: At point 2, the pressure inside the cylinder has become slightly greater than discharge line pressure. The resulting differential pressure across the discharge valve causes the valve to open, allowing gas to flow out of the cylinder. The volume continues to decrease toward point 3, maintaining a sufficient pressure differential across the discharge valve to hold it open.

Line 3-4: At point 3, the piston reaches TDC and reverses direction. At TDC, as the piston comes to a complete stop prior to reversing direction, the differential pressure across the valve becomes equal. This allows the discharge valve to close. The volume increases, resulting in a corresponding drop in pressure in the cylinder. The gas trapped in the cylinder expands as the volume increases toward point 4. At point 4, the gas pressure inside the cylinder becomes less than suction line pressure, creating a differential pressure that opens the suction valves. The cycle then starts over again. The shape of the re-expansion line (Line 3-4) is dependent on the same compression exponent that determines the shape of the compression line.

5.6.2 Suction Valve Leak

Figure 5.2 illustrates the PV diagram of a typical compressor cylinder with suction valve leakage. The difference between the theoretical PV diagram and the actual

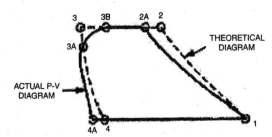

FIGURE 5.2 PV diagram illustrating the effects of suction valve leakage.

PV diagram will depend on the severity of leakage through the suction valves. The following is a step-by-step analysis of the PV diagram in Figure 5.2.

Line 1-2A: During compression, gas leaks out through the suction valve(s). Since gas is being pushed out of the cylinder during the compression stroke, the piston must travel further to reach the discharge valve opening pressure. If the leak is severe enough, the pressure within the cylinder will not reach discharge pressure. The cylinder volume at point 2A is less than point 2, resulting in a shorter effective discharge stroke or a loss in discharge volumetric efficiency (DVE).

Line 2A-3B: During the discharge stroke, gas is exiting through both suction and discharge valves. Should the leak be severe enough, the discharge valve will close prematurely at 3B instead of point 3.

Line 3B-3A: With the discharge valve prematurely closed, the piston is still moving towards TDC as gas continues to leak out of the cylinder through the suction valve. The internal cylinder pressure at point 3A is less than discharge line pressure at point 3. This effect may not be noticeable unless severe leakage is present.

Line 3A-4A: The cylinder's re-expansion slope occurs more quickly than normal due to the continuing gas leakage through the suction valve(s), thus causing the suction valve to open at point 4A.

Line 4A-1: The early opening of the suction valves causes the actual suction volumetric efficiency (SVE) to be greater than the theoretical SVE.

Symptoms:

1. Inlet temperature rises because of the re-circulation of the gas.
2. Leaking suction valve cap temperature will increase. Other valve cap temperatures may increase, but not as significantly.
3. Actual discharge temperature will increase (actual discharge temperature compared to theoretical discharge temperature).
4. Indicated horsepower may be lower than normal.
5. Compression ratio may decrease.

6. The calculated capacity based on the SVE will be higher than the calculated capacity based on the DVE, resulting in a capacity ratio greater than 1.0.
7. The compression and re-expansion lines will not match the theoretical PV curve.

5.6.3 Discharge Valve Leak

Figure 5.3 illustrates the PV diagram of a typical compressor cylinder which is experiencing discharge valve leakage. The difference between the actual PV diagram and the theoretical PV diagram will depend on the severity of leakage through the discharge valves. The following is a step-by-step analysis.

Line 3-4A: During re-expansion, the trapped gas in the cylinder is expanded as gas leaks through the discharge valve(s) into the cylinder increasing internal cylinder pressure. This increase in pressure causes the piston to move further down the stroke, re-expanding gas as it enters the cylinder through the discharge valve until it reaches a point where pressure is reduced, allowing the suction valves to open at point 4A. The result is a smaller effective suction stroke, thus reducing suction volumetric efficiency. If the discharge leak is severe enough, the internal cylinder pressure will not reach suction pressure.

Line 4A-1B: During the suction portion of the cycle, gas is entering the cylinder through the open suction valve and leaking discharge valves. The cylinder pressure can rise to a point causing premature closure of the suction valves at point 1B.

Line 1B-1A: The suction valve has closed, cylinder volume is increasing, and the internal cylinder pressure is rising, which results in a higher pressure at point 1A than suction line pressure at point 1.

Line 1A-2A: The actual compression line will not match the theoretical compression line since the pressure at 1A is not the same as the pressure at 1, and gas continues leaking into the cylinder through the discharge valves during the compression stroke. The discharge valve opens when cylinder pressure rises above discharge line pressure.

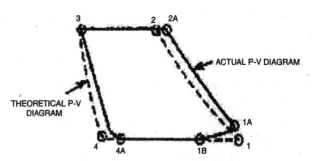

FIGURE 5.3 PV diagram illustrating the effects of discharge valve leakage.

Symptoms:

1. The actual discharge temperature will be higher than the discharge temperature observed in normal operation, or as compared to the theoretical discharge temperature.
2. The measured cylinder capacity will be less than the design cylinder capacity.
3. Capacity calculations based on DVE will be greater than capacity calculations based on SVE, resulting in a capacity ratio of less than 1.0.
4. Indicated horsepower may be lower than normal.
5. The actual compression and re-expansion lines will differ from a theoretical PV curve.

5.6.4 Piston Ring Leakage

Figure 5.4 illustrates the PV diagram of a typical HE compressor cylinder which is experiencing piston ling leakage. The shape of the actual PV diagram will depend on the severity of the leakage.

Line lA-2A: As the piston travels from point 1A to 2A, gas is initially leaking from the HE side of the piston into the CE, as would happen with a leaking discharge valve.

Line 2A-3B: Gas is exiting through the discharge valve and continues to leak past the rings. Should the leakage be severe enough, premature closing of the discharge valve could occur at point 3B.

Line 3B-3A: As the piston slows, and continues toward TDC, gas continues to leak past the ring, resulting in internal cylinder pressure drop to point 3A. This pressure at point 3A is lower than application pressure (point 3).

Line 3A-4A: During the re-expansion stroke, gas continues to leak past the rings, resulting in a much quicker drop to suction pressure until pressure equalizes on both sides, just like a leaking suction valve. After pressure equalizes fairly far down the stroke, pressure is now higher on the crank-end side of the cylinder, and gas

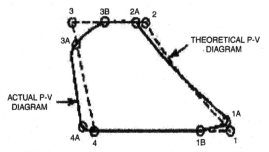

FIGURE 5.4 PV diagram illustrating the effects of ring leakage.

starts leaking into the head-end side, again looking like a leaking discharge valve. Usually this happens so far down the stroke that it is not noticeable.

Line 4A-1B: Gas is entering the cylinder through the suction valves and is leaking past the piston rings. This leakage results in premature closing of the suction valves at point 1B.

Line 1B-1A: The suction valves have closed and the cylinder volume is increasing. Pressure in the cylinder increases due to continued piston ring leakage into the cylinder. The pressure at point 1A is higher than design pressure (point 1).

Symptoms:

1. Measured capacity might be lower than the application capacity.
2. Discharge temperature will increase due to re-circluation of the gas. Compare actual discharge temperature to a normal value or theoretical discharge temperature. With severely leaking rings, discharge temperature may rise 80°F or more (double acting cylinder).
3. Leaking rings usually show up as a capacity ratio of greater than 1. However, leaking rings can also show up as a capacity of less than 1.
4. The measured compression and re-expansion lines will not match theoretical compression and re-expansion lines.

5.6.5 General Operation Limits

Generally, three common operational problems grouped together are: pulsation effects, valve losses, and cylinder gas passage losses. Their effect on compressor performance should be minimized as much as possible in the cylinder design and taken into consideration in the stated compressor horsepower and capacity figures. Even though they are taken into account in the compressor design, they are sometimes either underestimated or undefinable to the accuracy required and are responsible for performance problems. Figure 5.5 illustrates the cylinder losses for a typical PV diagram.

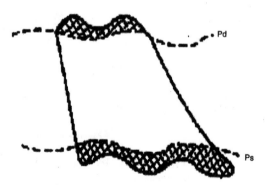

FIGURE 5.5 Operational and design problems.

Another area of fault is that of restricted passages. Passageways may be blocked for a number of reasons. Some of these include: incorrect cylinder design or sizing bottle restrictions such as plugged screens or broken diffuser tubes, valve restrictions such as plate lift decreased through improper machining processes or debris stuck in the valves. In the case of discharge, it may be plugged with melted piston ring debris. When passages are restricted, there may be excessive valve losses with the hump of the discharge line more pronounced towards the end of the stroke.

If a sharp rise should occur just before the end of the stroke, valves may be partially covered by the piston. DVE will be smaller than before, but due to the added valve losses, hp may not necessarily be less. In fact, total hp may be higher than normal.

These restrictions can also occur on the suction side for the same reasons, for the same type of action, but during the suction cycle. Suction terminal pressure may be less than the line pressure. Because enough gas cannot get into the cylinder, the slope of the compression line will be longer. This will mean less capacity and a lower than normal DVE. Horsepower may decrease somewhat and suction valve hp losses will be high.

5.6.6 Pulsation Effects

While the suction and discharge valves are open, the acoustic pulsation present in the system is passed into the compressor cylinder. Should the pulsation levels be of sufficient amplitude, the valve opening and closing times can be affected. Also, the average inlet and/or discharge pressures of the cylinder may be different than the design pressures with the net result being horsepower and capacity values which are different than the design values. These values may be greater or smaller, depending on the pulsation characteristics. The change in horsepower and flow may be proportional, resulting in actual BHP/MMSCF figures that are the same as design. However, the predicted loading curves will no longer be accurate.

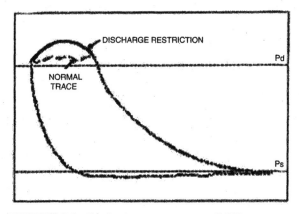

FIGURE 5.6 Discharge passage too small PV.

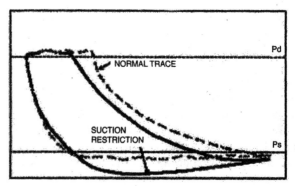

FIGURE 5.7 Suction passage too small PV.

5.6.7 Valve and Cylinder Gas Passage Losses

Valve horsepower loss is due to the pressure drop across the compressor valve. Cylinder gas passage loss is the pressure drop between the cylinder flange and the compressor valve. Should these losses exceed the cylinder design allowances, actual flow will be less than the design flow. (Note that these losses are also affected by gas pulsations.) A general rule of thumb is that valve and cylinder gas passage losses should not exceed 5% of the indicated horsepower for that cylinder end.

5.6.8 Excessively Strong Discharge Valve Springs

Strong discharge valve springs will be evident when evaluating a PV curve, usually identified by a normal trace until the start of the discharge stroke. Pressure will have to rise higher than normal to open the valve. A single hump may appear and then taper off until the cylinder reaches the end of the discharge stroke. With extremely stiff springs, there may be oscillations above and below the discharge line throughout the discharge stroke. Pressure pulsations can also show a similar pattern. In this case, it is necessary to look at the bottle pressure trace for indications of pulsations. Horsepower may not increase much, but excessive valve hp losses will be evident.

5.6.9 Excessively Strong Suction Valve Springs

The suction valve may have stiff valve springs as well. The same effects occur with suction springs as with discharge. For stiff springs, a single dip would appear at the beginning of the suction stroke. SVE will probably stay the same or a little less, but computed valve losses would be much higher.

An easy way to determine the difference between excessive spring forces and valve chatter created by weak or broken springs, with either suction or discharge, is:

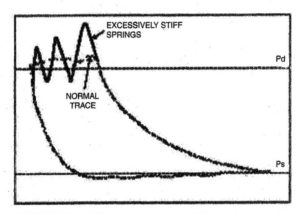

FIGURE 5.8 Stiff discharge spring PV.

- If valve hp losses are high, then the most probable cause is excessively stiff springs.
- If valve hp losses are low, then the most probable cause is weak or broken springs.

5.7 COMPRESSOR PRESSURE/TIME (PT) PATTERNS

5.7.1 Double Acting Compressor Cylinders

A double acting cylinder moves gas on both sides of the piston simultaneously. The furthest end from the crankshaft is referred to as the head-end (HE), and the cylinder end closest to the crankshaft is the crank-end (CE). A double acting cylinder requires suction and discharge valves on both ends of the cylinder.

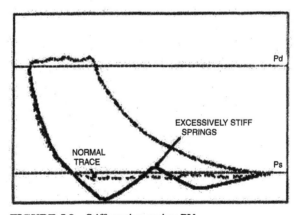

FIGURE 5.9 Stiff suction spring PV.

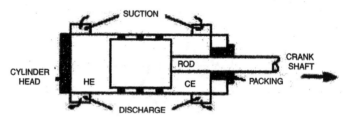

FIGURE 5.10 Double acting compressor cylinder.

There are three possible pressure measurement points:

- Suction nozzle/bottle
- Discharge nozzle/bottle
- Head and crank end cylinder measurements

While HE and CE cylinder pressure measurements are the most common, nozzle pressures also have value in determining causes of excessive valve and passage losses, or pulsation. The analyst must decide what is to be done with information obtained when determining the necessity to collect the above pressure readings.

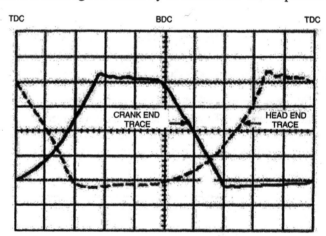

FIGURE 5.11 Double acting compressor cylinder PT.

The above diagram represent typical HE & CE cylinder pressure traces with the suction and discharge pressure traces overlaid.

5.7.2 Suction Pressure Time Trace

At line #1 (Fig. 5.12), suction line pressure, we see a line moving across the screen. Ideally, the line would be very steady. This represents the pressure, preferably at the suction inlet nozzle of the cylinder. The function of this line is to allow the analyst to evaluate the flow of gas entering into the compressor cylinder and its

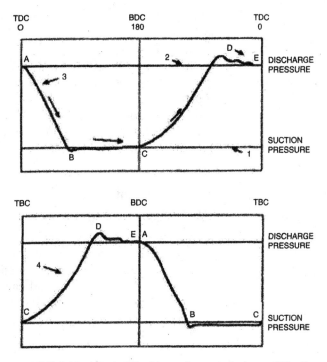

FIGURE 5.12 HE & CE with suction and discharge PT's displayed.

effect on the internal cylinder pressures. The area below the suction pressure line within the PV curve is considered valve and passage horsepower loss.

5.7.3 Discharge Pressure Time Trace

In a similar manner, the trace at the #2 position is that of the discharge line pressure collected from the discharge nozzle leading into the discharge bottle. Ideally, this line should be very steady. As with the suction pressure trace, this aids in the determination of internal cylinder pressure characteristics. The area above the discharge pressure line within the PV curve is considered valve and passage horsepower loss.

5.7.4 Head-End Pressure Time Trace (Internal Cylinder Pressure)

Trace #3 is a representation of the head-end pressure within the cylinder. Top dead center (TDC) starts at the far left of the pressure screen. At TDC, both the cylinder pressure and discharge line pressure should meet as the discharge valves close.

Line A-B: The cylinder pressure quickly drops to just below suction line pressure, allowing the suction valve to open.

Line B-C: The cylinder draws gas as the piston moves toward bottom dead center (BDC). As the piston slows and comes to a stop at BDC, pressure equalizes across the valve and the suction valve closes.

Line C-D: The piston moves from BDC toward TDC, compressing the gas within the cylinder until the pressure gets above discharge line pressure, allowing the discharge valve to open.

Line D-E: The cylinder discharges gas and continues moving towards TDC. The piston slows and comes to a stop at TDC. At TDC, pressure across the valve equalizes, allowing the discharge valve to close.

5.7.5 Abnormal Pressure Time (PT) Patterns

The PT trace is usually used to provide pressure reference points with overlaid vibration traces. Internal cylinder pressures are at given degrees of crank angle. From this it can be determined where normal vibration events will happen, and possible causes for other vibration events. While suction valve, discharge valve, and ring leaks affect the pressure time curve, and can be evaluated using the PT curve, they are more easily diagnosed using the PV curve.

5.7.6 Passage Restrictions

The next problem seen is that of restricted gas flow through the valves and piping. Gas flow restrictions could be caused by: the passageway being too small for the volume of gas; restricted suction screens; orifice plates that have been incorrectly designed; or other obstructions. Recognition of a restriction is easily made by the PT or PV curve. Figure 5.17 illustrates how a cylinder trace would appear with a restriction on the discharge side of the cylinder.

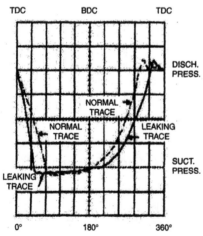

FIGURE 5.13 Discharge valve leaking.

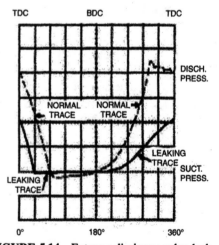

FIGURE 5.14 Extreme discharge valve leak.

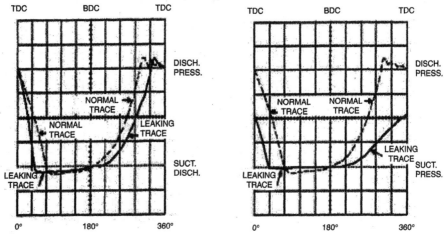

FIGURE 5.15 Suction valve leaking.

FIGURE 5.16 Extreme suction valve leak.

5.7.7 Rod Load Reversal

There are two types of reversals. The first, piston or crank angle reversal is the physical reversing in direction of the piston, which happens at both TDC and BDC positions. The second type of reversal is pressure reversal. Pressure reversal occurs as the internal cylinder pressure goes from more pressure on the head-end side to more pressure on the crank-end side of the piston. Without pressure reversal, lubrication of both sides of the crosshead pin may not take place. This lubrication is necessary and lack of lubrication will cause early failure.

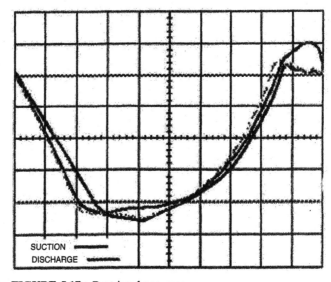

FIGURE 5.17 Restricted passages.

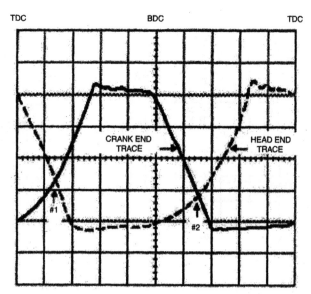

FIGURE 5.18 HE & CE PT trace.

The head-end trace begins with the re-expansion stroke. The pressure is higher than the crank-end side of the cylinder during the re-expansion until it meets the increasing pressure of the crank-end at point #1 (Fig. 5.18). When pressure is highest on the head-end side of the cylinder, the rod is said to be in compression, and the cross head pin is being pushed to the back of the bushing. At point #1, the pressure on both sides of the piston is equal.

Continuing to follow the head-end trace during the re-expansion cycle, the pressure becomes less than that of the crank-end which places the rod in tension. The clearance is changed in the bushing and the pin is forced to the other side of the bushing. In a similar manner, the pressure will reverse just after the head-end begins its compression stroke and the crank-end is on its re-expansion stroke. The pressure again equalizes (at point #2) with a change from rod tension to compression. This moving back and forth of the pin allows the clearance on both sides to open and accept oil and thus provide a lubricating oil film.

With improper cylinder unloading, it is possible to create a situation in which rod load reversal does not take place. Valve failures can also produce a non-reversal situation if the compressor is already running close to a non-reversal condition. It is necessary for the rod load to go from compression to tension for a short period of time. (API standard 618 provides additional information on the duration required to provide adequate lubrication.)

5.7.8 Partially Covered Valves

Partially covered valves are sometimes seen in cylinders with large piston/head clearance after an overhaul as a result of incorrect setting of the piston position.

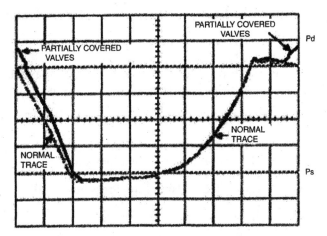

FIGURE 5.19 Partially covered valves using PT trace.

Generally, the clearances are set at one-third of total on the crank side and two-thirds on the head side. This allows for thermal growth in the rod and piston and the *operating* clearances should then be equal on each side. If improperly set, it can affect cylinder performance. Figure 5.19 indicates that a problem exists with either the way the piston was set, or the machining in the process of a cylinder refit, or in the dimensions of a new piston. The PT shows the possibility of partially covered valves. Note how the pressure rises suddenly at the end of the stroke. One way for this to occur is by covering valves so that the gas is restricted as it leaves the cylinder.

5.7.9 Compressor Rod Loading

Calculation of rod loads may be done with external suction and discharge line pressures. More accurate values will be obtained using actual internal cylinder pressures, taking into account the rise in pressure caused by valve restrictions or other factors that might be present within the cylinder.

5.7.10 Rod Load Calculations

Compression Rod Load (CRL) Internal = (HEA × HEPmax) − (CEA × CEPmin)

HEA = Area of HE of the cylinder
HEPmax = HE discharge pressure that yields maximum total differential pressure with CE pressure throughout cycle
CEA = Area of CE of the cylinder minus rod area
CEPmin = Suction pressure at point of maximum differential pressure with HE pressure

Tension Rod Load (TRL) Internal = (HEA × HEPmin) − (CEA × CEPmax)

HEPmin = HE suction pressure at point of maximum differential pressure with CE pressure
CEPmax = CE discharge pressure that yields maximum total differential pressure with HE pressure throughout cycle

Compression Rod Load (CRL) Using Line Pressures = (HEA × HEPd) − (CEA × CEPs)

HEPd = HE discharge measured line pressure
CEPs = CE suction measured line pressure

Tension Rod Load (TRL) Using Line Pressures = (HEA × HEPs) − (CEA × CEPd)

HEPs = HE suction measured line pressure
CEPd = CE discharge measured line pressure

Manufacturers will provide actual rod load limits for both compression and tension, as found in most compressor manuals.

Below is an example from real data gathered in the field on an actual unit. Calculated rod loads using line pressures and then rod loading using internal pressures.

Line pressures of the cylinder were found to be:

- Suction = 700 psi
- Discharge = 1733 psi
- Piston diameter = 4.625″
- Rod diameter = 2.500″

Compression = $(3.14 \times 2.31^2 \times 1733) - ((3.14 \times 2.31^2) - (3.14 \times 1.25^2)) \times 700 = 20793$ lbs
Tension = $(3.14 \times 2.31^2 \times 700) - ((3.14 \times 2.31^2) - (3.14 \times 1.25^2)) \times 1733 = 8848$ lbs

The compression and tension loads differ. Limits from the manufacturer may be different between each as well. The limits for this rod might be something like 25000 lbs. compression and 15000 lbs. tension.

The actual measured internal pressures of this cylinder as read from the PT curve are different from the measured line pressures. Both cylinders go over the measured discharge line pressures. The suction pressure also is different from the head to the crank-end.

The maximum pressure on the head-end is 2040 psi. At the same time, the crank-end pressure is 660 psi. These are the two pressures to be used for discharge and suction for internal calculations. The maximum pressure during the CE cycle is somewhat less at 1940 psi and the suction pressure is 600 psi. This represents the maximum differential experienced during the tension rod load cycle. This is what is used for the calculation of rod load tension.

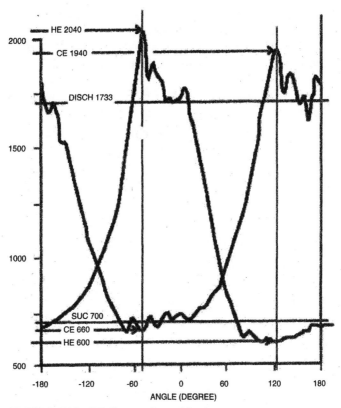

FIGURE 5.20 PT diagram for rod load.

Compression = $(3.14 \times 2.31^2 \times 2040) - ((3.14 \times 2.31^2) - (3.14 \times 1.25^2)) \times 660 = 26427$ lbs.

Tension = $(3.14 \times 2.31^2 \times 600) - ((3.14 \times 2.31^2) - (3.14 \times 1.25^2) \times 1940 = 12990$ lbs.

The results indicate that rod loads as predicted from line pressures would be less than those measured from actual operating data.

5.8 COMPRESSOR VIBRATION ANALYSIS

The basic types of vibration associated with compressors are the same as for engines. Recognition of these types is of concern for the analyst. Vibration caused from transients is the first group. When a valve slams open with a sharp mechanical impact, a high, straight line peak is created that exists only briefly. The second group is a flowing pattern. A vacuum or leak, where the velocity of the gas changes enough to produce a wedge-type formation, will be seen with the trace. The last group is scuffing or roughness. This is seen as a broadening of the trace throughout the area. The width of the base line determines when a cylinder has become scored.

With compressors, there are basically three events within the cycle that should cause normal vibration for a single cylinder end.

- One is suction valve opening. This event should represent a sharp mechanical impact. A gas passage noise may be present for a short period of time.
- Another is discharge valve opening. Here we see a sharp mechanical impact and a short period of flowing gas at the opening of the valve.
- Last is discharge valve closing. Because of the pressure present in the discharge line, the valve will quickly close. Therefore, there should be a small closing peak with a slight blow as the gas is being shut off.

Suction valve closing is a definite event, but the pressure drop during closing changes relatively slowly. Because the pressure slowly changes from suction to compression, the valve should gently close with low flow.

When analyzing compressors, the repeatability of patterns compared to base measurement is critical. Pattern repeatability and data collection consistency are important. While trending or during a first-time analysis, the analysis may be difficult.

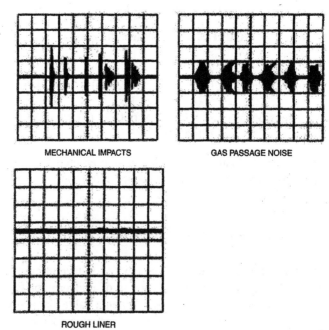

FIGURE 5.21 Types of vibration.

COMPRESSOR ANALYSIS   **5.21**

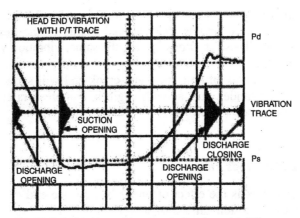

FIGURE 5.22 Normal head-end vibration with PT trace.

5.8.1 Use of An Analyzer

When using vibration instruments in the time domain, the acceleration should be set to 1.0 to 2.0 volts/division for most units. For others using computerized analysis equipment, the default scale should be set to 2.0 g's. What is most important is a standard baseline scale that is set for each type of equipment.

5.8.2 Normal Vibration Patterns

All the events previously discussed may be seen in Figs. 5.22 and 5.23. The opening events may change relative to crankshaft position due to operational changes. The

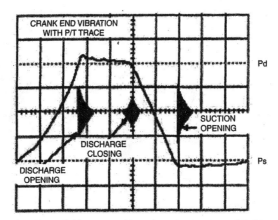

FIGURE 5.23 Normal crank-end vibration with PT trace.

actual position of the opening/closing events should be referenced to the pressure traces. The diagrams show normal events of the head-end and crank-end of the cylinder. These relationships are the same: vibration versus time, and pressure versus time.

An overlay of both head and crank-end will show complete cycle of both ends of the cylinder. Just as with engines, compressors exhibit cross talk and echoing as described below.

5.8.3 Compressor Cross Talk

Cross talk is the effect of one valve, on one end of the cylinder, presenting itself in the vibration/ultrasonic waveform of other vibration traces collected around the entire cylinder body. This is the reason that overlaying one pattern on top of another provides valuable information.

Figure 5.24 indicates normal valve action of both the head and crank-end as referenced to pressure.

Key:

HE denoted by caps
CE denoted in lower case

A,a Discharge Closed	A,a to B,b Re-expansion Stroke
B,b Suction Opening	B,b to C,c Suction Stroke
C,c Suction Closed	C,c to D,d Compression Stroke
D,d Discharge Opening	D,d to E,e Discharge Stroke
E,e Discharge Closing	1 & 2 Cylinder Pressure Equalization

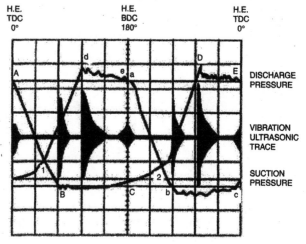

FIGURE 5.24 Normal PT and VT illustration.

5.8.4 Pressure Reversal

An event that should not be seen as vibration, except in very low amplitude or in magnified resolution, is pressure reversal. The pressure reversal does not change when the piston changes direction. The load on the wrist pin and rod components will be compression or tension depending on the pressure difference that exists between the HE and CE. The change in direction of this difference will cause the cross head pin clearance to shift from one side to the other.

The change in clearance from one side to the other allows penetration of oil into the clearance to lubricate the crosshead pin. The pressure reversal occurs just after the two pressures equalize. This can be seen in the Fig. 5.24 at points 1 and 2.

5.8.5 Normal Vibration Pattern Wrap-up

When looking at a single trace of either end, one must remember what other events are occuring and when the vibration/ultrasonic trace is being viewed. When examining ends of the cylinder individually, the size of the vibration events should be compared. Crank-end events will, or should, become smaller when measuring the head end valves. The converse is also true when measuring the crank-end valves. Figures 5.25 and 5.26 indicate what should be seen.

5.9 ABNORMAL VIBRATION/ULTRASONIC TRACES

5.9.1 Leaking Suction Valves

Beginning at BDC, which is where the suction valve closes, as the piston moves towards TDC, and as soon as compression pressure starts to build, the suction valve

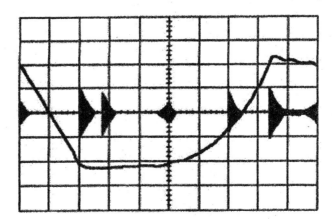

FIGURE 5.25 Head-end PT with vibration overlaid.

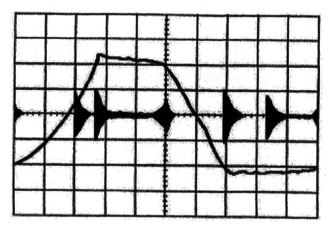

FIGURE 5.26 Crank-end PT with vibration overlaid.

may start leaking. This is not to say that all suction valves leak immediately after they close. There are many cases in which valves will close immediately after BDC and remain closed until there is enough pressure to force gas back out of the suction valve. If a suction valve is going to leak at all, it must leak during the discharge stroke, which is the area of highest differential pressure across the suction valve. With the capability of using a filter for both the vibration and/or ultrasonic transducers, this vibration will usually show up as high frequency in nature since this involves a leak rather that a mechanical event. There are three rules for identifying leaking valves:

1. Valves will not leak when they are open. This is because there is a free flow of gas through the valve.
2. A valve can leak from the time it closes until it opens again.
3. The greatest area of leakage will occur when the greatest pressure differential across the valve is present. Leakage of a suction valve will be greatest during the discharge stroke.

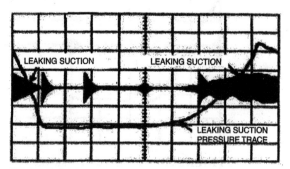

FIGURE 5.27 Head-end suction valve leak.

5.9.2 Leaking Discharge Valves

With discharge leakage, the trace remains similar in that the leak is again indicative of a high frequency rather than that of a mechanical low frequency. In this case, the use of a filter is highly recommended. The vibration pattern of a leaking discharge valve will:

1. Not indicate a leak when valve is open
2. Only show a leak from time valve closes until it opens again
3. Indicate most likely time of leakage will occur when greatest pressure differential across the valve is present. This would mean that leakage of a discharge valve will be greatest during the suction stroke.

5.9.3 Leaking Rings

Leakage across rings will occur when there is sufficient pressure differential across them. The key to identifying leaking rings is to determine when rings would *not* leak. This would occur when pressure is equal on both sides of the piston. Refer to points #1 and #2 of Fig. 5.29. Rings will leak the most when the greatest pressure differential is present. To help identify ring leakage, consider that:

1. Rings will not leak when the pressure is equal on both sides of the piston.
2. Rings are going to leak when the greatest pressure differential is present.

5.9.4 Cylinder Roughness

With cylinder roughness, the baseline of the trace will broaden in the roughest area. This is typically seen when abrasives enter the cylinder, when the cylinder surface

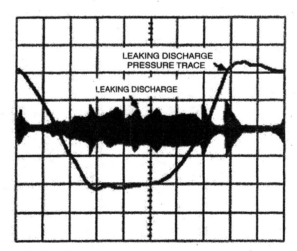

FIGURE 5.28 Head-end discharge valve leak.

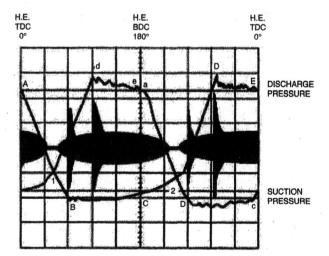

FIGURE 5.29 Leaking rings.

fails (scuffing or scoring), or when large wear particles come from the rings or piston. Although the rubbing, or scraping, is a mechanical action, the frequency is high in most cases. If analysis is not performed at least once every six weeks, the cylinder may become smooth, and even though the injury or wear may still exist, the vibration pattern may not show up. This does not mean the fault is gone, but that the components have worn each other to the point where lower friction is created. This may be difficult to see if the pressure trace is not monitored with the vibration. A look at the pressure trace will show the resultant leak. Figure 5.30 below indicates roughness in the cylinder. Note the difference in baseline trace during the rub.

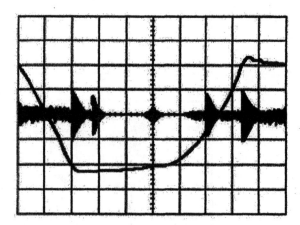

FIGURE 5.30 Head-end cylinder roughness.

5.9.5 Over Lubrication/Leaking Glycol System

If the glycol leak is relatively small, the symptoms will be similar to over lubrication. If the leak is large, there will be extreme changes during compression and re-expansion as the slope will be very steep. Over lubrication of a cylinder will create impact spikes seen in the trace. They will usually appear to be evenly spaced and show up as impact moments. Since this is mechanical in nature, the frequency will be low. Figures 5.31 and 5.32 shows a representation of over lubrication or presence of liquids.

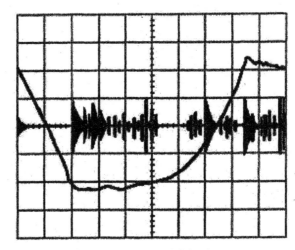

FIGURE 5.31 Head-end over lubrication.

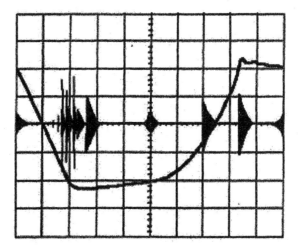

FIGURE 5.32 Head-end cylinder liquids.

5.28 CHAPTER FIVE

5.10 SYSTEMATIC COMPRESSOR ANALYSIS

The primary focus of a systematic analysis approach is to ensure a thorough compressor evaluation that consumes the least amount of time. Utilizing a form such as in Fig. 5.34, to record all the pertinent information makes it less likely to waste time or overlook important analysis infomation.

Follow the analysis format by completing items 1-17 as shown in Fig. 5.34. Use a check mark to indicate good condition; a check mark with a line through it to indicated marginal condition; an X to indicate poor condition; and a — to indicate not applicable or no data was taken or available.

This should only be considered a guideline, adjusted to meet specific needs and abilities. It is assumed that the analyst has a thorough grasp of the analysis concepts discussed here.

5.10.1 Data Validity (Basic PV/PT)

Check the basic PV/PT for accuracy.

Look to see that the head end PT curve pressure drops off immediately after TDC to ensure that phase angles are correct. Look to see that the crank end PT pressure curve drops off immediately after BDC.

Channel resonance can be identified by a jagged line during the compression and re-expansion strokes. If channel resonance is present, it should be corrected if possible, then this process started over again.

5.10.2 Corrections (VEs, VEd, CRC, Ps, Pd, MCA)

If any of the following corrections are made, note them with the below abbreviations:

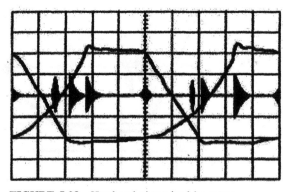

FIGURE 5.33 Head-end piston/rod looseness.

Compressor Systematic Analysis

Analysts Name_____ Unit_____
Date_____ Location_____

                                                                  # ___ # ___ # ___ # ___ # ___ # ___ # ___

1. Data Validity (basic PT/PV_____
2. Connections (VE, CRC, Pd, Ps, MCA_____
3. Theoretical PT/PV_____
4. Trends of Data_____
5. Compression Ratio_____
6. System Valve Losses (<5%)_____
7. Capacities_____
8. SVE & DVE (<25%)_____
9. Suct. Cap/Disch. Cap_____
10. Actual-Theor. Disch Temp (<40 deg.)_____
11. Rod Load/Reversal 12 Valve Cap Temp._____
12. Valve Cap Temp._____
13. Log P - Log Clearances N-Ratios_____
14. Cross Head Knocks_____
15. Gas Passage Noise_____
16. Mechanical Impacts_____
17. Cylinder Stretch_____

Notes

FIGURE 5.34 Systematic analysis work sheet.

VEs or VEd—Volumetric Efficiency(Suction or Discharge)

CRC—Channel Resonance Correction

Ps—Pressure Suction(Suction Terminal Pressure)

Pd—Pressure Discharge (Discharge Terminal Pressure)

MCA—Marker Correction Angle

It is important that any changes in the data be easily recognized by anyone that may review analysis.

5.10.3 Theoretical PT/PV

Many analyzers are now able to overlay collected PV's over theoretical curves based upon operating conditions. This display can prove helpful in identifying both subtle and obvious distortions in the PV curve. While this is a helpful tool, it should not be relied upon as the only indicator on which to make analysis calls.

It is important to make sure the information necessary to create theoretical curves is correct (gas analysis, clearances, geometry information).

Analysis of theoretical PT/PV—With the ability to overlay and display theoretical pressure volume and pressure time curves to actual curves, the analyst should evaluate curves looking for differences.

5.10.4 Trend (All Collected Data)

At start use alarm limits that are set per design operating conditions. Values for certain operating parameters change with operating conditions. Temperatures change with load and overall vibration changes with load. Because operating condition changes with load, it is not the individual readings themselves that are the helpful indicators of condition, but deviations from a baseline curve over an operating envelope. If the system is not capable of such calculations, utilizing statistical process control features may be a way of developing an alarm limit that takes into consideration changes in operating conditions. The most basic form of trends is a raw trend line of parameters. These can be helpful when looking at correlated parameters. Valve cap temperatures can be compared to each other on a single cylinder or like stages. Temperatures can be compared as well as ratio of capacities. Compare values across the unit in a single step or come up with a single efficiency number that can be used to determine compressor degradation (flow balance, actual capacity vs theoretical capacity, etc.).

Analysis of trends—Use whatever form of trending capabilities are available through the system. Note any anomalies.

5.10.5 Compression Ratios

This is the ratio of absolute discharge pressure (PSIA) to absolute suction pressure (Pd/Ps). Atmospheric pressure is added to both suction and discharge pressures read from the PV curve to convert Pd and Ps to absolute. In most cases, if a measured value is not put in for atmospheric pressure, the system will assume 14.7 psi (pressure at sea level) for calculations.

Analysis of compression ratios—Generally, compression ratio should not cause rod loads to exceed manufacturer limits. When compression ratio approaches 3.0, VE is very low, especially DVE. When VE is less than 25%, the analyst should question the calculations that rely heavily on accurate VE.

5.10.6 System (Valve) Losses

From pressures at inlet and discharge nozzle, cylinder system losses can be calculated. System losses refer to horsepower lost due to piping and valve pressure drop. If nozzle pressures are not taken, the area above Pd and below Ps are considered system losses. In most cases, pressure is taken at the neck of the bottle, just prior to gas entering the cylinder, or just after gas is leaving the cylinder. Valves are designed with losses in mind. (Generally, the more efficient the valve is, the less durable it is. Also a valve can be durable at the expense of high losses.) To measure actual compressor valve horsepower losses, collect a pressure trace from a tapped compressor valve cap and overlay on the PV curve.

Analysis of valve (system) losses—As a general rule, total system losses above 5% should be investigated further to determine root cause of the losses. By taking nozzle pressure, it may be possible to further identify the reason for losses—to determine if losses are piping or valve related. If valves are suspected to be the problem, it may be worth tapping a single valve to obtain a pressure trace to compare with the PV curve.

5.10.7 Capacity

Using calculated capacities for analysis purposes requires some foreknowledge of the compressor application and design specifications, or previous data collection with which to compare. Capacities are calculated from the SVE and DVE each independently. The results are suction capacity and discharge capacity. In theory, they should be equal or very close to equal. Usually when they are not, it is an indication of a problem occuring within the cylinder.

Analysis of capacity—Measured capacity should be compared to theoretical capacity. Also compare capacities to horsepower curves *generated for the unit*. Capacities can only be trended when operating conditions are constant.

5.10.8 Ratio of Capacities (ROC)

ROC is a single measure of cylinder condition. In simplest terms, ROC is suction capacity divided by discharge capacity or gas-in divided by gas-out. The ratio should be 1 or as near to 1 as possible. A range for determining acceptability is generally .95 to 1.10. The ROC should be considered when compared to historical readings for the cylinder and unit. ROC greater than 1.1 indicates leaking suction valves or rings. ROC less than .95 indicates leaking discharge valves or rings.

TDC reference must be accurate. VE must not be affected by pulsation or channel resonance. The smaller the VE, the more likely there exists an error within the resulted calculated values.

5.10.9 SVE and DVE

Suction volumetric efficiency and discharge volumetric efficiency are obtained from the pressure volume curve. They are also known as the effective suction and discharge stroke. They are read from the terminal pressure curve, crossing either the compression or re-expansion line.

Analysis of SVE and DVE—Note VE less than 30%: there may be calculation errors if VE is less than this.

5.10.10 Discharge Temperature Delta

This reading is the actual discharge temperature minus the theoretical discharge temperature. In most cases, theoretical discharge temperature does not take into consideration frictional heat, so the actual discharge temperature should be higher.

Analysis of DTD—High DTD generally indicates re-circulation of gas caused by leaking valves or rings. A general rule of thumb is that the DTD should range from 10 to 40°F higher than theoretical. This value should be trended. With ring leakage, DTD can go to 100°F or more. Valve leakage results, typically, in DTD of 40 to 50°F.

Negative numbers for the temperature difference identify a need to evaluate what goes into the theoretical calculation and the method of discharge temperature collection. Taking temperatures using skin temperature values typically give readings 10 to 20°F cooler than internal temperatures. Ambient temperature, sunlight, and whether the unit is inside or out, affect the readings.

5.10.11 Rod Load

The maximum rod load reached in compression and tension during the stroke should not exceed equipment manufacturer (EM) limits, and should go from compression to tension for a short period of time. Rod reversal allows lubrication to both sides of the cross head pin. Rod load can be based upon internal cylinder or line pressures and should consider effects of inertia. It is important to identify the method used by the EM to set the limits and make any comparisons on same basis.

5.10.12 Valve Cap Temperatures

Temperatures are the least reliable single indicator of valve leakage. Temperatures do help confirm leaking valves when applied with other information throughout the analysis process.

Analysis of valve cap temperatures—In addition to actual valve temperatures, discrepancies in temperatures between valves on the same cylinder can give a good indication of possible problems.

5.10.13 Cross Head Knock

Looseness associated with cross head knock comes at pressure reversal points. This reversal is also called the cross over point. This refers to the change from compression on the rod to tension. Valve vibration events should not be confused with knocks at the reversal points.

Analysis of cross head knocks—Match the rod load display with the vibration traces taken. Look for events that occur at the reversal points. Make sure the events are not valve related. Knocks identifed in this area should be of immediate concern.

5.10.14 Gas passage noise as expected

Analysis of gas passage noise as expected—Identify any pattern that exhibits unexpected gas passage noise. Concentrate on noise associated with leakage.

5.10.15 Impacts as Expected

Impacts as expected—Identify any abnormal impacts throughout the vibration traces.

5.10.16 Cylinder Stretch/Flap Analysis

Cylinder stretch is simply the movement in the horizontal direction from the frame to the end of the cylinder. Horizontal is perpendicular to the crankshaft. Movement should be very small and typically less than 3 mils. If this exceeds 3 mils, a reading should be taken on the distance piece to determine where the source of movement starts. A general rule is when the movement is greater than 3 mils in the horizontal direction, cylinder supports should be checked.

Flap analysis refers to the end of the cylinder flapping or having movement in all directions (horizontal, vertical, and axial). If this movement appears to be excessive, it can be assumed that bolts connecting the cylinder to the distance piece or frame are loose or broken.

5.10.17 Condensed Liquids Entrapped In The System

Similar results are obtained when condensation of liquids occurs in the system. These liquids are formed when pressure differential and temperatures are just right to cause the gas to condense. This typically occurs on the suction side due to lower temperatures. If condensation is heavy enough, vibration will become audible as fluid is slammed through the valves by the piston. If allowed to increase, the results can be almost as bad as detonation within an engine cylinder. If fluids in the cylinder are from a cooling system leak that has reached large enough proportions, the trace may be similar. The use of a filter may indicate both high and low frequency vibration.

5.10.18 Steps in the Cylinder, Ring Land, and Wear Band Looseness

Vibrations may occur at strange frequencies in line with steps in the cylinder. They will typically be present near the end or beginning of the cylinder stroke. Band or ring looseness will again appear at points of either the pressure or mechanical reversals.

5.10.19 Other Types of Looseness

These are types that exist in the mechanical train. Crosshead bushings/shoes, rods, bearings, pins, pistons and clearance plugs will all create vibrations that are similar to vibrations due to looseness of rings and wear bands. The amount of looseness will determine what amplitudes are seen. As discussed in the pressure section, the most likely places for this to occur are at the points of pressure and mechanical reversals. The vibration event will be seen from 10 to 25 degrees after the event. Exactly where it is seen is dependent on the acceleration of the rod, which is a function of speed of the unit. When viewed with a filter, the vibration typically is low frequency.

Figure 5.33 illustrates a cylinder with a piston that is loose on the rod. There is a difference in the valve and rod events in reference to crank angle position. With double acting cylinders, the event should be seen on both reversals. Some analysts have monitored this event in a logarithmic scale in order to track the looseness earlier.

5.10.20 Unloader Faults and Problems

Unloaders sometimes create more problems than is realized. The problem is that, if the unloader is not permitting the valve plate to return to its original position, it will usually look like a leaking suction valve. The typical causes of this problem includes the fingers not being dimensioned correctly or use of a compressed gasket. Unloader chairs or valve chairs may be loose in the cylinder, which can cause vibration.

CHAPTER 6
COMPRESSOR AND PIPING SYSTEM SIMULATION

Larry E. Blodgett
Southwest Research Institute

2.1 INTRODUCTION

Simulation of compressors and piping systems covers a broad range of disciplines Any discussion of simulation would be incomplete without an understanding of the nature of the process simulated and the simulation objective. The simulation of pulsations requires plane wave acoustics and occasionally three dimensional acoustics. The simulation of mechanical vibrations (excited by acoustic shaking forces) requires mechanical dynamics and finite element understanding. The simulation of stress requires mechanics of materials and fatigue theory. Reciprocating compressor pressure volume simulation requires fluid dynamics and thermodynamics. Reciprocating compressor valve simulation requires acoustics, mechanical dynamics and fluid dynamics. The central concept that relates these disciplines is the *dynamic* concept. Of course an understanding of statics is also required, although it is usually not at the heart of most efficient designs. A basic understanding of both statics and dynamics is required to recognize what is necessary in developing and using a particular simulation or model.

The term *simulation* or *model* will be used interchangeably to mean a tool which exhibits similar properties of an actual machine. The simulation is usually based on mathematically analogous processes. Therefore most simulations are mathematical ideas that respond in a similar enough fashion to predict the desired properties of the system to be designed or analyzed.

Another central issue in compressor and piping simulation is the realization that a system is an assemblage of compressor and piping that forms a unified system. Proper simulation must address itself to the system as a whole and not isolate processes which are interactive in the system. Statics and dynamics both influence a machine's performance, therefore they must both be included in an optimized machine design. Specialization that minimizes the overall character of the system, usually detracts from the success of a design effort.

6.1.1 Defining The Overall Task

The task of designing or analyzing a compressor and piping system includes:

- Piping acoustics (from compressor valve to acoustic termination)
- Piping mechanical dynamics (compressor manifold and external)
- Pressure drop analysis (efficiency considerations)
- Compressor valve dynamics (both performance and reliability)
- Compressor performance (cost efficiency)
- Piping mechanical statics (thermal expansion, etc.)

The major point to be made by addressing the overall design task is that the sub-projects are all influenced by each other. Mechanical piping changes can influence the acoustics and acoustics can influence both mechanics and performance. The cost-effective and technically sound acoustical design cannot be performed in a vacuum. The use of concurrent analysis[1] is without doubt the best approach.

6.2 GENERAL MODELING CONCEPTS

Normally, model processes are separated into areas which efficiently exhibit the desired properties. Many models are limited intentionally so that the designer will not make an effort to misuse the model. Therefore, it is not uncommon to see several seemingly isolated simulations being performed which are then applied simultaneously. The use of a simultaneous design philosophy is very beneficial. To this end, we will be illustrating simulations in a focused effort to indicate the nature and use of simulations realizing that they will all be combined in a unified effort to optimize machine reliability and efficiency.

6.2.1 Static Systems

Static analysis in piping systems is usually divided into two areas:

- Static fluid loss associated with pressure drop and fluid dynamic efficiency (fluid related)
- Temperature, weight and pressure forces which determine static integrity (mechanical related)

Pressure drop simulations vary considerably in use and complexity. They are based initially on the fundamental loss mechanism. For pipeline efficiency, this loss factor might be empirical such as Spitzglass, Babcock, Weymouth or Panhandle. The rational method of Darcy is more common in simulations in the last 20 years. The Darcy method is rationally developed from the physical properties of fluids and Bernoulli's general energy theorem. Bernoulli's theorem can be stated as follows:

$$Z_1 + \frac{P_1}{\rho_1} + \frac{v_1^2}{2g} = Z_2 + \frac{P_2}{\rho_2} + \frac{v_2^2}{2g}$$

Z_1 and Z_2 = potential head at condition 1 and 2
P_1 and P_2 = static pressure at condition 1 and 2
ρ_1 and ρ_2 = density at condition 1 and 2
V_1 and V_2 = velocity at condition 1 and 2
g = acceleration due to gravity

All practical formulas for fluid flow are derived from this theorem, with modifications to account for frictional losses.

Mechanically related models dealing with temperature, weight and static pressure forces are usually included in a thermal flexibility analysis. The major static issues are pipe stress, displacement, machinery forces and moments, and cooler nozzle forces and moments. These will be discussed in detail in the section reserved specifically for them.

6.2.2 Dynamic Fluid Transient Systems

The modeling of dynamic flow which is not acoustically related is generally achieved through solutions of the basic equations of energy, motion or continuity, plus equations of state and other physical property relationships. The most popular solution is the *characteristics method* (method of characteristics). This method converts the two partial differential equations of motion and continuity into four total differential equations. These equations are then converted to finite difference expressions using a method of specific time intervals. The resultant computational process is performed in the time domain and can yield very rigorous results. When large intermittent fluid flow problems are solved, this type of approach is necessary. It can also be applied to acoustic problems but is computationally intensive. Emergency shutdown and sudden machinery loading must be analyzed in the time domain using such techniques.

6.3 PREDICTING PULSATIONS, VIBRATIONS, AND STRESS

Pulsation, vibration and dynamic stress can best be understood in terms of a dynamic energy source and systems which can be resonant. Initially, the energy is generated by the machinery (reciprocating compressor). If the piping natural frequencies are frequency coincident, the energy is magnified through acoustic resonance. The unbalanced pressure forces in piping systems couples into the mechanical piping system causing vibration. If the mechanical natural frequency of the piping is frequency coincident with the pulsation energy, secondary magnification results. When large vibrational displacements occur in stiff systems, excessive stress results at the points of stress concentration. If the cyclic stresses exceed the endurance limit of the piping material, fatigue failure results.

6.3.1 Pulsations and Piping Acoustics

Dynamic energy is generated by the compressor in normal operation. The reciprocating process produces intermittent flow and pressure. These flow and pressure variations are conveyed into the gas in the piping. The dynamic energy (both pulsative flow and pressure) first transfers into the gas or piping acoustics. Figure 6.1 illustrates the mass flow versus time waves that commonly occur at compressor valves. Figure 6.2 illustrates the frequency content of the head end discharge flow pulse. It is readily apparent that the frequency content of the pulsative flow is limited to compressor rpm and multiples of compressor RPM. A single compressor end produces decreasing amplitudes moving from the first compressor order (rpm × 1) to the higher multiples. When the front (head end) and back (crank end) ends of the piston are used simultaneously, cancellation and reinforcement of compressor order occurs. Most notably, the odd orders (1×, 3×, 5× ...) tend to be reduced due to cancellation of the two ends. Reinforcement occurs on the even order (2×, 4×, 6× ...). Therefore, double acting compressors cylinders produce strong pulsative flow at even orders. This reinforcement and cancellation occur with significant acoustic involvement (on a single cylinders) due to the relatively close proximity of the head end and crank end valve in the cylinder passage. A much more complex case occurs when multiple cylinders (operating in parallel or series) are connected by piping elements. In such cases, the reinforcement or cancellation of energy

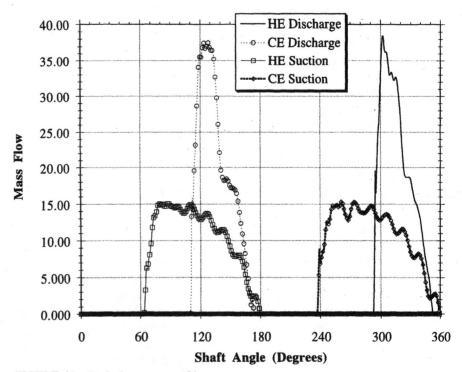

FIGURE 6.1 Typical compressor flow patterns.

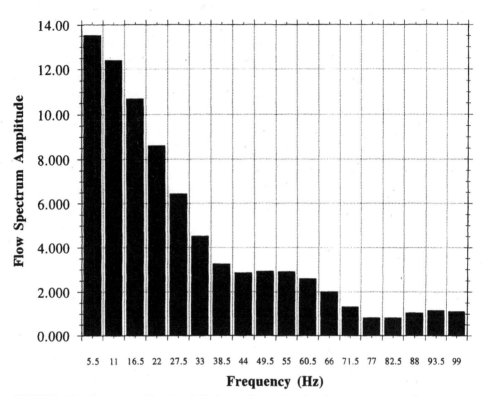

FIGURE 6.2 Spectrum of head end discharge flow pulse showing compressor orders.

occurs due to the crank shaft phase and the piping acoustics between the cylinders. The models to analyze simple systems are almost trivial compared to the level of sophistication required to analyze multiple cylinders with complex piping systems. It is always good to keep in mind the transfer of energy through a system is: cylinder excitation; acoustic transfer and amplification; mechanical transfer and amplification; acoustical to mechanical coupling; resultant shaking force; mechanical vibration; and eventual pipe material strain and stress.

The piping system can be viewed as a complex organ pipe network. The normal piping system will have several acoustic natural frequencies which, if excited, develop standing wave patterns (acoustic mode shapes). As the flow and pressure wave travel out from the compressor, they are transmitted and reflected in the piping system. Whether a wave is reflected or transmitted is determined by the change in impedance from element to element. The simple acoustic impedance (Z) is determined by the gas velocity of sound (c = ft/sec), the gas density (ρ lb/cu ft) and cross sectional flow area (A = sq ft) of the acoustic element.

$$Z = \frac{\rho c}{A}$$

This type of simplistic thinking is actually the basis for more complex models that are used in everyday acoustic analysis.

The design of piping systems related to compressors from an acoustic viewpoint was first developed in 1952. Forty-four years of advances in analytical dynamics, instrumentation, and computer systems have continuously improved the engineer's ability to develop low maintenance, cost-effective and efficient designs. For many years, the only available techniques were based on electro acoustical (analog) or simple mathematical models. With the advent of the desktop computer came digital acoustic techniques. The use of the computer to solve basic acoustic piping calculations was not new. It actually existed on mainframe computers for many years, but the man-to-machine interface was inefficient and cumbersome.

There are many different types of acoustic piping models in use today. The majority of digital models use the transfer matrix method. A fairly complete list of methods would include the following:

- Electro acoustical model (the analog)
- Transfer matrix
- Method of characteristics
- Simultaneous differential equations
- Acoustic finite wave
- Finite difference methods
- Spectral method
- Boundary-integral method
- Impedance methods (linear analysis)

An accurately modeled compressor and piping system requires both time domain and frequency domain calculations. The use of frequency-to-time domain transforms has led to a semi-rigorous approach in the frequency domain appearing to have true time domain interaction when in reality it does not exist. True time domain models include electro acoustic (analog), method of characteristics, or simultaneous differential equation solutions.

6.3.2 Time Domain Models In Reciprocating Compressors

The process of developing a reciprocating compressor and piping design involves the representation of the *compressor cylinder, pressure operated valves* and *a valid acoustic piping model*. The piston motion and the valve action produce a periodic intermittent mass flow from the suction piping and into the discharge piping. It is important to note the discontinuous nature of the flow pattern. If a single flow pulse is converted to the frequency domain, the flow can be viewed in terms of frequency multiples of the compressor speed (rpm/60). The nature of these discontinuous pressure functions results in pressure pulses being produced at the machine speed and multiples of one times machine speed.

The acoustic natural frequencies of the piping system can be excited by the piston pulse causing pressure and velocity magnification. The volumetric properties of the piping tend to introduce a smoothing process to the more severe interruptions characteristic of the opening and closing compressor valves.

The performance of reciprocating compressors can be generally inferred from the internal cylinder pressure and the manner in which it interacts with the pressures outside the suction and discharge compressor valves. The cylinder external pressures can be helpful or harmful to the overall cylinder compression and flow process. It is important to note that the piston motion, mechanical valve model, and outside pressures should be represented in the time domain to allow for proper interaction.

When acoustic standing waves are present in the piping system, they can couple through elbows and capped ends, resulting in significant shaking forces. The major contributor of acoustic shaking force is due to the standing wave which is a by-product of acoustic resonance. Therefore, acoustic resonance has two disadvantages: the amplitude of the pulsative is magnified; and the energy is concentrated in a form that efficiently couples to shaking forces. By limiting or controlling the pulsation amplitude, the coupled shaking force can also be limited. The control of shaking forces reduces vibration that can cause maintenance problems or fatigue failures.

Through design analysis, *non resonant acoustical* and *mechanical* systems can be designed which limit vibration, ensure efficiency and increase reliability of the machine and its piping system.

In simple systems, the design analysis approach can be closed form equations in combination with past successful experience. However, in most cases, the complexity associated with multiple cylinders and extensive piping configurations requires the use of Analog or digital techniques.

6.3.3 Frequency Domain Acoustic Models

The most popular model used in piping acoustics is based on the transfer matrix approach. The development of the equations used in constructing the model follows the following path:

- Plane waves in an inviscid stationary medium
- Plane waves in a viscous stationary medium
- Plane waves in an inviscid moving medium
- Plane waves in a viscous moving medium

Implicit in the development of the impedance in acoustical systems is the recognition of a direct analogy to frequency domain analysis of electrical transmission networks. This is the fact that inspired the first acoustic piping design tool which dominated piping design for many years, and continues to hold considerable advantage compared to existing digital computer applications. The use of inductors (coils), capacitors and resistance forms the basic analogous components which relate directly to fluid mass property, fluid resiliency and fluid resistance. The mass flow and fluid pressure are directly analogous to electrical current and electrical voltage. Even today the pro and cons of "analog" versus digital continues to be a matter of much debate.[2] The real test of any model is its ability to produce faithful results that allow a knowledgeable piping designer to produce safe efficient compression systems.

6.3.4 Piping Mechanical Models

The single most important factor in dynamic mechanical models is determination of accurate natural frequency calculations. Accurate natural frequency calculations allow for the proper separation of pulsative energy (both incident and resonant) and mechanical natural frequencies. Should coincidence occur, mechanical resonance results and almost certain problems will ensue. High vibration due to resonance can result in one or more of the following problems.

- Loosing nuts and bolts associated with valves, piping restraints or other bolted elements. This results in general high maintenance cost.
- Vibration induced fatigue of smaller lines such as instrument lines
- Vibration induced fatigue of major piping elements

Where piping is the primary moving element, the vibrational mode shapes are dependent on the model possessing the distributed stiffness and mass properties of the pipe. When valves or concentrated masses are present, it is also important that these elements have specified rotational inertia properties. It is very important that restraints are modeled with proper stiffness values. As a general statement, the mass and stiffness distribution and magnitude must be properly modeled to ensure accurate natural frequency calculations. A knowledge of vibrational mode shape can help in determining when a piping geometrical configuration is susceptible to pulsation energy. The transfer of pulsation energy to the mechanical system generally occurs due to area coupling at pipe closed ends and piping elbows. In piping systems with pulsation energy, the more elbows the greater probability of a vibration problem.

6.3.5 Compressor Immediate Mechanical Analysis

Reciprocating manifold systems are composed of crosshead guides, distance pieces cylinders, cylinder supports, suction nozzles, discharge nozzles, suction manifold bottles, discharge manifold bottles, discharge bottle restraints and attached piping on suction and discharge. The vibration patterns associated with such elements are most accurately viewed as lumped masses connected with generally massless springs. Therefore, the approach required to model a compressor manifold is quite different than that required to model the distributed properties of pure piping systems. Calculating proper natural frequencies and mode shapes for compressor manifolds requires a very specialized understanding of such elements as:

- Crosshead guide flexibility
- Distance piece flexibility
- Cylinder rotary inertia
- Nozzle branch connection flexibility
- Manifold bottle rotary inertia
- Discharge bottle restraint added stiffness
- Attached piping dynamic effects

6.3.6 Dynamic Stress Models

Dynamic stresses can be calculated in both piping systems and manifold systems with proper attention to the element properties and the forces and moments at each end of the element. In most cases, a finite element type of approach can be used to calculate the dynamic stresses. Experience has shown that a distortion energy theory algorithm correlates well with practical field failure experience. The improvements of the distortion energy theory over the total strain energy theory account for the experimental observation that hydrostatic states of stress must be properly assessed. The later contributions of Von Mises and Hencky have led to the best overall techniques. The effort associated with a complete FEA analysis is not necessary and would be prohibitive (if performed correctly) from a time and cost viewpoint. Practically, the most conservative and reliable dynamic stress criteria is to simply ensure the maximum dynamic peak-to-peak stress is less than 6,000 psi. This accounts for worse case mean stress, stress concentration, surface effect and size effects.[3]

6.4 RECIPROCATING COMPRESSOR PRESSURE VOLUME ANALYSIS

Compressor system models are composed of both the compressor and the piping system. Therefore, when a compressor simulation analysis is performed, a PV (pressure vs. volume) and PA (pressure vs. crank angle) display of the cylinder internal pressure is available. Figure 6.3 illustrates the PA display along with actual pressure levels at the suction and discharge valves. Figure 6.4 illustrates a typical PV card. The advantage of this PV card is the inclusion of the pulsation effects. Ideal PV calculations yield four basic components.

- Suction volumetric efficiency
- Discharge volumetric efficiency
- Compression line
- Re-expansion line

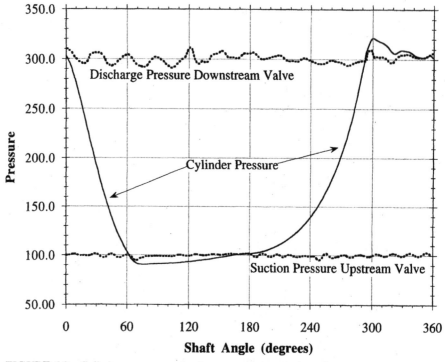

FIGURE 6.3 Cylinder pressure and valve pressure vs. crank angle.

During the period of time the suction or discharge valve is open, the internal cylinder pressure is influenced by the pressure beyond the valve in the piping. The changes in acoustic impedance cause pulsative energy to reflect back upon the compressor valve and to actually enter the valve port. This influence is very significant. The nature of the pressure profile on the PV card during these time periods is very similar to the pressure immediately outside the valves. A primary influence is on the area of the card which is proportional to the work performed for each rotation of the shaft. The work combined with the rpm yields the horsepower of the compressor. High frequency pulsative energy tends to produce numerous waves during the inlet or outlet flow time. Low frequency pulsative energy tends to cause the PV card to balloon or swell. A ballooning card usually suggests the horsepower is increased with a corresponding increase in flow. Therefore, the efficiency of the compressor is deteriorated. The compression and re-expansion lines can also be displaced, causing very significant increases in required horsepower with a small increase in flow. Displaced compression and re-expansion lines in many cases are symptoms of increased valve impact velocities and limited valve life.

6.5 *VALVE MOTION MODELS*

The suction and discharge valve motion is determined by the dynamic properties (mass, stiffness and damping) of the valve elements and the differential pressure

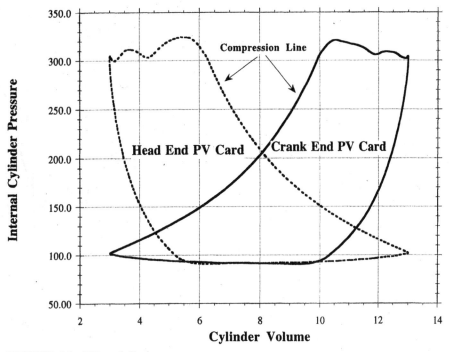

FIGURE 6.4 PV card display.

across the valve. The differential pressure and the effective pressure area determines the force that operates the valve. The differential pressure is composed of both static and dynamic components. The valve motion cannot be properly predicted without including the pulsative energy present in the system. Figure 6.5 illustrates the pulsative pressure at the exit of the discharge valve and also in the common nozzle. This data shows the complex nature of pulsative energy and why this would surely influence the valve motion. The energy content is quite different as you move from the valve exit to the common nozzle. Figure 6.6 illustrates the spectral content of the energy in the nozzle. This spectral energy content shows the dominance of the basic double acting cylinder (dominate rpm × 2 energy) and acoustic response associated with the cylinder internal passage at approximately 64 to 69 Hertz. This is typical and illustrates the requirement of the modeling process.

An adequate compressor model will include a mechanical valve model coupled into the driving pulsative energy and the open and closed limits of valve element travel. It is important that this model be evaluated in the time domain. The results of the model should yield valve spring and weight parametric analysis capabilities. The valve displacement, velocity and acceleration are directly available, allowing for direct evaluation of impact velocities and forces. At present, several valve manufacturers have impact velocity criteria which are used to screen valve reliability. These criteria have not proven totally reliable up until now, and are used as a simple criteria which should not be over emphasized.

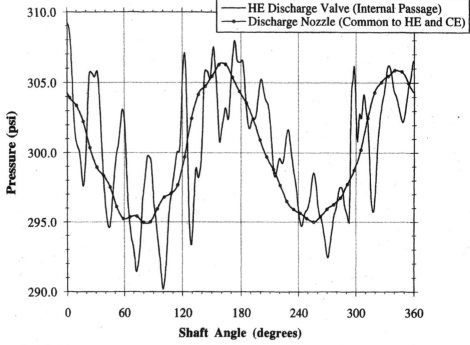

FIGURE 6.5 Pulsation at valve and nozzle.

Ultimately, the dynamic valve simulation should perform accurately enough to allow for optimization of the valve motion so that the valve operates ideally.

6.6 THERMAL FLEXIBILITY MODELS

The modeling of thermal expansion and piping flexibility has grown considerably since its introduction with the Mare Island Project. Several commercial programs have been developed over the years to calculate pipe stresses, displacements, forces and moments on piping systems as well as external loads on machinery components. The most important aspect of thermal, pressure and weight modeling is to properly define the end conditions or limitations of such elements as stops, supports, hangers, clamps and anchors. The common but false assumption that using large values will ensure a conservative design is simply not true. Where piping systems are large and interactive, the restraining effects actually determine if a model is valid or simply an educational exercise. Here again the real value of static analysis is dependent more on engineering expertise than on the model per se. This is true since most of the commercial models are adequate from a technical viewpoint.

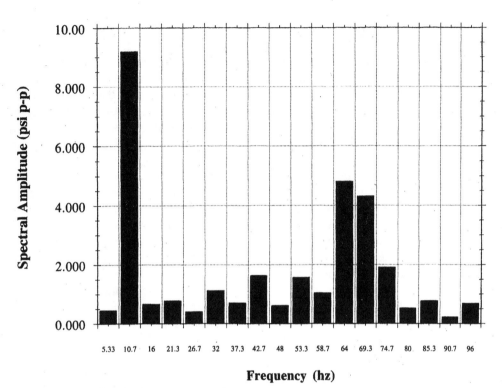

FIGURE 6.6 Pulsation spectrum at nozzle.

6.6.1 Piping Thermal Effects—Forces, Moments, Stresses

Depending on the program used, several force drivers can be included. A list might include the following:

- Thermal expansion
- Pressure
- Thermal bowing
- Weight
- Wind
- Earthquake
- Support and restraint displacement
- Restraint friction
- External loads

Such forces are combined to determine total forces which, through the geometry of the system, result in moments. These forces and moments induce stress in the piping which is compared with various piping codes in the generation of a code compliance report. The codes also specify the calculation of stress intensification to be used in the compliance report.

In thermal flexibility analysis, it is important to point out one very common modeling procedure which yields incorrect results. It is common to model extremely high stiffnesses (anchors) as boundaries of the system and also to simulate the effect of piping restraints. Stiffness characteristics of piping restraints such as base elbows, piping clamps and directional guides should be properly evaluated and applied to the flexibility model. This is important to achieve proper results and also to allow for economically viable designs. The use of large or infinitely rigid values is defended on the basis that it is always conservative. This may be true regarding calculated stresses, but in terms of economics, mechanical stability (susceptibility to vibration) and overall design quality, it is often not conservative. Such modeling techniques often result in unnecessary design changes (i.e. additional elbows, expansion loops, deletion of dynamic restraint) which give the system excessive flexibility. Without the use of proper restraint flexibilities, most analyses are of questionable value.

As with most models, it is important to understand that the results are based on the ideal behavior of a perfectly constructed computer model. In addition, numerous assumptions are made during the modeling process, all of which affect the calculated loads. Therefore, computer models used in this type of analysis should serve as design tools to provide general characteristics of a piping system and to avoid significant thermal-related problems. The application of codes for pipe stress and compressor loading are meant to serve as guidelines, and strict adherence to or violation of these codes neither guarantees success or failure of a piping system.

6.6.2 Forces and Moments on Machinery

The forces and moments due to the static piping effects are combined according to codes or manufacturer's guidelines to determine the forces and moments on the machinery component (centrifugal compressor case or reciprocating cylinder). The calculated values are then compared with allowable values to determine if the piping system is exerting an excessive load on the machinery. Large forces and moments on high speed rotating equipment tend to produce case distortion and misalignment. Such distortion produces excessive wear and bearing failures as well as excessive vibration at rpm and rpm $\times$ 2. Where such analyses are performed, all legitimate flexibility should be included to ensure the forces and moments are reasonable and consistent with actual conditions. It is advantageous to understand that the most commonly overlooked property is hidden flexibility. The actual systems tend to be more flexible than expected. In most cases this tends to produce lower forces and moments, but there are some exceptions.

6.6.3 Forces and Moments on Coolers

The application of thermal modeling to predict forces and moments on cooler nozzles has become a common practice. There are several factors which must be considered to ensure proper modeling yields relevant results. In cases where oversimplified assumptions are made, the risers to coolers are made extremely flexible and become candidates for severe vibration when significant pulsative energy is present. The proper balance of static and dynamic considerations is the only real assurance of true reliability.

6.6.4 Buried Piping Models

Buried piping models have been developed based on soil properties and mechanics. The soil stiffness is simulated through the use of discrete restraints distributed along the buried pipe. In some cases the restraint distribution is evenly spaced, and in other cases the distribution is variable. The ability to analyze buried pipe is a very desirable capability and is much needed due to the fact that a great deal of piping is buried. Although, these models are a valiant effort to simulate soil, they have not been totally proven in terms of accuracy. Perhaps the most obvious question arising in the mind of the analyst is the assumption that soil is homogenous in nature. Common knowledge suggests that soil components vary greatly over a given area and depth. Therefore, the results derived from such models should be closely monitored and evaluated to ensure the results do not violate common engineering understanding.

6.7 REFERENCES

1. Drummond, Rick, "Concurrent Analysis of Compressor Piping Systems," Southwest Research Institute, San Antonio, Texas, 1994.
2. Blodgett, Larry E., "Using Analog and Digital Analysis For Effective Pulsation Control," American Gas Association, Operations Conference, Operating Section Proceedings, 1996.
3. Young, Warren C., *Roark's Formulas for Stress and Strain* New York, N.Y.: McGraw-Hill, 1989).

CHAPTER 7
VERY HIGH PRESSURE COMPRESSORS (over 100 MPa [14500 psi])

Enzo Giacomelli
General Manager Reciprocating Compressors
Nuovo Pignone

Alessandro Traversari
General Manager Rotating Machinery
Nuovo Pignone

7.1 DESIGN PROCEDURE

7.1.1 General Information on High Pressure Services

The mechanical difficulties connected with high pressures were encountered by compressor manufacturers during the development of ammonia synthesis processes after 1920, as pressures of about 100 MPa (14500 psi) were reached.

Subsequently, the ammonia synthesis process was developed by reducing the final pressure and today's values are around 32 MPa (4640 psi). Other applications such as Urea production required pressure up to 35 MPa (5075 psi), but today, due to improvement in the process, reaction can be obtained at pressures of 15 to 20 MPa (2175-2900 psi). The storage and reinjection of natural gas also use high pressures ranging between 15 and 60 MPa (2175 to 8700 psi). Each application has particular difficulties connected to gas compression, liquid carry over, lubrication and corrosion.

The use of high pressure processes increased after World War II with industrial demand for low density polyethylene (LDPE),[1] whose polymerization is achieved by bringing the gas up to 350 MPa (50750 psi) by using special types of reciprocating compressors (Fig. 7.1). Compressors in this service have been given the name *hypercompressor*.

FIGURE 7.1 12 cylinder compressor for very high pressures (*courtesy of Nuovo Pignone*).

Plant capacities have considerably increased. In the early 1960s, the normal production per single line was on the order of 7,000 t/yr,* while lines producing 200,000 t/yr LDPE are now in operation.

These plants are dangerous in case of erroneous or unsafe operation, because of the high flammability of the gas and the very high pressures involved in the production process.

One of the sources of risk is to be found in the secondary hypercompressor, where two main areas called for special care in the design and manufacturing stages: parts subjected to pressure and the crankgear.

The cylinders are subject to pressure fluctuations, which can cause fatigue failure if design, manufacture and material selection are not adequate.

The crank mechanism, driving the plungers, must perform its task with great accuracy, because any misalignment might cause failure[2] with possible risk of fire. Plungers are in fact brittle items owing to the very hard metal used, generally solid tungsten carbide.

As a consequence, the design of these machines must be based on sound fundamental choices[3] and supported by the most up to date analysis methods and experimental techniques.[4]

---

*The abbreviation t denotes the metric ton of 1,000 kg.; yr = year.

Polyethylene types, differing in molecular weight, density, and degree of crystallization, can be divided into two categories depending on the process used to polymerize the ethylene:

- High Density Polyethylene (HDPE) is obtained with a low-pressure process, working at pressures below 5 MPa (725 psi) and giving a product with density ranging from 0.945 to 0.970 g/cm^3 (.034–.035 lb/in^3).
- Low Density Polyethylene (LDPE), produced by high-pressure processes, operating between 100 and 350 Mpa (14500–50750 psi), gives a product with a little lower density (0.91 to 0.93 g/cm^3) (.033–.034 lb/in^3).

The introduction of low pressure processes producing Linear Low Density Polyethylene (LLDPE) with operating pressures below 2 MPa (290 psi) began to erode the sectors served by LDPE, reaching today plant capacities up to 500,000 t/yr of polyethylene.

The possibility of obtaining linear polyethylene with high pressure (HPLLDPE) was then investigated and practical applications were found in some existing plants with final pressures from 80 to 170 MPa (11600 to 24650 psi), in order to provide some operating flexibility. Since the polymerization of these products requires the machine to operate in very difficult conditions, the process is now applied only in existing plants.

Although extremely good results were obtained with plastic sealing elements on the packings, the difficulty of controlling the process, in general, the severe lubricating conditions and the presence of aluminalkyle were resulting in unreliable performance.

About 50% of the new plants for Low Density Polyethylene today are built to use compressors to bring ethylene to high pressure and then subject the gas in a reactor to temperature and a catalyst. The polymer formed has good mechanical and optical properties. Furthermore, these processes have no environmental impact, as they contain no metals and involve only ethylene and energy.

This technology is simple to use, competitive and offers possibilities for development of catalysts such as the "metallocenic" ones, well fitting with homogeneous systems (like the high pressure LDPE).

At present, high pressure polyethylene production covers the following sectors:

Product	Density g/cm^3 (lb/in^3)	Gas	Pressure MPa (psi)
LDPE	0.920 (.033)	ethylene	120–350 (17400–50750)
Copolymers 1	0.927–0.935 (.033–.034)	ethylene + vinylacetate (5 to 40%)	120–350 (17400–50750)
Copolymers 2	0.927–0.935 (.033–.034)	ethylene + acrylates	120–350 (17400–50750)
HPLLDPE	0.880–0.920 (.032–.033)	ethylene + butylene (10 to 40%)	80–150 (11600–21750)

Although low pressure processes are continuously improving, with the help of special catalysts, the high pressure process will still be needed in future.

7.1.2 Features of the Machines for LDPE Services

These pressures are reached by reciprocating compressors, having special design and using advanced materials such as NiCr-Mo steels vacuum melted and sintered tungsten carbide.

For safety and reliability of operation, the following aspects need to be faced:

- The fatigue of the components subject to fluctuating pressure between suction and discharge
- The sliding seal, between piston and cylinder
- The definition of a crankmechanism capable of giving perfect linear movement of the pistons
- The risks connected with the gas in the polymerization of ethylene

The proper design of very high pressure machines requires investigation of thermodynamic, mechanical, process, operational and safety aspects.[5] As an example, the motion work (Fig. 7.2), should drive the plungers with absolute precision.

The crosshead movement should be perfectly straight and coincide with the cylinder axis under both cold and warm conditions. The thermal expansion of the crosshead should not affect alignment. Both planes of the vertical and horizontal guides should pass through the centerline of the plunger, which is centered to the crosshead without clamping and is supported by the packing bush in the cylinder. The cylinders (Fig. 7.8), which are directly connected to the frame, require particular care in the design of parts subject to gas pressure and of the seal arrangement.

Poppet type valves should be sized to minimize fluttering phenomena that would damage both the seat and the poppet.

Solid tungsten carbide or plated plungers are used with bronze or very special plastic packing, depending on the process and pressure.

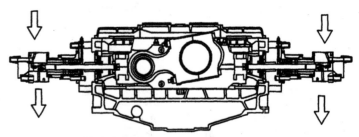

FIGURE 7.2 Special crank mechanism with self-aligned high pressure cylinders directly connected to the frame.

For reasons of safety, a distance piece is inserted in the cylinder arrangement. A first recovery vent connected to flare or safe area, an oil barrier used to cool the plunger, and an emergency vent allow for proper gas leakage control without entering into the crankcase. Should gas enter the frame in emergency conditions, relief valves are provided. Nitrogen purging of the crankcase to eliminate oil vapor mixtures ensures safe operation.

7.1.3 Crankgear Arrangements

For structural reasons, it is preferred the cylinders be of the single acting type. The thrust mechanism of each plunger is subjected to a unidirectional load with difficulty of lubricating the load bearing connections, where surfaces always remain in contact, without the possibility of establishing an oil film (for example, between the small end of the connecting rod and the crosshead pin.) In the design of the crank-mechanism for polyethylene compressors, various compressor manufacturers have used different solutions to resolve the problem of the linear transmission of movement and the unidirectional load.

- The conventional type of crosshead drives through a spherical coupling an auxiliary crosshead (Fig. 7.3), in order to reduce the oscillation transmitted to the plunger, but does not provide reversal of load on the small-end bearing. This bearing must be sized with low specific loading to allow for conditions of precarious lubrication and the transient state between liquid film lubrication and dry friction.
- An external auxiliary crosshead (Fig. 7.4) is driven by the main crosshead through tie bolts. Opposed plungers are attached to this auxiliary crosshead (yoke). With this construction, the thrust reversal is met and oscillation due to the low friction force acting on it is reduced by the auxiliary crosshead. The problem of thermal expansion might still exist, but the solution for this is to locate the main and auxiliary crossheads in different frames, at some distance apart. Accessibility of the inside cylinder may still be somewhat restricted in this type machine.

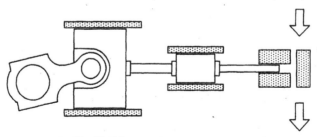

FIGURE 7.3 Crankgear using conventional and auxiliary crossheads.

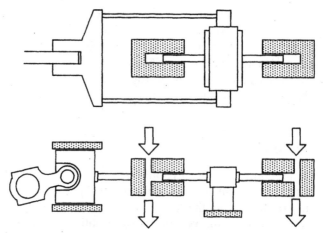

FIGURE 7.4 Crankgear with main crossheads connected to auxiliary crossheads in separate frame.

- Another solution is to utilize a hydraulic device for the transmission of force to the piston. This is able to provide capacity variations, give perfect linear movement to the pistons, and also may have a complete oil control circuit.
- Two special crossheads (Fig. 7.5), one directly attached to the connecting rod, the other positioned on the opposite side of the crankshaft, connected by tie bolts, satisfy the thrust reversal requirement. Auxiliary crossheads are required to avoid transmitting to the pistons oscillations and to absorb the thermal expansion of the main crosshead. This expansion is a function of the distance between the crosshead pin and sliding surfaces and causes a displacement between the axis of the crosshead with respect to the axis of the cylinder.
- Another special arrangement (Fig. 7.6) has a crosshead, with sliding surfaces and guides in the lower part of the frame. Two connecting rods are used on each side. Thrust reversal is ensured by having two plungers on opposite sides of the main crosshead. In order to obtain adequate linear movement during operation, an auxiliary crosshead is necessary to eliminate the thermal expansion resulting from the distance between plunger and main crosshead guides.

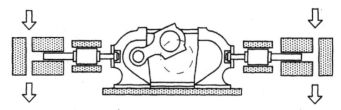

FIGURE 7.5 Crankgear using two special crossheads with tie bolts and auxiliary crossheads.

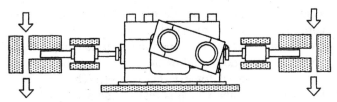

FIGURE 7.6 Special crosshead with two conrods and with auxiliary crossheads.

- A special and precise solution (Fig. 7.7), with two opposed plungers, is driven through a special crosshead, within which rotate crankshaft and connecting rod. The crosshead guides are plain and their axis is placed on the centre-line. The distance between crosshead slides is long, with a small clearance, and thus misalignment due to thermal expansion is avoided as the plunger axis and the slides are on the same plane. This patented solution[5] meets the above requirements most satisfactorily.

As a result, the movement of this crosshead is perfectly straight-line and coinciding with cylinder axis under both cold and warm conditions. In fact:

- Pitching is mechanically prevented by the very small clearance ratio between guide and wing and distance between the guides and also because the friction force on the slides practically coincides with the gudgeon pin axis;
- Thermal expansion of the crosshead does not affect alignment, as the two planes of the vertical and horizontal guides pass through the plunger center-line.

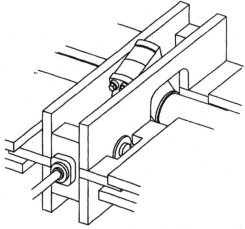

FIGURE 7.7 Special crosshead self-aligned assembly, adopted for secondary compressors in LDPE Plants.

This solution avoids the use of auxiliary crossheads which are necessary when a conventional main crosshead is used, to correct its rocking movement. Auxiliary crossheads make the machine longer, more complicated, more difficult to align and, above all, they do not allow the cylinders to be directly connected to the frame, to eliminate the need for external support. This condition is in fact fundamental to building a machine which is really "factory aligned" and needs no realignment in the field.

The last two solutions are the only ones still applied in new plants.

7.1.4 Characteristics of Cylinders

As they work at very high pressures, hypercompressors should incorporate features to make them safe, reliable, and efficient.[5]

The cylinders, connected by long hydraulically-tightened tie rods, consist of three major parts: a flange connecting the cylinder to the motionwork frame, the main packing housed in a cartridge assembly, and the cylinder head containing suction and discharge valves.

The cylinder with radial valve arrangement (Fig. 7.8) is a typical structure with a compression chamber, and a plunger with packing for gas sealing. Packing cup geometries are complex, as housings are provided for guide bushes and seal rings, and also because of the presence of lube oil holes.

Similar cylinder arrangements are used with high pressures over 230–250 MPa (33350 to 36250 psi), where the cylinder head will have axial valves (Figs. 7.9 and 7.10).[36]

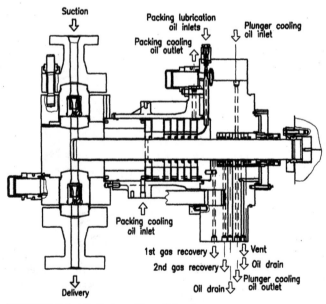

FIGURE 7.8 Cylinder with radial valves.

VERY HIGH PRESSURE COMPRESSORS **7.9**

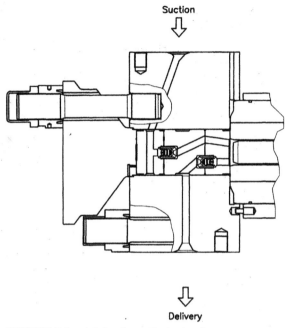

FIGURE 7.9 Axial valve cylinder head.

Construction with vertical or radial pipes-inlet and outlet (Fig. 7.9)-allows better fastening to the foundation in order to obtain a very low vibration level. An arrangement with axial and radial pipe arrangement (Fig. 7.10) is also used in relation to plant layout requirement.

Valves of radial type (Fig. 7.11) are used up to 250 MPa (33350 psi). Axial combined suction-delivery valves cover the entire operating pressure range with

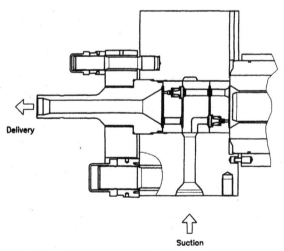

FIGURE 7.10 Multipoppet axial valve cylinder head.

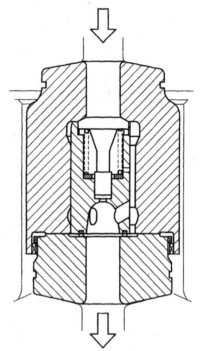

FIGURE 7.11 Radial valve assembly.

monopoppet (Fig. 7.12) or multipoppet (Fig. 7.13) solutions. All valves are of the poppet type with gas velocity usually between 30 and 60 m/s (98 to 197 ft/sec). Experience indicates outstanding performance of radial valves with over 30,000 hours of operation, due to the following advantages:

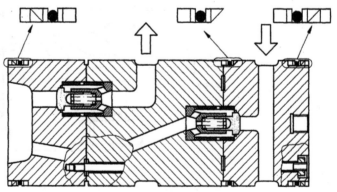

FIGURE 7.12 Axial combined valve assembly.

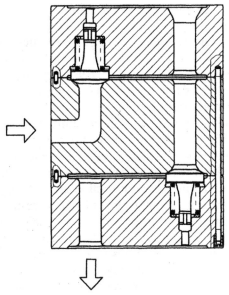

FIGURE 7.13 Axial multipoppet valve assembly.

Greater reliability

- A very simple valve structure
- Less possibility of polymer formation since the gas velocity is a little higher
- Very low sticking effect on the poppet during operation because of higher clearance and higher drag force on the poppet
- Larger diameter of springs gives greater endurance

Better sealing

- The machining of the seats is more difficult on multipoppet valves
- The higher clearance on a monopoppet valve allows for a better contact between seat and poppet with less possibility of damage

Monopoppet valves have greater allowance for reconditioning after long operation and are easy to assemble.

On the axial valves, special seals separate the two sections from each other and from the outside. This solution serves a dual purpose by:

Allowing precompression of the valve bodies with pressure outside them, so that their tensile stresses are minimized.

Eliminating high stress fluctuations in the cylinder head at the suction and discharge crossbores.

The areas of heads subjected to high stress work in a precompression state, when "autofrettage" is applied.

Two-piece construction for compression chamber and packing cups is obtained by shrink-fitting to precompress the most highly-stressed areas.

The cylinder heads have large holes intersected by lateral ducts for gas suction and discharge. Cylinder design is important in order to attain the goal of safety-reliability-performance.[6] It is necessary to solve problems connected with high fatigue stresses due to high fluctuating pressures and the variations in geometries causing stress raisers.

Variations in geometry are minimized to reduce the effects of stress concentration. Surface roughness is very important from the fatigue point of view and special machining procedures are needed to obtain very high levels of finish. In case of a cylinder head with radial valves, where the whole crossbored cylinder is subject to pressures varying between suction and delivery, stress fluctuations must be evaluated at the critical points and adequate prestressing applied.

The highest pressures require a solution with a combined suction-discharge valve, enabling reduction of stress concentration effects on the head by setting the intersection points at places where pressure fluctuation is low.

In a cylinder for very high pressures, the material must also be efficiently exploited by the use of prestress methods such as shrink-fitting and autofrettage, which ensure more uniform stress distribution across the cylinder wall. These two methods may also be applied together when a high compressive prestress level is necessary in order to operate safely.

7.1.5 Safety Aspects

Safety is of great importance in LDPE plants on account of the flammability of the gas and the high pressures involved. The discharge temperature of the gas is limited today to 100 to 110°C* to reduce the risk of dissociation of ethylene in the cylinder, while in the past 120°C was reached in single stage compressors. In this case, to ensure low final temperature, the gas was cooled before entering the cylinder.

As a further precaution, the plant is usually installed in the open air, sometimes under roofing, or in well-ventilated enclosures to prevent any accumulation of gas due to accidental leakage.

For the safety of personnel, it is the practice of engineering and manufacturing companies to position the instrumentation and control equipment at some distance from the most hazardous parts of the plant, and to ensure appropriate management and use of modern monitoring and automation systems.[8] Centralized control and operation of the plant also allows optimization of process parameters and shorter periods of transient operation, with even more outstanding performance of the whole plant.

Additional benefits derive from the application of monitoring systems with maintenance based on predictive rather than preventive criteria.

---

*In 1st stage 130°C shutdown, 115°C alarm. In 2nd stage 120°C shutdown, 100°C alarm.

With modern materials and design techniques, reciprocating compressors including very high pressure units now reach very high safety levels with reliability and availability factors over 99.5%.

Special safety concepts are applied to compressors, as the plungers are very brittle and misalignment may produce failures. An intrinsically safe crank mechanism is of prime importance.[7] The plungers must always be kept free to rotate and to move axially and a special connection to crosshead (Fig. 7.14) is required.

Furthermore, the cylinder bolts must be long enough to permit elastic elongation in case of decomposition or overpressure in the cylinder. The head is lifted by the elastic stretching of the tie rods and the gas is released through special relief channels in the head itself. A separation chamber between cylinder and frame is designed to vent any excess ethylene leakage from packings to the atmosphere, thus preventing gas from entering the frame. Nitrogen purging is applied to the frame as in internal combustion engines. The material used for high pressures zones is selected ductile enough to "leak before breaking."

7.2 STRESS CONSIDERATIONS

7.2.1 Heavy-walled Cylinders Under Pressure

Typical arrangements of high-pressure cylinders of the plunger and packing types (Figs. 7.8, 7.9, and 7.10) consist of cylindrical bodies, nearly symmetrical, such as

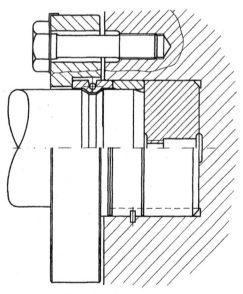

FIGURE 7.14 Plunger—crosshead special connection.

the cylinder head, where suction and delivery valves are housed, formed by the intersection of the central duct and the lateral ducts feeding the valves, the compression chamber and the packing cups.

As all these areas are under pressures ranging from the value of suction to that of delivery, they are susceptible to fatigue. For the highest pressures, a solution utilizing a combined suction and discharge valve is employed. Stress concentration in the head is reduced by arranging these points of intersection at a position of low pressure fluctuation (Figs. 7.9 and 7.10).

At the design stage, the target is to simplify the shape by reducing it as nearly as possible to a cylindrical form or, to the intersection of cylinders. This facilitates calculation and tends towards a safer design.

Thick cylinders under internal pressure have distribution of stresses in the wall such that the radial stress (always in compression) has a value equal to the internal pressure and decreases to zero at the outside diameter, while the circumferential stress (always tensile) has a maximum value at the inside diameter and decreases to a value which remains above zero. Circumferential and shear stress cannot be reduced below the value of the internal pressure at the inner bore.

This distribution is a disadvantage, as a small part of the cylinder wall is under high stress, the outer fibres being less loaded as evidenced by the trend of the shear stress per unit pressure $(\sigma_c + p)/p$ versus the OD/ID (outside diameter/inside diameter) ratio K (Fig. 7.15). For values of K in excess of 2.5 there is little advantage in increasing the wall thickness, and over 5 the inside ideal stress is practically that of a cylinder with infinite wall thickness.

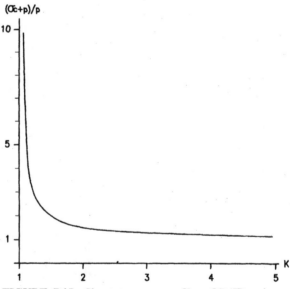

FIGURE 7.15 Shear stress versus K = OD/ID ratio on heavy-wall cylinders.

To design cylindrical bodies for very high pressures, it is necessary to utilize the material more efficiently by using prestressing methods such as shrinkage or autofrettage, which result in a more uniform stress distribution across the wall.

The simplest method of prestressing is compound shrinkage. It makes it possible to have the critical parts working at points that are more favourable on the endurance limit diagram, while the stress range will remain unchanged, depending on the total radial thickness only.

Tungsten carbide liners (with a normal modulus of elasticity about three times that of steel) have been used inside the cylinder. These offer resistance to fatigue when the tungsten carbide is prestressed so that it is always in compression even under the most severe operating conditions.

This construction was applied with piston rings as sliding seals, between piston and cylinder (Fig. 7.16). Tungsten carbide has a low friction coefficient, great resistance to wear, and good resistance to corrosion. It is a suitable material for sliding seal elements under heavy contact pressure.

The application of brittle wear resistant materials is important, although they are undesirable if highly stressed in high-pressure equipment because of the risk of fragmentation if they should fail.[9] The failure criterion for brittle materials is the maximum tensile stress, i.e., when the liners are subjected to internal pressure greater than external pressure, the circumferential stress at the inner diameter has to be considered.

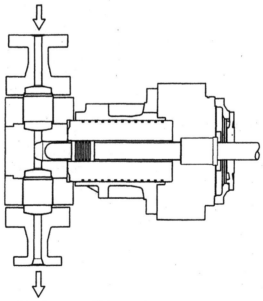

FIGURE 7.16 High pressure cylinder with WC liner and piston rings.

The state of stress in any cylindrical body is worsened by the presence of radial holes, necessary in the cylinder for the suction and delivery of gas (in the heads), as well as for bringing lubricating oil to the sealing elements in the packing cups.

In the intersection of the radial holes with the bore, the stress system[10] is composed of:

σ_c = circumferential stress at intersection of radial hole, with the bore due to the internal pressure in the bore and within the radial hole

σ_a = axial stress due to preload on the cylinder and effects of internal pressure within the radial hole

σ_r = radial stress caused by the internal pressures acting on the points of intersection of the radial hole and the bore

The presence of cross-bores and the consequent stress concentration has a strong effect on the fatigue strength of cylinders. To reduce the risk of failure, the material has to be prestressed, inducing compressive stresses by shrinkage or autofrettage.

The latter process consists of stressing one part of ring beyond the yield point of the material. A greater plastic deformation occurs on the inside of the cylinder. When the autofrettage pressure is released, a compressive stress results in the inner zone and a residual tensile stress in the outer part. This method is preferred to shrink-compounding in the presence of crossbored cylinders such as those used for high-pressure cylinder head and packing cup.

Some tests[11] made on a cylinder with a ratio of 2.25 between OD and ID diameter, with the cylinder wall overstrained up to the geometric mean radius, showed it to be as strong after autofrettage as non-autofrettaged cylinders without stress concentration. These results were obtained after applying pressure at least one million times. It is possible to use shrinking and autofrettage, thus obtaining the advantage of a more uniform stress distribution across the total cylinder wall thickness, ensuring a safer component under working conditions.

7.2.2 General Design Criteria

Most of the analytical design considerations on the components of the cylinders are based on the traditional equations of stresses and strain related to a cylinder under pressure. Therefore, the formulas needed to obtain the stress level and the determination of the point of failure can be found in the contributions of many authors.

- Materials used in high pressure services are normally steel alloys with mechanical properties ranging from 1000 to 1500 MPa (145000–217500 psi) UTS (Ultimate Tensile Strength).
- Selection of materials is generally related to the stress level, as the corrosion effects are negligible, apart from very special applications, where stress corrosion may occur.

- When high stress levels are involved, Vacuum Remelted (Electroslag or Vacuum Arc Remelted) or superclean steels have to be used.
- These features are very important because the inclusion level is very low and the risk of fatigue failures is strongly reduced. The characteristics of such materials with relevant endurance limits are reported on Fig. 7.17a and 7.17b.

The design of cylinder components has to make reference to thick walled cylinders under pressure, with the use of shrink-fitting construction,[11] autofrettage technique,[12,13,14,15] often applied to cross-bored cylinders[16,17,18] generally under fatigue.[19,20,21]

The initial contact pressures, between the various components of the cylinders, arising after preloading of the main bolts, should have a safety factor of 1.2 against yield stress of the material. The preload is usually established by assuming contact pressure on two adjacent parts will come to a level 10 to 20% above the internal pressure.

Stress risers have to be accounted for when holes or changes in geometry are involved. The stress concentration factors are generally applied to the range part at the stress, without considering the mean stress level.

Various methods could be applied to determine factor or safety. The basic approach is to compare in a consistent way the stress level in operation with an allowable value, usually endurance or yield strength, for the material being used.

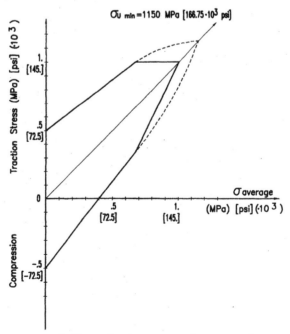

FIGURE 7.17a Traction-compression Smith diagram for NiCrMo steel.

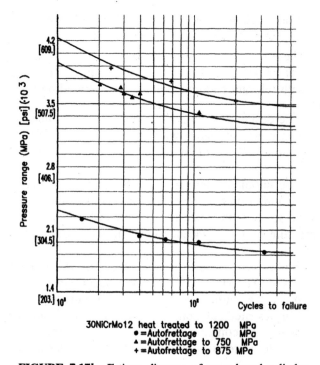

FIGURE 7.17b Fatigue diagram of cross-bored cylinder ($K = 3$) with and without autofrettage, tested with internal pressure.

A combined stress calculated for a cylinder component is usually based on the Von Mises criterion since almost always a triaxial stress state is generated by the complex loading of the cylinder components.

Considering the fact that cyclic stress is made up of *mean* (*m*) and *range* (*R*) stresses, the conditions related to suction and discharge are reduced to a mean stress and a range stress and then combined in accordance to the criterion of Von Mises

$$\sigma_{mises\ m} = (1/\sqrt{2})\ [(\sigma_{c(\text{mean})} - \sigma_{r(\text{mean})})]^2 + (\sigma_{r(\text{mean})} - \sigma_{a(\text{mean})})^2 \\ + (\sigma_{a(\text{mean})} - \sigma_{c(\text{mean})})^2]^{1/2} \quad (7.1)$$

$$\sigma_{mises\ R} = (1/\sqrt{2})\ [(\sigma_{c(\text{range})} - \sigma_{r(\text{range})})]^2 + (\sigma_{r(\text{range})} - \alpha_{a(\text{range})})^2 \\ + (\sigma_{a(\text{range})} - \sigma_{c(\text{range})})^2]^{1/2} \quad (7.2)$$

Safety factors are thus defined in respect to the *Mises* (*mean*) and the *Mises* (*range*) stresses as:

$$V_m = \text{yield strength}/\sigma_{mises\ m} \qquad (7.3)$$

$$V_R = \text{endurance strength}/\sigma_{mises\ R} \qquad (7.4)$$

A combined safety factor, under fatigue conditions, is defined in order to make a comparison of the stress level during operation relative to the allowable limits.

$$V_f = \frac{1 + V_m \cdot V_R}{V_m + V_R} \qquad (7.5)$$

As the stress level is generally very high, the prestress level is determined in such a way as to put the most critical parts, i.e., those subject to high variable pressures, always under compression. As fatigue failures occur when tensile stresses are present, the possibility of operating under precompression stresses practically eliminates this phenomenon.

The thickness of various components is very high because they are sized for fatigue and not for steady state conditions. Considering the geometry of such components, they are usually oversized against failure under steady pressure conditions.

The Results of the Repeated Pressure Fatigue Tests. Fatigue strength of cross-bored cylinders made from forged materials have been investigated for the components used on packing cups. Thick-walled cylinders with $OD/ID = K = 3$, internal bore diameter 12 mm (.47 inch) and radial holes 2.5 mm (.10 inch) were subjected to cyclic internal pressure.

Different NiCrMo steels with different heat treatment and autofrettage conditions were tested.

The non-autofrettaged fatigue strength of all cross-bored specimens was around 175 to 189 MPa (25375–27405 psi), which is a typical value. With material having a UTS 50% higher the endurance limit was 10% lower because of a higher notch sensitivity.

Autofrettage at 750 MPa (108750 psi) increases the endurance limit of all materials tested up to about 350 MPa (50750 psi) and 6% fatigue strength increase can be obtained raising the autofrettage pressure 11%. When autofrettaging at the same pressure, increasing the ultimate tensile strength of the material does not improve the fatigue limits of cross-bored cylinders. Only radiusing corners and improving surface finish, increases fatigue limits.

7.2.3 Formulas for LDPE Cylinder Components

Calculation of Packing Cups with Axial Holes. On a generic cup idealized in Fig. 7.18, the effects of preload and operating pressures will influence the stress level. This can be determined by the following formulas.

The cylinder bolts provide preload, on area $A + B$.
This preload is reduced by internal pressure acting over the area within D_1.

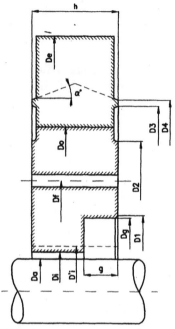

FIGURE 7.18 Sketch of packing cup with axial holes.

$$A = (\pi/4)(D_2^2 - D_1^2) \text{ and } B = (\pi/4)(D_4^2 - D_3^2)$$

It is desirable, for sealing purposes, to maintain contact pressure higher than the discharge pressure p_d by a coefficient x, so

$$F = [(\pi/4)D_1^2 + x(A + B)] p_d \tag{7.6}$$

Bolt load F will increase over the initial preload and has to be determined by analysis of all the cylinder components in (Fig. 7.21). From this the initial preload P required from the bolting and thus the initial face contact pressure p_{SAB} can be determined.

The contact pressure due to preload on the mating surface of two adjacent cups without internal pressure is

$$p_{SAB} = -P/(A + B) \tag{7.7}$$

The value of axial stress during operation, on the internal part of the cup is

$$\sigma_a = -\frac{P - \left(\dfrac{k_c}{k_c + k_b}\right)\dfrac{\pi}{4} D_1^2 p}{\dfrac{\pi}{4} \cdot ((D_4 + \lambda h)^2 - D_g^2)} \tag{7.8}$$

where p = generic value of the pressure that will assume the value of suction or discharge for fatigue analysis considerations
k_c = spring constant for cups
k_b = spring constant for bolting
λ = tang $\alpha°$ allows to calculate the conical volume where the axial stresses are acting (generally $\alpha° \cong 20°$)

The contact pressure on the mating surface s of two adjacent cups, taking into account of the service pressure (suction/discharge) effect, is

$$p'_{SAB} = \left[P - \left(\frac{k_c}{k_c + k_b} \right) \frac{\pi}{4} D_1^2 p \right] / (A + B) \tag{7.9}$$

The load value P may be estimated or calculated with accuracy using the model shown in Fig. 7.21.

The equation that correlates the interference fit pressure at diameter D_o with the Young's Modulus (E_e, E_i) of the external and internal materials involved and the dimension of outer and inner rings, is

$$p_s = E_i \, I/Z \tag{7.10}$$

where I = interference fit (per diam. D_o) where the ratios between diameters, and outer ring and inner ring modulus of elasticity are

$$K_o = (D_e/D_o) \qquad K_i = (D_o/D_i') \qquad \alpha = (E_i/E_e) \tag{7.11}$$

where average ID of the cup, considering the groove size is

$$D_i' = D_i + [(D_g - D_i) \cdot g/h] \tag{7.12}$$

and the geometrical characteristics of the shrink-fit cylinder are given by the co-efficient

$$Z = \alpha \, (K_o^2 + 1)/K_o^2 - 1) + (K_i^2 + 1)/(K_i^2 - 1) + .3 \, (\alpha - 1) \tag{7.13}$$

The radial stress (i.e., the contact pressure between external and internal part of the cup) at interference fit diameter due to internal pressure is

$$p_s' = (2p)/[Z(K_i^2 - 1)] \tag{7.14}$$

The total radial stress at diameter D_o due to shrink fitting and pressure is

$$p_o = p_s + p'_s \tag{7.15}$$

The circumferential stress on the bore of the external part is

$$\sigma_{ce} = [(K_o^2 + 1)/(K_o^2 - 1)] \cdot p_o \tag{7.16}$$

As the service pressure is variable, it is necessary to obtain the *mean* (m) and the *range* (R) values of axial, circumferential and radial stresses, between *suction* (s) and *discharge* (d).

$$\sigma_{ce(m)} = \frac{\sigma_{ce_d} + \sigma_{ce_s}}{2}, \quad \sigma_{ce(R)} = \frac{|\sigma_{ce_d} - \sigma_{ce_s}|}{2} \qquad (7.17)$$

$$\sigma_{re(m)} = \frac{-p_{o_s} - p_{o_d}}{2}, \quad \sigma_{re(R)} = \frac{|-p_{o_s} - (-p_{o_d})|}{2} \qquad (7.18)$$

$$\sigma_{ae(m)} = \frac{\sigma_{ae_d} + \sigma_{ae_s}}{2}, \quad \sigma_{ae(R)} = \frac{|\sigma_{ae_d} - \sigma_{ae_s}|}{2} \qquad (7.19)$$

On the most heavily loaded point (the bore) of the external part of the cup, the mean Mises stress and the amplitude of the range of the Mises stress are calculated with formulas (7.1) and (7.2). The compound safety factor under fatigue condition calculated with formulas (7.3), (7.4) and (7.5).

Inner ring circumferential stress level at the hole is

$$\sigma_{ri} = \frac{p - K_i^2 p_o}{K_i^2 - 1} + \frac{(p - p_o)D_o^2}{K_i^2 - 1}\left[\frac{1}{D_f^2}\right] + p \qquad (7.20)$$

where D_f is the diameter where the hole is located.

The circumferential stresses at the hole in the inner ring are

$$\sigma_{cin} = [(\sigma_{cis} + \sigma_{cid})/2] \mp [k_t (\sigma_{cid} - \sigma_{cis})/2] \qquad (7.21)$$

where k_t is the stress concentration factor due to notch or hole.

The radial stress at the hole is

$$\sigma_{ri} = \frac{p - K_i^2 p_o}{K_i^2 - 1} - \frac{(p - p_o)D_o^2}{K_i^2 - 1}\left[\frac{1}{D_f^2}\right] + p \qquad (7.22)$$

The total concentrated circumferential stress, due to the hole, is

$$\sigma_{ci} = \sigma_{cin} - \sigma_{ri} + p \qquad (7.23)$$

The mean and range values of circumferential, radial and axial stresses at the notch at discharge are

$$\sigma_{ci(m)} = \frac{\sigma_{cid} + \sigma_{cis}}{2}, \quad \sigma_{ci(R)} = \frac{|\sigma_{cid} - \sigma_{cis}|}{2} \qquad (7.24)$$

$$\sigma_{ri(m)} = \frac{-p_s - p_d}{2}, \quad \sigma_{ri(R)} = \frac{|-p_s - (-p_d)|}{2} \qquad (7.25)$$

$$\sigma_{ai(m)} = \frac{p_{SAB_s} + p_{SAB_d}}{2}, \quad \sigma_{ai(R)} = \frac{|p_{SAB_s} - p_{SAB_d}|}{2} \qquad (7.26)$$

where p_{SAB} has been calculated with (7.9).

The mean and the amplitude of the range of Mises stress are given by formulas (7.1) and (7.2).

The compound safety factor for the inner ring V_{fi}, can be calculated as before, using formulas (7.3), (7.4) and (7.5).

Typical Pressures for a Thick-Walled Cylinder Subject to Autofrettage. Considering the idealization of a cylinder (Fig. 7.19) and that $K = D_e/D_i$ and $K_c = D_c/D_i$ and $K_x = D_x/D_i$ the following formulas apply:

Autofrettage pressure, i.e., the pressure necessary to reach the yielding conditions at the diameter D_c of the cylinder, having a yield point σ_y is

$$P_A = (\sigma_y/\sqrt{3})\,[1 - (K_c^2/K^2) + 2\ln K_c] \tag{7.27}$$

The pressure producing yielding at inner diameter, obtained by putting $K_c = 1$, is

$$p_y = (\sigma_y/\sqrt{3})\,[1 - (1/K^2)] \tag{7.28}$$

The full autofrettage pressure, i.e., the value of internal pressure capable to plastically strain the whole thickness of the cylinder is

$$p_{fA} = (2/\sqrt{3})\,\sigma_y \ln K \tag{7.29}$$

The bursting pressure, according to Faupel,[35] considering the ultimate tensile strength σ_U, is

$$p_b = (2\,\sigma_y/\sqrt{3}) \ln K\,[2 - (\sigma_y/\sigma_U)] \tag{7.30}$$

The stresses in the plastic zone, i.e., when $1 \leq K_x \leq K_c$, are calculated with the following formulas (characterized by index p = plastic)

- Stresses during autofrettage (index A)

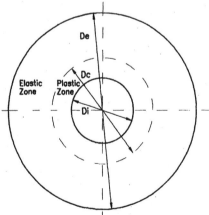

FIGURE 7.19 Thick-walled cylinder with plastic deformation.

Circumferential stress

$$\sigma_{cpA} = 1{,}155\,\sigma_y\,\{1 - \ln(K_c/K_x) - [K^2 - K_c^2]/2K^2]\} \quad (7.31)$$

Radial stress

$$\sigma_{rpA} = -1{,}155\,\sigma_y\,\{\ln(K_c/K_x) + [K^2 - K_c^2]/2K^2]\} \quad (7.32)$$

Axial stress

$$\sigma_{apA} = 1{,}155\,\sigma_y\,\{0{,}5 - [(K^2 - K_c^2)/2K^2] - \ln(K_c/K_x)\} \quad (7.33)$$

- Stresses after autofrettage (Residual stresses) (index R_{es})

Circumferential stress

$$\sigma_{cpRes} = \sigma_{cpA} - \{[1/(K^2-1)][(K^2 + K_x^2)/K_x^2]\,p_A\} \quad (7.34)$$

Radial stress

$$\sigma_{rpRes} = \sigma_{rpA} + \{[1/(K^2-1)][(K^2 - K_x^2)/K_x^2]]\,p_A\} \quad (7.35)$$

Axial stress

$$\sigma_{apRes} = \sigma_{apA} - [1/(K^2-1)]\,p_A \quad (7.36)$$

where P_A has been calculated with (7.21).

- Stresses during operation (index OP)

Circumferential stress

$$\sigma_{cpOP} = \sigma_{cpRes} + \{[1/(K^2-1)][(K^2 + K_x^2)/K_x^2]\,p_i \\ - [K^2/(K^2-1)][1 + (1/K_x^2)]\,p_e\} \quad (7.37)$$

Radial stress

$$\sigma_{rpOP} = \sigma_{rpRes} - \{[1/(K^2-1)][(K^2 - K_x^2)/K_x^2]\,p_i \\ + [K^2/(K^2-1)][1 - (1/K_x^2)]\,p_e\} \quad (7.38)$$

Axial stress

$$\sigma_{apOP} = \sigma_{apRes} + [1/(K^2-1)]\,p_i \quad (7.39)$$

where p_i are the suction or the discharge pressure and p_e is the external pressure during operation.

The Mises stresses in the plastic zone during operation is

$$\sigma_{misesOP} = (1/\sqrt{2})\,\sqrt{(\sigma_{cpOp} - \sigma_{rpOP})^2 + (\sigma_{rpOP} - \sigma_{apOP})^2 + (\sigma_{cpOP} - \sigma_{apOP})^2} \quad (7.40)$$

Stresses acting into the elastic zone ($K_c < K_x \le K$) are calculated with the following formulas: (characterized by index e = elastic)

Stresses during autofrettage

$$\sigma_{ceA} = 0{,}577\,\sigma_y\,(K^2 + K_x^2)K_c^2/(K^2\,K_x^2)] \tag{7.41}$$

$$\sigma_{reA} = -0{,}577\,\sigma_y\,[(K^2 - K_x^2)K_c^2/(K^2\,K_x^2)] \tag{7.42}$$

$$\sigma_{aeA} = 0{,}577\,\sigma_y\,(K_c^2/K^2)$$

Stresses after autofrettage (residual stresses)

$$\sigma_{ceaRes} = \sigma_{ceA} - \{[1/(K^2 - 1)][(K^2 + K_x^2)/K_x^2]\,p_A\} \tag{7.43}$$

$$\sigma_{reaRes} = \sigma_{reA} + \{[1/(K^2 - 1)][(K^2 - K_x^2)/K_x^2]\,p_A\} \tag{7.44}$$

$$\sigma_{aeaRes} = \sigma_{aeA} - [1/(K^2 - 1)]\,p_A \tag{7.45}$$

Stresses during operation

$$\sigma_{ceOP} = \sigma_{ceaRes} + \{[1/(K^2 - 1)][K^2 + K_x^2)/K_x^2]/K_x^2]p_i$$
$$- [K^2/(K^2 - 1)][1 + (1/K_x^2)]\,p_e\} \tag{7.46}$$

$$\sigma_{reOP} = \sigma_{reaRes} - \{[1/(K^2 - 1)][K^2 - K_x^2)/K_x^2]\,p_i$$
$$+ [K^2/(K^2 - 1)][1 - (1/K_x^2)]\,p_e\} \tag{7.47}$$

$$\sigma_{aeOP} = \sigma_{aeaRes} + [1/(K^2 - 1)]\,p_i \tag{7.48}$$

The Mises stresses in the elastic zone during operation

$$\sigma_{miseseOp} = 1/\sqrt{2}\,\sqrt{(\sigma_{aeOP} - \sigma_{reOP})^2 + (\sigma_{reOP} - \sigma_{aeOP})^2 + (\sigma_{aeOP} - \sigma_{ceOP})^2} \tag{7.49}$$

7.2.4 Advanced Analysis of Stresses on Compressor Cylinder Components

Compressor Cylinder Idealization by FEM. Computer programs can be used to evaluate stress and strain in compressor cylinders (Figs. 7.8, 7.9, and 7.10) by means of the finite element method (FEM). Especially subjected to the effects of cyclic pressure are the compression chamber and the first packing cups, bolts and valve elements and radial valve cylinder heads. Due to the pressure drop generated by the seal rings, the packing cups downstream of the first seal element undergo lower fluctuation.[23] The axial valve cylinder head, owing to the layout of the suction and delivery ducts, is subject to practically static suction pressure along the outboard section of the valve housing bore and practically static delivery pressure along the inboard one.

Proper selection of the tie rod axial preload is also very important, to avoid excessive stress concentrations at the point of discontinuity between the compressed and non-compressed areas of each packing cup.

Different configurations are used to reduce this stress concentration (Fig. 7.20) with the purpose of smoothing the peak of contact pressure. If the axial load is too great, the risk of packing cup fatigue failure is high. The packing cup profiles should be maintained during reconditioning.

Fretting can occur between the mating surfaces, whereby the initiation point of a fatigue failure may be reached as two adjacent cups tend to deform in a different way. The upstream cup has a larger inner diameter (Figs. 7.20, and 7.22) and is subjected to higher pressure because of the pressure drop generated by the seal ring. These conditions cause relative movement of the contact faces, and then fretting, if the preload is not such as to make the two cups behave as a single piece. Another factor to be considered in determining optimum load is the temperature at which the cylinder is going to operate, since this affects the length of the elements, and thus the axial load itself. The compressor cylinder in a polyethylene plant is a complex pressure vessel, due to its geometry and to the variation in the pressure distribution along the packing which makes it necessary to evaluate all the most severe conditions.

FEM analyses carried out on high pressure cylinder components,[3] studied as separate entities, leads to assumptions in determining the actions that the rest of the cylinder transmits to the part examined. By considering the whole cylinder, interactions, taking place among all of the components, can be determined with greater accuracy to establish the stress and deformation states.

The cylinder is considered to consist of elementary components, and each component is idealized as a spring having an equivalent stiffness (Fig. 7.21).

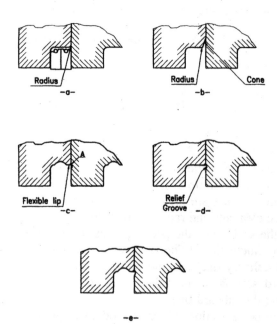

FIGURE 7.20 Different profiles of packing cup mating surfaces.

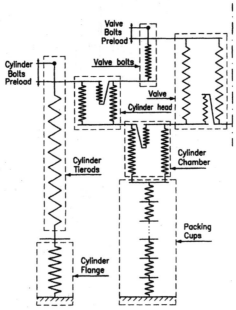

FIGURE 7.21 Cylinder elastic simulation.

In the calculations, each cylinder component is considered under different load conditions, simulating the actions simultaneously occurring during operation, so that elastic behaviour can be determined (Fig. 7.22).

The elementary load conditions acting on the packing cups are the effects of preloads, gas pressure (viewed as superimposition of a constant and a differential pressure), shrink-fitting and temperature. The total load and deformation states of each cylinder component are obtained under preload, suction and delivery conditions.

In order to make a comparison of calculation results with experimental measurements, deformation was measured on a cylinder under both static and dynamic conditions by dial gages on the cylinder and strain gages on the tie rods. Static measurements were made of total elastic deformation resulting from the application of preloading on both cylinder and valve tie rods. (Total elastic deformation is equal to the interference produced between the tie rods and the rest of the cylinder during tightening.) The dynamic measurements of deformation (by strain gages) on the cylinder tie rods show that the oscillation of tie rod deformation passing from suction to delivery is equal to 52.6 $\mu\varepsilon$ (microstrains). This means that the fraction of gas thrust acting on the piston section affecting the tie rods is 32%.[3]

Optimization of Tie Rod Preload. Optimum preload can be obtained considering axial stress σ_a (contact pressure) and τ/f ratio along the cup radius (Fig. 7.23). τ is the shearing stress that may cause sliding of adjacent cups and $f = 0{,}2$ (assumption) is the friction coefficient. Sliding, and thus fretting, occurs when

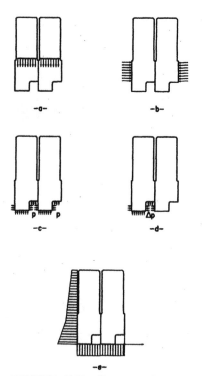

FIGURE 7.22 Forces acting on packing cups: *a*) Shrink-fitting; *b*) Tie rod preloading; *c*) Uniform internal pressure; *d*) Pressure drop; *e*) Temperature effect.

$$|\tau/f| > |\sigma_a|$$

The pressure distribution along packing cups was considered with a delivery pressure of 300 MPa (43500 psi) in the compression chamber and a differential pressure of 100 MPa (14500 psi).

When the differential pressure increases, the contact pressure is attenuated at the inner diameter. Moreover, different internal pressures in two adjacent cups result in different deformations. This causes shearing actions between the mating surfaces that increase when the differential pressure increases, thus facilitating fretting (Fig. 7.23). The gas can infiltrate inside and separate the surfaces. When pressure decreases during the compression cycle, the surfaces mate once again, passing through a sliding stage. The resulting "hammering" and sliding leads to fretting on the surfaces. These phenomena are eliminated when axial stress simultaneously exceeds τ/f and the pressure along the seal surface. The axial preload ensuring this represents an optimum, taking into consideration an adequate safety factor.

By knowing the main stresses on the cup mating surfaces during suction and delivery, it is possible to make a strength evaluation (Fig. 7.24).[3]

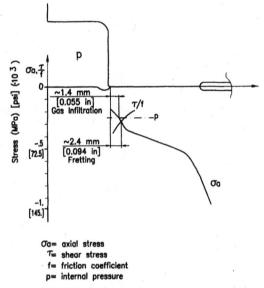

FIGURE 7.23 Axial and shearing stress on mating surfaces of cups.

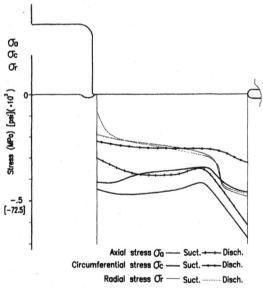

FIGURE 7.24 Main stress distribution on a packing cup contact surface.

Fatigue Reliability of Packing Cups. The discontinuity between adjacent packing cup pressed and unpressed areas (Fig. 7.20) is one of the most highly stressed regions in high pressure cylinders. It can be affected by high differential pressures through a single cup as the annular shoulder supporting the seal ring is subject to pronounced bending, which produces tensile stresses (radial and circumferential) on the high pressure side. Assuming the geometry of Fig. 7.20d, high stress concentrations may result on the bottom of the relief groove. Calculations provide detailed knowledge of the stress state at this point, by superimposing effects. The main stresses, determined for suction and delivery, were combined by the Von Mises criterion for each condition, thus obtaining the mean Mises stress and its variation. These stresses can be increased if stress raisers are present, due to accidental notches on the relief groove (Fig. 7.25).[3]

7.2.5 Axial Valves Design

General Solutions. In suction-discharge multipoppet valves (Fig. 7.13), particularly suitable for large diameters, operating pressures produce very high stresses. Considering the most highly stressed part of the valve, the piece is idealized as a rectilinear axis cylindrical body with 6 holes, having two different diameters.[24] When operating, this part is subjected to constant pressure on the outer edge (except for the gas pulsations in the line) and to fluctuating operating pressure on the inside, i.e., on the hole edges. There is also a compression in the direction of the axis due to gas pressure and the axial preload created by tightening the flange tie rods holding the valve in position.

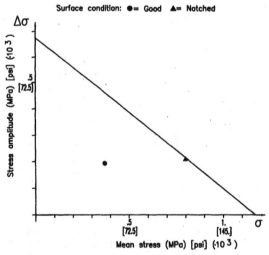

FIGURE 7.25 Ideal stress of packing cup relief groove in different operating conditions.

By the principle of superimposition of effects, the stress conditions generated by external pressure, internal pressure and axial preload can be considered separately.

The holes are assumed to be of the through type and have a diameter which is constant, with the geometry of the valve section unchanged in any plane perpendicular to the valve body axis. Without axial stress, the calculation approach brings up the problem of an elastic body in a plane stress condition. Consequently, the problem consists of establishing the stress condition due to external and internal pressure in a plate geometrically schematized in Fig. 7.26.

The plate has three axes of symmetry, 60° apart, which correspond to the diameters through the hole centers. In this structure, the greatest stresses are on the inner edges of the holes, particularly on the points lying on the axes connecting two adjacent holes and on the axes of symmetry. The most interesting points (Fig. 7.26) are used to compare different calculations.

The stress condition of this elastic body could be determined through an exact procedure, i.e., analytically, by solving the elastic problem, or through approximate procedures using:

- Existing formulas for comparable geometrical bodies
- The finite element method[25,26]
- Strain gages on the piece boundary
- Photoelastic models

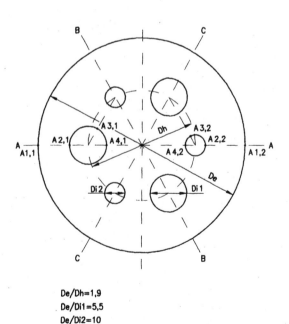

De/Dh=1,9
De/Di1=5,5
De/Di2=10

FIGURE 7.26 Cross-section of multi-bored plate.

Solving a plane problem using the elasticity theory,[27] means finding the Airy function.

The stress function is complex due to the presence of several boundaries inside the plate and consequently the resolution of the equation system defining the elastic problem will also be very troublesome. An analytic solution of a similar case has been found by Kraus.[28]

Evaluation of the stress distribution on the valve body can also be made using equations for thick-walled cylinders under external and internal pressure.[28] In the case of cylinders with a central hole, the formulæ are to establish the stress distribution in any point of the radial thickness. More complex are the equations for cylinders having eccentric holes,[28] giving circumferential stress in any point of the external and internal boundary. A further evaluation of circumferential stress can be made, (only for the points in Fig. 7.26) by utilizing existing studies on stress concentration factors in plates, whose notches are represented by holes.[29] In this case, the plate is assumed to be compressed uniformly, as in a solid cylinder, with the pressure acting on the outside. Variations in the circumferential and radial stresses on the required points referring to the center of the valve body being known, the circumferential stresses, resulting from the presence of the holes, can be determined. Furthermore, holes of different diameters require further simplifying assumptions.

Strain Gage Method. A model of the plate was equipped with strain gages on external and internal surfaces to measure the trend of the circumferential stresses on the boundaries, with pressure acting inside and outside.[24] The model was bigger than the valve, to allow positioning of the strain gages on the internal surface and because of seal problems in the passage area of the connecting wires to the strain gages. The test pressure value was kept under 30 MPa (4350 psi). To minimize effects of systematic and accidental errors of the measuring instruments, the value of the microstrains undergoing measurement was increased, by adopting a light alloy model instead of steel, having a normal modulus of elasticity $E = 72500$ MPa (10,512,500 psi) (about 1/3 that of the steel used for the valve). To eliminate uncertainties as to the elastic properties of this material, some specimens were taken from the piece the model was made from, to obtain the Young's modulus and Poisson's ratio for converting the microstrains into stresses.

FEM Application. The calculations were made with pressure acting separately on the external and internal peripheries. It was assumed, according to the symmetry of the system, that there was no rotation in the nodes determining the diameters of the half-plate, and that displacement would occur only in the direction parallel to the circumference. The procedure used for calculation involved finite elements with triangular elements having three nodal points, with the general element having 6 degrees of freedom and a linear shape function,[24] whose trend of stresses is shown in the graphs in Fig. 7.27 in relation to pressure.

The trend of circumferential stress with pressure acting on the outside is similar on hole edges. In fact, its lowest values comply with those predicted in points A2.1, A3.1, A2.2 and A3.2. The lowest value ($\sigma_c/p_e = -2.9$) is assumed to be at point

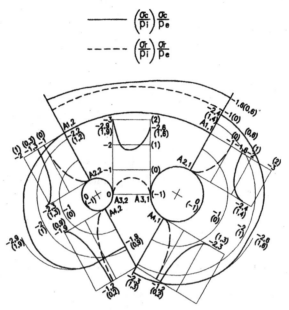

FIGURE 7.27 Circumferential and radial stresses on plate edge and symmetry axes.

A.3.2, i.e., the internal boundary point of the hole having the smallest diameter and also related to the straight line joining the centers of two adjacent holes. The highest value ($\sigma_c/p_e = -1.9$) is at point A4.2, i.e., at the smallest hole, toward the plate center and along a symmetry axis. Furthermore, with internal pressure, the curves of circumferential stresses on the inner edge of the holes show a similar trend, the highest value being point A3.2. The trend of circumferential and radial stresses is alike (Fig. 7.27), both in the case of external pressure and that of pressure in the holes.

The sum of circumferential or radial stresses in the case of external pressure and unit internal pressure is constant and equal to -1, i.e.

$$(\sigma_c/p_e + \sigma_c/p_i) = -1$$

The foregoing can be proved analytically for thick cylinders with centered or eccentric holes, as formulæ exist for stresses along the thickness and at the boundary respectively. In any case, if unit pressure exists inside and outside a cylinder, the stress condition is the same at any point of the thickness and the hoop and radial stresses are:

$$\sigma_c/p = \sigma_r/p = -1$$

This is the result of two different loading conditions, with external and internal pressure; the above equation can thus be obtained by the superimposition effect. These statements apply to any type of stress (hoop, radial or direct, according to the reference axes) involving multiconnected domains, regardless of boundary shape, provided the internal pressure is considered on all internal profiles at the same time.

Comparison of Results.   In a polar-type representation (Fig. 7.28), the values are compared with different methods. The stresses due to internal pressure are bracketed.[24] The trends of the curves determined according to the finite element method and the experimental measurements are similar, and the stress values are very near. The experimentally determined values, except for the central zone of the small hole, are slightly higher than those calculated with the finite elements. At the area of greatest concentration (points A3.1 and A3.2), the results practically coincide.

The use of conventional equations led to results sufficiently in accordance with one another and generally lower than those obtained through the finite element method. This occurs especially at the point of greatest concentration when the thick cylinder formulæ are used. At the same points, according to the theory of notches, the results practically coincide with those obtained through the finite element method and experimental measurements.

Knowledge of the effective stress condition, proper choice of materials and obtaining a high degree of finite elements in the zones of greatest stress concentration makes it possible to arrive at the actual safety coefficient and thus ensure reliability against fatigue failure.

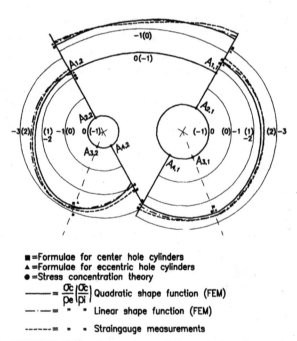

■ = Formulae for center hole cylinders
▲ = Formulae for eccentric hole cylinders
● = Stress concentration theory
——— = $\dfrac{\sigma_c}{\rho_e}\left(\dfrac{\sigma_c}{\rho_i}\right)$ Quadratic shape function (FEM)
—·— = " " Linear shape function (FEM)
------ = " " Straingauge measurements

FIGURE 7.28   Comparison of theoretical and experimental results on multi-bored plate.

7.3 PACKING AND CYLINDER CONSTRUCTION

7.3.1 Technical Solution for Cylinder Components

Two solutions have been used for this special pressure vessel:

- A hard metal liner (sintered tungsten carbide with 9 percent cobalt binder), shrink-fit into a steel cylinder, on which a piston equipped with special piston rings (Fig. 7.16) was sliding
- A packing arrangement cup housing the seal rings, with a hard metal plunger (Fig. 7.8)

Although the first solution was providing fairly good results, it was more affected by plant conditions, low polymers and catalyst carrier as the lubrication was obtained by injecting oil into the gas suction stream. The packed plunger solution is less influenced by such factors, considering that the lubricant is injected directly onto the sealing elements through holes and grooves on the packing cups.

The technological development of sintering WC (11 to 13 percent Co) plungers of large size in one piece, the lower quantity of oil consumed, the excellent performance, and other process considerations[21] led to preferring packed plungers over liners on the compressors manufactured in the last 25 years.

The selection of materials for components under pressure is very important.

Mechanical properties must always be carefully analyzed and, when extreme fatigue conditions exist, aircraft-quality electroslag or vacuum arc remelted steels should be utilized. To obtain adequate fatigue strength of pressure components, it is necessary to use autofrettage when operating pressures are very high.

Sealing surfaces between cylinder components play an important role in achieving good cylinder performance. These are normally flat annular surfaces lapped to a finish of 0.2 microns CLA* and pressed together by tie rods so that their resulting load provides sufficient contact pressure to achieve seal. Since little can be done to modify the actions the cups are subjected to during operation, care should be taken to prevent the consequences of accidental surface defects by performing local precompression treatments, such as cold rolling, shot peening, ionitriding etc.

Special attention is required for the surface finishing of elements in direct contact with the fluid subjected to pulsating pressure. In order to eliminate superficial faults as much as possible, which could cause fatigue failure, very high grade finishes are required. Tungsten carbide plungers and liners have surfaces with 0.05 microns CLA; with the additional advantage of reducing to a minimum the coefficient of friction between the moving parts. It is difficult to obtain these low roughness values on the gas passages in the cylinder heads and on the surfaces of steel cylinders in general, without the use of special machinery.

---

*CLA = Center Line Average.

With a surface finish of 0.8 microns (32 microin.), fatigue life is reduced by 15% as compared to that of a finish of 0.025 microns (1 microin.) It is not necessary to obtain perfectly smooth surfaces, as it has been proved that finishes of 0.1 microns (4 microin.) have no greater fatigue resistance than surfaces with roughness of 0.025 microns (1 microin.).

7.3.2 Sliding Seals Between Piston and Cylinder at Very High Pressures

The contact between the sliding parts for adequate sealing is severe for packing and particularly for piston rings. Under normal pressure, as the relative movement is parallel and does not allow perfect lubrication, only a transient condition of film lubrication and dry friction exists. Oil particles, between the contact points, prevent galling, but to keep the friction coefficient within allowable limits, and to avoid excessive heat generation, a correct selection of materials (chemical and physical properties) is necessary. Experience has shown that the most suitable materials for sealing elements are bronzes, having good wear resistance and mechanical properties.

Cast iron and bronze or various combinations of these metals were used in the past for piston rings. Special bronzes are still utilized for packing sealing elements, although plastic elements can be used up to 250 MPa (36250 psi) when the process requires low heat generation to avoid decomposition in the cylinder. Relating to the plunger material, in the past, nitrided steels were used for plungers in ammonia compressors up to 100 Mpa (14500 psi). Usually, today piston rods are made of steel coated with tungsten carbide (11 to 13% Co) up to pressure of 60 MPa (8700 psi).

In polyethylene plants, with more severe pressure conditions and more precarious lubrication by white oils, liners or plungers are made of tungsten carbide with cobalt bonding. When the cobalt content is increased, the hardness decreases, but the toughness increases, and this quality is more important for plungers than for liners. Today, the steel plunger coated with tungsten carbide can be used up to 140 MPa (20300 psi), usually on the first stage of secondary compressors.

The sliding surface of plungers and liners should be machined to the maximum degree of finish obtainable in order to reduce the friction coefficient to a minimum. Values of 0.025 to 0.05 microns (1 to 2 microin.) CLA of roughness are normally achieved. In case of WC coated plungers, the surface roughness is 0.1 microns CLA (4 microin.). The surfaces of sealing elements do not require the same high quality, since they are softer and on the plunger they are polished during operation, but still need lapped mating surfaces and more accurate geometry to prevent leaks and failures.

The life of the sealing elements is influenced by other factors. The stroke and revolutions per minute (RPM) determining the average piston speed influence the life, since heat generation increases with speed. The RPM are limited by compressor size and arrangement, dynamic loads on the foundation, operation of the cylinder valves, and pressure pulsation in the gas pipes.

The stroke is selected to have a mean piston speed between 2.7 and 3.3 m/s (530 to 650 ft/min). A long stroke is generally desirable since this exposes a longer part of the plunger out of the packing, for more effective cooling. The life of sealing elements is influenced by the system supplying the oil to the cylinder, the amount and quality of oil, the shape of sealing elements, and the linearity of plunger movement. A continuous film of oil must be applied to the sliding surfaces. The type of oil is selected mainly for process reasons (i.e., the need to keep the product pure), and also its lubricating properties. It is current practice to use white oil.

The shape of the sealing elements used is similar to those used in conventional machines.

The piston rings solution, with lubricating oil entrained by the gas, needs only few rings for efficient sealing, but also to enable the one most distant from the gas-oil mixture to be lubricated. Each combined piston ring is made of two rings in the same groove, with a further ring mounted beneath. The ring gaps are positioned out of alignment to give a complete seal effect. On the top of the rings there is a bronze insert, improving the anti-friction properties and the running-in.

A packing arrangement is usually composed of 5 elements, for pressures up to 350 MPa (50750 psi). In the past, solutions with 3 to 8 sealing elements were also applied. The ring nearest the pressure is a breaker ring of special shape, suitable for damping the high pressure fluctuations but not designed to provide effective seal, as this function is performed by the following ring couples, whose life is consequently increased.

The amount of oil applied must be controlled accurately, since trouble can arise from either excessive or insufficient lubrication. If excessive oil is injected and the seal rings are providing perfect seal, the oil pressure can rise to a value above that of normal conditions and the contact pressure between rings and plunger could cause seizure. Of great importance is the linearity of the piston movement, since it ensures that the sealing elements will not be subjected to irregular operating conditions and thus forced to assume an incorrect position in their housing, with consequent overstressing and reduction in life.

It is necessary to keep the temperature low by cooling the plunger with oil around it, outside of the main packing. This is important mainly to reduce the risk of thermal cracks on the plunger surface.

7.3.3 Autofrettage of Various Cylinder Components

General Aspects. The use of autofrettage, applied to tubular and vessel-reactors, has been extended to pumps[18] and to machines operating particularly in tubular-reactor plants, as it is effective where the probability of fatigue failure is high. This technique allows components to be built using materials with lower mechanical properties.

Autofrettage is performed on cylinder heads with combined axial valves, when high pressures are involved, as gas pulsations are still present and fatigue must always be taken into consideration. Cylinder chambers and packing cups are excluded, as they can reach adequate prestress levels through shrink-fitting. Packing cups with axial holes and oil distribution cups require additional prestress only inside the lube oil hole. The distribution cup has no shrinkage, and generally has curved holes normally obtained through a special procedure, such as electro discharge machining (EDM). In this case, a proper polishing procedure should be applied to fully remove the surface modified by local defects.

Autofrettage of injection quills and check valves operating on ethylene secondary compressor second stages is also common practice when pressures are very high. Cylinder heads with radial valves are shrink-fit and are autofrettaged only when differential pressure between suction and discharge is very high. Autofrettage pressure is determined by operating conditions, geometry, presence of prestresses (due to shrink-fitting), and properties of the material. Autofrettage pressures for hypercompressor cylinder parts range between 500 MPa and 1300 MPa (72500 to 188500 psi).[30] Autofrettage of axial holes is performed after shrink-fitting of the cup on the finished piece, only upon completion of machining before final lapping of the mating surfaces. In this case, autofrettage pressure has been applied up to 1100 to 1300 Mpa (160000 to 188500 psi).

Test Rigs and Seals Arrangement. Few types of seals withstand very high pressure applications, due to the fact that the geometries of the cylinder components to be autofrettaged are often complex. On polyethylene compressor cylinder parts, seals are restricted to conical seating surfaces, metal gaskets, plastic O-rings and special arrangements:[30]

- The cone solution (Fig. 7.29), typical of high pressure tubing, has been applied up to 1300 MPa (188500 psi).
- Annealed copper gaskets are used up to 1300 MPa (188500 psi) (Fig. 7.30).
- Viton O-rings are employed for small-diameter seats, tapered (Fig. 7.31) or flat (Fig. 7.32), protected against extrusion by the metallic contact between the parts.

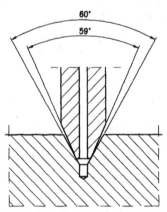

FIGURE 7.29 Surface seal with conical seat.

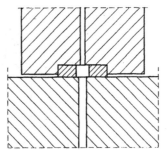

FIGURE 7.30 Metal seal.

Positive results were obtained on diameters up to 76 mm. (3 in.) and up to 900 MPa (130500 psi) for the latter solution.

- Self-sealing arrangements (Fig. 7.33) are used for wider diameters, in order to follow the bore, subject to considerable strain under high pressures.

  These seals are made as follows:

  - A seamless plastic O-ring with hardness between 75 to 90 Shore A, with good surface finish
  - Hard plastic (a polyamide resin) and geometrically precise shoulder rings. Dimensions have to be carefully checked, as plastics are subject to alteration with the passage of time.
  - Bronze antiextrusion rings with a 45° angle
  - Bronze rings to preload the seal assembly and to guide the inner core of the device

In autofrettage of radial valve cylinder heads, similar seals are used and internal mandrels are applied to reduce fluid volume. Axial valve cylinder heads are autofrettaged (Fig. 7.34) with special seals (Fig. 7.33) to achieve seal on the large inner diameter which can be accomplished by providing a smooth surface finish and

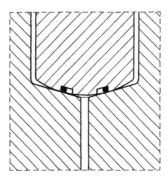

FIGURE 7.31 Plastic O-ring with conical seat.

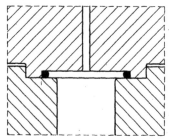

FIGURE 7.32 Plastic O-ring with flat seat.

using great care in assembling the rig to avoid local damage in the seal zone. An internal bar reduces fluid volume. The seals are preloaded, the assembly is balanced and no additional support is required for the inner core. Lateral (suction and discharge) holes are plugged by flanges using combined metallic and O-Ring seals (Fig. 7.32). Autofrettaged packing cup axial holes (Fig. 7.35) use metal seals (Fig. 7.30). The test rig for the oil distribution cups uses axially-directed seal (Fig. 7.30) and radial seal (Fig. 7.31). Autofrettage of injection quills utilizes cone seals (Figs. 7.29 and 7.31).

Autofrettage Procedure. In equipments operating at very high hydrostatic pressures, the fluid must be able to transmit pressure without undergoing freezing effects, related to fluid properties, operating temperatures and tubing size. Pressure may increase at the pump and, due to solidification problems within the tubing, may be much lower inside the piece to be autofrettaged.

Brake oils have been used up to 500 MPa (72500 psi) with some drawbacks (i.e., corrosion on pump seal rings caused poor performance). Prexol 201 overcomes solidification problems and gives adequate intensifier plunger seal life, up

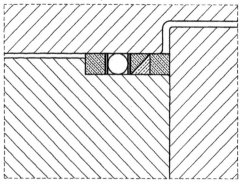

FIGURE 7.33 Special seal with O-ring.

VERY HIGH PRESSURE COMPRESSORS 7.41

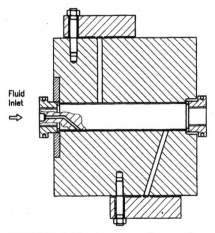

FIGURE 7.34 Apparatus for autofrettage of axial valve heads.

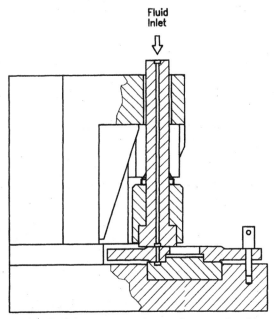

FIGURE 7.35 Apparatus for autofrettage of packing cups.

to 1300 MPa (188500 psi). As oil properties are altered by use at the highest pressures, oil should be changed frequently.

The whole autofrettage process is controlled by resistance strain gage-type transducers to check pressure at pump discharge, close to the piece undergoing autofrettage and, if critical conditions exist, at the end of the circuit or at the far side of the cylinder component.

Strain gages on the outer surface of the piece are used when an autofrettage procedure must be defined for the first time, or in case of complex shapes, and may detect internal pressure or deviations in mechanical properties. For safety reasons, dimensional checks are performed after autofrettage.

The inner diameter of axial valve cylinder heads should be checked, to assess the amount of metal to be removed, which should be as small as possible in order to preserve the benefits of prestressing. At plastic strain conditions, duration of tests appreciably affect the final results. Pressure should rise slowly to allow strain to take place completely during each loading condition. In short tests, yield point and ultimate tensile stress are increased while strain decreases. In the case of steel, a pressure rise of 10 MPa (1450 psi) per second is on the safe side. Generally, the test requires a pressure rise of 5 minutes minimum. Pressure increase is related to the volume of the fluid in the whole system and its components (tubing and the intensifier). The autofrettage pressure is maintained for 15 minutes (5 minimum) and a slow pressure decrease takes place in about 5 minutes. Slow return to final conditions eliminates errors in dimensional measurements, allows time to check the autofrettage effect, and allows the special seals to return to their original positions in their housings after having undergone severe strain, thus reducing disassembly problems.

Very high pressure systems have potential hazards, although risks are not as great as when gases are handled, due to the great energy involved (the fluid possesses compressibility and can be trapped inside the system). If gaskets in the hydraulic system fail, the jettisoned particles could cause injury to people or damage objects. Fluid leak at high speeds, reduced by the small volumes involved, is another risk. To prevent air from being trapped in the hydraulic circuit during test rig assembly, a vent valve is temporarily opened at the highest point of the circuit and oil is allowed to drip out, prior to tightening. To reduce risks from stored energy, the volume of the system is reduced: the piping is made as short as possible and suitable inner cores are used in large components like cylinder heads. The compact system is positioned in a safe area (bunker with fencing around the equipment to protect the surroundings). Steel shield between assembly and pump and metal sheets around the pressure tubes are added protection. The operator's work station is separate. Before disassembling any part of the test rig, the pressure is relieved from the circuit.

Some authors,[31,32] advise heat-treating the material at about 250°C for an hour to allow component dimensions (i.e., eliminating flexural stresses without affecting residual body stresses)[33] and the material elasticity to be restored. (Others recommend higher temperatures.) At the same or higher temperatures, decarburizing problems might arise on the surfaces. This is not common practice with polyethylene compressors, as components have proven successful field operation. In any case, this heat treatment cannot be performed when the tempering temperature of the material is lower than the heat treatment temperature.

Axial valve cylinder heads, requiring accurate inner bore dimensions, must be machined after autofrettage. Remachining is also performed in the seat area quills (oil distribution cup side) and thus the modified prestress level area is quite limited. Appropriate allowances must be considered, and material removal must take into account the reduction in the prestress level.

It is generally advisable to perform autofrettage on finished parts. The combination of autofrettage and shrink-fitting, especially when high ultimate tensile strength materials are used, is complex. Autofrettage before shrink-fitting is normally carried out on radial valve cylinder heads, allowing use of lower autofrettage pressure, with advantages. Lube oil holes of packing cups are autofrettaged at a pressure of 1100 MPa (159500 psi). Autofrettage contributes to increasing the availability of secondary ethylene compressors which operate in plants with tubular reactors or in general when pressures exceed 200 MPa (29400 psi).

7.3.4 Typical Behaviour of Packings

Packings today consist typically of one (or two) split breaker rings and five radial tangential sealing rings (Fig. 7.36). The rings are made of special bronze alloys, usually with high lead content, uniformly distributed, so as to guarantee sufficient strength, low friction coefficient and high thermal conductivity, for a rapid dissi-

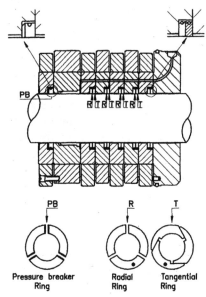

FIGURE 7.36 Packing assembly.

pation of the friction heat through the packing cups. The hardness of the rings varies from 55 to 80 Brinell (measured with a 10 mm. ball and 500 kg. load).

The plunger on which the sealing elements slide is made of solid tungsten carbide, with surface finish of 0.05 microns [2 microin.] CLA. The synthetic lube oil of the cylinders has lower lubricating properties than oils used for normal services, since for ethylene polymerization, pollution of the final product must be reduced to a minimum.

Packing performance is greatly influenced by the above parameters and by the efficiency of the breaker rings, whose action is very important, as can be seen by analysing the operating conditions of a packing. The pressure inside the cylinder can be considered as consisting of a constant portion (suction pressure) and a fluctuating portion (the difference between discharge and suction). The static pressure distribution tends to overload the last ring (frame side), which has to handle almost the whole load.[23] This is similar to packings, operating at constant pressure, for example on ammonia synthesis compressors. The variable pressure increases due to polytropic compression, and then decreases due to the expansion of the gas remaining in the clearance volume, and assumes constant values during discharge and suction effect.

Breaker rings oppose a rapid pressure increase in the cylinder, limiting gas leakage and reducing the propagation of the pressure wave towards the seal rings. Their most important function, however, is to delay the "backflow" from the packing rings towards the cylinder chamber, when the plunger begins its back stroke. If this action is inadequate, the pressure upstream of the first sealing element will suddenly drop to the suction value, due to the steep slope of the expansion curve. The resultant of the forces acting on the first sealing ring is suddenly inverted, causing rapid expansion of gas under the radial and especially under the tangential ring, which exerts a stronger sealing action, with the following problems:

- Breakage of the dowel pin between radial and tangential ring
- Breakage of the lips of the tangential cut rings
- Damage to the garter springs of the sealing element

When, after a certain period of operation, the first sealing pair no longer performs its function, the problems occur in the second pair and the process of progressive damage continues through the various rings of the packing. To analyze the operating conditions, behaviour and performance of packings, measurements were taken at the lube oil injection quills and in the compression chamber (Fig. 7.37) of a first and second stage cylinder on a compressor having a capacity of 53,000 kg/hr (1945 lb/min), operating in a plant with a vessel reactor. Packings had a three piece pressure breaker ring, with small circumferential clearance and five grooves of radial tangential seals (with axial clearance of about 0.15 mm. [.006 in.]). The distribution of the pressures along the packing in relation to the crank angle (Fig. 7.37) and during the suction and discharge strokes (Fig. 7.38) is quite similar on first and second stage.[23]

In general, the first three sealing elements are affected by the pressure fluctuation of the cylinder, while the last two are subjected to an almost steady pressure (Fig.

VERY HIGH PRESSURE COMPRESSORS   **7.45**

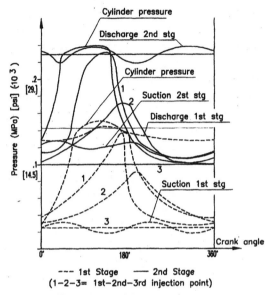

FIGURE 7.37 Operating pressures on a 1st and on a 2nd stage cylinder.

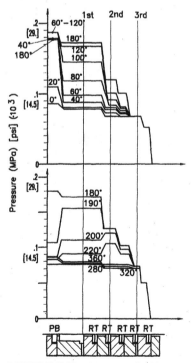

FIGURE 7.38 Pressure distribution on the 2nd stage packing during discharge and suction stroke.

7.38). The pressure variation occurring at the first sealing pair is propagated to the following elements in the proportion of 70% on the second pair, 30% on the third and a negligible amount on the two final sealing elements. The steady pressure, at the external oil injection quill, does not significantly change and therefore about 60% should be supported by the last sealing pair, in the hypothesis of labyrinth behaviour.[34] Pressure pulsations, upstream of the suction valve and downstream of the delivery valve, may have an influence on the cycle pressures in the cylinder (Fig. 7.37). The pressure breaker rings behaviour appears good since the delaying action is evident during the compression period and a considerable sealing effect is evidenced in both stages during compression and expansion. The breaker ring, in fact, withstands about 80% of the pressure fluctuation, (100 MPa [14500 psi] in the second and 70 MPa (10150 psi) in the first stage). The efficiency is higher in the first stage, due to the greater variability of the specific volume of the gas (5% in the second and 16% in the first stage). This may be partly explained by the difference between the polytropic coefficient in first and second stage. It should be recalled that when the physical conditions of a gas are close to those of a liquid, the task of the breaker ring is more difficult and its effect is lower.

The fluctuating part of the pressure affects the first three seal rings, with the second and third withstanding a differential pressure of 50% and 30% as compared to the first sealing couple (Figs. 7.37 and 7.38). The steady part of the pressure is mainly supported by the last two sealing elements.

Some packings were dismantled and analyzed after 10,000 to 20,000 hours of operation. The wear of each radial and tangential element was compared (Fig. 7.39)

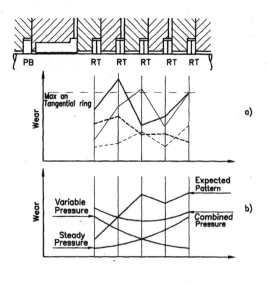

FIGURE 7.39 Wear distribution on 1st and 2nd stage packings.

and there was a similar wear pattern curve for radial and tangential rings. For first or second stage, the trend for higher wear is on the first and last elements. The wear rate is higher in tangential rings as compared with radial ones. On first stage, the first pair did not wear completely, as the dowel broke, due to "backflow" and then the pressure loaded the second pair, causing accentuated wear. Wear on the tangential ring higher than the amount allowed by the butt gap is frequently observed due to non-uniform wear on the rings, resulting from the high pressures and forces acting on them. On the second stage, the "backflow" caused breakage of the springs (of the coil type) of the first pair and later breakage of the dowel pin of the second pairs. The work of withstanding the variable pressure was then carried out by the third sealing element.

The maximum wear on the frame side elements of both stages is due to the constant pressure to which they are subjected, considering that lubricating conditions are not optimal. Wear on the radial ring of the last pair of the first stage packing is an exception, encountered in other compressors, which can be explained as follows: The radial rings, subject to steady pressure, tend to remain in their position without effecting an appreciable sealing action towards the plunger, but simply creating a barrier to the pressure at the cuts of the tangential rings. In the zone subject to variable pressure, the first radial rings are forced to exert a sealing action on the gas that tends to re-enter the cylinder during the suction phase. The sealing effect is not complete, since the radial cuts allow the gas passage.

A general wear pattern can be derived connected with the pressure distribution along the packing (Fig. 7.39). The steady portion of the pressure causes a type of wear with maximum values reached on the frame side sealing ring. The fluctuating portion of the pressure causes wear with an opposite trend, with the highest values on the first sealing pair. The resultant wear will be a curve with its maximum values at the extremities of the packing. Generally, the theoretical maximum value is either towards the first ring (pressure side) or towards the last (frame side) depending on the predominance of the fluctuating or the steady portion. The practical wear pattern is different as the "backflow" can make some sealing elements inefficient. The performance of the sealing elements is strongly influenced by operating conditions, lubrication and alignment. The normal plunger runout is within 0.075 mm. (.003 in.), as easily measurable by proximity probes, with alarm 0.15 mm. (.006 in.) and trip 0.2 mm (.008 in.). Long life of packing rings has been reported up to 65,000 operating hours, with 180 MPa (26100 psi) final pressure.

7.4 BIBLIOGRAPHY

1. Crossland, B., K. E. Bett, and Sir Hugh Ford: *Review of Some of the Major Engineering Developments in the High-Pressure Proc. Polyethyene Process,* 1933–1983, Institute of Mechanical Engineering, 1986, Vol 200, Ne A4.

2. Andrenelli, A., "Reciprocating Compressors for Polyethylene Production at Pressures Higher Than 3000 Atmospheres," *Quaderni Pignone* 13.

3. Traversari, A., M. Ceccherini, and A. Del Puglia, *Advanced Elastic Analysis of Compressor Cylinders for H.P. Low Density Polyethylene Production,* ASME Joint Conference of the PVP, Materials, Solar and Nuclear Engineering Division, Denver, Colorado, June 21–21, 1981, Session 0A-6.

4. Traversari, A., and F. Bernardini, "Aspects of Research on Secondary Compressors for Low Density Polyethylene Plants," *Quaderni Pignone* 25, June 1978, pp. 123–124.

5. Vinciguerra, C., *U.S. Patent 3, 581.583 to Nuovo Pignone S.p.A.,* January 15, 1969.

6. Andrenelli, A., "Special Features in Reciprocating Compressors for Polyethylene Production," *Proceedings of the Industrial Reciprocating and Rotary Compressors: Design and Operational Problems,* Institution of Mechanical Engineers, Vol. 184, Part 3R, October 13–16, 1970, pp. 106–113.

7. Traversari, A., P. Beni P., *Approaches to Design of a Safe Secondary Compressor for High Pressure Polyethylene Plants,* High Pressure Symposium: Safety in High Pressure Polyethylene Plants, Tulsa, Oklahoma, March 12–13, 1974.

8. Giacomelli, E., and M. Agostini, *Safety, Operation and Maintenance of LDPE Secondary Compressors,* ASME PUP Division Conference, New Orleans, Louisiana, 1994.

9. Manning, W. R. D., "Ultra-high-pressure Vessel Design, Pt. 1," *Chem. Proc. Eng.,* March, 1967.

10. Morrison, J. L. M., B. Crossland, and J. S. C. Parry, "Fatigue Strength of Cylinders with Cross Bores," *J. Mech. Eng. Sci.* 1959 1 (N. 3).

11. Parry, J. S. C., "Fatigue of Thick Cylinders: Further Practical Information," *Proc. Inst. Mech. Engrs* 1965–66 180 (Pt. 1), 387.

12. Chaaban, A., K. Leung, and D. J. Burns, "Residual Stress in Autofrettaged Thick-Walled High Pressure Vessels," *PVP,* Vol. 110, 1986, pp. 56–60.

13. Kendall, D. P., "The Influence of the Bauschinger Effect on Re-Yielding of Autofrettaged Thick-Walled Cylinders," ASME Special Publication, *P.V.P.,* Vol. 125, July, 1987, pp. 17–21.

14. Yang, S., E. Badr, J. R. Sorem, Jr., and S. M. Tipton, "Advantages of Sequential Cross Bore Autofrettage of Triplex Pump Fluid End Cross Bores," *P.V.P.,* Vol. 263, High Pressure—Codes, Analysis and Applications, ASME, 1993.

15. Manning, W. R. D., *Design of Cylinders by Autofrettage,* Engineering (April 28, May 5 and May 19, 1950).

16. Chaaban, A., and N. Baraké, "Elasto-Plastic Analysis of High Pressure Vessels with Radial Cross Bores," *P.V.P.,* Vol. 263, High Pressure—Codes, Analysis and Applications—ASME, 1993.

17. Chaaban, A., *"Static and Fatigue Design of High Pressure Vessels with Blind-Ends and Cross Bores,"* Ph. D. Dissertation, University of Waterloo, 1985.

18. Chaaban, A., and D. J. Burns, *"Design of High Pressure Vessels with Radial Cross Bores,"* Physical 139, 140B, pp. 766–772, North-Holland, 1986.

19. Rees, D. W. A., "The Fatigue Life of Thick-Walled Autofrettaged Cylinders with Closed Ends," Fatigue Fract. Eng. Mater. Struct., Vol. 14, pp. 51–68, 1991.

20. Rees, D. W. A., "Autofrettage Theory and Fatigue Life of Open-Ended Cylinders," *Journal of Strain Analysis,* Vol. 25, pp. 109–121, 1990.

21. Parry, J. S. C., "Fatigue of Thick Cylinders: Further Practical Information," *Proc. Inst. Mech. Eng.,* 1965–66, 180 (Part I).

22. Kendall, D. P., and E. H. Perez., "Comparison of Stress Intensity Factor Solutions for Thick-Walled Pressure Vessels," *P.V.P.*—Vol. 263, High Pressure—Codes, Analysis and Applications, ASME, 1993.

23. Traversari, A., and E. Giacomelli, "Some Investigation on the Behaviour of High Pressure Packing Used in Secondary Compressors for Low Density Polyethylene Production," *Proceedings of the 2nd Int. Conf. on H.P. Engineering,* University of Sussex, Brighton, England, July 8–10, 1975, pp. 57–58.

24. Giacomelli, E., *"Finite Element Method on Polyethylene Compressor Valves Design,"* Quaderni Pignone 26, January 1979, pp. 19–25.

25. Zienkiewicz, O. C., "Axi-Symmetric Stress Analysis," *The Finite Element Method in Engineering Science,* (London, Eng.: McGraw Hill, 1971), pp. 73–89.

26. Tottenham H., and C. Brebbia, *Finite Element Techniques in Structural Mechanics* (Southampton, Eng.: Millbrook).

27. Muschelisvili, *Some Basic Problems of the Mathematical Theory of Elasticity,* Moscow, 1949.

28. Kraus, H., "Pressure Stresses in Multibore Bodies," *Int. J. Mech. Sci.* (Pergamon Press Ltd., 1962), Vol. 4, pp. 187–194.

29. Peterson, R. E., *Stress Concentration Design Factors* (New York, N.Y.: John Wiley and Sons, 1974).

30. Giacomelli E., P. Pinzauti, and S. Corsi, *Autofrettage of Hypercompressor Components up to 1.3 GPa: Some Practical Aspects,* ASME PUP Division Conference, Orlando, Florida, 1982.

31. Vetter, C., and H. Fritsch, "Zur Berechnung und Gestaltung von Bauteilen mit Beanspruchung durch schwellende Innendruck," *Chemie Ingr. Tech.,* 1958, 40 (n. 24).

32. Morrison, J. L. M., B. Crossland, and J. S. C. Parry, "Strength of Thick Cylinders Subjected to Repeated Internal Pressure," *Proc. Inst. Mech. Eng.,* 1960, 174 (no. 2).

33. Giacomelli, E., and P. F. Napolitani, *"Ricerca Sperimentale sul Comportamento degli Accoppiamenti Forzati Albero-mozzo,"* Thesis, Dept. Mech. Eng., University of Pisa, Italy, 1969.

34. Cosimi, L., "Il Compressore a Pistone a Secco con Tenuta a Labirinti," *Il calore,* 1961—N. 3.

35. Faupel, J. H., and F. E. Fisher, *Engineering Design,* (New York, N.Y., Chichester, Brisbane, Toronto: John Wiley and Sons, 1981).

36. Whiteley, K. S., *Ullmann's Encyclopedia of Industrial Chemistry,* Vol. A21, Section 1.5.1, Polyofins, 1992.

CHAPTER 8
CNG COMPRESSORS

Mark Epp
Jenmar Concepts

8.1 INTRODUCTION

The introduction of natural gas as a fuel for automotive and mass transportation has provided an entirely new application for the compressor. The problems of energy supply shortages, poor air quality and high energy costs have contributed to the importance of natural gas as an alternative to crude oil based fuels.

Natural gas is a mixture of gases in which the primary constituent is methane, typically at 85.0 to 95.0 mole percent.

As a transportation fuel, stored natural gas must be compressed for an increase in energy density. The compressor is used to boost the pressure of natural gas and is the primary equipment of the compressed natural gas (CNG) refueling station.

8.2 CNG COMPRESSOR DESIGN

The compressor type used is the multi-stage reciprocating piston compressor. Compressor size commonly ranges from 25 to 250 brake horsepower (BHP). The design of the CNG compressor resembles the high pressure air compressor but with some important differences.

8.2.1 Suction And Discharge Pressures

Discharge pressures of 3600 to 5000 psig preclude the use of the multi-stage reciprocating piston compressor. Suction pressures are site specific and dependent on the operating pressures of the local gas utility distribution pipeline. Suction pressures can range from inches water column to 1000 psig. Most often pressure regulation and metering is supplied by the gas utility providing stable suction pressures to the compressor. To minimize energy consumption, CNG compressor manufac-

FIGURE 8.1 Compressed natural gas refueling station (*courtesy of IMW Altas Inc.*).

turers configure compressors specific to the application and suction gas pressure available (refer to Fig. 8.2).

The total pressure ratio from suction pressure to discharge pressure determines the number of compressor stages used. For the same pressure ratio, a natural gas compressor will generate lower discharge temperatures than an air compressor. This is due to the lower specific heat ratio property ($k = C_p/C_v$) of natural gas relative to air. For this reason, natural gas compressors of similar technology can operate at higher pressure ratios than air compressors. The gas discharge temperature of a compressor stage is one of the limiting factors determining maximum stage pressure ratio. The maximum discharge temperatures allowable are a function of the acceptable operating temperatures of the sealing materials used, including piston rings, rod rings, o-rings and gaskets. Avoiding high discharge temperatures also decreases compression horsepower. To maintain satisfactory discharge temperatures a suitable number of compression stages must be selected. Table 8.1 provides a guide to the number of compressor stages required for a given suction pressure. There is some overlap of suction pressure ranges. Some compressors, such as those with oil lubricated and cast iron piston rings can operate at higher pressure ratios than compressors using nonlubricated and special material piston rings such as the filled Teflons™.

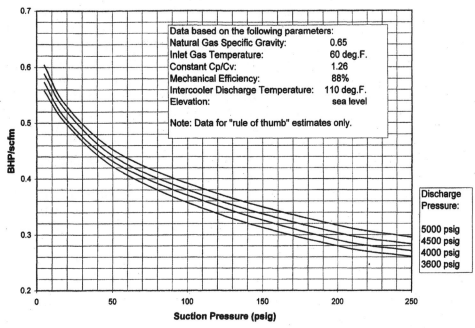

FIGURE 8.2 Compressor brake horsepower vs suction pressure.

8.2.2 Compressor Sealing

Natural gas is a flammable gas. It has also been identified as an atmospheric greenhouse gas. CNG compressors are designed to eliminate or severely restrict gas leakage emissions. Uncontrolled leakage may result from random leaks that occur in the piping system or compressor caused by static seal failures. Controlled leakage is expected leakage from compressor rod packings and seals. As industrial emissions standards tighten, consideration must be given to CNG compressor manufacturer's gas leakage rate data.

TABLE 8.1 Compressor Stages vs Suction Pressure

Suction pressure (psig)	No. of stages	Discharge pressure (psig)
6" H_2O-10	5	3600–5000
6" H_2O-100	4	3600–5000
80–350	3	3600–5000
250–1200	2	3600–5000
1000+	1	3600–5000

Pressurized Crankcase. Compressors with pressurized crankcases collect seal leakage gas in the crankcase and recycle it into the suction pipe. No leakage gas is lost to the atmosphere. The crankcase is pressurized at suction pressure. Special rotating shaft seals prevent gas leakage from crankshaft drive end extensions. Most pressurized crankcases are limited in use to suction pressures below 250 psig, however compressors with higher seal operating pressures are available.

Pressurized crankcases are most often used on trunk piston type compressors. Trunk pistons have a linear guide and piston as one integral part. There is no rod sealing between the piston and linear guide. Without a linear traveling piston rod and seals (see Atmospheric Crankcase, below), piston ring leakage flows into the crankcase. To hold leakage gas at suction pressure, the crankcase must be designed as a pressure vessel with heavy rounded walls and internal or external structural ribs (see Fig. 8.3). Some pressurized crankcase compressors use a cantilevered shaft to eliminate one shaft seal. Other components including oil lubrication systems, static seals on inspection plates and cover seals must withstand the elevated pressures.

Atmospheric Crankcase. Compressors with atmospheric crankcases commonly use double acting cylinders and crossheads. Crossheads allow the use of a piston rod which moves linearly and compresses both to the head and crank end (see Fig. 8.4). The piston rod is readily sealed using a series of rod packings. Rod packings are assembled in a packing case with gas leakage vented and piped for discharge

FIGURE 8.3 Pressurized crankcase compressor (*courtesy of Compair Reavell Limited*).

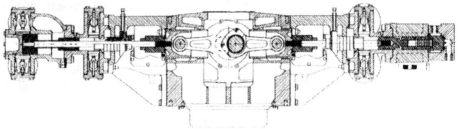

HPSS COMPRESSOR

FIGURE 8.4 Atmospheric crankcase compressor (*courtesy of Gemini Engine Company*).

to atmosphere. New rod seal leakage can be very low and commonly less than 0.1% of total cylinder mass flow rates. Most compressors of this type vent the gas at source rather than allowing the gas to leak into the crankcase. The crankcase operates at atmospheric pressure, eliminating the need for special shaft seals, gaskets, and elevated pressure lubrication systems.

The atmospheric crankcase is most suitable for large compressors where design for pressure containment is difficult. Atmospheric crankcase type compressors using rod sealing also allow compressors to be designed for high gas suction pressures beyond what is practical for pressurized crankcase type compressors. In addition, maintenance procedures are less onerous, allowing crankcase inspections without depressurization.

Blow Down Gas Recovery. Similar to air compressors, the natural gas compressor must be depressurized for start up. This necessitates that on shutdown, gas entrapped in the compressor and piping system must be vented. Unlike an air compressor which can be vented to atmosphere, the natural gas compressor must be provided with a blow down gas receiver tank. This tank must be adequately sized to allow the compressor to depressurize and reach a pressure equilibrium sufficiently low for compressor start up.

Upon compressor depressurization, valves may operate under sonic or choke flow conditions during blow down. Sonic gas velocities will quickly erode the seats and seals of some valves. Sonic flow can be managed by using line orifices, special valve seat materials, or specially designed valves which protect seats and seals from direct flow impingement.

8.2.3 Lubrication

Compressor lubrication has become an issue of debate within the CNG industry. Lubricated compressors require lubrication of piston rings, rod packings and valves. Nonlubricated or oil free compressors use special materials for these components, eliminating the need for additional oil injection. Proponents of nonlubricated compressors claim that they achieve the highest discharge gas quality. Proponents of lubricated compressors maintain that with well engineered lubrication and filtration systems, similar discharge gas quality is attainable. A lubrication oil carry over limit maximum of 0.5 lb/mmscf at compressor discharge has become a common industry standard. This standard can be met using nonlubricated compressors or lubricated compressors with filtration.

In deciding lubricated versus nonlubricated, other factors to consider are outlined in Table 8.2.

8.2.4 Piston Ring and Seal Performance

The extreme gas pressures exerted in the final stages of a natural gas compressor present some unique design problems. Piston ring wear rates increase dramatically with increasing stage pressures. High pressure differentials across piston rings contribute to ring extrusion between the piston and cylinder clearances. Lowering clearances reduces ring extrusion, but increases the possibility of piston contact with the cylinder wall as the piston wear bands deteriorate. Extreme pressures also contribute to high operating PV (the product of surface pressure and velocity) values of piston rings. The result can be high piston and cylinder wear rates. High PV also generates high ring surface contact temperatures. These temperatures can be higher than measured gas discharge temperatures resulting in piston ring material creep and extrusion. Another less understood factor in piston ring wear is the apparent loss of oil viscosity at high operating pressures.

High pressure static sealing using compliant and porous o-ring materials can result in seal failure upon rapid decompression. O-ring materials including Buna-N and Viton™ are porous and allow high pressure natural gas to permeate the material. If the o-ring is operating at high pressure for some extended time period and then the compressor shuts down and rapidly decompresses, the gas entrained in the o-ring will rapidly expand. Failure of the seal is caused by rapid expansion of entrained gas causing bubbles and lacerations in the o-ring material as the gas

TABLE 8.2 Lubricated vs Nonlubricated Compressors

	Lubricated	Non-lubricated
Advantages	—increased piston ring life —allows use of metallic rings —air cooling or noncooling systems sufficient —higher pressure ratios and discharge temperatures allowable —fewer stages necessary in some cases —higher operating speed —reduced capital cost —longer overhaul intervals	—low to nil oil contamination of discharge gas —reduced lubrication requirements —less filtration needed —reduced routine maintenance —reduced noise levels with liquid cooled units
Disadvantages	—oil contamination of discharge gas stream —oil deposits in pressure vessels reducing capacity to store gas —oil contamination of on board vehicle equipment —increased vehicle emmissions —higher compressor oil consumption —increased maintenance requirements on lubrication systems —increased noise levels with air cooled compressors	—increased cooling requirements —lower maximum discharge temperatures —reduced piston ring and rod packing life —lower pressure ratios and discharge temperatures required —more stages may be necessary —increased capital cost —lower operational speeds —reduced valve life —shorter overhaul intervals

escapes. High operating temperatures, and pressures along with the presence of compressor lubricating oils, seems to exacerbate the problem. The higher durometer seal materials provide some improvement likely due to their lower porosity. If suitable, the use metallic gaskets should be considered.

8.2.5 Other Compressor Design Considerations

Adequate gas aftercooling is key to the satisfactory performance of high demand CNG refueling stations. The cooler the natural gas is when it enters the storage vessels or refueling vehicle, the more dense the fuel, enhancing storage capacity. Final gas discharge temperatures should be a maximum of 20 to 30°F above ambient air temperatures. Further reductions in temperature will greatly increase aftercooler cost.

Compressors used in time filling applications (see Section 8.4.2) operate with a large variation in discharge pressure. Some compressors require operation above some discharge pressure minimum. This minimum pressure is critical to the proper operation of piston rings. Without sufficient back pressure the rings will not seat adequately against the cylinder walls and in some lubricated compressors, oil consumption will increase as oil is drawn into the compressor cylinder. To address the problem, a back pressure regulator can be installed at compressor discharge and set to the manufacturer's minimum required pressure.

Compressor package noise levels are often critical to the acceptance of an installation by approval authorities. Noise level requirements are often site-specific and set by city and municipal bylaws. Reduced noise levels also enhance fuel marketing efforts. A commonly specified noise level is 75 dba measured at 10 feet from the perimeter of the compressor skid. Most compressor packagers can meet this noise level with an enclosed and sound attenuated compressor skid package.

8.2.6 Compressor Electrical Systems

Natural gas, being flammable, requires that all electrical equipment and wiring within a code specified distance from natural gas compressors and gas containing equipment be explosion-proof. The explosion-proof classification most used in the CNG industry is class 1, division 1 or 2, group D in accordance with National Fire Protection Association (NFPA) standard 70, the National Electrical Code. Codes such as NFPA 52 define the boundaries of explosion-proof areas.

Explosion-proof electrical enclosures and junction boxes are designed to withstand internal explosions without flame propagation out of the enclosure. They are metallic and of heavy wall construction making them expensive and cumbersome to access. The compressor skid of Fig. 8.5 shows an explosion-proof disconnect panel standing to the right of the instrumentation panel. Some alternatives to the use of explosion-proof equipment include:

- Conventional contact closure instrumentation with intrinsically safe circuitry
- Standard electrical enclosures with air purge systems and failure shutdown interlocks
- Impenetrable gas tight electrical rooms within the hazardous area with gas detection
- Locating electrical equipment remote and outside the hazardous area

8.3 CNG STATION EQUIPMENT

In addition to the compressor, other items of equipment are required to complete the CNG refueling system. The descriptions of equipment that follow are representative only. The CNG industry has few system design standards. Most installations are uniquely designed to meet a performance specification.

FIGURE 8.5 Compressor skid package (*courtesy of Gemini Engine Company*).

8.3.1 Priority Panel

The Fast Fill CNG type refueling station has the compressor filling one or several large pressure vessels. The natural gas vehicle (NGV) then refuels its on board tank from the pressure vessels. In this way, refueling flow rates are maximized and independent from compressor capacity.

The Three Bank Priority Panel uses a single multi-port valve or multiple valves to control the flow of discharge gas from the compressor to a series of storage pressure vessels. The pressure vessels are divided into three banks with each bank having one or more pressure vessels joined by a manifold. The three storage banks are designated the high, medium and low banks, with the high and low banks having the highest and lowest filling priority respectively. The compressor operates to maintain maximum pressure in the storage banks through the priority panel by filling the high bank, medium bank and low bank in turn.

If all storage vessel banks are depleted in pressure, most priority panels shift all compressor discharge flow to feed the refueling vehicles directly. At this stage, unless the compressor has a very large flow capacity, refueling flow rates are slow and equal to the rate of compressor flow.

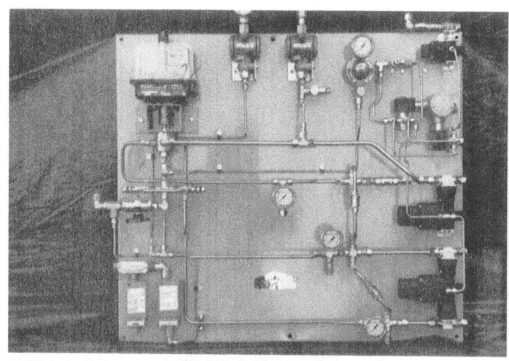

FIGURE 8.6 Priority panel (*courtesy of IMW Industries Limited*).

8.3.2 CNG Dispenser

While the compressor is maintaining storage bank pressures in accordance with priority, NGV's are drawing gas from the CNG dispenser. CNG dispensers resemble liquid fuel dispensers and often accompany them at the same retail service station.

The CNG dispenser includes a special compressed gas meter and two-way sequencing valves. Upon connecting the special refueling nozzle to the NGV and authorizing the dispenser, the first sequencing valve S1 opens (see Fig. 8.7). High pressure gas from the low bank drains into the NGV's fuel tank which is initially at a low or "empty" pressure (e.g. 200 to 500 psig). As the difference in head pressure between the low storage bank and NGV fuel tank reduces, the flow rate decreases. At a minimum flow rate, as measured by the dispenser meter, sequencing valve S2 opens. The gas flow rate increases with the increase in head pressure. Again, as the flow rate drops with the head pressure, a minimum flow rate signals the opening of sequencing valve S3. When the NGV fuel tank reaches full pressure, the sequencing valve closes, terminating the fill. In this way, multiple vehicles can be filled consecutively or simultaneously, depending on the number of dispensers provided.

Refueling flow rates are dependent on how restrictive the piping system is between the pressure vessels and the NGV fuel tank. Commonly the most restrictive piping is on board the NGV. Since there is no industry standard for on board NGV piping systems, predicting refueling performance is difficult.

FIGURE 8.7 CNG dispenser (*courtesy of Fueling Technologies Inc.*).

Typical refueling flow rates for passenger cars and light trucks average 300 to 500 scfm with maximum flow rates reaching as high as 900 scfm. Refueling flow rates for large commercial, industrial and public transportation vehicles can average 1500 to 2500 scfm with maximum flow rates reaching 5000 scfm.

8.3.3 Emergency Shutdown Systems

The CNG station Emergency Shutdown (ESD) System uses fail safe closed valves, strategically located to shut off gas flow in an upset scenario. The ESD valves are located in piping systems near pressure vessels, compressor suction and discharge lines, and inside CNG dispensers. Most ESD valves use either pneumatic, spring return operators or spring return electric solenoid actuators. The means of initiating valve closure include:

- Loss of electric power
- Manually depressing ESD push buttons located in various station locations
- Seismic activity detection devices
- Proximity or limit switch devices sensing motion of impacted equipment
- Vibration sensing equipment
- Gas, fire and heat detectors
- Loss of pneumatic control pressures from plastic air line ruptures caused by fire

8.3.4 Pressure Vessels Storage

Fast fill type CNG stations store compressed gas in one or more pressure vessels. The size, design pressure and configuration of the pressure vessels determine how much stored gas can be used for refueling before replenishment by the compressor is required. The CNG industry commonly uses pressure vessels configured in three banks. Each bank has at least one pressure vessel. Higher capacity installations use multiple pressure vessels joined by a manifold and perform as one larger volume bank. The CNG dispenser uses a set of valves which open in sequence during refueling. Each sequencing valve allows the flow of gas to the vehicle from one bank. By opening the valves in sequence, a greater percentage of gas can be withdrawn from the pressure vessel storage than if all the pressure vessels were joined by a manifold as a single bank.

Pressure vessel storage full pressures range from 3600 to 5000 psig. Industry standard NGV fill pressures are 2400, 3000, and 3600 psig at a standard gas temperature of 70°F. To complete NGV refueling at least one bank of a pressure vessel storage must have a higher pressure than the NGV fuel tank final fill pressure.

Storage Utilization. Only a fraction of the total stored weight of gas from full storage pressure vessels can be dispensed. At some point, there will be insufficient pressure in the storage vessels and the NGV fuel tank fill cannot be completed. At this point, the amount of gas remaining in the pressure vessels is substantial and may be 50 to 75% of the full amount. Storage utilization is defined as

Total gas weight dispensed/Storage full gas weight × 100%

Storage utilization increases with the number of storage banks used. The increase in storage utilization becomes marginally less with each additional bank. The industry has found the use of three banks to be the best compromise in terms of cost, complexity and performance.

In addition, storage utilization is highly dependent on storage full pressure and final vehicle fill pressure (see Table 8.3). Considerable gains in storage utilization occur when maximum storage fill pressures are increased. This allows the use of

TABLE 8.3 Storage Utilization (SU)*

Maximum storage vessel pressure % (psig)	SU —single bank —3600 psig fill	% SU —single bank —3000 psig fill	% SU —3 bank —3600 psig fill	% SU —3 bank —3000 psig fill
3600	n/a	12	n/a	36
4000	6	17	24	43
4500	10	21	34	49
5000	13	25	40	53

Note: The SU values listed were generated with the following assumptions: minimum differential pressure between the storage bank and NGV fuel tank is 100 psid; isothermal compression and expansion; no compressor replenishment during fill; three equally sized storage bank volumes for the three bank SU values; gas critical temperature of 366°F.; gas critical pressure of 669.84 psia; 70°F. gas temperature; 53 psig NGV fuel tank start of fill pressure.

*SU values generated with the assistance of Ralph O. Dowling, P.E., Christie Park Industries, using the Institute of Gas Technology "Cascade" computer program.

smaller storage vessel volumes and vessel cost savings. The benefits of higher storage fill pressures are offset, however, by an increase in compression energy costs and higher compression equipment and maintenance costs.

Other variables affecting storage utilization include

- NGV fuel tank empty pressures
- Dispenser sequencing valve switching set points
- Initial fuel tank gas temperatures
- Heat transfer rates from storage tanks, piping and NGV fuel tanks
- Successive vehicle filling versus periodic filling
- Relative volumes of each storage bank
- Gas critical pressure and critical temperature

8.3.5 Gas Dehydration

New standards for compressor discharge gas quality are making the use of gas dehydration equipment mandatory in many CNG installations. The Society Of Automotive Engineers standard SAE J-1616 specifies maximum allowable compressor discharge gas water content. This standard ensures that NGV fuel tanks will be safeguarded from corrosion and will operate safely for the life of the vehicle.

Gas dehydration equipment is most commonly installed on the suction or final discharge line of the compressor or compressor system. Gas dehydration equipment installed on compressor suction lines can incur significant piping line losses. A minimum pressure loss specification should be included with other process parameters when sizing and selecting equipment.

8.4 CNG STATION SYSTEM DESIGNS

Most CNG refueling station systems are custom designed to meet specific cost and performance criteria.

8.4.1 Three Line Fast Fill System

The three line fast fill system is commonly used for retail sale of CNG. It is among the most costly, but provides the maximum performance, with CNG dispensers providing fuel at flow rates comparable to liquid fuel dispensers. Fuel flows directly from the pressure vessel storage and is independent of compressor capacity as long as head pressure remains at the pressure vessels.

With the three bank storage, storage utilization is high. This enables the system to best meet random surges in demand. High storage utilization also allows the compressor to operate for extended periods of time, with a minimum of stops and starts.

8.4.2 Time Fill System

The time fill system requires the least amount of equipment and is the least expensive, and no pressure vessels are required. The compressor discharges directly

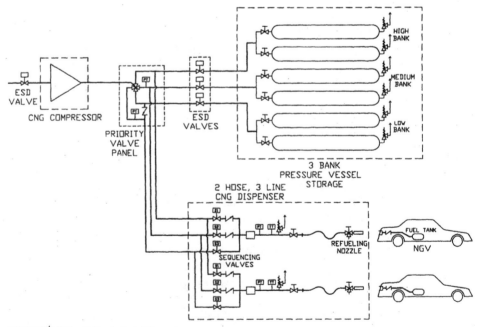

FIGURE 8.8 3 Bank fast fill system.

to the refueling vehicles. Refueling rates are dependent on the number of vehicles refueling at once. The system finds application with private vehicle fleets where fast refueling times are not important and vehicles are parked for an extended period of time during an off shift.

Upon connection of the refueling nozzle to the NGV, a pressure loss in the feeder line will be sensed by a pressure switch at the compressor. The compressor will start to compress gas through a time fill control panel. This panel has a pressure regulator and temperature instrumentation with shut off valve to stop the fill when the fill pressure is reached.

8.4.3 Single Line Fast Fill System

The single line system is similar to the three line fast fill system, but uses only a single bank storage. The dispenser is without sequencing valves and the priority panel has only one priority.

The use of more than one dispensing hose is not recommended. If two vehicles with different fuel tank pressures are connected to the same supply line, the refueling of one of the vehicles will halt until pressures equalize.

The system is less costly than the three bank system, but storage utilization is low. Surges in demand cannot be met as effectively. Low storage utilization results in frequent compressor start and stop operation.

8.4.4 Diverter System

The diverter system is similar to the single line fast fill system, but provides two hose refueling capability. A diverter valve directs all compressor flow to the first hose authorized. Upon authorization of the second hose, refueling will begin directly from the single bank storage without robbing compressor discharge flow

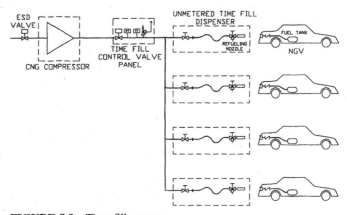

FIGURE 8.9 Time fill system.

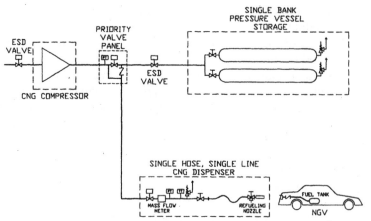

FIGURE 8.10 Single line fast fill system.

from the first hose authorized. As soon as the fill is complete at the first hose, the diverter valve switches all compressor flow to the second hose.

This system finds application in the refueling of fleet vehicles one after another. The toggling of the diverter valve spaces vehicle filling so that at least one vehicle is always connected to a hose. While one vehicle is moved into position and connected to the fueling hose, the second vehicle is nearing fill completion. In this way, refueling is nonstop and the compressor operates continuously with a minimum of starts and stops.

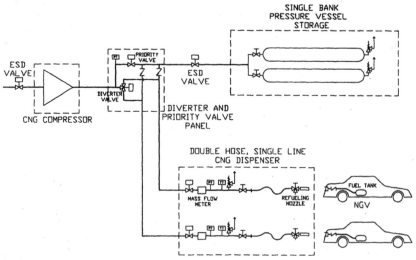

FIGURE 8.11 Diverter system.

8.5 EQUIPMENT SELECTION AND SYSTEM PERFORMANCE

The following example will demonstrate how to select a compressor and pressure vessel storage for a three line fast fill station. For other refueling system types a similar approach can be adopted.

Example:

A proposed retail CNG refueling station site will have 25 psig regulated and metered gas available by the local gas utility. Maximum compressor discharge gas pressure will be 3600 psig. The station will be of the three line fast fill type, with maximum NGV fill pressures of 3000 psig. The proposed CNG refueling installation is forecast to have a filling frequency distribution as per Table 8.4. The average fuel consumption of each vehicle is also determined to be 4.6 gallons gasoline equivalent per day.

8.5.1 Compressor Selection

Table 8.4 indicates the number of NGVs that will fill up each business hour. Total amount of gas dispensed each day will be

$$75 \text{ NGVs} \times 4.6 \text{ gal equiv.} \times 108.7 \text{ scf/gal equiv.} = 37,500 \text{ scf}$$

This is also the amount of fuel that must be compressed daily. A maximum number of compressor operating hours per day is arbitrarily set at 8. Fewer operating hours increase compressor size, cost, energy demand charges and interrupted operation. More operating hours reduce compressor size and the ability to meet random increases in refueling demand. The required compressor flow capacity is calculated.

$$37,500 \text{ scf/8 hrs/60 min/hr} = 78 \text{ scfm}$$

From Fig. 8.2, a 25 psig suction pressure and 3600 psig discharge pressure requires 0.48 BHP/scfm. At 78 scfm, the required compressor horsepower is 37.4 BHP. A 40 BHP, 4 stage compressor is selected, providing a flow capacity of 83 scfm.

TABLE 8.4 Daily Refueling Frequency Distribution

Hour end time (a.m.)	1	2	3	4	5	6	7	8	9	10	11	12
No. of Vehicles	0	0	0	0	0	0	2	20	5	2	3	8
Hour end time (p.m.)	1	2	3	4	5	6	7	8	9	10	11	12
No. of Vehicles	6	2	2	3	10	6	2	2	2	0	0	0

8.5.2 Pressure Vessel Storage Sizing

The filling frequency distribution is critical to the selection and sizing of the pressure vessel storage. The hours from 7am to 8am is the busiest with 20 fill ups. Gas withdrawn from the storage pressure vessels during this hour is calculated.

$$20 \text{ NGVs} \times 4.6 \text{ gal equiv.} \times 108.7 \text{ scf/gal equiv.} = 10{,}000 \text{ scf}$$

In this time, the compressor has also operated to refill the storage. To maximize compressor running duration, control systems do not start up the compressor until there is a substantial drop in pressure of the storage. To account for this, it is assumed that the compressor operates for 45 minutes of the hour. Storage replenishment flow is calculated.

$$45 \text{ min.} \times 83 \text{ scfm} = 3735 \text{ scf}$$

In addition, it is assumed that the two vehicles between 6am and 7am did not initiate compressor operation. This accounts for an additional 1000 scf depleted from storage.

The total amount of gas that must be provided from the storage pressure vessels is calculated.

$$10{,}000 \text{ scf} - 3735 \text{ scf} + 1000 \text{ scf} = 7265 \text{ scf}$$

The size of the pressure vessel storage can now be estimated. From Table 8.3, the storage utilization factor is 36%. The size of the pressure vessel storage is calculated.

$$7265 \text{ scf}/36\% = 20{,}181 \text{ scf natural gas @ 3600 psig}$$

Three pressure vessels each having an internal volume of 22.8 cu. ft are selected. Each vessel provides a storage capacity of 7,082 scf at 3600 psig working pressure, or a three vessel total of 21,246 scf.

8.5.3 Other Equipment

A priority panel is selected and sized for the compressor discharge capacity of 78 scfm. Pressure losses through the panel are kept below 50 psid so that the storage pressure vessels are filled to a maximum pressure without compressor shutdown. A two hose metered CNG dispenser is specified so that two vehicles can be filled simultaneously during peak demand times.

8.6 CODES AND STANDARDS

CNG refueling equipment and installations must comply with numerous codes and standards. In North America, work is currently underway to further develop codes

TABLE 8.5 Industry Codes and Standards[t]

AMERICAN NATIONAL STANDARDS INSTITUTE (ANSI)	
ANSI/AGA NGV1	Standard for Compressed Natural Gas Vehicle Refueling Connection Devices
ANSI/AGA NGV2	Basic Requirements for Compressed Natural Gas Vehicle (NGV) Fuel Containers
ANSI/ASME B31.1	Power Piping
ANSI/ASME B31.3	Chemical Plant and Petroleum Refinery Piping
ANSI/ASME B31.8	Gas Transmission and Distribution Piping System

AMERICAN SOCIETY OF MECHANICAL ENGINEERS (ASME)	
ASME Section VIII, Division 1 and 2	Boiler And Pressure Vessel Code
ASME Section IX	Boiler And Pressure Vessel Code: Qualifications Standard for Welding and Brazing Procedures, Welders, Brazers, and Welding and Brazing Operators

AMERICAN PETROLEUM INSTITUTE (API)	
API Specification 11P	Specification for Packaged Reciprocating Compressors for Oil and Gas Production Services
API Standard 618	Reciprocating Compressors for General Refinery Service
API Standard 661	Air-cooled Heat Exchanger for General Refinery Services

NATIONAL FIRE PROTECTION ASSOCIATION (NFPA)	
NFPA 37	Standard for the Installation and Use of Stationary Combustion Engines and Gas Turbines
NFPA 52	Compressed Natural Gas (CNG) Vehicle Fuel Systems
NFPA 54	National Fuel Gas Code
NFPA 70	National Electrical Code
NFPA 496	Purged and Pressurized Enclosures for Electrical Equipment

SOCIETY OF AUTOMOTIVE ENGINEERS (SAE)	
SAE J1616	Recommended Practices for Compressed Natural Gas Vehicle Fuel

[t]Industry codes and standard compiled with the assistance of Ray Benish, CNG Systems Inc.

and standards for the industry. The National Fire Protection Association (NFPA) Standard 52 is one of the most important industry standards and invokes other important documents including the ASME code and ANSI standards. Beyond the codes and standards listed in Table 8.5, state regulations, building codes and municipal bylaws are routinely enforced.

CHAPTER 9
LIQUID TRANSFER/VAPOR RECOVERY

William A. Kennedy Jr.
Blackmer/A Dover Resource Company

Gas compressors are often used in the bulk transfer of liquefied gases from rail cars or truck transports into a storage vessel. Liquefied gases are products that are contained at a vapor/liquid equilibrium in closed systems above atmospheric pressure. Typical examples are hydrocarbons (propane, butane, propylene, etc.), carbon dioxide, refrigerant gases, chlorine, hydrogen chloride, and some solvents. Most of these products must be kept free of contaminants like moisture, oil and non-condensables such as air or nitrogen.

While the use of rail cars to transport compressed liquefied gases is a widespread and safe practice, the process engineer is faced with several system design problems because of the following:

1. The system NPSHA (Net Positive Suction Head Available) is less than required by a liquid pump. Top mounted control valves with "dip tubes" are used on these types of rail cars.
2. The liquid vapor pressure is above atmosphere at ambient temperatures.
3. The system must not be contaminated with moisture, oil, or air.

Unloading a rail car can be handled by a liquid pump, air padding (or other non-condensable gas), gas compressor or a combination system using a liquid pump and gas compressor. The problems and merits of each method are discussed below.

9.1 TRANSFER USING A LIQUID PUMP

Cavitation and loss of pump prime are common problems when using liquid pumps to unload rail cars. Since liquefied gases are stored at their vapor/liquid equilibrium point, any reduction in pressure (caused by fluid friction in the pump inlet line), or increase in temperature (caused by heat gain in the pump inlet line), results in vapor forming in the inlet piping and/or internal pump cavitation. Either condition reduces the transfer rate and damages internal pump parts.

Another problem when using a liquid pump is the amount of product left in the rail car when it appears to be empty since the dip tubes do not reach all of the way to the bottom of the tank (see Fig. 9.1). In addition, there is product in vapor form remaining in the rest of the rail car. For example, in a typical 11,000 gallon tank car, with the dip tube 3 inches from the bottom, there would be about 115 gallons of liquid left below the dip tube. Usually there is even more liquid remaining because a liquid pump will lose its prime before the liquid level reaches the bottom of the dip tubes. The amount of vaporized product left would depend on the product being transported. A propane tank car on an 80°F day (vapor pressure = 144 psia) would still contain 1465 pounds of propane (or 344 gallons when liquefied) in the vapor space of an 11,000 gallon tank car *after* all of the liquid had been removed.

The use of a liquid pump to unload a compressed liquefied gas from a rail car with top mounted outlets presents the following problems:

1. Extreme difficulty in priming and maintaining the prime due to poor NPSH conditions and likely cavitation problems
2. Low and unpredictable transfer rates due to cavitation in the inlet line and pump
3. Excess noise and internal pump damage due to cavitation
4. Failure to remove all the liquid from the tank car
5. Failure to remove any vapor from the tank car

9.2 AIR PADDING

Another unloading method that will overcome some of the liquid pump problems is to use compressed air to "pad" the car. A dedicated system designed to supply oil free air at a pressure greater than the maximum vapor pressure of the liquid in the tank car is required. Usually the air must be dry (−40°F dew point is typical) to prevent moisture contamination. Nitrogen is sometimes used in place of dry air. While air padding solves the system NPSH problem by pushing the liquid out, it leaves several problems unaddressed and creates others.

1. There is still liquid left below the dip tube.
2. The vaporized liquid is still in the tank car.
3. If the air drying system fails, the resulting moisture contamination can result in product quality or system corrosion problems.
4. The most serious problem may be the dilution of the vaporized liquid with air. The presence of a non-condensable gas (air or nitrogen) will cause tank pressures greater than the vapor pressure of a pure product. When the rail car is

LIQUID TRANSFER/VAPOR RECOVERY **9.3**

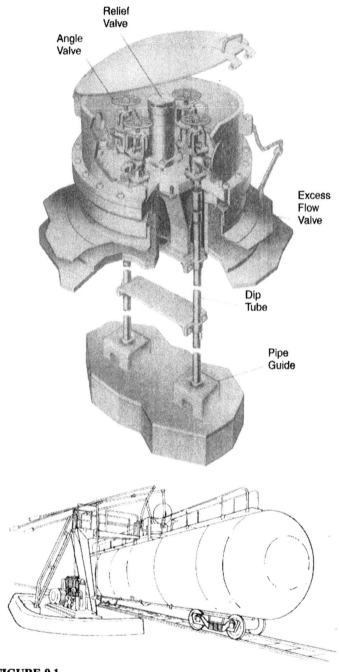

FIGURE 9.1

later filled with liquid, this may cause relief valves to open unexpectedly releasing product to the atmosphere.

9.3 TRANSFER USING A GAS COMPRESSOR

The best solution for unloading rail tank cars is the use of a gas compressor rather than a liquid pump or air padding. Figure 9.2 shows a typical schematic of a liquid transfer operation using a gas compressor. The vapor section of the receiving tank is connected to the compressor suction while the vapor section of the rail tank car is connected to the compressor discharge. A separate liquid line connects the liquid sections of the two vessels. Liquid transfer begins as the compressor transfers vapor from the receiving tank to the vapor section of the rail car. In a well designed system, the pressure differential developed between the two vessels will be 30 psi or less. The process is continued until the liquid level falls below the liquid dip tube opening in the rail car. This phase of the operation requires that enough gas be transferred to displace the volume of liquid leaving the rail car plus the amount of gas condensing into liquid in the rail car. Since the increase in gas temperature caused by the heat-of-compression helps keep the gas from condensing, the compressor discharge line (leading to the rail car) should be insulated and the compressor should be installed near the rail car, ensuring minimum heat is lost from the compressor discharge gas. In order to get the remaining product out of the rail car, the system is changed to a "vapor recovery" mode, which is possible with a gas compressor but not with a liquid pump or air padding.

Figure 9.3 illustrates the vapor recovery mode. The "liquid" line is closed. The vapor section of the rail car is connected to the compressor suction and the compressor discharge is connected to the liquid section of the receiving vessel. The

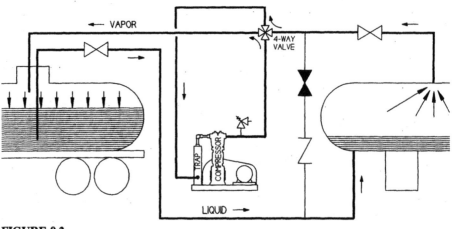

FIGURE 9.2

VAPOR RECOVERY

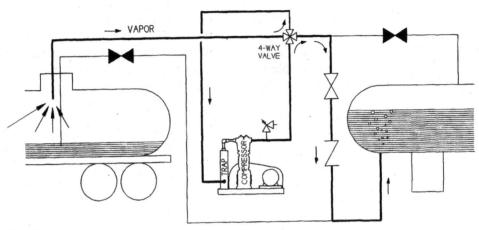

FIGURE 9.3

connection change can be accomplished easily with the use of a multi-port selection valve in the piping to and from the compressor. During the early stages of vapor recovery, the rail car pressure will decrease slightly, causing the liquid below the dip tube in the rail car to vaporize. Once vaporized, it is transferred by the compressor into the receiving tank. Bubbling the gas through the liquid phase in the receiving tank ensures rapid condensation with little increase in receiver tank pressure. In extreme conditions, a separate condenser may be required.

After all of the liquid in the rail car has vaporized, the pressure will begin to decrease as more gas is removed. The degree of vapor recovery is usually dependent on economics—the value of the gas, the power required to operate the compressor and time available to hold the rail car. A rough rule of thumb is to reduce the rail car pressure to 25% of the product vapor pressure.

The use of a gas compressor for unloading rail cars addresses all the problems noted above when using liquid pumps or air padding. A well designed compressor system will provide years of safe operation with limited down time. The increased product recovery (the liquid below the dip tubes and the vapor in the rail car) will actually reduce the number of rail cars required over a period of time. The NPSH problems and related cavitation go away and there should be no product contamination.

When designing gas compressor systems for liquid transfer/vapor recovery operations, some unique system requirements must be considered. An often overlooked, but important system feature is the rated liquid flow of the excess flow valve on the rail car. The excess flow valves located in the dip tubes, shown in Fig. 9.1, are designed to automatically close at a given flow rate to prevent spillage due to a major line break. Once the allowable flow rates are determined, the size of the compressor can be established. An oversize compressor will cause the excess flow valves to close, stopping the unloading operation. Another critical decision involves how low the rail car pressure is to be reduced. This determines whether a single- or multi-stage compressor is required. Careful analysis of the compressor operating temperature is required. Normally, every effort is made to keep the operating temperature of a compressor low, which increases the life of the machine, so water cooling is common in the compressor industry. However, except in extreme cases, an air cooled compressor is best used in a liquid transfer operation because it is desirable to keep the gas temperature up to help reduce condensation of gas in the rail car during the liquid transfer mode.

The compressor manufacturer can supply performance data to help determine the time required for liquid transfer, time needed to reduce rail car pressure to various levels, power requirements and operating temperatures. These are rather complex calculations since the compressor is not operating at steady state conditions, and compressor performance varies with product, ambient temperature, tank sizes, piping losses, location of the compressor, etc.

9.4 COMBINATION COMPRESSOR/PUMP SYSTEMS

While the use of a compressor solves many problems with unloading rail cars of compressed liquefied gases, there are some limitations that must be observed. High pressure differentials will decrease the transfer rate. This is usually caused by poor piping design; great separation of rail car from the storage tank; elevation difference of two tanks; or metering liquid flow with a high pressure drop meter, i.e., positive displacement type meter.

If any of these conditions exist, a combination compressor/liquid pump system may be the best solution (see Fig. 9.4). This is the same as Fig. 9.2, except a liquid pump has been inserted in the liquid line. The NPSH problems caused by the poor pump suction conditions (dip tube and long suction line) are solved by using the gas compressor to increase the rail car pressure with gas from the receiving vessel. With improved suction conditions, the liquid pump can now provide the high differential pressure required by the system. Once the liquid level drops below the dip tube, the liquid pump is turned off. Removal of the remaining liquid and vapor recovery take place as shown in Fig. 9.3.

9.5 COMPRESSORS FOR LIQUID TRANSFER/VAPOR RECOVERY

While there are many variables, typical rail car liquid transfer/vapor recovery compressors (Fig. 9.5) will be rather small (up to 60 CFM piston displacement), air cooled and single stage. Larger compressors may be used for multiple rail car unloading. The following design features are common:

1. Cylinder working pressure greater than the maximum expected product vapor pressure plus design system differential pressure

LIQUID TRANSFER/VAPOR RECOVERY 9.7

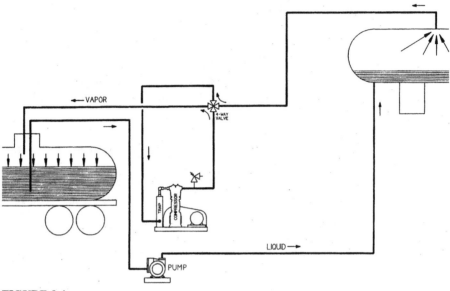

FIGURE 9.4

2. Atmospheric vented crankcase design with a filtered pressure oil system to lubricate the main bearings and connecting rods
3. Crosshead/piston rod construction allowing for separation of the crankcase from gas compression area

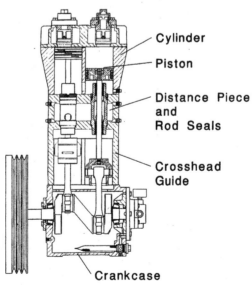

FIGURE 9.5

4. Non-lubricated gas compression construction for oil-free transfer of gas
5. Piston rod seals to control gas leakage and prevent crankcase oil from entering the cylinder area. While a single seal per rod may be used, the preferred construction is two seals per rod with a closed distance piece allowing for venting, purging or padding between seals. For certain gases (chlorine, vinyl chloride, hydrogen chloride, etc.), a third seal per rod and a second distance piece is recommended.
6. High temperature switches, high/low pressure switches, high liquid level switches in liquid traps, etc., help prevent unexpected equipment problems.

Since the gas is being handled at, or very near, its vapor-liquid equilibrium point, there is always concern with condensation in the piping or in the compressor cylinder. To help protect against liquid entering the compressor (which may result in severe damage), several system designs must be considered:

1. A liquid trap should be located as near the compressor inlet as possible. The trap must be sized to accommodate any anticipated condensation that could occur in the inlet piping. Liquid level alarms, shut-off devices (mechanical or electrical), mist pads and/or sight gages may be part of the system. Figure 9.6

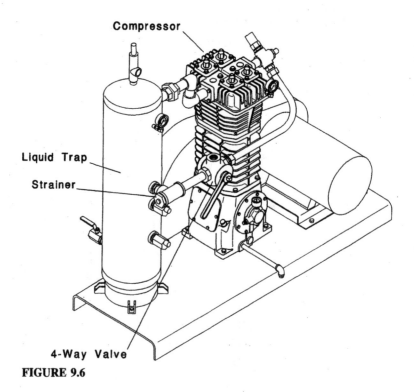

FIGURE 9.6

shows a typical package including a liquid trap, 4-way valve, strainer and drive system.
2. Piping should be designed to prevent condensate from draining into the compressor during shutdown.
3. Operating procedures must be established to drain any condensate prior to start up.

CHAPTER 10
COMPRESSED NATURAL GAS FOR VEHICLE FUELING

Adam Weisz-Margulescu, P. Eng.
FuelMaker Corporation

Air quality is a major public concern. Motor vehicles are a major source of air pollutants that have negative effects on the environment. Faced with unacceptable air quality and growing public concern, governments and industry have taken a number of initiatives to reduce motor vehicle emissions. The trend is clear: in order to achieve acceptable air quality motor vehicle emissions standards will become more stringent. The natural gas vehicle industry has become the leader in the drive for clean air. At the same time, due to stringent indoor clean air mandates, fuel pricing and supply problems, the natural gas forklift market has become the fastest growing market niche in the natural gas vehicle industry.

Until the development of appliances for compressing natural gas to pressures required for fueling vehicles, the slow-growth in public fueling infrastructure has made it difficult for motor vehicles to readily access natural gas fuel. The high cost of natural gas fueling equipment is prohibitive for a small number of vehicles. These compact units give a small fleet operator the ability to perfectly size their fueling requirements to the exact number of vehicles they service. At the same time, the cost of the fuel station and conversion will be recovered through the fuel price differential. These appliances can be used as an independent slow-fill gas refueling appliance to provide compression for up to two vehicles simultaneously, or they can be configured with multiple number of vehicle refueling appliances coupled together to provide fuel to the fast-fill storage system. The storage system in turn will provide fast-fill to the natural gas vehicle. For fleets that park their vehicles indoor, an indoor remote panel is used (Fig. 10.1).

10.1 REFUELING APPLIANCE

The refueling unit is generally a self-contained, oil-free outdoor appliance that will fill a 26.4 U.S. gallon gas cylinder to a pressure of 3000 psig @ 68°F within 8

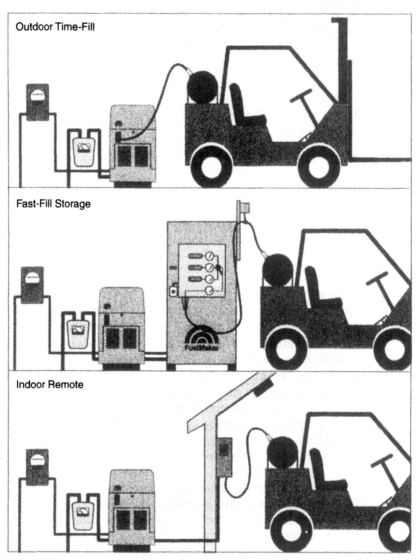

FIGURE 10.1 CNG fueling systems.

hours, which corresponds to an average flow rate of 1.8 SCFM. The flow rate is roughly the energy equivalent to about 1.1 U.S. gallons of gasoline per hour, depending on the energy content of natural gas. Average power consumption is only about 1.3 kW using an electrical supply of 230/208 Volts @ 60 Hz (Fig. 10.2). The appliance is connected to low-pressure gas system from 7 in. water column to 2 psig at rated flow. It is usually supplied with one fibre-reinforced high pressure fill hose (second hose can be connected) connected to the unit via a breakaway fitting which allows the hose to be disconnected without damage should the user drive the vehicle away without disconnecting. The refueling nozzle supplied is

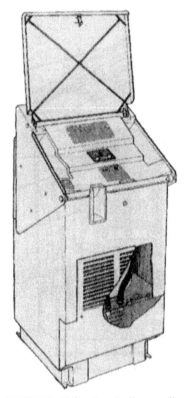

FIGURE 10.2 A refueling appliance.

suitable for natural gas "slow-fill" applications. At the completion of each refuelling cycle, the high pressure gas contained downstream of the compressor is returned to a blowdown vessel, thus reducing the pressure in the fill hose to approximately 29 psig.

A typical appliance is composed of the following modules (Fig. 10.3):

1. Compression
2. Controls
3. Electronics

The compression module and control module form one compact, integrated unit (Fig. 10.4). The gas flow with the unit "on" is illustrated in Fig. 10.5. When the unit is turned "off" the gas is recirculated as shown in Fig. 10.6.

The blowdown vessel is part of the controls module and will accommodate the volume of gas contained by the fill hose, refueling nozzle, and the space between the vehicle receptacle and check valve only. This limits the maximum length of the fill hose. The maximum length of the fill hose is limited by the NFPA 52 Code as well, to 25 ft.

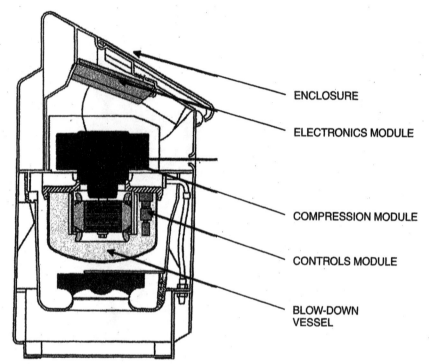

FIGURE 10.3 Modules within the refueling appliance.

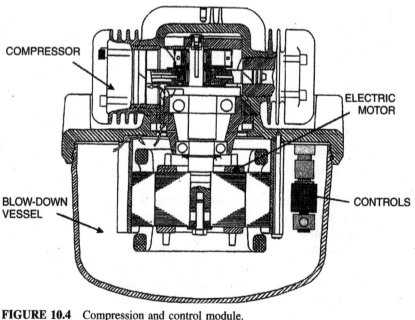

FIGURE 10.4 Compression and control module.

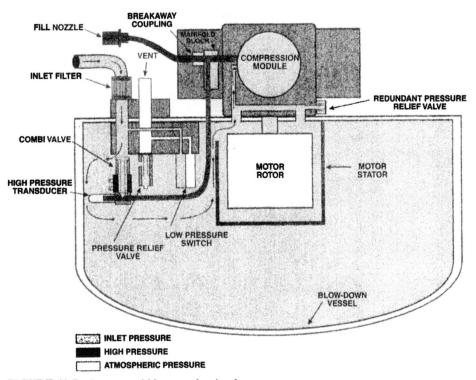

FIGURE 10.5 Pressure within control unit when on.

The electro-mechanical controls are mounted on the convection plate, which represents the interface between the blowdown vessel and the compressor (Fig. 10.7). The low pressure switch is set to shut down the unit if the inlet pressure drops below 5 in. of water column (Fig. 10.8). The low pressure relief valve will release the pressure into the vent line in case the blow-down vessel is over-pressurized (Fig. 10.9). The high pressure transducer monitors the high pressure output from the compressor (Fig. 10.10). It is calibrated for 2900 psig. The pressure is temperature compensated. The temperature sensor is mounted in the inlet air stream and determines the allowable fill pressure for a particular ambient temperature. The temperature/pressure compensation feature attempts to fill the storage tank with a constant mass of gas, regardless of the ambient temperature. This prevents the vehicle tank from being over-pressurized if the ambient temperature rises. The convection plate temperature sensor will shut down the unit at 167°F and the motor temperature switch will turn off the motor at 275°F.

The electronics module controls the operation of the unit. The schematic diagram is shown in Fig. 10.11. Some parameters can be changed by the installer or service personnel in the field via a programming device available. The electronics module is interfaced with the user panel. Starting, stopping and monitoring of the unit takes place at the user panel. It has separate Start and Stop buttons and three indicator lights.

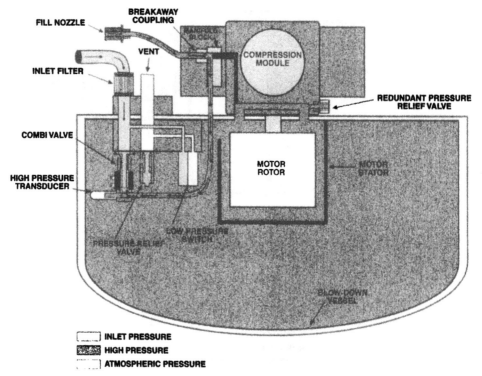

FIGURE 10.6 Pressure within control unit when off.

10.2 COMPRESSOR

The compression module as shown in Fig. 10.12 is a reciprocating motion type, four-stage, four-cylinder non-lubricated arrangement with the direct-drive rotor mounted directly on the drive-shaft. As all reciprocating compressors, this unit operates on an adiabatic principle: the gas is drawn into the first stage cylinder via the crankcase, is compressed in the individual cylinders, is moved from stage to stage via integrated gas passages through inlet and discharge valves, and finally is passed through the fourth stage discharge valve against discharge pressure into the vehicle storage tank. The gas is cooled between stages through the integrated passages by conduction and radiation. The geometry of the compressor and the design of the gas passages between stages will facilitate the dissipation of the heat generated by compression into the surrounding aluminum structure. In turn, the finned cylinder heads are cooled via forced convection with a separate two-speed fan mounted in the enclosure below the blowdown vessel. In case of blocked vent lines, a pressure relief valve mounted on the compressor housing will release to atmosphere at 145 psig. A high pressure burst disc installed in the fourth stage cylinder head will provide protection if the pressure rises above 3335 psig.

The gas is drawn into the blowdown vessel through the inlet pipe, passing through an inlet filter and a combination valve. The blowdown vessel is an integral

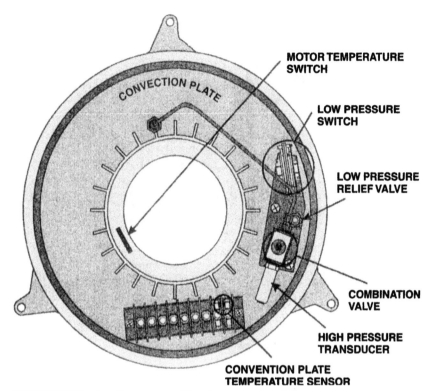

FIGURE 10.7 Interface between blowdown vessel and the compressor.

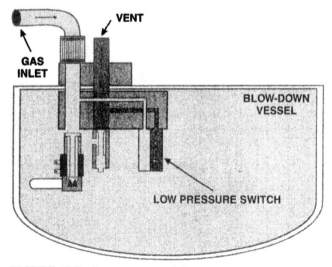

FIGURE 10.8 Low pressure switch.

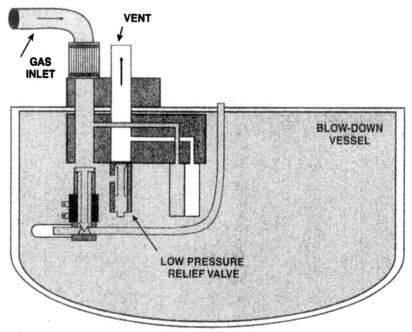

FIGURE 10.9 Low pressure relief valve.

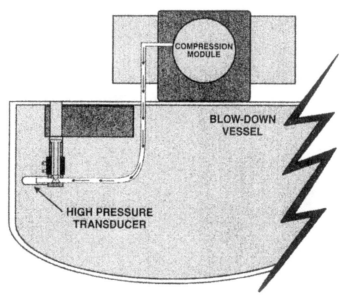

FIGURE 10.10 High pressure transducer.

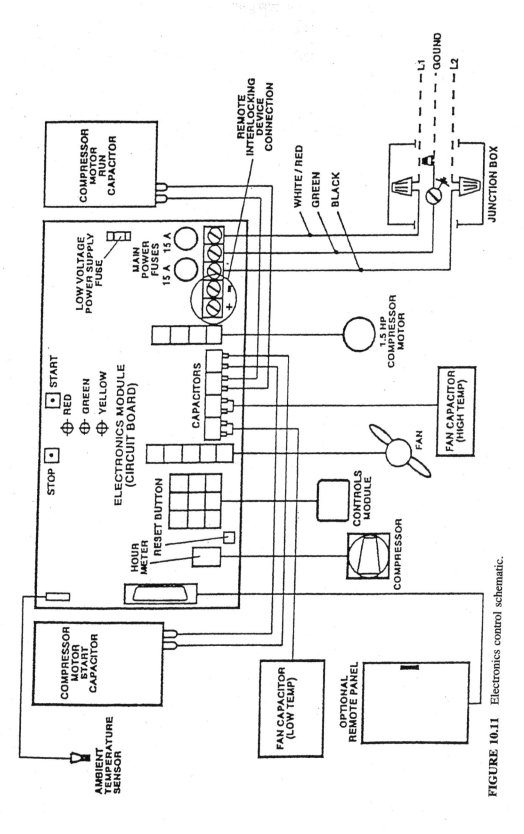

FIGURE 10.11 Electronics control schematic.

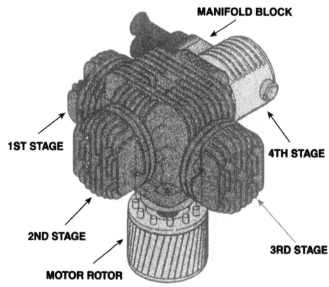

FIGURE 10.12 Compressor.

component system of the unit, designed to reduce the delivery side pressure (in the fueling hose and the space between vehicle receptacle and check valve) from operating level to approximately 29 psig. "Blowdown" allows the nozzle to be disconnected from the vehicle.

Via two holes in the compressor housing the gas enters the back end of the first stage cylinder. The first stage piston is fitted with six valves kept closed by disc springs and activated by differential pressure in the system. During the downstroke of the piston, the valves open and the gas rushes into the first stage cylinder cavity. As the piston reverses direction, the pressure increases and the valves close, and the compression cycle is completed. On top of the cylinder, identical valves are installed into a valveplate. At the end of the stroke these valves will open due to the differential pressure, and the gas is pushed through passages in the compressor housing to stage 2 and subsequently to stages 3 and 4. In the second and third stages, the gas enters the compression chamber via valves installed in the cylinder sleeves and exits through one similar but larger valve placed in the centre of the cylinder head. Before entering the fourth stage compression chamber, the gas is filtered again. In order to minimize pulsation, the gas is passed through a damper before exiting through the high pressure block. The fill hose has one end attached to the breakaway coupling and the fill nozzle end connected to the vehicle receptacle (Fig. 10.13). During the "off" cycle, the gas is being recirculated into the blowdown vessel through the high pressure block and tube connected to the convection plate (Fig. 10.14).

The piston-head clearance is kept to a minimum. But due to clearances necessary to permit operation and allow valve passages to be incorporated, the piston does

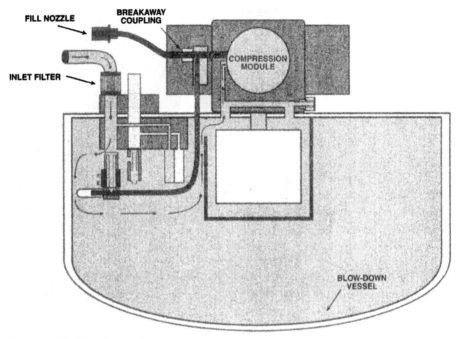

FIGURE 10.13 Flow during on cycle.

not sweep the entire volume of the cylinder. Hence the actual cylinder capacity is lower than the displacement. The volumetric efficiency of the cylinder is:

$$E_v = Q/C_{dis}$$

where

E_v is the volumetric efficiency
Q is the total volume through-put per unit of time at suction conditions in CFM
C_{dis} is the volume swept by all pistons per unit of time in CFM

To improve efficiency, the compression ratio per each stage is kept as close to constant as possible. To achieve the discharge pressure necessary, the approximate ratio/stage is four.

10.3 COMPRESSOR BALANCE

The compressor is driven by a constant speed (1750 RPM) 1.5 HP electric motor. The rotor is mounted directly on the drive-shaft, while the stator is mounted on the convection plate. Constant air gap is being maintained to minimize temperature increases and eddy current losses.

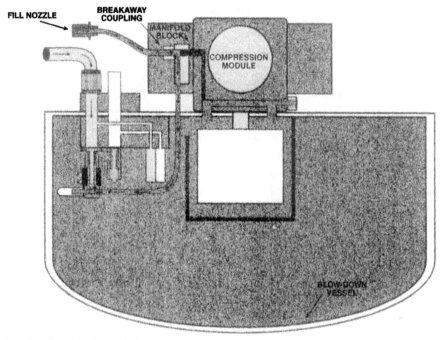

FIGURE 10.14 Flow during off cycle.

The drive-shaft is supported by two ball bearings. The crank end of the crankshaft (drive-shaft) facilitates the conversion of the rotary motion into the reciprocating motion of the compressor.

The pistons for stages 1 and 3 are mounted in opposite direction on the yoke. The same is true for stages 2 and 4. The two piston/yoke assemblies are installed on the pin end of the crankshaft 90° off to each other in the horizontal plane of the compressor. The reciprocating motion is achieved through a set of linear bearings riding on a sliding block inside the yoke assembly.

Two counterweights are installed on each side of the yokes to balance the crankshaft assembly. Both synchronous inertia forces originating in masses with rotating motion and inertia forces originating in masses with purely reciprocating motion must be considered. Since the sum of the reciprocating masses in one direction (stage 1 + stage 3) are equal to the sum of the reciprocating masses in the other direction (stage 2 + stage 4), the oscillating forces of this compressor are identical to a V2/90° engine. For this kind of engine, the oscillating forces can be compensated in the counterweights. The following equation has to be satisfied:

$$M_{osc} = M_{cw}$$

where we can approximate:

$$M_{osc} = (m_{rot} + k\, m_{rec}) \times d_{cor\text{-}cgosc}$$

and

$$M_{cw} = m_{cw}(\text{top}) \times d_{cor\text{-}cgcw}(\text{top}) + m_{cw}(\text{bottom}) \times d_{cor\text{-}cgcw}(\text{bottom})$$

where:

M_{osc} is the moment produced by the oscillating masses
M_{cw} is the moment produced by the counterweight masses
m_{rot} is the sum of the rotating masses
m_{rec} is the sum of the reciprocating masses
$d_{cor\text{-}cgosc}$ is the distance from the centre of rotation to the centre of gravity of the oscillating masses
m_{cw} is the counterweight masses
$d_{cor\text{-}cgcw}$ is the distance from the centre of rotation to the centre of gravity of the counterweight masses
k is a constant percentage factor of compensation for reciprocating masses and is 0.5 for V2/90° engine mathematical model

10.4 COMPRESSOR COMPONENTS

The main components of the compressor are (Fig. 10.15):

1. Housing
2. Drive unit
3. Piston assemblies
4. Cylinder head assemblies
5. High pressure unit

The housing is made of cast aluminum and has integrated stainless steel tubes to provide the passages from one stage to the next stage. All static seals are natural gas compatible nitrile elastomers. The geometry and layout of the fins help air circulation to optimize cooling.

The drive unit has the crankshaft housed by a flange unit via two ball bearings. The flange unit facilitates the attachment of the drive unit to the housing. The two yoke assemblies are mounted on the pin end of the crankshaft, sandwiched between the two counterweights. The separated rotor is mounted at the free end of the crankshaft (Fig. 10.16). The bearings in each yoke assembly are well protected from dirt and are lubricated with a high viscosity synthetic grease suitable for a service temperature range of −40°F to +300°F.

The four piston assembly units are mounted on the two yokes and provide the sequential compression of the gas. The compression of the gas takes place in the

10.14 CHAPTER TEN

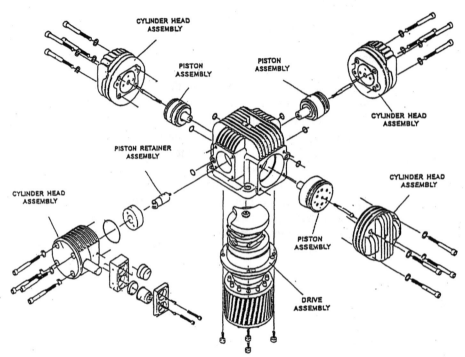

FIGURE 10.15 Exploded view of compressor.

four cylinder sleeves mounted in their respective cylinder heads. Sealing is achieved via non-lubricated sealing elements conceived as integral part of the pistons. Differential pressure valves for each stage are integrated at both the entry and exit of each individual stage to regulate the flow of the gas. All internal and external seals are made of natural gas compatible elastomers.

These refueling appliances would typically have a service interval of 2250 hours. At that time, the sealing and guide rings are checked and replaced when necessary, all "O" rings are replaced and the drive system is verified and repacked.

10.5 NATURAL GAS AS FUEL

Compressed natural gas as a vehicle fuel has been accepted all over the world. From Canada and U.S.A., where the majority of the installations exist, to Europe, Australia, South-America and Japan, the concept of natural gas as an alternate fuel for motor vehicles of all kinds has been embraced by more and more people. Considering the overwhelming environmental benefits, cost and availability, this system will expand as we approach the 21st century.

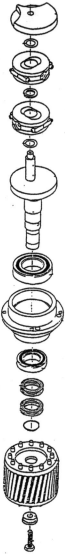

FIGURE 10.16 Exploded view of drive assembly.

CHAPTER 11
GAS BOOSTERS

Karl-Heinz Bark
MaxPro Technologies

Gas boosters are an alternative to high pressure stationary-type compressors. These boosters offer a compact, lightweight design that requires no electrical power or lubrication, thereby providing a more flexible and efficient source for delivering high presssure gas.

Gas boosters will compress gases such as nitrogen and argon up to 15,000 psi, while oxygen can be compressed up to 5,000 psi using special seals and cleaning procedures. A wide variety of other gases can be compressed including hydrogen, natural gas, ethylene, nitrous oxide, neon, carbon dioxide, carbon monoxide and breathing air.

In applications where high output pressures are required and the gas supply pressure is low, gas boosters can be operated in series. To achieve higher gas flows, two or more boosters can work in parallel as a unit.

11.1 APPLICATIONS

- Low pressure gas reclaim from storage bottles
- Breathing gas systems for scuba and fire department tanks
- Gas pressure and leak testing
- Charging of accumulators and high pressure inflation bottles for helicopter pop floats
- Boosting gas pressures from oxygen and nitrogen generators
- Nitrogen injection for molding machines
- Leak detection systems
- Low pressure autoclaving
- Cleaning petroleum tanks
- Glass blowing with oxygen

11.2 CHAPTER ELEVEN

- Typical gases—air, nitrogen, helium, oxygen, nitrous oxide, neon, argon, krypton, carbon monoxide, methane, ethylene and natural gas

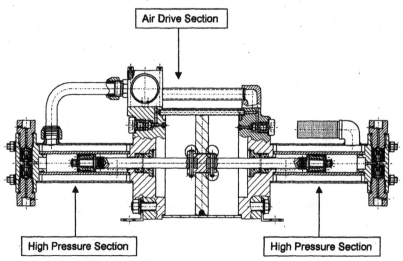

FIGURE 11.1 Gas booster cross section.

11.2 CONSTRUCTION AND OPERATION

A gas booster consists of a large air driven piston directly connected to a smaller area gas piston. The gas piston strokes in a high pressure gas section. The gas section contains inlet and outlet check valves. The air drive section includes a spool valve and pilot valves that cycle the pistons in both directions. Gas seal assemblies in the high pressure section are vented on the back side to prevent gas from getting into the air drive section. Cooling of the gas is provided by routing the cold exhausted drive air over the gas barrel section.

11.2.1 Air Drive Head

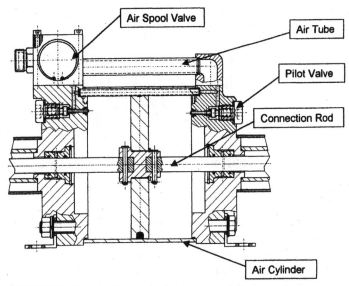

FIGURE 11.2 Air drive cross section.

11.2.2 Air Spool Valve

The air spool valve is a pilot operated spool that channels the compressed air to either side of the air drive piston, depending on the position of the spool. In certain operating conditions with high air consumption, it is possible that the regulated air drive pressure drops in the back chamber. For this reason, it is important that full air pressure is available to an unregulated air connection.

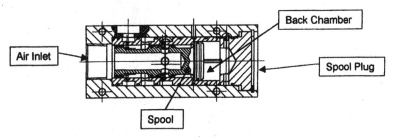

FIGURE 11.3 Spool value cross section.

11.2.3 Working Principle

After turning on the air drive, the spool moves to its upper position. Thereby the control line (Sx1) is released. The drive air is now at the pilot value (Vp1) and at the bottomside of the air piston which now makes a suction stroke.

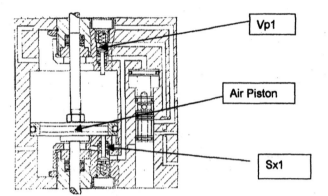

FIGURE 11.4 Air drive logic cross section.

Reaching its upper end position, the air piston switches the pilot valve (Vp1). The spool moves to its start position and the control line (Sx2) is released. The air piston switches the pilot valve (Vp2). By aid of a logical switching of the control lines, the volume (x) can bleed into atmosphere and the cycle returns. The booster will cycle as fast as it is able. To control cycle speed, an air speed valve may be installed at the air exhaust connection.

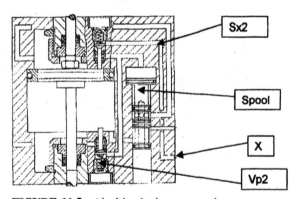

FIGURE 11.5 Air drive logic cross section.

Low Pressure Check Valve Assembly (under 1,000 psi)

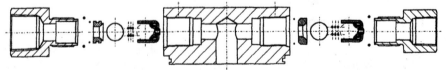

FIGURE 11.6 Low pressure check valve cross section.

High Pressure Check Valve Assembly (above 1,000 psi)

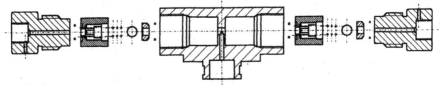

FIGURE 11.7 High pressure check valve cross section.

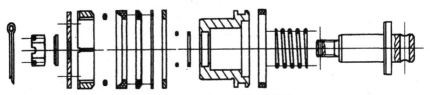

FIGURE 11.8 High pressure piston seal assembly (30:1 booster).

The seal assembly is one of the booster wear parts. This seals the gas barrel without letting any gas into the air drive section. The materials of some seals may change, depending on the gas, the pressure, and the temperature.

11.2.4 Dead Volume

Dead volume is that which does not displace but which must be put under pressure for the function of the compressor. This volume results, for instance, through bores, tubes, or valve cross sections. The high pressure plunger can completely stroke (in the pressure direction) and not eject the total gas volume. During suction stroke, the gas expands into the gas barrel until the pressure is equal to or less than the gas supply pressure (p_s), at which point only new gas enters the booster.

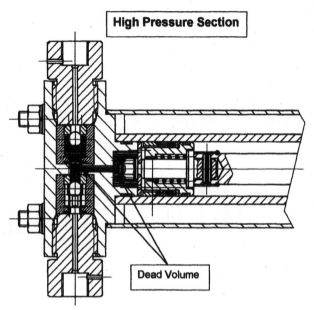

FIGURE 11.9 High pressure cross section.

11.2.5 Exhaust Air as Cooling for the High-Pressure Section

The exhaust air expands after the pressure stroke from the air drive section and the air is guided over the high pressure cylinder for cooling. On two-stage boosters, the connection line between the low and high pressure sections adds cooling.

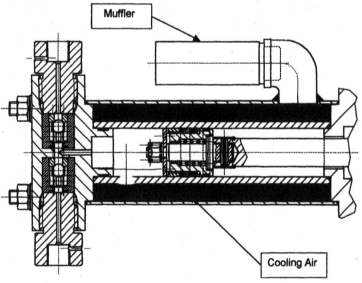

FIGURE 11.10 Gas end cooling.

11.2.6 Intercooler

The exhaust air from the air drive, used in boosters with a relatively high compression ratio, is used for cooling the high-pressure cylinders. Under certain operating conditions, it is sometimes necessary to reduce the stroke rate in order to avoid overheating the seals, or connect an additional Intercooler. This intercooler provides a longer lifetime of the high pressure piston seals in continuous duty applications.

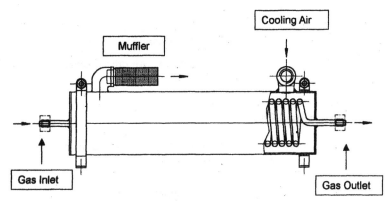

FIGURE 11.11 Inter-stage cooler.

11.2.7 Air Preparation

Version 1

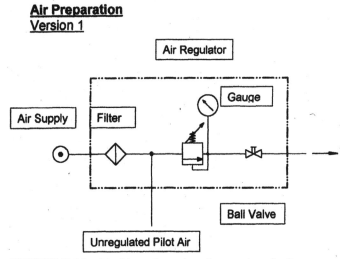

FIGURE 11.12 Air drive schematic to booster (standard).

Version 2. For applications with very dry air, when the remaining moisture of the grease is lost, it is necessary to use an additional oiler.

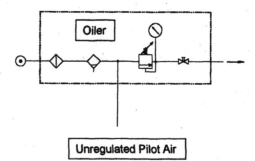

FIGURE 11.13 Air drive schematic to booster (very dry air).

11.2.8 Pressure Control Filter

The pressure control filter regulator and the ball valve are accessories and not considered part of the booster.

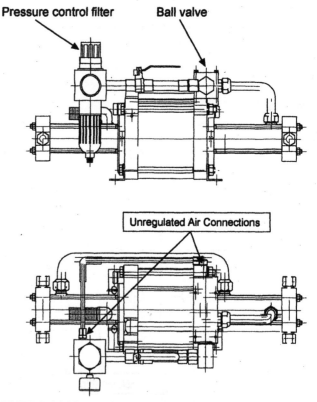

FIGURE 11.14 Typical plumbing arrangement for air supply.

11.2.9 Unregulated Pilot Air

The external pilot port must be connected with a separate supply line, otherwise the fill level of the back chamber is too small. It is possible that the controlled air drive pressure drops during the operation.

11.2.10 Basic Pressure Relation

Pressure Ratio (p_R) $\qquad p_R = p_O/p_A$

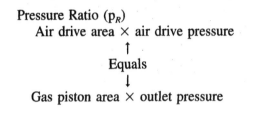

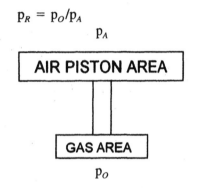

Example
Pressure ratio 30:1 and air drive pressure
90 PSI = 30 × 90 PSI = 2700 PSI outlet pressure
for single acting/single stage booster.

Compression Ratio $\quad C_R = p_O/p_S$

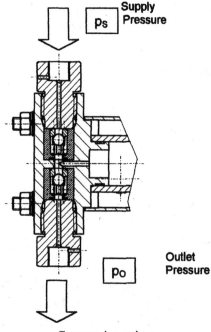

Compression ratio =
gas outlet pressure/gas supply pressure
FIGURE 11.15 High pressure check valve assembly.

The compression ratio is the ratio between gas supply pressure and gas outlet pressure, while the booster is still able to produce a gas flow. If the compression ratio is exceeded, the booster only works into the dead volume.

Stage Pressure Ratio $\quad A_1/A_2$. This is the actual area ratio between the first and the second stage gas plungers. This only applies to double acting/two stage boosters.

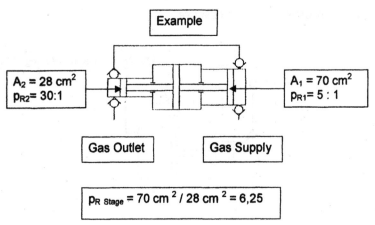

FIGURE 11.16 Example of 2-stage pressure ratios.

11.2.11 Suction Pressure Minimum/Maximum

If the suction pressure falls below the minimum stated suction pressure, a two stage booster would be required. For the single acting/single stage booster, the maximum suction pressure is lower, so that the air piston does not hammer against the top cap. The maximum suction pressure for double acting/single stage boosters is identical to the maximum gas outlet pressure. In case of double acting/two stage boosters, the maximum gas supply pressure depends on the air drive pressure. When the gas supply pressure is exceeded, the first pre-booster stage cannot boost the gas higher and cannot push it to the second stage.

11.2.12 Calculation of Outlet Pressure

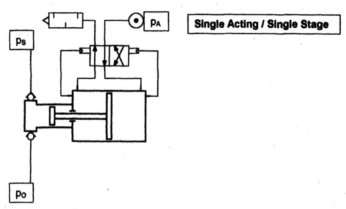

Outlet pressure (p_O) =
pressure ratio (p_R) × air drive pressure (p_A)

FIGURE 11.17 Typical installation diagram single acting/single stage booster.

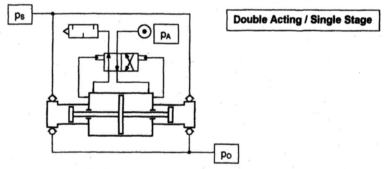

Outlet pressure (p_O) = pressure ratio (p_R) ×
air drive pressure (p_A) + supply pressure (p_S)

FIGURE 11.18 Typical installation diagram double acting/single stage booster.

Calculation of Outlet Pressure

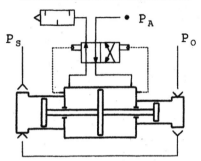

Outlet pressure (p_O) = pressure ratio (p_R) × air drive pressure (p_A) + (p_{R2}/p_{R1}) × p_S

FIGURE 11.19 Typical installation diagram single acting/2-stage booster.

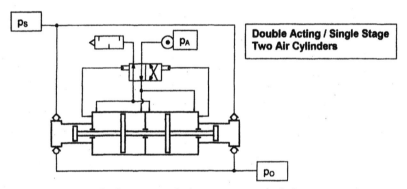

Outlet pressure (p_O) = pressure ratio (p_R) × air drive pressure (p_A) + supply pressure (p_S)

FIGURE 11.20 Typical installation diagram double acting/single stage/double air head booster.

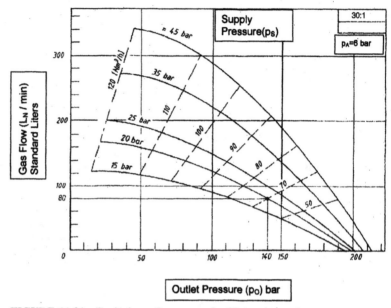

FIGURE 11.21 Typical gas flow curve for 30:1 ratio gas booster.

To find the gas flow, refer to a flow chart similar to Fig. 11.21 of the selected booster. The gas flow depends on gas supply pressure (p_O) and the air drive pressure (p_A). The flow chart above is only for 6 bar air drive. For 4 or 8 bar air drive, another flow chart would be used.

Example: Outlet Pressure = 140 bar and Air Drive = 6 bar

Gas Supply Pressure: 20 bar Gas Flow: 80 L_N/min

11.2.13 Selection of a Booster

Boosters are selected based on gas outlet pressure required, gas supply, and the air drive pressures available. Flow capacity should be checked from a flow chart, as the example in Fig. 11.21. If the flow capacity is not sufficient, a double acting booster will be required. If the supply pressure is not sufficient, it would be necessary to go to a two stage booster.

Example:

Air drive pressure $\qquad p_A$ = 6 bar (88 psia)
Working pressure $\qquad p_O$ = 140 bar (2058 psia)

GAS BOOSTERS 11.15

Gas supply pressure p_S = 15 bar (218 psia)
Gas flow required F = 40 L_N/min (1.41 SCFM)

Using technical data, a 30:1 ratio unit is the booster that fits these conditions. Refer to flow curve to verify flow capability.

11.2.14 Filling a Storage Tank (in a specific time)

Medium	Shop Air
Air Drive Pressure (p_A)	6 bsar (88 psi)
Supply Pressure (p_S)	6 bar (88 psi)
Working Pressure (p_O)	80 bar (1176 psia)
Tank Volume	20 Liter (.71 Cu Ft/Min)
Filling Time (t)	20 min

1. Pre-selected booster according to the pressure ratio
   $p_R = p_O/p_A = 80/6 = 13$
2. Required volume in tank at 80 bar.
   $V_N = V \times p_O = 20$ Liter $\times$ 80 bar = 1600 L_N
3. Volume in tank at supply pressure.
   $V_{N1} = V \times p_S = 20$ Liter $\times$ 6 bar = 120 L_N
4. Volume to be supplied by booster.
   $V_N - V_{N1} = F_{Fill} = 1600$ L $-$ 120 L = 1480 L_N

11.2.15 Average Gas Flow

$F\phi = V/t = 1480\ L_n/20$ min = 74 L_N/min

A booster would be selected with this capacity, based on the performance curves.

CHAPTER 12
SCROLL COMPRESSORS

Robert W. Shaffer
President
Air Squared, Inc.

Initial patents for the scroll concept date back to the early 1900's. Unfortunately the technology to accurately make scrolls did not exist and the concept was forgotten. In 1972, the scroll concept was re-invented.

The potential and advantages of the scroll compressor over reciprocating compressors were immediately recognized by the refrigeration industry. Because of the tremendous pressure for better efficiency of refrigeration compressors in the early '70s, there was a strong incentive to pursue the scroll: the balanced rotary motion reduced noise and vibration; there were no valves to break; and valve noise and valve losses were eliminated; fewer parts were needed; and rubbing velocities, along with associated frictional losses were lower. Not only did the scroll compressor offer improved efficiency, it also had the added benefit of greater reliability, smoother operation and lower noise. Today, scroll compressors are used extensively for residential and automotive air conditioning by many well known companies.

The development of scroll type compressors for air has not been as rapid. Air is much more difficult to compress than refrigerant, especially when oil is not used for sealing and cooling. By the '90s, machine tool technology had progressed to the point where scrolls could be accurately made and the first dry, oilless scroll compressor was introduced in January, 1992. The oilless scroll air compressors had the same inherent features as the scroll refrigeration compressor when compared to reciprocating oilless air compressors, durability, reliability, lower noise and vibration.

Currently scroll refrigerant compressors are well established as the standard of the industry. Scroll air compressors are extending from the initial three and five horsepower models into larger and smaller sizes from one to ten horsepower. Figure 12.1 shows typical scroll air compressors ranging in size from 1/8 to 1.0 hp.

Recently introduced technology is expected to make scroll air compressors practical in the fractional horsepower sizes.

FIGURE 12.1 Scroll air compressors from 1/8 to 1.0 HP.

12.1 PRINCIPAL OF OPERATION

The fundamental shape of a scroll is the involute spiral. The involute is the same profile used in gear teeth. An involute is a curve traced by a point on a thread kept taut as it is unwound from another curve. The curve that the thread is unwound from, that is, used for scrolls, is a circle. The radius of the circle is the generating radius.

A scroll is a free standing involute spiral which is bounded on one side by a solid flat plane, or base.

A scroll set, the fundamental compressing element of a scroll compressor, vacuum pump or air motor, is made up of two identical involutes which form right and left hand components. One scroll component is indexed or phased 180 degrees with respect to the other to allow the scrolls to mesh, as shown in Fig. 12.2. Crescent shaped gas pockets are formed bounded by the involutes and the base plates of both scrolls. As the moving or orbiting scroll is orbited about the fixed scroll, the pockets formed by the meshed scrolls follow the involute spiral toward the center and diminish in size (the motion is reversed for an expander or air motor). The orbiting scroll is prevented from rotating during this process to maintain the 180 degree phase relationship of the scrolls.

The compressor or vacuum pump's inlet is at the periphery of the scrolls. Air is drawn into the compressor as the inlet is formed as shown in Fig. 12.2. *b, c,* and *d.* The entering gas is trapped in two diametrically opposed gas pockets, Fig.

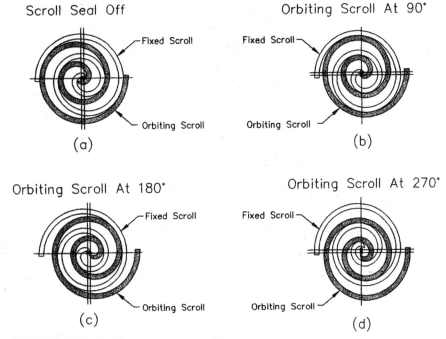

FIGURE 12.2 Scroll compressor operation.

12.2 *a*, and compressed as the pockets move toward the center. The compressed gas is exhausted through the discharge port at the center of the fixed scroll. No valves are needed since the discharge is not in communication with the inlet. Figure 12.2 shows the scroll positions as the line connecting the centers of the two scrolls is rotated clockwise, illustrating how gas pockets diminish in size as the orbiting scroll is orbited.

12.2 ADVANTAGES

Scroll refrigerant compressors, air compressors and vacuum pumps have the following advantages:

- Scroll compressors can achieve high pressure. The pressure ratio is increased by adding spiral wraps to the scroll. Pressures as high as 100 to 150 psig can be achieved in a single-stage air compressor.
- Scroll compressors are true rotary motion and can be dynamically balanced for smooth, vibration-free, quiet operation.
- There are no inlet or discharge valves to break or make noise and no associated valve losses.
- Scroll compressors can be oil flooded, oil lubricated, or oil free.

- Due to the unique orbital motion, the rubbing velocities of the sliding seals are significantly less than piston rings or vanes for comparable speeds. Rubbing velocities are typically 30 to 50% less, resulting in greater durability.
- Air is delivered continuously, therefore there is very little inlet or discharge pulsation and associated noise.
- The scroll compressor has no clearance volume that gets re-expanded with associated losses. The compression is continuous.
- Noise levels 3 to 15 dBA lower than other compressor technology are typical.

Table 12.1 gives some typical performance for scroll compressors operating on air.

12.3 LIMITATIONS

Although scroll compressors continue to expand into larger and smaller sizes, there are limitations. Since the scroll has a leakage path at the apex of the crescent shaped pockets, there are limits to how small a scroll compressor can be as a function of discharge pressure. Large displacement scroll compressors become large in diameter and the moving or orbiting scroll becomes massive. The maximum centrifugal force generated by the orbiting scroll gives a practical maximum size in a single-stage scroll.

12.4 CONSTRUCTION

Minimizing leakage of compressed gases within the scrolls is the key to performance in a scroll compressor.

There are two primary leakage paths in a scroll compressor. There is a leakage path at the apex of the crescent shaped air pockets where the scroll involves are

TABLE 12.1 Typical Performance Data of Scroll Air Compressors

Nom. power (HP)	Speed (RPM)	Disch. press. (PSIG)	Air flow (ACFM)	Sound power (dBA@ 1 m)
5.0	3050	120	15.0	59
3.0	2630	120	9.0	59
1.0	3450	100	4.0	NA
0.3	1750	30	3.0	49
0.02	3000	10	0.3	39

in closest proximity. This leakage is minimized by running the scrolls with a very small gap at these points. The size of the gap at the apex of the air pockets is a function of scroll geometry, and the scroll geometry is a function of the scroll manufacturing process.

There is also a leakage path between the tip of the involute and the opposite scroll base. Since the involute is relatively long if stretched out, this path is of primary importance. This leakage path is sealed by either, running the scrolls very close together and using oil to seal the remaining gap or using a floating tip seal as shown in Fig. 12.3. The floating tip seal acts much as a piston ring in a piston type compressor and bridges the running gap between the scrolls. For oil-free compressors, the tip seals are made of self lubricating materials.

Driven by a demand from the refrigeration industry, machine tool builders have improved the speed and accuracy of scroll manufacture. These new machine tools can produce finished scrolls in one to five minutes with involute accuracy of 0.0002 to 0.0005 inch and with good surface finish. Spindle speeds as high as 30,000 RPM are typical machining scrolls made of aluminum or cast iron. Most of the major machine tool manufacturers have standard scroll machining centers.

12.4.1 Lubricated Scroll Compressors

Typically scroll compressors used as refrigerant compressors are oil lubricated. Lubrication greatly simplifies the compressor design. Design features include:

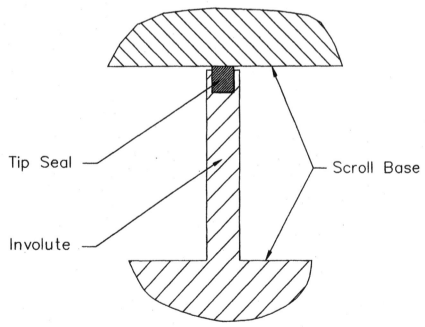

FIGURE 12.3 Section through involute showing tip seal.

FIGURE 12.4 Typical scroll showing idler shafts.

- Cast iron or aluminum scrolls with no special coatings or surface treatment required
- A simple eccentric drive at the center of the orbiting scroll
- A flat plate thrust bearing to support and locate the orbiting scroll axially

Since refrigerant compressor are hermetically closed systems, no special oil clean up is needed at the discharge. The oil can simply circulate through the refrigeration system and return to the compressor to seal and lubricate.

12.4.2 Oilless Scroll Compressors

Oilless or oil-free scroll compressors are typically used for air and other gases where the cost of oil clean up is a factor, or where zero oil can be tolerated in the discharge. Design features include:

- Cast iron or aluminum scrolls coated to improve corrosion and wear resistance
- Tip seals are required for good performance and are made of a self lubricating material.
- Idler shafts supported by sealed rolling element bearings are used to support the axial thrust load, locate the orbiting scroll axially and maintain the 180 degree scroll phase relationship. See Fig. 12.4.

12.5 APPLICATIONS

Scroll compressors can primarily be used in those applications where its advantages are of benefit, specifically low vibration and noise, and durability. Although scroll compressors can be cost competitive, if cost is the most important factor, alternative technology should also be considered.

Some possible applications are given below.

- Residential air conditioning
- Automotive air conditioning
- Process controls
- Pneumatic controls
- Laboratory
- Home health care
- Medical and hospital
- Computer peripherals
- Optical equipment

Scroll compressors can be used where vane or reciprocating compressors are used. They can be dry or oil lubricated.

CHAPTER 13
STRAIGHT LOBE COMPRESSORS

A.G. Patel, PE
Roots Divison
Divison of Dresser Industries Inc.

13.1 APPLICATIONS

Straight lobe compressors are used for pneumatic conveying of materials, aerating liquids, extracting gases and vapors, providing low pressure air/gas, supercharging engines and drying materials, etc.

13.1.1 Operating Characteristics

Capacity range: 5 cfm to 60,000 cfm

Pressure range: 15 psi

Vacuum range:

15 in. hgv for conventional compressor
27 in. hgv for externally aspirated compressor or liquid sealed
0.5 micron as a vacuum booster

Higher pressure and vacuum levels could be achieved through staging.

13.2 OPERATING PRINCIPLE

The more prevalent straight lobe compressors usually have rotors with two or three lobes. The operating principal for a two-lobe compressor is described below.

In the two-lobe compressor, two figure eight rotors are mounted on parallel shafts within an elongated cylinder. A set of timing gears keeps the rotors in synchronization. In Fig. 13.1, the lower rotor is presumed to be the drive rotor. As it rotates clockwise, the inlet is created on the right hand side and the discharge is created on the left hand side. The driven rotor turns counter clockwise through the action of the timing gears (not shown). In position 1, the drive rotor is delivering volume A to the discharge, while the driven rotor is trapping the volume B between the housing and itself. In position 2, the driven rotor has sealed off volume B from the inlet and the discharge. Volume B is basically at inlet conditions. In position 3, volume B is being discharged by the driven rotor, while the drive rotor is in the process of trapping volume A. The two-lobe compressor discharges four equal

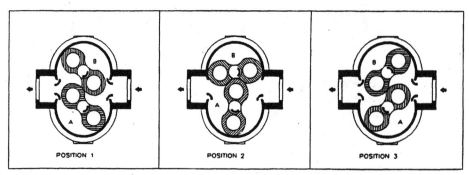

FIGURE 13.1 Two lobe compressor pumping schematic.

volumes of the medium per one rotation of the drive shaft. There is no internal compression of gas involved. The system resistance determines the head.

13.3 PULSATION CHARACTERISTICS

In Fig. 13.1 position 3, the trapped volume A, which is primarily at the inlet pressure condition, is being exposed to the discharge pressure. The higher pressure discharge medium suddenly rushes in to occupy the volume A. This sudden inrush produces a pressure pulse.

$f = 2 \times N \times K$
f = pressure pulse frquency, hz
N = compressor revolution per second
K = number of lobes

In a two-lobe compressor, the pulse frequency is four times the compressor rpm. The amplitude of the discharge pulse is controlled by controlling the rate of pressure change of the trapped volume A. In their Whispair™ design, Roots Operations, a division of Dresser Industries, uses discharge gas to pre-charge the volume A in a controlled manner to reduce pulsations.

In most of the applications, the use of a discharge pulsation dampner is mandated to avoid pulsation damage on the equipment down stream of the compressor.

13.4 NOISE CHARACTERISTICS

The discharge pressure pulse is one of the main contributors to noise in a straight lobe compressor. The other contributing factors are gears, bearings and flowing gases. The impeller actions generate varying loads on the bearings. These loads not only vary in magnitude during rotation of the impeller, they also change direction causing shock loading that the bearings transfer to the mounting structures.

The vibrations of these structures radiate noise. For controlling noise, it is not uncommon to enclose the straight lobe compressor in an acoustic housing.

13.5 TORQUE CHARACTERISTICS

The straight lobe compressors have a pulsating shaft torque. The torque pulsations could be in the range of ± 10% of the mean. Stiff couplings like gear type couplings are not recommended as they would transmit the torque pulsations possibly causing damage to the driver.

13.6 CONSTRUCTION (FIG. 13.2)

The main components of a straight lobe compressor are the rotors ②, the casing ①, the timing gears ⑦, the bearings ⑤, and the seals ④.

13.6.1 Rotors

The rotors (2) are nothing more than a set of two toothed gears. The common profile for the rotor lobes is involute; cycloidal profile is also used sometimes. The

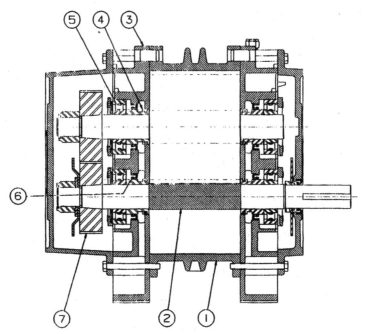

FIGURE 13.2 Cross section through straight lobe compressor.

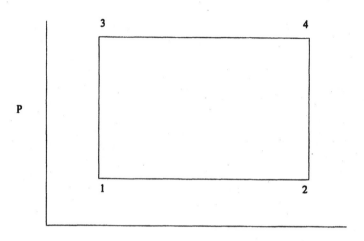

FIGURE 13.3 PV Diagram For Single Stage Compression

shafts are either integrally cast, pressed through, or bolted stub shafts. The clearances between the rotors and between the casing and the rotors is held to a minimum to reduce leakage flow, the main source of compressor volumetric inefficiency. The rotors are generally hollow; for dusty environments they are plugged to prevent rotor imbalance.

13.6.2 Casing

The casing consists of cylinder (1) and end plates (3) also known as head plates. The casing is normally designed for 25 psig rating.

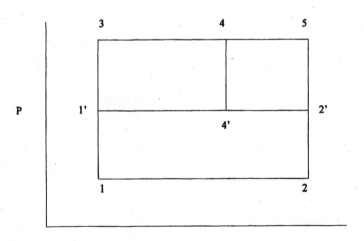

FIGURE 13.4 PV diagram for two stage compression.

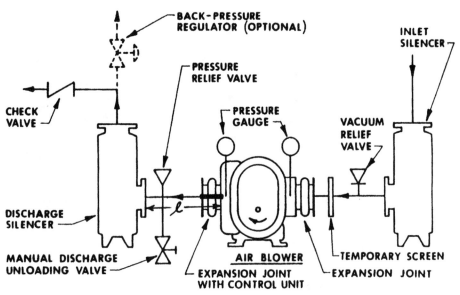

FIGURE 13.5 Suction and discharge arrangement for straight lobe compressor.

13.6.3 Timing Gears

Timing gears maintain the rotor phasing without contact. Timing gears are generally mounted on the shafts with some form of keyless interference fit to permit easy timing of the rotors.

FIGURE 13.6 5000CFM straight lobe blower driven by variable frequency driver. Installed in a wastewater treatment plant. (Roots Division, Dresser Industries Inc.)

FIGURE 13.7 5000CFM straight lobe blower driven by a Waukesha engine. Installed in a wastewater plant. (Roots Division, Dresser Industries Inc.)

FIGURE 13.8 1760CFM, 250PSI acetylene product blower (Roots Division, Dresser Industries Inc.)

13.6.4 Bearings

Antifriction bearings are generally the bearings of choice.

13.6.5 Seals

The head plate seals are generally labyrinth or piston ring. In air compressors, the area between the headplate seals and the oil seals is vented to the atmosphere to prevent pressure build-up in the end covers. In gas applications, the vents are closed and mechanical face seals are used in place of oil seals. There are two types of mechanical seals available: splash lubricated and pressure lubricated. Pressure lubricated seals use oil pressure higher than the gas pressure and hence have better gas sealing ability than the splash lubricated.

13.7 STAGING

Two main reasons for staging straight lobe compressors are for achieving higher compression ratios, and for reducing power consumption.

13.7.1 Higher Compression Ratios

Generally, single-stage straight lobe compressors are limited to a compression ratio of 2.0, or about 15 psi rise on air from an ambient of 14.7 psia. Above this pressure rise, the temperature rise across the compressor becomes excessive. Higher compression ratios could be achieved by staging the compressors where discharge gas from the each stage is cooled before sending it to the next stage.

13.7.2 Reduction of Power

The straight lobe compressors do not have any internal compression. The power required by a single-stage compressor is represented by a rectangle 1-2-4-3 on PV diagram as shown in Fig. 13.3.

The air is drawn into the compressor at a constant inlet pressure represented by line 1-2. The trapped volume is instantly compressed to discharge pressure as represented by line 2-4. The air is discharged at a constant discharge pressure represented by line 4-3. The cycle is completed by line 3-1.

By adding one more stage, power reduction represented by area 2'-4'-4-5 in Fig. 13.4 is realized. The first stage draws in air at a constant inlet pressure. The air is compressed to the intermediate pressure 2'. Since the first stage discharge air volume has been reduced to 4', the second stage need to have a displacement of only 4'. So the work required by the first stage is 1-2-2'-1' and the second stage is 1'-4'-4-3.

13.8 INSTALLATION

Figure 13.5 shows the recommended installation for a straight lobe air compressor. The location of discharge silencer with respect to the compressor flange is very critical. If distance "*I*" is not properly selected, there exists a possibility of setting up resonance of discharge gas column. To avoid resonant situation,

$$I <> \frac{nc}{4f}$$

where

$n = 1, 3, 5, \ldots$
c = velocity of sound in ft/sec = $\sqrt{kgRT}$
f = excitation frequency in Hz = 4 × rotational speed in rev/sec for two-lobe compressor = 6 × rotational speed in rev/sec for three-lobe compressor
k = specific heat ratio
g = gravity constant = 32.16 ft/sec^2
R = gas constant in lb.ft/°R
T = gas temperature in °Rankine

For positive displacement compressors, use of relief valves is very important. Any piping blockage could overload the compressors beyond design pressures.

CHAPTER 14
THE OIL-FLOODED ROTARY SCREW COMPRESSOR

Hasu Gajjar
Weatherford Compression

The rotary screw compressor has attracted increased attention in the gas industry during the last decade as an ideal compressor for low pressure/high capacity operations, with low pressure defined as suction pressure at or near zero psig and discharge pressure at less than 400 psig. Developments are underway by screw compressor manufacturers to go to higher discharge pressures.

The rotary screw is a positive displacement compressor, like its better known relative, the reciprocating compressor. In a comparison between the two, the rotary screw gleans honors for its simplicity, low cost, easy maintenance and almost pulsation-free flow. It takes a back seat to the reciprocating compressor, however, in handling high pressure (see Fig. 14.1.)

Benefits offered by rotary screw compressors include:

- Simple maintenance
- Low maintenance costs
- Long compressor life
- Full use of driver horsepower
- Low operating expense
- Low purchase price
- High compression ratios (Rc) up to 16 Rc per stage
- Operation at low suction pressure up to 26 inches of vacuum
- Light weight
- Compactness

14.1 TYPES OF COMPRESSORS (see Fig. 14.2)

Rotary compressors may be either positive displacement or dynamic compressors. The positive displacement rotary utilizes either vanes, lobes or screws to literally

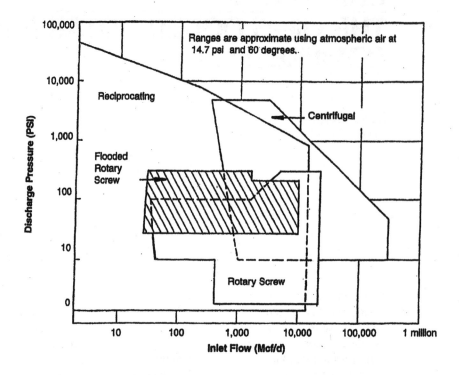

FIGURE 14.1 Reciprocating/rotary screw compressors.

pack the gas into the discharge line. Dynamic compressors, on the other hand, operate on an entirely different principle. Instead of reducing the volume of the gas to increase its pressure, the dynamic compressor works on transfer of energy from a rotating set of blades to a gas, and then discharges the gas into a diffuser where the velocity is reduced and its kinetic energy is converted to static pressure.

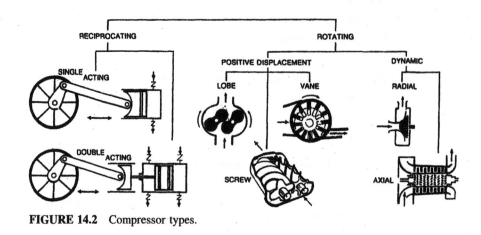

FIGURE 14.2 Compressor types.

Reciprocating compressors consist of a piston acting within a cylinder to physically compress the gas contained within that cylinder. They may be either single-acting or double-acting, and can be designed to accommodate practically any pressure or capacity. For this reason, the reciprocating compressor is the most common type found in the gas industry. Each compressor is designed to handle a specific range of volumes, pressures, and compression ratios.

Compared to the rotary compressor, the reciprocating compressor is more complex, and may cost more to maintain. However, its higher efficiency and ability to handle greater pressures outweigh these disadvantages.

In the selection of a compressor unit, one of the primary considerations, besides pressure-volume characteristics, is the type of driver. Generally, small rotary compressors are driven by electric motors, while the larger rotary compressors are usually turbine driven. Reciprocating compressors may be driven by electric motors, turbines (gas or steam) or engines (gas or diesel).

In some types of reciprocating compressors, the power cylinders and compression cylinders are integrated into one unit, and share the same frame and crankshaft. These compressors are referred to as **integral units.** The power and compression cylinders of an integral unit may be either horizontally opposed or in a V-configuration with the power cylinders on one bank and the compression cylinders on the other.

Another type of reciprocating compressor is the **separable unit.** In this type unit, the prime mover is separate from the compressor, thereby allowing the user to choose the driver best suited to the application. Although this design may be slightly more complex than that of the integral unit, its inherent flexibility often gives the separable unit an advantage over the integral unit.

A wide variety of compressor designs can be used on the separable unit including horizontal, vertical, semi-radial and V-type. However, the most common design is the horizontal, balanced-opposed compressor because of its stability and reduced vibration.

14.2 HELICAL ROTORS

The rotary screw compressor consists of two intermeshing helical rotors contained in a housing (see Fig. 14.3). Clearance between the rotor pair and between the housing and the rotors is .003 in to .005 in. The main rotor (male rotor) is driven through a shaft extension by an engine or electric motor. The other rotor (female rotor) is driven by the main rotor through the oil film from the oil injection; there is no metal contact.

The length and diameter of the rotors determine the capacity and the discharge pressure. The longer the rotors, the higher the pressure; the larger the diameter of the rotors, the greater the capacity.

The helical rotor grooves are filled with gas as they pass the suction port. As the rotors turn, the grooves are closed by the housing walls, forming a compression chamber. Lubricant is injected into the compression chamber after the grooves close

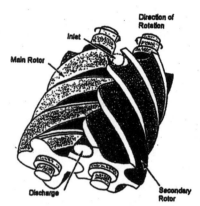

FIGURE 14.3 Typical rotary screw compressor.

to provide sealing, cooling and lubrication. As the rotors turn to compress the lubricant/gas mixture, the compression chamber volume decreases, compressing the gas/lubricant toward the discharge port. The gas/lubricant mixture exits from the compressor as the compression chamber passes the discharge port. Each rotor is supported by anti-friction bearings held in end plates near the ends of the rotor shaft. The bearings at one end, usually the discharge, fix the rotor against axial thrust, carry radial loads, and provide for the small axial running clearances necessary.

After compression, the gas/lubricant mixture enters a multi-stage separator which removes the lubricant from the gas. From the separator, the gas flows to the aftercooler. The lubricant is also cooled, returning to the compressor through a thermostatically-controlled valve.

Oil is the lifeblood of a rotary screw compressor. The limited clearance inside the rotary screw means that without proper lubrication, the screw may experience higher than normal wear. Rotary screw compressor operators use a synthetic hydrocarbon oil of ISO 100, 150 or 220 viscosity. Viscosity is selected based on the specific gravity of the gas. The gas analysis is very important in oil selection.

During the initial start-up of a unit, gas will dilute the viscosity of the oil. It should not be allowed to drop below the minimum recommended value.

Since the rotary screw has a closed oil system, it should use minimal oil. Most packagers design the package with oil carryover of five (5) parts per million. If a rotary screw compressor *loses* oil and no leak can be found, then the oil is going down the sell line. These conditions mean a scavenging line orifice is plugged or a coalescing filter has collapsed.

The oil filter for the oil injected into the rotary screw is a 10-micron filter. The fine mesh is needed to protect the bearings and shaft seal. An oil change is recommended only every 8,000 operating hours, unless the oil is contaminated. A regular oil sampling will help the operator determine when an oil change is needed.

Many rotary screw models are available with internal capacity control and variable volume ratio systems, which permit efficient variable load and pressure op-

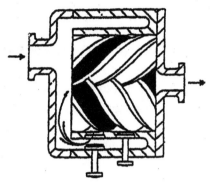

FIGURE 14.4a Lift valve unloading mechanism.

eration. Such systems are particularly desirable when constant speed electric motors are used and varying pressure conditions exist. (see Fig. 14.4a and 14.4b).

14.3 ADVANTAGES OF THE ROTARY SCREW COMPRESSOR

In many applications, the rotary screw compressor offers significant advantages over reciprocating compressors.

1. Its few moving parts mean the elimination of maintenance items such as compressor valves, packing and piston rings, and the associated downtime for replacement.

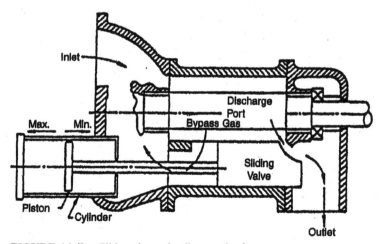

FIGURE 14.4b Slide valve unloading mechanism.

2. The absence of reciprocating inertial forces allows the compressor to run at high speeds, which results in more compact units.
3. The continuous flow of cooling lubricant permits much higher single-stage compression ratios.
4. The compactness tends to reduce package costs.
5. Rotary screw technology reduces or eliminates pulsations, resulting in reduced vibration.
6. Higher speeds and compression ratios help to maximize available production horsepower.

14.4 APPLICATIONS FOR THE ROTARY SCREW COMPRESSOR

A rotary screw compressor package is ideal for numerous compression applications, including:

1. Fuel gas boosting
2. Casing head gas boosting
3. Vapor recovery
4. Landfill and digester gas compression
5. Propane/butane refrigeration compression
6. Compression of corrosive and/or dirty process gases

A rotary screw compressor package can also be used to upgrade existing reciprocating compressor installations. By boosting low suction pressure, capacity may be increased at minimum cost with continued use of existing reciprocating equipment.

If an application requires large volume/low suction pressure, but discharge pressures are greater than the screw can provide, a combination screw/reciprocating unit with a common driver can be the solution.

14.5 VAPOR RECOVERY

One example of rotary screw utilization is for vapor recovery. Vapor recovery is the gathering of stock tank vapors and the compression of these vapors into the gas sales line. Capturing these vapors is profitable and environmentally advisable.

The gas vapors are gathered into a common header and fed into a vapor recovery unit (VRU), which usually includes a suction scrubber, a compressor, a driver, a discharge cooler and separator, and controls for unattended operation. The vapors are usually rich and wet, conditions which lead to condensate and the washout of lubricant. Washout causes excessive wear in vanes or piston rings. A rotary screw compressor does not suffer from these problems for two reasons: there is enough lubricant injected to take care of washout; and the rotary screw does not require the rotors to make contact with the stator.

The sizing of a rotary screw compressor for vapor recovery is extremely important. An oversized unit will operate in the partial load stage or shut down and start up too often. An undersized unit will not be able to keep up and the vents will emit vapors into the atmosphere, defeating economy and the effort to maintain clean air.

14.6 SIZING A ROTARY SCREW COMPRESSOR

To size a rotary screw compressor, one needs to know the suction and discharge pressures, the desired capacity, the gas analysis, temperature and the elevation. (See Eq. 1 for formula to determine capacity and Eq. 2 for formula to determine horsepower.)

EQUATION 1

Rotary Screw Compressor Capacity

$$ICFM = D^3(L/D)(GR)(RPM)(E_v/C)$$

Rotor Diameter	D
Rotor Length	L
Gear Ratio	GR
Driver Speed	RPM
Volumetric Efficiency	E_v
Rotor Profile	C

EQUATION 2

Rotary Screw Compressor Power

$$W_R = P_1 Q \times \frac{K}{K-1} \times \frac{(P_2/P_1)^{K-1/K} - 1}{E_a} + WL$$

Pressures	P_1, P_2
Gas Flow Rates	Q
Gas Properties	K
Adiabatic Efficiencies	E_a
Mechanical Losses	W_L

Figures 14.5A and 14.5B show adiabatic efficiency at different pressure ratios, while Fig. 14.6 shows efficiency change with variable volume ratio.

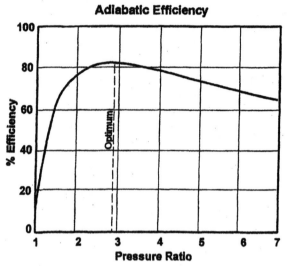

FIGURE 14.5a Adiabatic efficiency (pressure ratio to 7).

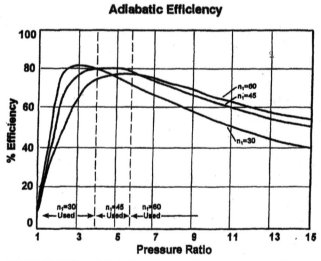

FIGURE 14.5b Adiabatic efficiency (pressure ratio to 15).

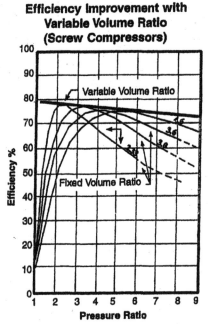

FIGURE 14.6 Efficiency improvement with variable volume ratio.

CHAPTER 15
DIAPHRAGM COMPRESSORS

G. Reighard
Howden Process Compressors, Inc.

15.1 INTRODUCTION

A diaphragm compressor is a specialized piece of equipment used to compress gases when little or no leakage is tolerable. The difference between an ordinary piston compressor and a diaphragm compressor is in the method of compression and the accompanying seals. A piston compressor compresses the gas with a moving piston; the piston has a dynamic gas seal in the form of piston rings which are not leakproof. A diaphragm compressor also has a piston with piston rings, but the piston moves a volume of hydraulic oil. The oil bends a set of diaphragms up and down, and the diaphragms compress the gas. Because only static seals are involved, compression is achieved without the escape of gas through a dynamic seal.

15.2 THEORY OF OPERATION

To understand the operation, refer to Fig. 15.1, which shows a typical drive arrangement, crankcase, hydraulic system and compression head.

A diaphragm compressor usually has an electric motor as the prime mover, with flexible belts for power transmission. The belts turn a compressor pulley or flywheel, which then rotates a crankshaft. Similar to other reciprocating compressors, the crankshaft provides reciprocating motion through a connecting rod on an eccentric journal. The connecting rod is attached through a wrist pin to a crosshead, which rides in a distance piece or cylinder.

A hydraulic piston is attached to the crosshead. This piston rides in a hydraulic cylinder and is sealed with piston rings. The piston pulses a fixed volume of hydraulic oil back and forth against the diaphragm set. The oil forces the diaphragm set against the gas head; it is the diaphragms which actually compress the gas.

It is noteworthy that there is a dynamic seal in a diaphragm compressor, but it has two distinct advantages over a piston compressor:

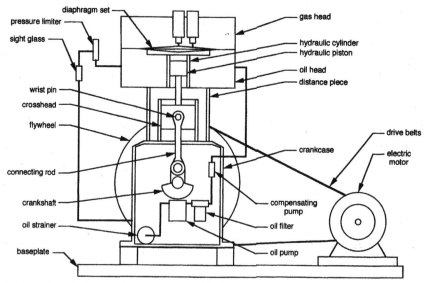

FIGURE 15.1 Main components of a diaphragm compressor.

1. Because it is on the oil side and does not contact the gas, there is no dynamic gas seal to leak.
2. It is lubricated with oil.

In a diaphragm compressor, the hydraulic oil serves several purposes: in addition to lubricating the running gear, it pulses the diaphragms to provide gas compression, and it provides a cooling effect.

The hydraulic circuit begins in the bottom of the crankcase, which acts as a reservoir for lubricating oil. Oil flows into the circuit through a strainer. Where required, the oil flow is cooled, usually through a water-cooled heat exchanger. The flow enters the main oil pump and is discharged through a filter. Next the oil is split into two streams, with the bulk of the oil flow going to lubricate crankshaft bearings, connecting rod journals, wrist pins, and sliding surfaces of the crosshead. A small portion of the flow is diverted to the compensating circuit.

The purpose of the compensating circuit is to make up any oil that leaks past the piston rings on the hydraulic piston. In this circuit, oil flows from the main pump to a check valve, through a low-volume reciprocating compensating pump, through another check valve, and into the oil head. However, the makeup oil flow is very large compared to the actual amount of oil lost through piston ring leakage.

During each stroke, the hydraulic oil pushes the diaphragm set into full contact with the gas head. If leakage past the hydraulic piston rings were not made up, the pulsed oil volume would decrease, and the diaphragm set would not completely contact the gas head. Continued hydraulic piston ring leakage would have the same effect as increasing the volumetric clearance in a piston compressor, with resulting losses in performance. Therefore the proper operation of this compensating circuit is critical to the performance of the compressor.

To regulate the makeup oil, a valve is mounted on the oil head to maintain proper oil pressure internally, but also allow any excess oil to flow back into the crankcase. This valve is known as the hydraulic pressure limiter. It is adjusted to develop the desired gas discharge pressure in the compression head. The limiter is opened by the development of a peak oil pressure (the limiter pressure), and is closed by a spring.

On some small models, the function of the compensating pump is provided by differential displacement, where the piston displacement is slightly larger than the cavity volume between the gas and oil heads. This allows the diaphragms to touch the oil head; the additional travel of the piston draws oil into the oil head through the oil check valve. However, this approach is limited in its application.

The compression head consists of an oil head bolted to a gas head, a diaphragm set, seals, suction valve(s), discharge valve, valve retainers and bolting. The oil head contains the hydraulic oil; the gas head contains the gas being compressed, and the diaphragms separate the gas from the oil. The suction and discharge valves are self-actuating, opening and closing due to differential pressures, allowing gas to enter and leave the head at the proper times.

To understand the relationship of the dynamic pressures inside the compression head, refer to Fig. 15.2. This figure shows the pressures developed inside the oil head during one typical revolution.

At Point A, the hydraulic piston is at top dead center (TDC) and the pressure has reached its peak. The piston begins its downward stroke, away from the gas head. The oil pressure drops rapidly, as does the gas pressure on the other side of

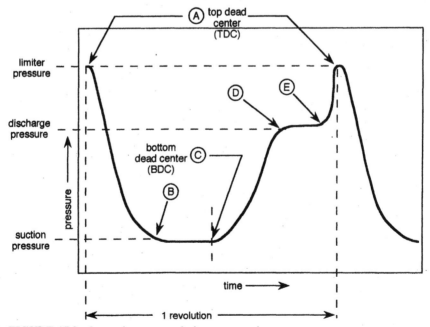

FIGURE 15.2 Internal pressures during compression.

the diaphragm set. Because the external gas pressure is higher than the internal gas pressure, the discharge valve closes. The remaining gas in the compression head then expands from discharge pressure to suction pressure. When the internal gas pressure is slightly less than suction pressure (Point B), the suction valve opens and admits fresh suction gas into the head.

As the piston approaches bottom dead center (BDC), an important function occurs: the compensating pump injects a small amount of oil into the oil head to allow for piston ring leakage. When the hydraulic piston reaches BDC, the flow of suction gas into the head stops (Point C). The diaphragms are now at their closest position to the oil head.

As the piston begins its upward stroke toward the gas head, internal gas pressure exceeds external gas pressure and the suction valve closes. Gas is now trapped inside the compression head causing the oil pressure and the gas pressure to increase simultaneously. When the internal gas pressure exceeds external gas pressure (Point D), the discharge valve opens and gas flows from the head into the discharge piping. This flow of gas stops when the diaphragm set contacts the gas head, and the discharge valve closes (Point E). However, the piston continues to travel a small amount further. As a result, the pressure in the oil head continues to rise although the trapped gas remains at discharge pressure. The hydraulic pressure limiter then opens and discharges a small amount of oil back to the crankcase. At TDC (Point A), the piston begins to travel away from the gas head. The limiter closes, maintains the oil pressure inside the compression head, and the next compression cycle begins.

For proper operation of the compressor, the hydraulic pressure limiter must provide a peak oil pressure higher than gas discharge pressure. As shown in Fig. 15.2, gas compression is intimately tied to the proper operation of the hydraulic system.

15.3 DESIGN

Model numbers used by various manufacturers commonly denote the basic crankcase (with a given stroke and rod load); the configuration; and the maximum pressure rating of the head(s). Strokes run the range from 1.25 inch through 9 inches; the most popular strokes are 2.5 inches through 5 inches. Configuration of diaphragm compressors can be varied. Single-stage designs are common, with the head mounted either vertically or horizontally. Two-stage designs are also common, built in "L," "V," and horizontally opposed styles. Three-stage designs are less common, and four-stage systems are unusual. Pressure ratings usually run from 150 psig to 15,000 psig. However, compressors have been built outside this range, with some operating at vacuum conditions, and others at 45,000 psig and higher.

Drivers for diaphragm compressors are usually electric motors, ranging from 1 HP up to 200 HP. Although purchasers often specify high efficiency motors, they have lower slip and higher currents than standard motors. As a result, standard efficiency motors are usually the better choice for reciprocating compressor applications.

Belt drives are the most common means of power transmission. A small sheave is mounted on the motor shaft and a large sheave on the compressor shaft. The large sheave usually doubles as a flywheel, with a heavy rim to provide rotational inertia for reducing peak motor current.

Crankcases are very similar to those used for piston compressors. The one significant difference is the provision for shaft-mounted compensating pumps, which provide makeup oil to the oil heads to counteract piston ring leakage.

Proper oil filtration is mandatory for successful operation of a diaphragm compressor. As with all rotating equipment, bearings may be damaged if fed with dirty oil. In addition, the diaphragms are susceptible to cracking by bending around dirt or metal particles if the oil is not kept clean continuously. To keep the oil clean, it first passes through a strainer as it leaves the crankcase. After leaving the pump, it passes through a filter, usually of 25-micron rating.

In the past, splash lubrication with slingers or dipping extensions has been provided on some models. However, it has generally been replaced by forced lubrication.

Forced lubrication is typically provided by a shaft-driven gear pump. The pump may be driven by direct coupling onto the crankshaft, or by a gear attached to the crankshaft. Auxiliary pumps with electric motors may be provided as an option. Relief valves are fitted to the discharge of these pumps with any overflow being returned to the sump in the crankcase.

An oil heater is often required for low-temperature operation. The heater is generally an immersion type, installed directly into the crankcase. Oil coolers are often provided as well; these are typically installed next to the oil pump. In some wide operating temperature ranges, a compressor may have both an oil heater and an oil cooler.

The compensating pump is driven by an eccentric on the crankshaft, and returned by a spring. The pump has a small plunger sealed to a body with either piston rings or a tightly-toleranced fit. Check valves are mounted at the suction and discharge ports to prevent backflow.

The peak hydraulic pressure is established by the hydraulic pressure limiter. The limiter is opened by the peak oil pressure and is returned by spring action. It is usually externally adjustable and has a rugged design, opening and closing up to 500 times a minute. Limiters usually have a bypass valve (sometimes external) that allows the oil head to fill with oil without developing internal pressure. This bypass valve is designed for use only after the head has been taken apart.

An optional feature recommended for large heads is a set of valves used for filling the head with hydraulic oil. Filling is done with a separate pump which can be a hand pump or an auxiliary motor-driven pump.

Selection of the proper lubricant is vital in a diaphragm compressor. For most applications, a standard hydraulic or general purpose oil is used, with an ISO viscosity range between 46 and 100 centistokes. Anti-wear and anti-foam additives are generally used.

For certain applications, synthetic lubricants such as halocarbon oils are specified, especially in the compression of oxidizers such as oxygen and fluorine. When these synthetic oils are furnished, care must be taken to ensure that all materials in the hydraulic circuit are compatible with the oil.

For a typical cross-section view of a diaphragm compressor head assembly, refer to Fig. 15.3. At the lower end of this assembly, the piston rod is attached to the crosshead. The piston is assembled into a hydraulic cylinder, guided by rider rings and sealed with piston rings. While piston rings at one time were made exclusively of cast iron, they are now commonly made of specialized filled plastic materials.

The oil head is a rugged component, designed to withstand the peak hydraulic pressure in the system. The hydraulic cylinder is mounted on the crankcase end of the head. As the piston pushes oil up through the cylinder, the oil head distributes the oil evenly underneath the diaphragms through a series of holes or grooves. These holes or grooves must be limited in size, or the diaphragms can be overstressed by bridging these features. On the side of the oil head facing the diaphragms is a shallow cavity which limits the deflection and stresses in the diaphragm set.

The oil head has an inlet port to receive oil from the compensating pump, and an outlet port to discharge oil through the limiter. Because they are pressure-retaining components, thickness of the compression heads (both gas and oil) is critical; U.S. designs follow the ASME Boiler and Pressure Vessel Code for flat heads. The oil head is sometimes water cooled to limit seal temperature.

The gas head is rugged as well, since it also must withstand the peak hydraulic pressure in the system. On the side of the gas head facing the diaphragms is a cavity that matches the one in the oil head. The diaphragms make contact with this

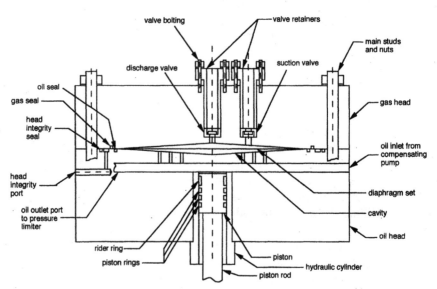

FIGURE 15.3 Diaphragm compressor head assembly.

cavity at top dead center, and the cavity limits the deflection at this point. In the cavity are radial grooves leading to the discharge port at the center; these grooves help to sweep gas out of the head during discharge. The suction port is offset radially from the discharge port. As noted previously, all of these holes or grooves must be limited in size. On the external face of the gas head, ports are machined to receive the suction and discharge valves. The gas head is usually water-cooled.

As noted above, both the gas head and the oil head have a shallow cavity machined into them. Although other designs exist, two of these cavity designs are well known; they are the "two-radius" and the "free-deflection" approaches. However, a feature common to all cavity designs is a very small center deflection with a large cavity diameter. This ensures that diaphragm stresses are kept within the elastic range. At low pressures, a large cavity is used; such a cavity may be 36 inches in diameter with a deflection of 0.5 inch at the center. For high pressures, though, cavities can be quite small. A cavity with a 3-inch diameter may have a center deflection of only 0.030 inch.

In U.S. design practice, ASME Code allowable strengths and guidelines are used for the selection of bolting. High-strength studs are often specified as SA193 Grade B7, with nuts made from SA194 Grade 2H material. Fatigue resistance is of prime importance due to the high cyclic application of these fasteners.

Many of these compressors are located in corrosive environments such as salt spray from the ocean or in a caustic atmosphere in a chemical plant. Because of this, a well designed compressor head will use either plated or coated bolts to prevent rust. While bolting previously made use of cadmium or zinc plating, recent trends are toward the use of PTFE coatings. These coatings provide lubricity as well as corrosion resistance.

Although bolted head assemblies are the most common, other methods of head closure exist. Several of these methods make use of a large body, usually forged, and a large internal nut to retain the compression heads. The internal nut may be closed with either set screws or by hydraulic pressure.

Diaphragms are flat, circular pieces of sheet metal with a smooth finish. A stack of three diaphragms is commonly used, with each of them serving a different purpose. Because it is in constant contact with the process gas, the gas diaphragm must provide corrosion resistance. The middle diaphragm has slots, grooves, or holes to conduct any leakage from a broken gas or oil diaphragm to the periphery where such leakage can be detected. The oil diaphragm, which rarely has a corrosion problem, transmits hydraulic pressure to the other diaphragms and provides a barrier to the oil.

Due to tiny differential displacement between the diaphragms as they bend up and down, provision must be made for lubricating the interfaces between diaphragms. To this end, the middle diaphragm may be coated with a dry film lubricant. Alternatively, the use of dissimilar metals can prevent wear on these surfaces.

Seals are critical to the proper operation of these compressors. While earlier versions used metal-to-metal seals with rather high bearing pressures between mating surfaces, the predominant seal in use today is an elastomer o-ring. O-rings are very forgiving in their ability to seal minor imperfections, and they are available in many different materials to suit varying process gases. Less common are metal seal rings and metal o-rings, which can be used to seal aggressive process gases.

Because leakage is such an important issue in the application of diaphragm compressors, most of them are fitted with a head integrity system as shown in Fig. 15.4. This system, also known as a leak detection system, will sense any one of four leakage sources:

1. A broken gas diaphragm
2. A broken oil diaphragm
3. A leaking gas seal
4. A leaking oil seal

In the event of a broken diaphragm, the gas or oil will find its way through the slotted or grooved middle diaphragm to the outside of the diaphragms. With a ruptured gas or oil seal, the leaking gas or oil will also travel to the diaphragm periphery. There the leak is contained inside another seal to prevent leakage to the atmosphere. The leak is then conducted to an external port for monitoring by the head integrity system. This system is composed of four parts: a manual vent valve, a relief valve, a pressure gauge and a pressure switch. The manual vent valve allows resetting of the switch; the relief valve prevents overpressurization; the pressure gauge allows verification of the leak; and the pressure switch will shut the compressor down.

As a rule, valves used in diaphragm compressors are self-actuated, opened by the flow of suction or discharge gas, then returned and sealed by differential pressures. Springs are used to control the valve sealing element and damp out unwanted motion. The sealing elements are usually flat discs or guided poppets; occasionally

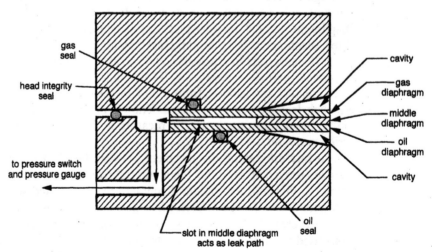

FIGURE 15.4 Head integrity system.

balls are used. Although some designs require multiple suction valves, most heads use a single suction valve and a single discharge valve. Seals are usually elastomer o-rings, metal gaskets, or metal seal rings. Where high heat is generated during compression, valve retainers are fitted with Belleville springs to accommodate thermal expansion.

Numerous special configurations of diaphragm compressors have been built. Compressors have been designed with remote head assemblies; the motive hydraulic oil is sent from the crankcase to the head through piping. Certain compressors have been built from light materials such as aluminum where low weight is a requirement. In other variations, air cylinders are prime movers rather than electric motors. There are many other possible modifications to the basic design.

15.4 MATERIALS OF CONSTRUCTION

Materials used for the crankcase, crankshaft, bearings, connecting rods, wrist pins, and crossheads are very similar to those found in an ordinary piston compressor. However, materials used to manufacture compression heads and valves can vary greatly.

Due to the limited availability of materials, the earliest versions of compressor heads were made from intricate and robust iron castings. However, recent designs make use of carbon steel plate or forgings. A typical plate specification for oil heads is SA516 Grade 70.

Because gas heads are in intimate contact with the process gas, there are several different materials used for their manufacture. For inert gas service and lower cost, gas heads may be made of carbon steel plate. For more corrosion resistance, many gas heads are made from stainless steel plate such as SA240 type 304 or 316. For severe environments, the selection may range to Monel or Alloy 20. Many other metals could be suitable providing they meet the structural requirements of the head design. Along with the oil heads, gas heads are typically machined from plate; forgings are used when dictated by strength or thickness requirements.

For ordinary inert gas applications, carbon steel sheet is adequate as a diaphragm material; however, the majority of diaphragms are made of high tensile-strength stainless steel, usually type 301 or 316. Other stainless steels such as 17-4 PH are also used. For added corrosion resistance, Monel or Alloy 20 may be used. While other materials are possible, they are sometimes impractical due to limited commercial availability in the proper thickness, width, or strength.

Valves for diaphragm compressors are often produced from corrosion-resistant materials such as alloy steels and 300 or 400 series stainless steels. As with the other gas-contacting parts, the valves may also be made from Monel or Alloy 20. Sealing elements are sometimes made of specialized plastics such as PEEK or Vespel. Springs are commonly stainless steel; Inconel is also a good choice for springs.

15.5 ACCESSORIES

Because many diaphragm compressors are customized in design, it is common to have customized accessories as well. These accessories are mounted on a steel baseplate, which also holds the motor and belt guard.

Because dirt and debris can be very harmful to the diaphragms, a process gas suction filter should be installed. An ideal filter is made of pleated stainless steel mesh with a 10-micron rating. Double suction filters with changeover valves are useful where production time is at a premium.

Where there is no significant volume of gas in the suction and discharge piping, accumulators are recommended to smooth out compressor operation. Pulsation dampeners tailored to the application are recommended where the pulsation could be detrimental to either the process or equipment.

When the process gas may contain moisture, a separator and trap should be installed at the suction to each stage and any point where condensation may occur, especially after a cooler. Liquid ingestion can damage not only the diaphragms, but the compression heads as well.

Gas coolers are usually installed at the discharge of each compressor stage. These may be shell-and-tube types, spiral tubes, or tube-in-tube types for smaller flow rates. Intercooling of gas is required not only for protection of seals and equipment, but for efficient operation.

Flow rates on many diaphragm compressors are low enough that tubing can handle the main flow of gas. At lower pressures, compression fittings are well suited. At pressures greater than 5000 psig, however, coned and threaded tubing provides a very safe joint with a very low leak rate.

When pipe is used, screwed joints may be employed but only with certain precautions. Due to the nature of gases handled (including those that are flammable, pyrophoric, and toxic), only very low leak rates may be acceptable. In those cases, welded joints are preferable. Although socket welds are used frequently, butt welds are used to meet the most stringent requirements where radiography can verify the soundness of a joint.

Maintenance joints for piping are very often standard raised-face flanges; however, flat-faced o-ring unions are used in many applications.

To monitor compressor performance, a pressure gauge should be installed at the suction port and discharge port of each compression head. Very often pressure switches are also required. These are installed at suction or discharge but rarely at an interstage section. Where final product temperature is critical, a discharge temperature switch or transmitter should be used.

For overpressure protection, a relief valve or rupture disc should be installed at the discharge of each stage, whether interstage or final discharge. Failure to do this could result in serious harm to personnel or property damage.

Capacity control is usually provided by some form of gas bypass, which returns the excess flow from discharge back to the suction piping. Common bypass methods maintain a constant discharge pressure, provided by an air-operated control valve, or a manual back-pressure regulator. If justified by the demand cycle, on-off control can be supplied. With on-off control, there should be only a few starts per hour, and process gas should be vented from the discharge to allow unloaded restarting.

Variable speed drives are usually not used for capacity control; the varying speed could play havoc with oil pressure and flow from shaft-driven oil pumps. Mechanical valve unloaders are generally not used; the location of valve inlet and outlet ports directly over the valves would complicate this design.

For protection of the hydraulic system, an oil pressure gauge and pressure switch are usually provided. For added protection, an oil level switch may also be used. These switches are usually wired to shut the compressor down. Water flow switches are often installed; these are good insurance against accidental loss of coolant.

15.6 CLEANING AND TESTING

Because cleanliness is of extreme importance in the reliable operation of a diaphragm compressor, its components must be carefully cleaned whether they contact the gas side or the oil side. Dirt and debris on either side can cause diaphragm failure. Diaphragm compressor manufacturers have established high standards of cleanliness for this reason. However, there are some applications that call for even stricter control of contamination. One of these is oxygen or oxidizer service, where all gas side components must be free of both debris and oil. The mere presence of these contaminants can cause a reaction with the process gas. In addition, the crankcase and all hydraulic equipment must be free of hydrocarbon oils; ignition can occur if the oxidizer contacts these hydrocarbons.

A manufacturer of high-quality compressors will complete not only a mechanical run test but a performance test before shipment. The majority of these compressors can be run at full suction and discharge pressures during test. The actual flow rate is measured, using inert gas (typically nitrogen) to simulate the process gas. The entire compressor system should be checked for leaks on the gas side, the hydraulic side, and the cooling circuits. In addition, all required nondestructive examination such as liquid penetrant tests and radiographic tests for piping should be completed before shipment. These tests should be specified in writing by the purchaser and reviewed by the manufacturer early in the production of the machine.

15.7 APPLICATIONS

Diaphragm compressors are frequently applied where an ordinary piston compressor could experience problems due to leakage, gas contamination or pressure limitations. Some of the applications where diaphragm compressors are ideally suited are listed below:

15.12 CHAPTER FIFTEEN

Type of gas	Examples
Toxic	Boron trifluoride
Corrosive	Fluorine
Flammable	Hydrogen
Pyrophoric	Silane
Oxidizer	Chlorine
Radioactive	Uranium hexafluoride

In addition, the applications can provide cost-effective compression where low contamination is required, high pressures are developed, or high temperatures will be handled.

Many different markets have applications for diaphragm compressors. One of the largest users is industrial gas production and distribution, where these compressors are used to fill cylinders with common gases such as helium, nitrogen and argon. This is a well-established segment where diaphragm compressors are a standard, non-contaminating, reliable method of compression.

Some other areas that make good use of diaphragm compressors include the following mentioned below.

15.7.1 Automotive Air Bag Filling

Gas mixtures are compressed to high pressures to pressure test and leak test gas canisters. The compressed gas then remains in the gas canister to provide actuation of the airbag during an accident.

15.7.2 Tank Car Unloading

Vaporized product from a bulk liquid tank car can be used to pressurize the car and offload it to a storage vessel.

15.7.3 Petrochemical Industries

Gases are produced and distributed throughout a plant to various processes after compression. Gases are also recirculated at low compression ratios where only piping losses must be overcome.

15.8 LIMITATIONS

Diaphragm compressors typically operate in the range of 300 to 500 rpm. While speeds as low as 100 rpm are possible, provision must be made to maintain oil pressure. Speeds of 750 rpm have been achieved in modified diaphragm compressors.

Atmospheric suction pressure is common for diaphragm compressors. Low suction pressures such as 2 or 3 cm Hg absolute are also possible, although special provisions must be made for degassing the oil in the crankcase sump. In extreme conditions, the suction pressure could be as high as 5000 psig.

Discharge pressures of 2,500 psig are very common; this is typical of many cylinder-filling operations for industrial gas handlers and producers. There are many current applications where discharge pressures of 5,000 to 10,000 psig are routine. Diaphragm compressors have been built with discharge pressures of 45,000 psig and higher. However, such designs must be carefully executed when the process pressures exceed the yield strength of ordinary metals.

Diaphragm compressors can be successfully applied at very small flow rates of less than one standard cubic foot per minute (SCFM). Typical flow rates lie in the range of 5 to 100 SCFM, although flow rates of 700 SCFM and higher have been achieved.

Physical size is sometimes a constraint in the design of a diaphragm compressor. While some compressor heads may be only 10 inches in diameter, certain low pressure heads may have a diameter of 40 inches which can limit the installation in tight quarters. Such large heads may weigh nearly 2000 pounds, which can provide a challenge when removing the head for maintenance. In addition, the choice of available diaphragm materials is limited when the diameter exceeds 36 inches.

15.9 INSTALLATION AND MAINTENANCE

Installation of a diaphragm compressor should follow the same good practices used for any reciprocating compressor. Foundation design should account for vibration due to unbalanced forces and moments. These values are available from the manufacturer.

Piping design should account for flexibility, vibration isolation and thermal expansion. For large flow rates, pulsation dampeners should be evaluated to avoid acoustic resonance. Where personnel have routine access to the compressor, consider guards to prevent contact with discharge lines. These frequently are hot enough to cause burns on unprotected skin.

One of the primary goals of a good diaphragm compressor installation is to prevent contamination of the process gas. Filters should be used to prevent particulate entry; separators are required when liquid could present a problem; and components must be kept clean during any maintenance work.

Maintenance plans for the compressor should consider: lifting equipment for the gas head and oil head; ability to clean all exposed components on both the gas and oil side; adequate lighting; and accessibility to the crankcase and compression heads.

15.10 SPECIFYING A DIAPHRAGM COMPRESSOR

The purchaser of a diaphragm compressor should specify sufficient detail so that the manufacturer can provide equipment which meets the requirements safely and efficiently. A thorough specification will contain these key points:

- Required flow rate
- Composition of gas handled
- Special gas conditions such as corrosives, condensibles, and contaminants
- Suction pressure and temperature
- Required discharge pressure and temperature
- Type of cooling available
- Electrical area classification
- Power supply available
- Control voltage and type of controls required
- Site conditions such as ambient pressure, temperature, and humidity
- Preferred materials of construction
- Limitations on envelope size

Because of the current global equipment market, the user should specify applicable national or international standards. A design that is approved for one site

PHOTO 1 Burton Corblin two stage 2000 psi diaphragm compressor.

PHOTO 2 Burton Corblin two stage 6000 psi diaphragm compressor.

PHOTO 3 Burton Corblin two stage 8000 psi diaphragm compressor.

may be inadequate for the same service but in a different country. Typical standards invoked for the United States markets are:

- ASME B&PV Code for pressure vessels
- ASME B31.3 for piping
- National Electrical Code for wiring
- API 618 for basic compressor construction

In addition, the purchaser should list any of its own in-house standards that apply, such as those for motors, pressure vessels, and piping.

CHAPTER 16
ROTARY COMPRESSOR SEALS

James Netzel
Chief Engineer, John Crane Inc.

16.1 INTRODUCTION

The selection of a sealing system for a compressor is critical if satisfactory performance and reliability are to be realized. The type of compressor and the method of lubrication used will determine the type of sealing technology to be applied. For some compressors a liquid lubrication system is required, while others will require a gas lubricated system.

Sealing technology is an evolutionary process. Design concepts and improvements in material of construction are some of the most notable achievements in this field. Sealing systems can be divided into four classes, based on the type of lubrication.

1. **Contacting**
   a. Liquid lubricated
   b. Gas lubricated

2. **Non-contacting**
   a. Liquid lubricated
   b. Gas lubricated

16.1.1 Contacting Liquid Lubricated

Seals, both mechanical face-type seals and lip seals, are cooled and lubricated by the lubricating oil in the compressor. This system is a condition of mixed lubrication where the load or contact pressure is partly carried by a fluid film and partly carried by the mechanical contact between the sealing surfaces. This is the most common sealing system found in industry for all types of rotating equipment.

16.1.2 Contacting Gas Lubricated

The contacting surfaces are designed to run dry with very light contact loads. Cooling and lubrication is achieved from the gas being sealed. This is a condition of boundary lubrication where the sealing surfaces are in contact, though separated from hard contact by material transfer films. This sealing concept is only applied to very light duty services.

16.1.3 Non-Contacting Liquid Lubricated

This method of sealing is dependent on a geometry change at the seal interface. Spiral grooves or similar features are incorporated into one of the sealing surfaces to generate hydrodynamic lift to separate the seal faces. This system is generally used to move a small quantity of liquid lubricant from a low pressure source to a high pressure side of a seal. This non-contacting concept is applied to specialized sealing applications to eliminate hazardous and toxic leakage and on those applications where abrasives are present.

16.1.4 Non-Contacting Gas Lubricated

This method of sealing an industrial compressor has become very popular over the years for it has eliminated expensive oil lubrication equipment. This design is also based on the concept of hydrodynamic lubrication and the incorporation of geometry changes to one of the sealing surfaces such as spiral grooves. The only heat that is developed is that of shearing gas at the seal interface. Therefore, it is the **most energy efficient** sealing system available to industry. This type of system can run on the process fluid being sealed or a neutral barrier fluid like nitrogen, purified air, or steam. This type of sealing system, which was designed for compressors, is now being applied to all types of difficult sealing applications on rotating equipment.

To successfully apply any of these sealing concepts, the following information must be considered.

1. **Process gas being compressed**
2. **Operating condition**
   - Pessure
   - Tmperature
   - Seed
   - Buffer fluid (if required)
3. **Space available for the seal**
4. **Utility reliability**
   - Steam

- Water
- Electricity—number of sources
- Plant air

5. **Disposal of buffer liquids**
6. **Disposal of gas sealants**
7. **Auxiliary equipment**
   - Controls, alarms, and trips
   - Direct
   - Closed loop pneumatic
     - Electronic
     - Dedicated mainframe
     - Computer

A detailed description of the gas to be compressed is necessary to ensure the proper selection of the sealing system to be used. This includes such items as whether or not the gas is hazardous or toxic, the effect on materials of construction, and whether contaminants are present. If a buffer fluid is required, then consideration must be given to availability, quality, backup, pressure level and cleanliness.

16.2 TYPES OF SEALS

16.2.1 Labyrinth Seals

Labyrinth seals represent the simplest method of sealing a rotating shaft. A series of knife edges are designed into either the housing or shaft. The clearance between the knife edge and its mating surface is a closely controlled value to limit leakage from the compressor. There is no limit on speed and labyrinth seals can be used at high temperature. The pressure limit is low and typically limited to 5 psi per knife edge. Leakage from this device is high. When applied to non-hazardous process gases, leakage can be vented to atmosphere. When the gas being compressed is hazardous to the environment, a buffer gas at higher pressure than the process gas, is injected between two labyrinth seals as illustrated in Fig. 16.1.

16.2.2 Carbon Ring Seals

Carbon ring seals are close clearance sealing devices similar to labyrinth seals. These types of seals are primarily used for low pressure, low temperature applications. Seal leakage is lower than a labyrinth seal. To prevent the leakage of process gas to atmosphere, a buffered gas or steam is injected between sets of carbon rings. This type of seal may also be used as a pressure breakdown device. There is no limit on shaft speed. This type of seal is illustrated in Fig. 16.2.

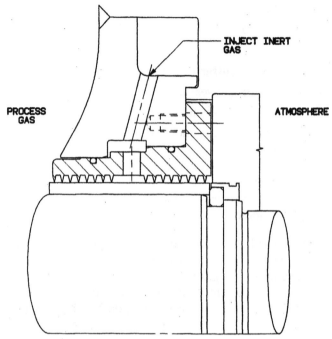

FIGURE 16.1 Typical labyrinth seal installation.

16.2.3 Bushing Seals

Bushing seals are always used with a buffered oil system to contain gas within a compressor. The oil must be maintained at a pressure of at least 0.3 to 1 bar above the process pressure. The oil is always injected between the inner and outer bushings, creating the seal. The amount of leakage is always dependent on the operating

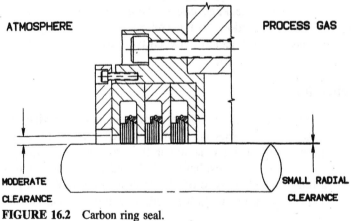

FIGURE 16.2 Carbon ring seal.

conditions of the compressor and the shaft size of the unit. Leakage of oil to the process gas can be 40 to 75 liters (10 to 20 gallons) per day. This leakage can be degassified and returned to the oil reservoir. Leakage to atmosphere is returned to the reservoir. Shaft speed for this seal is limited to less than 115 m/s (375 ft/sec). This type of seal may be used to a pressure of 3000 psig. A bushing type seal is illustrated in Fig. 16.3.

16.2.4 Pump Bushing Seal

This type of seal must also be used with a buffer liquid. The buffer liquid may be oil or water. Special designs may allow this sealing device to be used to pressure as high as 5000 psig. Speeds are limited to 115 m/s (375 ft/sec).

Buffer pressure is normally 0.3 to 1 bar greater than the process gas. Leakage is dependent on the operating conditions of the compressor and the shaft size of the unit. Static leakage can be high until the shaft begins to rotate. Oil leakage to the trap can be 4 to 20 liters (1 to 5 gallons) per day. Leakage is degassified and returned to a reservoir. A typical pump bushing seal is shown in Fig. 16.4.

16.2.5 Circumferential Seals

This type of sealing device uses segmented carbon rings held together with a spring. There may be one or multiple rings within the seal. Each ring is normally capable of pressure to 100 psig. This seal may operate directly in sealing the process gas

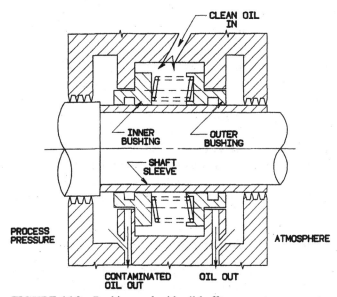

FIGURE 16.3 Bushing seal with oil buffer.

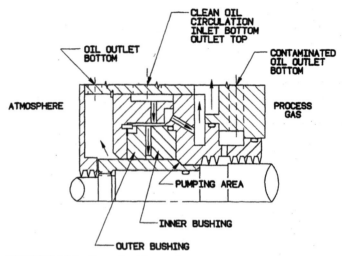
FIGURE 16.4 Pump bushing seal.

or a buffer gas supply. Leakage range is 2.8 to 29 l/min (0.1 to 1.03 SCFM), and the shaft speed is limited to 190 m/s (600 ft/sec). This type of seal is illustrated in Fig. 16.5.

16.2.6 Contacting Seal

Many small oil flooded compressors use conventional contacting seals to contain the oil and the gas being compressed. Refrigeration compressors typically use bellows seals and occasionally, o-ring seals. A bellows seal may take the form of an elastomeric bellows or a metal bellows as illustrated in Fig. 16.6 and Fig. 16.7 respectively. Most contacting seals of this type are used at speeds to 3600 rpm and pressures of 250 psig. As speeds increase on small oil flooded machines, the seal is held stationary to the unit as illustrated in Fig. 16.7.

On some larger centrifugal compressors a contacting face that requires an oil buffer is used to seal the compressor. This seal is shown in Fig. 16.8. Here, clean oil is injected over the rotating mating ring for cooling and lubrication. Oil leakage past the seal face is captured at an internal drain and separated from the gas. A labyrinth may be used to break down the pressure at the inboard side of the seal. The oil buffer pressure must be higher than the gas pressure. Leakage of oil can be as high as 8 gallons per day, and the speed is limited to 115 m/s (375 ft/sec).

16.2.7 Dry Running Non-Contacting Face Seals

Dry running non-contacting seals, as shown in Fig. 16.9, have been used to seal large industrial compressors since the early 1980's. This type of seal has become

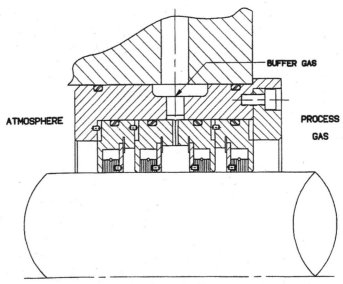

FIGURE 16.5 Circumferential seal with segmented carbon rings.

the most popular way to seal a rotary gas compressor. The success of this seal depends on the development of a fluid film at the faces. The non-contacting feature and film development is accomplished by incorporating a lift mechanism into the seal faces. Spiral grooves, Fig. 16.10, have proven to be the most efficient and stable way to achieve a non-contacting seal design. As the shaft begins to rotate, gas is compressed within the seal faces and then allowed to expand across the

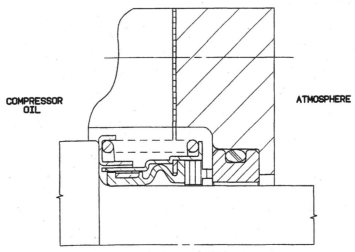

FIGURE 16.6 Elastomeric bellows seal (John Crane Inc.).

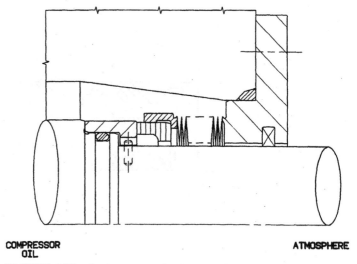

FIGURE 16.7 Stationary metal bellows seal (John Crane Inc.).

sealing dam. This generates enough opening force to separate the seal faces by a few nanometers during operation. Since there is no frictional contact, wear is eliminated and seal life is essentially infinite. A dry running non-contacting seal is designed to leak. The small amount of flow helps to remove the heat developed from constantly shearing the gas at the seal faces. The amount of leakage is significantly smaller than other types of seals. The effect of seal size and speed on leakage is shown in Fig. 16.11. Pressure and temperature also have an effect on leakage, as illustrated in Fig. 16.12. Leakage from the seal is vented to a vapor

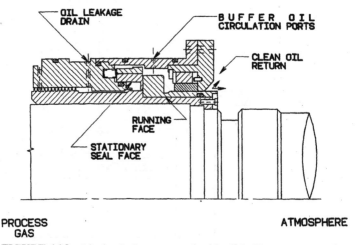

FIGURE 16.8 Mechanical contact seal with oil buffer.

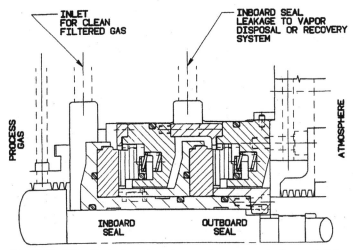

FIGURE 16.9 Typical tandem dry running non-contacting seal (John Crane Inc.).

disposal system when the gas is hazardous or toxic. Seal arrangement is an important part of any dry running non-contacting seal installation. Operating conditions such as the type of gas sealed, pressure, temperature, and speed, as well as abrasive contaminants, are considered in the selection of the seal arrangement. Typically when the process fluid is inert or non-toxic, a single seal is selected, Fig. 16.13. When the process gas contains abrasives, a steam flush may be used to provide a clean environment for the seal. Single seals are limited to 400 psi, 260°C, and 152 m/s.

Tandem seals, as shown in Fig. 16.9, are being used on hydrocarbon mixtures. These types of applications are found on pipeline, chemical, and refinery applications. Typically, operating conditions are limited to 1200 psig, 260°C, and 152

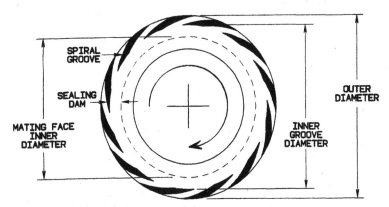

FIGURE 16.10 Spiral groove seal face.

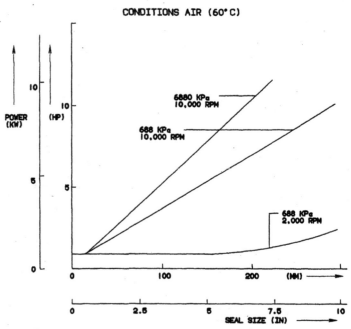

FIGURE 16.11 Dry running non-contacting seal performance: effect of seal size and speed on leakage.

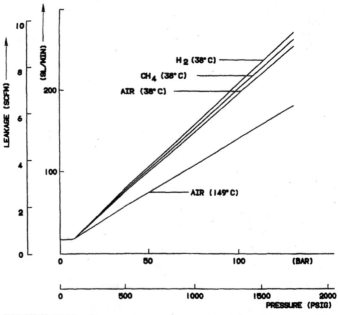

FIGURE 16.12 Dry running non-contacting seal performance: effect of pressure and temperature on leakage.

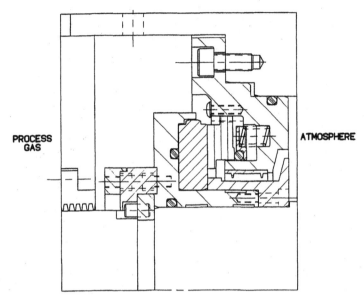

FIGURE 16.13 Single dry running non-contacting seal (John Crane Inc.).

m/s. Here, two seals are facing in the same direction in the seal chamber. Leakage from the seal is vented to a vapor disposal system. A typical control system for a tandem seal is shown in Fig. 16.14.

Triple tandem seals have been used to seal hydrogen recycle compressors in refinery service, with pressures to 2000 psi and temperature of 71°C. Shaft size and speed are 92 mm in diameter and 10,250 rpm, respectively. This unit was operated continuously for 24 months and the user estimates a 1.7 million dollars per year savings.

In some cases when a vapor disposal system or vapor recovery system cannot be used, a hazardous or toxic application may require a double seal arrangement, as shown in Fig. 16.15. Here an inert gas is used between the seals. Typically, plant nitrogen may be used. Nitrogen will leak to the process through the inboard and outboard seals. Leakage rates are normally less than 0.028 m^3/min. Double seals are normally used on services with pressures to 250 psig and temperatures from −60°C to 260°C, and speeds to 152 m/s.

Dry running non-contacting seals offer the user considerable savings over other types of sealing systems.

Sealing technology continues to evolve in solving complex problems defined by industry. The demands for higher operating pressures required the solution of explosive decompression of O-rings, the solution of secondary seal friction, and the deflection of seal face materials. The result is a high pressure non-contacting gas seal for pressures to 3000 psig and speeds to 180 m/s, as shown in Fig. 16.16. Here, spring energized polymer seals used with the supporting seal structure to control seal deflection, achieved the intended results.

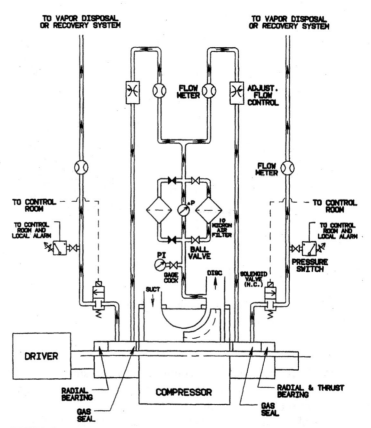

FIGURE 16.14 Typical emissions control system.

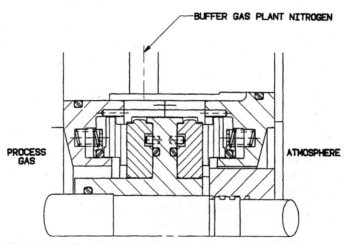

FIGURE 16.15 Typical double seal arrangement (John Crane Inc.).

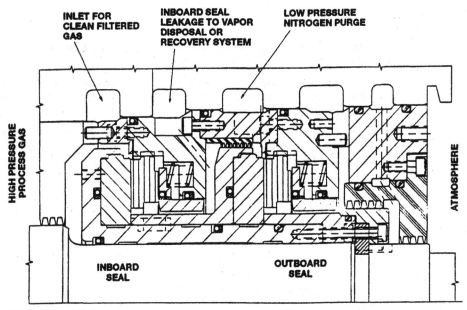

FIGURE 16.16 Tandem dry running non-contacting high pressure seal (John Crane Inc.).

In certain applications, there is a recognized need for the seal to work in the reverse rotation for an extended period. There are two specific circumstances that can cause a compressor to run in the reverse direction after shutdown.

- A leaky or "stuck open" discharge valve
- A large volume of gas between the compressor and its discharge valve

This has resulted in optimized bi-directional groove profile shown in Fig. 16.17. Even though this design has optimized bi-directional seal performance, the spiral groove design still provides a superior level of performance across the operating envelope.

Dry running non-contacting seals offer the user considerable savings over other types of sealing systems through increased mean time between maintenance and improved equipment reliability.

FIGURE 16.17 Bi-directional grooved seal face (John Crane Inc.).

16.3 FURTHER READING

1. Netzel, J. P., *High Performance Gas Compressor Seals,* 11th Internationa Conference on Fluid Sealing, BHRA Cranfield Bedford, UK (1987).
2. Shah, P, *Dry Gas Compressor Seals,* 17th Turbomachinery Symposim, Turbomachinery Laboratory, Department of Mechanical Engineering, Texas A&M University, College Station, Texas, November 1988.
3. Carter, D. R., *Application of Dry Gas Seals on a High Pressure Hydrogen Recycle Compressor,* 17th Turbomachinery Symposium, Turbomachinery Laboratory, Department of Mechanical Engineering, Texas A&M University, College Station, Texas, November 1988.
4. Atkins, K. E., and R. X. Perez, *Influence of Gas Seals on Rotor Stability of a High Speed Hydrogen Recycle Compressor,* 17th Turbomachinery Symposiu, Turbomachinery Laboratory, Department of Mechanical Engineering, Texas A&M University, College Station, Texas, November 1988.
5. Pecht, G. G., and D. Carter, *System Design and Performance of a Spiral Groove Gas Seal for Hydrogen Service,* 44th Annual Meeting, Society of Tribologists and Lubrication Engineers, Atlanta, Georgia, May 1989.
6. Dugar Jr., J. R., B. X. Tran, and J. F. Southcott, *Adaptation of a Propylene Refrigeration Compressor with Dry Gas Seals,* 20th Turbomachinery Laboratory, Department of Mechanical Engineering, Texas A&M University, College Station, Texas, September 1991.
7. Morris, J. R., C. G. Stroh, J. F. Southcott, *Retrofit of a Steam Turbine with Dry Gas Seals,* 22nd Turbomachinery Symposium, Turbomachinery Laboratory, Department of Mechanical Engineering, Texas A&M University, College Station, Texas, September 1993.
8. Mayeaux, P. T. and P. L. Feltman Jr., *Design Improvements Enhance Dry Gas Seal's Ability to Handle Reverse Pressurization,* 25th Turbomachinery Laboratory, Department of Mechanical Engineering, Texas A&M University, College Station, Texas, September 1996.

CHAPTER 17
RECIPROCATING COMPRESSOR SEALING

Paul Hanlon
C. Lee Cook, A Dover Resources Company

The first seals for reciprocating rods or pistons were of plastic-like materials, forced to conform to the shape of the seal cavity or stuffing box. This was replaced by "V" rings, "U" rings or similar shapes when it was realized that gas pressure could help supply the force necessary to cause soft materials to seal. This type of seal is still used in many small compressors.

As machines were built for higher temperatures, pressures and speeds, it brought out the need for stronger, more wear-resistant materials, and the logical choice was metal. Soft ones, such as babbitts, were chosen and used in configurations which, like the plastic materials, were forced to yield or deform to effect a seal. These were still, essentially, slow speed, low-pressure types of seals. In general, they required a large spring or force from the gland to push even the relatively soft metals down into contact with the rod or out against the cylinder.

Following this were pressure-actuated segmental or cut rings of harder metals, such as bronze, cast iron, or hard plastics, which could be made with narrow contact at the rod or cylinder and thus generate less total load and frictional heat. These also "float" with the rod or piston accommodating considerable misalignment or lateral motion.

17.1 COMPRESSOR PACKING

A packing is a seal around a shaft passing through a cylinder head. It consists of one or more rings contained within a case that is typically bolted to that head. Packing may be as basic as one ring in a case, or it may be an assembly consisting of a number of different type rings in a case that might have provision for lubrication, venting, purging, cooling, static sealing, temperature and pressure measurement, leakage measurement, and rod position detector.

A typical compressor packing consists of a series of seal rings in which each ring is meant to stop or restrict flow of gas to atmosphere or out into the distance piece. The rings are held in separate grooves or "cups" within a packing case. Each ring seals against the piston rod and also against the face of the packing cup at a right angle to the rod axis. Rings are free to move laterally along with the rod, and free to "float" within the grooves.

A basic packing arrangement consists of: 1) a pressure breaker that functions as a flow restricter (rather than a true sealing ring); 2) several seal rings that are meant to stop flow or leakage into the vent; and 3) a vent control ring that prevents gas from leaking into the distance piece from the vent.

Nomenclature for compressor and packing is given in Fig. 17.1. The packing illustration shows some elements that are not in all packing. Where the packing and other seals are located in a compressor cylinder is shown in Fig. 17.2.

In addition to the cylinder pressure packing is a wiper packing, which prevents oil from escaping along the rod out of the crankcase, and then a partition packing, in double distance piece machines, to prevent leakage between the distance pieces. The partition and wiper packing are under light pressure differential, but in compressors where very low loss of gas to atmosphere is required, these packings must provide a tight seal.

A packing assembly which incorporates wiper and distance piece all in the pressure packing is shown in Fig. 17.3. This type of packing allows for a very compact arrangement of the cylinder, but with somewhat greater opportunity for gas leakage into the crankcase, or crankcase oil leakage into the packing case. It would be most suited to lubricated or possible semi-lubricated cylinders.

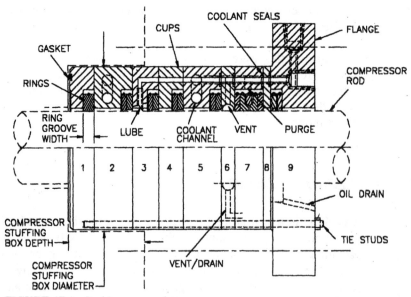

FIGURE 17.1 Packing nomenclature.

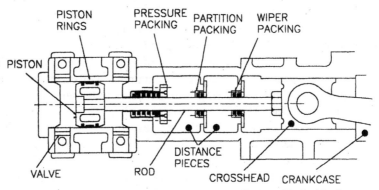

FIGURE 17.2 Seals in a typical compressor cylinder with double distance piece.

Even though seal rings do not provide a direct leak path, a vent to carry leakage away from the packing is necessary in most applications, since slight imperfections in the rings, or misalignment of ring, case, or shaft mating faces may allow some small volume of gas to blow-by.

Gas flow past a series of rings creates varying pressure differentials across individual rings, but because of rapid rise and fall of pressure within the compressor cylinder and because of low flow by the best sealing rings, this "labyrinth" effect in packing rings is usually insignificant.

Pressure drop is highest across rings nearest the pressure when the set is new. As packing wears, the downstream rings see more and more pressure drop as the leak paths of rings increase with age. A "reverse" drop exists across some rings

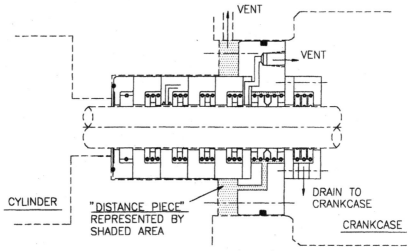

FIGURE 17.3 Compact packing incorporating distance piece and wiper into the cylinder packing case.

during the suction stroke; that is, gas will flow back out of the case toward the cylinder during that portion of the stroke. All this is illustrated in the "computer generated" curves shown in Fig. 17.4.

Pressure drop through a series of rings is highest across the one ring that forms the best seal. In packing sets containing well-made rings which fit the rod and grooves properly, the ring nearest the cylinder will, initially, carry almost the full pressure drop. As this ring wears, or changes for any reason, the load will shift to one of the other rings, usually the next ring in line. The pressure drop may distribute across a series of rings in almost any manner. The individual rings in a packing set wear, essentially, one-at-a-time, so the number of rings in the set influences packing life but has little effect on leakage.

17.2 BREAKER RINGS

Breaker rings, the simplest form of packing rings, are designed to restrict or control flow rather than effect a tight seal. In type P, the flow controlling orifice is the gap

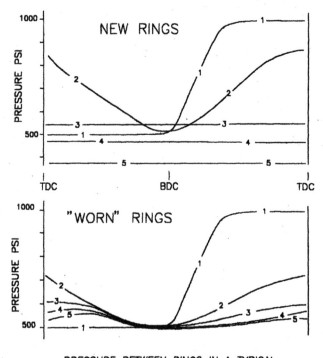

PRESSURE BETWEEN RINGS IN A TYPICAL (5) GROOVE PACKING (NUMBERS ON CURVE INDICATES PRESSURE UPSTREAM OF THAT PARTICULAR RING – 1 BEING PRESSURE WITHIN THE CYLINDER)

FIGURE 17.4 Plot of instantaneous pressures within a packing case.

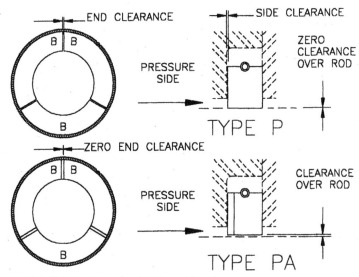

FIGURE 17.5 Type P and PA packing rings.

that is formed at the segment ends. In a type PA breaker, the orifice is formed by the clearance between the ring bore and rod. Type P and PA are illustrated in Fig. 17.5. The performance of both rings is very similar. However, a type PA breaker provides lower rod loading than a P, at the expense of a more difficult to control orifice area.

The most important function of breakers is to retard rapid expansion of gas from within the packing case back into the cylinder during the suction stroke. On the intake portion of the stroke, gas contained within the packing case tends to reverse direction and flow toward the cylinder where pressure is dropping rapidly to suction pressure levels. Without some restriction to this flow, an "exploding" action of the rings may occur causing premature ring failure and damage.

Pressure breakers are not required in all packing. They are generally not needed at pressures below 300 psi and are not required when the seal rings themselves act to restrict backflow.

Pressure breakers may be manufactured from almost any material, but those materials with a low thermal coefficient of expansion and high stiffness are preferable for a stable orifice area. Metal is usually superior to plastic in this respect; however, breakers can be made satisfactorily from some of the high strength plastics.

17.3 PACKING RING TYPE BT

The BT packing ring (Fig. 17.6) is a true sealing ring which is made up of a radially cut ring (facing the high pressure side) and a second, butt/tangent cut ring. The individual rings are doweled together so that the inner butt cuts of the tangen-

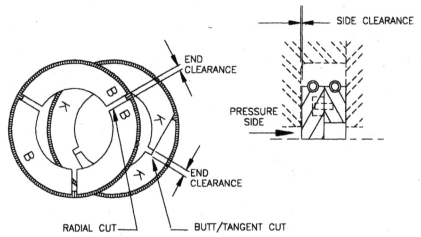

FIGURE 17.6 Type BT packing ring.

tially cut ring are offset, or staggered, between the radial cuts of the first ring in the pair. In this manner, gas passage by or through the ring is prevented.

In order for these rings to operate satisfactorily, it is necessary that both fit the rod perfectly. They must contact each other at their mating faces and the tangent ring must lie flat against the groove side face.

In addition to this, all sealing edges at the bore must be sharp and square. If these edges are beveled or rounded, they form a path for gas flow from the radial cut in the first ring to the radial portion of the cut in the tangent ring. If the tangential edges are not square, they form a passage from the outside toward the radial cut at the bore. All other edges are generally given small bevels to assist lubrication between the contact faces. Under ordinary circumstances, lubrication serves to help the seal function because it fills up minute crevices. This type of ring is single acting, or directional, in that it seals pressure from one side only.

17.4 PACKING RING TYPE BD

The BD packing ring (Fig. 17.7) is double acting, meaning it will seal pressure from either direction. It consists of two butt/tangent cut rings doweled so the radial portion of the cuts are staggered, blocking any flow path for leakage. All features important for establishing a tight seal with the BT ring, such as square sealing edges and proper face contact between rings and the rod and/or cup, are also important for sealing with the BD ring.

17.5 COMMON PACKING RING CHARACTERISTICS

The P, BT, and BD are the most common types of packing ring in general use. Numerous variations of these three exist, but sealing principles involved are gen-

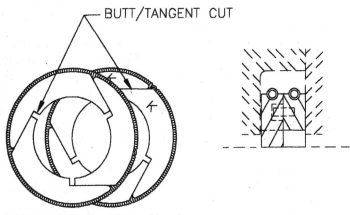

FIGURE 17.7 Type BD packing ring.

erally the same and the conditions affecting proper operation are common to all types.

Packing rings are made in segments for two reasons: 1) for installation over the rod; and 2) to allow free radial movement down against the rod. Free radial movement allows for slight size variations and also provides a means to accommodate ring wear. Both spring and gas pressure loading cause rings to contract or move radially inward toward the rod.

All types of packing rings are manufactured with an initial end clearance of sufficient size that no adjustment is required throughout their useful life. When they have worn to a point where the ends butt, they are normally discarded. In most rings, end clearance could be adjusted to maintain the proper opening. However, it is seldom practical to rework such a ring after it has operated in a butted condition. The bore will be worn out of round, and merely recreating end clearance will not correct poor rod contact.

17.6 PACKING RING MATERIALS

Lubrication and gas pressure are two factors that influence selection of ring material. Packing may operate at conditions that vary from full or normal lubrication all the way to dry or nonlube service. The common material used in packing fall into two categories:

1. *Metallic*—(typically bronze, cast iron, or to a diminishing extent, babbitt) which requires some lubrication, but may be used at very high pressure
2. *Nonmetallic*—(typically carbon-graphite, TFE (polytetrafluoroethylene), or one of the other plastics). The important advantage of nonmetallics is their ability to run in poorly lubricated applications.

Carbon-graphite and TFE blends work well in the complete absence of lubrication, but are somewhat limited insofar as the pressure at which they can be used due to their low strength (compared to metallics). Filled TFE blends are good to approximately 800 psi; but, with the addition of a backup, or anti-extrusion ring, can be used at much higher pressures. Carbon-graphites are moderately strong, but somewhat brittle, and therefore limited to approximately 1500 psi.

Plastic is the preferred material for tangent cut rings which require a high degree of conformability. All rings must conform at their joints and around the rod in order to establish a good seal, but some types will begin to leak slightly as they wear or as they change size due to thermal expansion. A good ring material is one that is not excessively stiff relative to the pressure acting on it.

Nonmetallic materials, including the new "high performance" plastics and their blends with fillers, are the most commonly used materials in the manufacture of modern packing rings. Metallics are principally useful at very high pressure, or as anti-extrusion rings, where strength or rigidity is an important requirement.

17.7 LUBRICATED, SEMILUBRICATED AND NONLUBRICATED PACKING

Packing rings made with TFE filled or blended with other materials (including other plastics) can now be used in many applications including those with less than full lubrication. By definition, *lubricated packing* receives just the amount of lubricant required for long life. *Semilubricated packing* is that which receives either reduced lubrication through a lubricator, or, in the event there is a short distance piece, minimal lubricant carry-over from the crankcase. *Nonlubricated packing* is packing installed in systems where double distance pieces, rod oil slingers, or some other means is used to prevent any lubrication from reaching the rod rings.

TFE is the material used most often for packing installed under conditions where lubrication is impaired or less than normal. Generally, this would be in the presence of corrosives, diluting liquids, or high temperatures. The sealing principles of all TFE ring types are virtually the same as those previously described. Due to their inherent flexibility, TFE rings have the added advantage of immediate sealing without requiring break-in.

Ring types TR and BTR (Fig. 17.8) are designs that include nonmetallic rings plus a rigid back-up ring. The back-up ring is made slightly larger than the rod. It has butted ends and does not grip (or grips lightly) the rod under pressure loading. Its function is to prevent extrusion of the softer nonmetallic ring and also, under some conditions, aid in conducting heat away from the rod surface through the light contact. Usually back-up rings are made of metal, but they can also be manufactured from the stronger plastics.

At pressures above 800 psi, type TR rings are the preferred alternative to the standard BT sealing ring. To obtain a better seal where there is insufficient pressure

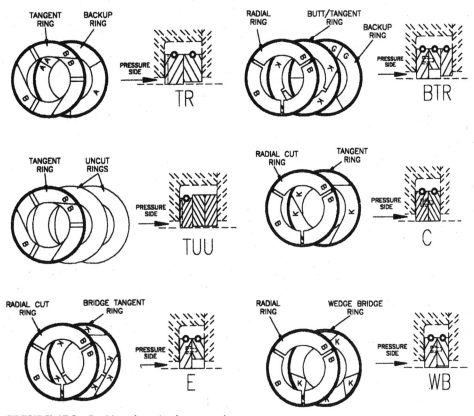

FIGURE 17.8 Packing rings (various types).

to properly actuate the TR ring, type BTR rings may be used. The BTR is limited to a pressure of about 2500 psi. Above this pressure, deformation or extrusion may occur into the gap of the butt/tangent ring.

To overcome sealing problems with the TR and deformation of the BTR, a tangent cut may be combined with an upstream radial cut ring. This is the type C (Fig. 17.8) or when a backup is added, the type CR. These styles are most useful when applied above 2500 psi.

The basic TFE ring set for pressures below 800 psi consists of BT rings to seal cylinder pressure with a double-acting BD ring downstream of the vent. This arrangement is the same for nonlubricated or fully lubricated cases. When the pressure exceeds 800 psi and the application is full or semi-lubricated, the arrangement should be as follows:

1. One breaker ring type P
2. BTR or TR rings in intermediate grooves
3. A BD ring beyond the atmospheric vent

Unless some special condition exists, the P and BD rings should be metal, and the BTR or TR rings a combination of metal-TFE. For best sealing of vent gas, the BD ring should be TFE.

The type P ring in this case will have an added function. Along with preventing backflow, garter spring breakage, etc., the fact that it is in contact with the rod around the entire circumference (without heavy pressure loading) allows it to function as a means of conducting heat. BD rings of metal may serve the same purpose.

17.8 PACKING RING TYPE TU

The TU ring is essentially the same as the TR, except the radially cut ring in the TR has been replaced with an uncut ring, which can be either plastic or metal.

When stiffness of the U ring is low compared to the pressure acting against the ring, it will collapse inwardly, maintaining a seal even as it wears. With proper choice of material in the TU ring, it can be made to seal low pressure (suction) with the tangentially cut ring, and high pressure (discharge) with the uncut ring. The TR ring can function in the same manner, but lacks the almost perfect seal characteristics of the uncut ring in the TU.

An added benefit with an uncut ring is reduced contact pressure between the ring bore and the rod. Compressive stress in the ring acts counter to gas pressure against the ring outside diameter, so there is lower ring-to-rod loading than there would be with a tangentially cut ring.

To prevent extrusion of the soft plastic sometimes used in the U ring, an additional back-up ring can be added to form the TUU style (Fig. 17.8). Typically the second U ring would be metal, or a very rigid plastic.

With proper choice of materials for the three rings in the TUU, each ring can be made to seal only over a particular pressure range. Thus wear and heat generation can be divided over the individual rings to give better performance and longer life.

The lower leakage and reduced wear that are advantages of the TU or TUU ring are somewhat offset by difficulty of installation. The rings must be installed over the rod end, and with an entering sleeve that is no larger than rod diameter as shown in Fig. 17.9.

17.9 THERMAL EFFECTS

When rings expand circumferentially due to temperature increase, there are usually leak paths created at the joints. The effect is similar whether this occurs in a true tangent-cut T ring such as in the TR, or in a butt/tangent ring such as in the BT.

A T ring is less likely to leak when its joint is covered by another upstream ring. This is the construction used in the Type C.

Contraction due to *lower* temperature rarely occurs, but if it does, it is much like having an undersized ring or oversized rod.

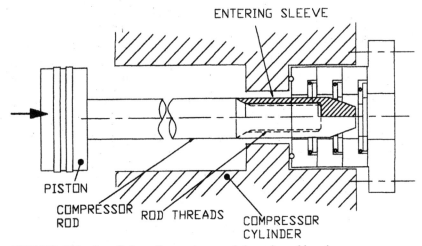

FIGURE 17.9 Installation of compressor rod through packing rings.

All the high temperature effects can be minimized by adding an upstream radially cut ring to block leak paths through the T ring. This addition is found in the C or CR ring arrangement. These ring styles are free of unsupported areas where extrusion can occur, so they're suitable for higher pressures than a BT or BTR.

One method to minimize the tendency toward breakage and lack of conformability of both butt-tangent and true tangent, or T rings, is to use "bridge" joints as illustrated in WB and E style (Fig. 17.8). The shorter, sturdier segments make the WB style suitable for service in situations where the elevated temperature and pressure may cause other style of rings to leak or break. Handling and installation of WB or E rings may present problems due to the greater number of segments involved. As with other style of rings, the E and WB may be used with back-up rings to prevent extrusion along the rod.

17.10 UNDERSIZED RODS

If a rod is undersize, but true in circularity and without taper, most rings still form an effective seal because bore contact will be in the center of each segment and at this point, the cut in each ring is overlapped by its mate. There is normally not a problem if undersize does not exceed .002 inch per inch of rod diameter. When a rod is undersized, some additional time is required for break-in before a packing will give the best seal.

17.11 OVERSIZED RODS

When packing rings have a somewhat smaller bore than rod diameter, segments touch only at each end, leaving the center away from the rod. This is directly in line with the cut, or joint, of the mating ring so there is a passage for gas flow along this line. This condition, if not too severe, will also be corrected by wear-in.

One problem that may occur with either an undersized or oversized rod is that lubrication may be blown off the rubbing surfaces by the passage of gas. The subsequent "dry contact" may result in high friction, high temperature or rapid wear.

17.12 TAPERED RODS

In the presence of lubricating films, which help fill (or block) very small leak paths, a packing can function adequately with some slight rod taper. Generally, taper is found at one, or both, ends of the ring travel area, the rest of which is relatively straight. The effect of the rod passing through the packing ring from the tapered to the straight portion and back, is wear at ring edges. This is due to the ring bearing directly on one edge while over the tapered section of rod and on the other while over the straight section. This dual edge wear leaves a gas passage along the bore from one radial cut to the other. The effect of taper or misalignment of the rod surface on leakage is illustrated in Fig. 17.10.

Tapered rods also cause cyclic flexing of garter springs. If this flexing is excessive, it can lead to spring breakage.

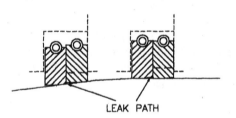

LACK OF CONTACT AS A RESULT OF A TAPERED OR BOWED ROD

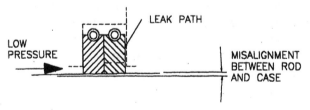

LEAKAGE AS A RESULT OF CASE TO ROD MISALIGNMENT

FIGURE 17.10 Effect of rod misalignment and taper on leakage.

17.13 PACKING LEAKAGE

If the complete packing, that is case plus rings, is considered there are three elements past which leakage can occur: gasket, cup faces and rings.

Conventional gaskets, which are generally soft metal, such as copper, aluminum, or pure iron, will basically give zero leakage if proper dimensions and smooth surfaces are applied. However, there have been some improvements in this recently using spiral wound gaskets and in a few instances, o-rings. In general though, to accommodate misalignment and discontinuities in the surface at the bottom of the stuffing box, the solid metallic gasket has been satisfactory.

The second element, case faces, can be made 100% leak free by the use of either a sealant or an o-ring applied between the mating surfaces. With this, and with proper uniform torque applied to the flange bolting, cup faces can be brought to a point where essentially no flow occurs between them.

The third element, rings, are probably what receive the most attention when trying to control leakage. Rod rings, since they do leak slightly, can act as a labyrinth or more likely as a labyrinth through which most of the pressure drop occurs across just a few rings. Changes in the rings due to pressure and temperature, and also the cyclic nature of the pressure, make it very likely the pressure drop will occur, for the most part, across only *one* ring. The pressure level, as it typically might occur across each ring in a 5-ring set, would be as illustrated in Fig. 17.4.

This was calculated on the basis of small orifices in new rings, larger leak paths for used rings, and equal volume between the rings. Pressures have been measured in actual packings and it has been determined that pressure distribution can occur in almost any fashion, but in general, will agree with what is illustrated by these curves. Efforts could then logically be directed at reducing leakage by 1) increasing the number of rings in the case or 2) designing and manufacturing individual rings that present the smallest orifice size.

Packings operating satisfactorily will leak at an average rate somewhere between zero and .2 scfm (standard cubic feet per minute). In general, packings that are considered to be unsatisfactory will be above .5 scfm. Current rings in use are, theoretically, very close to zero leak seals. The leakage that does occur by them, when they are properly manufactured and applied, by actual measurement falls in that zero to .2 scfm range, with the average generally about .1 scfm.

Most of the leakage that occurs by rings, flows through the joints due to "misfit" within the ring itself, or past the side or bore of the ring due to misalignment at these surfaces. Leakage past the ring due to less than perfect surface finish does also occur, but this is a minor leak path, except in those instances where the rings *cause* an increase in roughness or finish of the mating surface. No doubt some leakage also takes place due to manufacturing errors or a lack of quality. The primary reasons for leakage by rings usually is 1) misalignment between packing case and rod; and 2) temperature changes that distort the relative position between the ring and the rod, the ring and the case, or the ring segments themselves.

Misalignment, or as it is sometimes described, "run-out," between the rod and case, is generally measured by slowly moving the rod back and forth through its stroke and measuring lateral movement with an indicator. This can also be done electronically when the machine is running by the use of an analyzer and transducers to instantaneously pick up the rod position. Packing rings are able to accommodate lateral motion in that they can float within the case and, in general, exhibit little leakage as a result of this motion. However, angular motion (or the motion caused by a bent or temperature deflected rod) is detrimental and causes the sealing surfaces to open up leak paths as in Fig. 17.10.

There is one other type of misalignment that is important and that is the *angular* relation between case and rod. Usually this would be checked by measuring the squareness between the face of the flange and the rod. Correction to bring the case into alignment would be made by proper torquing of the flange bolting.

A rise in temperature of the rings, which is usually the result of frictional heat, can cause the rings to lift away at the bore or can cause separation at tangent cuts. High, uneven heating of the rod can also cause "bowing" of the rod, which will have the same affect as misalignment. Typically, the way high temperature results in leakage at the cuts and bore is shown in Fig. 17.11.

High temperature is probably the most destructive condition that can occur in a packing. Aside from the distortion that it causes, it may also lead to breakdown of the mating faces between the ring and rod, or the ring and case. All of these things ultimately lead to leakage and short packing life.

17.14 RING LEAKAGE AT LOW PRESSURE

Relative to the potential sealing problems that occur in higher pressure compressors, the sealing problems in low pressure (50 psi or less) compressors would seem insignificant. Wear and packing life may not be major problems, however, obtaining a good seal between packing ring and rod, and especially between the ring and cup face, may be difficult.

There are two reasons for this. The first is because of the relative stiffness of conventional seal rings. As the pressure differential increases across a ring, some of the leak paths will be decreased as the ring deforms. Using materials with low modulus of elasticity, such as plastics, will improve the seal. However, even plastics are relatively rigid when sealing against very low pressure, such as when suction is close to atmospheric pressure.

The second reason for leakage at low pressure is loss of ring contact with the flat cup face. The pressure and frictional forces acting on a ring is as illustrated in Fig. 17.12. Frictional forces acting in a direction away from the cup face may cause the rod to drag a packing ring away from the cup. When this occurs depends on ring size, coefficient of friction, and the pressure level. For "normal" size rings, which are well lubricated, approximately 50 psi differential pressure is required to overcome the effect of friction, and hold them against the cup.

In a compressor, pressure at the packing vent is normally well under 50 psi, so good practice dictates the use of a double-acting ring to minimize leakage from the vent into the distance piece. This ring functions by forming a reasonable seal

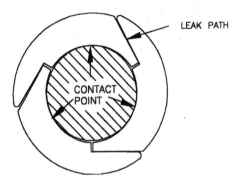

FIGURE 17.11 Leak paths in packing rings.

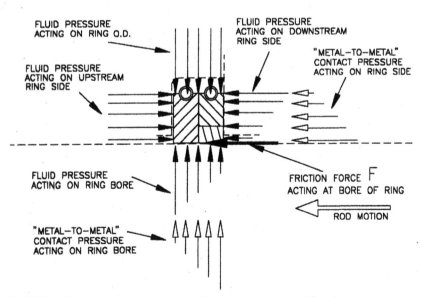

FIGURE 17.12 Pressure and friction forces acting on a packing ring.

regardless of which face it rests against. However, there is an instant during stroke reversal when even the double-acting ring will leak, and in order to improve on this it is necessary to use rings which are spring loaded in the axial direction, as illustrated in Fig. 17.13.

The advantage of ring types AT or WAT is the almost equal axial and radial load provided by the spring. If these loads are not equal, or very nearly equal, the ring can either "hang" in the groove and thus leak, or be pulled from one face to the other by the frictional force.

17.15 PROBLEMS ASSOCIATED WITH LOW SUCTION PRESSURE

In a compressor handling gas which would be undesirable to contaminate with air, and in which suction pressure is below atmospheric pressure (pressure *within the cylinder* falls to vacuum), air may pass back through the packing and into the cylinder. Either double-acting rings or those spring loaded against the cup face will

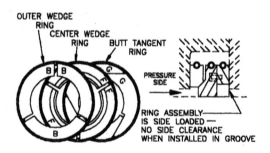

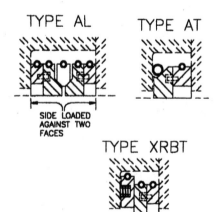

FIGURE 17.13 Side loaded packing rings.

reduce leakage, but neither will eliminate it completely due to minor leak paths at ring surfaces and insufficient gas pressure to make the ring conform.

One solution is to introduce *gas* at low pressure at the vent. Gas that enters the cylinder will be drawn from this rather than air from the atmosphere. Gas pressure in the vent need be only a few psi above atmospheric pressure.

Further loss of vent gas to the distance piece can be minimized with double vents, that is, a pressured vent, plus a vent to atmosphere with a ring between them. This double vent arrangement will control gas leakage into the distance piece to at least the same degree as with a single atmospheric vent.

When it is impractical to use a source of gas slightly above suction pressure, then gas at full discharge pressure may be used. This will increase leakage both toward the cylinder as well as out the atmospheric vent. It also puts a higher pressure load on the ring between the vent and the distance piece, or between the two vents, and may cause more frictional heat. This is the result of two rings loaded at high pressure, when normally only one ring would carry a high differential.

Everything discussed to this point assumes the gas used to exclude air was the same as that going through the compressor cylinder. There are circumstances where it is more practical to use a gas other than the process gas being compressed. The usual choice of a vent gas is an inert gas such as nitrogen. If nitrogen is used and pressure at the vent is less than discharge pressure, then gas in the vent or the distance piece will probably be a mixture of inert gas and process gas. Gas discharged from the compressor cylinder may also contain a small quantity of the inert gas.

17.16 PROBLEMS ASSOCIATED WITH LOW LEAKAGE REQUIREMENTS

In addition to low suction compressors, where entry of air should be avoided, there are a number of applications where any loss of gas to the atmosphere must be prevented. This is true for toxic or dangerous gases, very expensive gases, or in processes where it is desirable to maintain a constant quantity of gas in the system, such as within a refrigeration cycle.

There are several methods that will accomplish the goal of minimizing leakage. One way is to use a double-vented packing, with buffering gas introduced at constant pressure to the *outer* vent instead of the inner vent. The "buffering" gas is usually inert or at least can be tolerated as emission. Leakage from the main sealing rings flows out the inner vent in addition to some flow of buffering gas. The *loss to atmosphere* of the gas being compressed is almost zero and the mixture of the two gases from the inner vent can be recovered.

Another way to prevent leakage is much like the method used to prevent air from entering the cylinder during vacuum conditions. Buffering gas is introduced through the packing vent at a pressure which exceeds the compressor discharge pressure. The result is a gas mixture in the cylinder as opposed to one in the recovery vent. Depending on buffering pressure, one or more rings may be used between the buffer inlet and the distance piece. A second vent may be used to recover buffer gas or minimize leakage into the distance piece.

17.17 EFFECT OF RING TYPE ON LEAKAGE CONTROL

In each of the low pressure vent arrangements previously described, double-acting (BD) rings are normally used. When the requirements call for very low leakage, one of the axially loaded type of rings would normally be substituted. Axial loaded rings may be used for control of gas flow toward the cylinder, or into the vent or distance piece, as well.

When confronted with low pressures, another possible control method is to use "soft" packing or a lip seal. The result will be a tighter seal than with conventional rings. However, packing life may be somewhat shorter due to a limited resistance to, and compensation for, wear. In many reciprocating machines, lubrication is marginal, and when this is coupled with significant rod "float," there may be rapid wear on soft plastic seals. This soft seal arrangement, though not 100% effective, might be chosen for gases such as ammonia, propane, or methane. In general, the escape to atmosphere of these gases may be tolerated to a greater degree compared to gases such as vinyl chloride, chlorine, or hydrogen sulfide.

The quantity of leakage with any of the ring types or with lip seals can only be estimated based on experience and empirical data. Leakage paths through packing rings occur not by design, but due to tolerances, alignment, or various other characteristics of the compressor and ring itself. Although it would appear that flow rate would be directly related to pressure differential, this is not the case as rings often conform and effect a better seal at high pressure.

Under normal conditions and with most gases, leakage from the cylinder or out any of the vents will be on the order of 0.1 scfm. As factors which cause leakage progressively get worse (misalignment, dirt, poor finishes, poor lubrication, etc.), or with lighter gases such as hydrogen or helium, leakage rates can be expected to increase by two to four times. This is just a guide based on measured flow rates at various conditions. If a more accurate flow rate is required, it must be obtained by test and measurement.

17.18 LEAKAGE CONTROL WITH DISTANCE PIECE VENTING

Another method to prevent both air entry into the cylinder and/or gas leakage is to close, but vent, the distance piece rather than the packing case. To accomplish this, the conventional packing arrangement, less the atmospheric vent, may be used. It is also necessary to provide a positive seal at the partition packing in double distance piece machines, or for the wiper packing in a single distance piece machine. To reduce air contamination, the distance piece vent is used to *supply* gas, whereas when it is required to control leakage, the distance piece vent will be used to *recover* gas. This type of buffering for both single and double distance pieces is illustrated in Fig. 17.14.

In either case, one type of partition or wiper packing seal arrangement consists of side-loaded rings between which a buffering liquid (oil) or, more commonly,

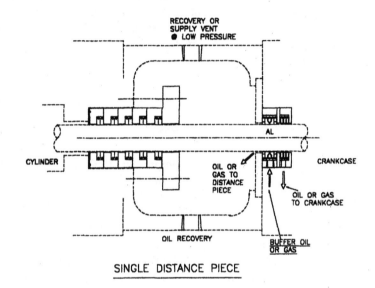
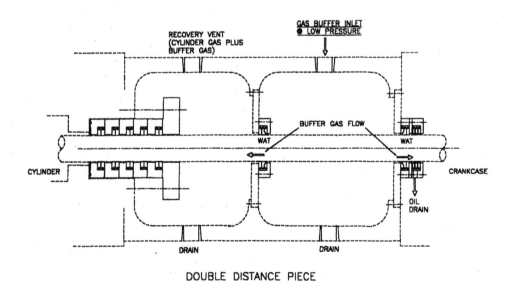

FIGURE 17.14 Distance piece venting to control leakage to atmosphere.

gas, is injected. When oil is used, it is generally taken from the crankcase. Part of the leakage past this seal goes back into the crankcase, while part must be recovered from the distance piece to also be re-piped to the crankcase. In the event gas is used, it must be one that is tolerated in the crankcase and suitable for mixing with the recovered gas. When two distance pieces are used, the second one can be pressurized to serve as the buffer.

In nearly every instance where gas is used in the distance piece to control leakage, the process pressure is very low which usually requires the use of axially loaded rings. The basic principle remains the same in all applications: the prevention of gas flow in one direction along some particular path is accomplished by establishing pressure conditions that cause gas to flow in the opposite direction along that same path.

17.19 STATIC COMPRESSOR SEALING

In some applications, after a compressor stops, it is desirable to maintain gas, under pressure, within the cylinder. Rod packing will generally leak slightly more when rod motion is stopped as compared to when it is moving. This is due to a number of factors: loss of oil which filled the leak paths; changes in the ring shape as the ring cools; and changes in rod alignment as the temperature changes.

The cylinder may be sealed by essentially an uncut, conformable ring which is forced against the rod by a piston that encircles the rod. Actuation occurs when pressurized gas is admitted to the piston. The piston, when actuated, wedges the seal ring inward against the rod. The shape of the seal is such that pressure from within the cylinder will cause the seal to move away from the rod when actuating pressure is lowered. A typical packing case containing this static seal is illustrated in Fig. 17.15. The pressure required to actuate the seal is one-half, up to full cylinder pressure.

After the compressor has stopped, a valve would be opened, admitting pressurized gas behind the internal piston. Pressure must be maintained on the piston for as long as it is desirable to seal the cylinder. De-actuation would take place when pressure on the piston is reduced, allowing the springs to push the piston back away from the seal. The seal itself would then lift from the rod surface. A vent to atmosphere (or some other low pressure area) must be located downstream of the seal in order for seal actuation to occur or for seal release. This can be adapted for pressures up to approximately 2000 psi. The seal itself is made of relatively soft synthetic rubber for pressures up to 700 psi or TFE for pressures above that.

17.20 COMPRESSOR BARRIER FLUID SYSTEMS FOR FUGITIVE EMISSIONS CONTROL

Although far down the list of offenders for total air pollution emissions, compressors have higher leak rates than most other industrial equipment. There is no federal

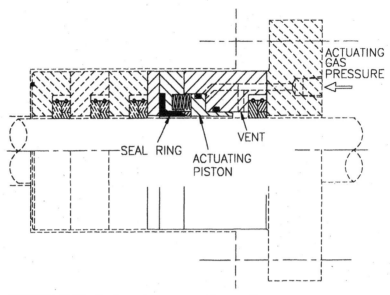

FIGURE 17.15 Static seal arrangement in a packing case.

regulatory limit for compressor leakage. It is indirectly established by the limit on the allowable concentration of VOC (volatile organic compounds) measured near the leak source.

Federal regulations consider 500 ppm (parts per million) to be "no detectable emissions." The regulations require that any compressor not in compliance with that emission level must have a barrier fluid (gas or liquid) system installed on its seal.

A "barrier" system is one in which a non-VOC liquid, or gas, is forced to flow into the seal in a direction opposite to that of the leakage. The barrier fluid or gas blocks the escape of leakage to atmosphere from the compressor seal or packing. The system can be installed in one of two ways:

1. A fluid, usually inert, is injected to form a barrier seal between the process gas, and atmosphere. The fluid pressure must be maintained slightly above the pressure upstream of the seal.
2. More commonly, a barrier gas is applied between a set of "barrier seals", usually WAT or AL rings as illustrated in Fig. 17.16. The barrier gas is held at a pressure exceeding the pressure in the vent which carries leakage away from the compressor seal.

On reciprocating equipment the motion of the rod will carry fluid, usually oil, in the form of a thin film under the seal face and out to atmosphere where emissions of gas carried in the oil may be released. The oil film on the rod surface absorbs gas while in the cylinder due to the relatively high gas pressure, and then releases this gas to atmosphere when the rod moves out of the packing. On non-lube ap-

17.22 CHAPTER SEVENTEEN

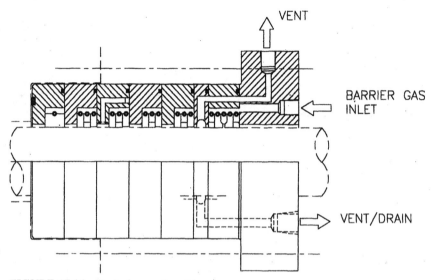

FIGURE 17.16 Typical purged packing.

plications, gas molecules can be carried out of the packing in the rod surface irregularities.

In either case, a small quantity of gas may escape the barrier seal and become fugitive emissions outside the packing flange. These may measure as much as 200 ppm, but generally are much lower. Gas transported by the rod surface becomes important only when allowable emissions approach zero.

17.21 WIPER PACKING

In addition to wiping oil from the rod surface and returning it through a drain back to the crankcase, wiper packing also has a sealing function. Pressure pulses from the crosshead, acting as a piston, may cause a "breathing" action through the wiper that will cause it to leak oil over into the distance piece. There can also be leakage of gas from the distance piece into the crankcase. The seal rings, as indicated in Fig. 17.17, must minimize both these leakages.

The seal and wiping function can be combined into one groove if it is necessary to seal only the crosshead pressure pulse. This single groove would normally contain a butt tangent cut ring paired with two wiper rings. Compared to the three wiper ring combination, this ring sacrifices some ability to scrape oil effectively from the rod, but it is sufficient for slow speed compressors.

To wipe oil from the rod surface requires an apparent contact pressure between ring and rod of about 50 psi (controlled by the garter spring). Lower pressure tends to leave a thicker film, while high pressure may allow rapid wear of ring edge or rod surface. Ring to rod fit and contact is the most important factor in wiper packing performance.

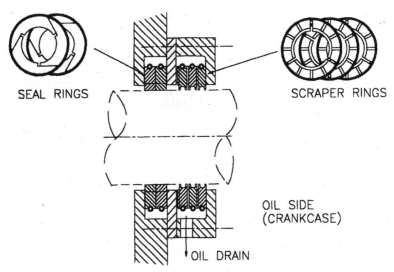

FIGURE 17.17 Typical wiper packing arrangement.

From a wiping standpoint, metal or hard plastic works best for the wiper ring. However, *seal* rings, if made of a softer, more conformable plastic (typically filled TFE), offer the best performance from a wear, friction, and sealing standpoint.

The wiper packing case should have a drain(s) shielded from oil spray or splash that results from crosshead motion. The force of this oil thrown up by the crosshead can, in some instances, block free drainage away from the wiper rings. Establishing good open drains as well as proper ring/rod contact becomes more important on the smaller, higher speed, compressors.

17.22 HIGH PRESSURE (HYPER) PACKINGS

"Hyper" is generally taken to mean over 10,000 psi. (Compressor discharge pressure might go above 100,000 psi.) At these pressure levels, fluids are usually acting very much like liquids in that compressibility is low. The type of rings used to seal these pressures are similar to those used in lower pressure compressors, but ring designs and materials must be selected to withstand high "compressive" cyclic pressures.

The problems (wear, friction and heat) caused by high contact pressure between ring and plunger are normally overcome by using ring sets that act partly as a labyrinth, and thus spread the pressure drop across several rings. Life of hyper packings is increased also by high lubrication rates plus, in some cases, coolant (oil) flow across the rod downstream of the packing.

The other principle problem in hyper packings is the containing of very high cyclic pressure within the case, which is essentially a thick-walled pressure vessel. Particular attention has to be paid to stress concentrations, such as holes or notches,

17.24 CHAPTER SEVENTEEN

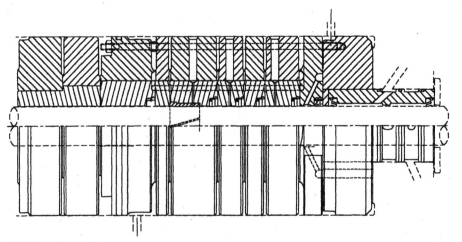

FIGURE 17.18 High pressure packing with compounded packing cups and oil circulating around plunger and case.

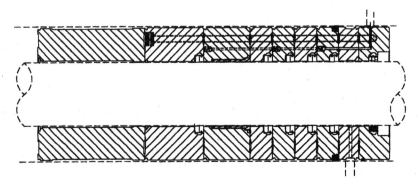

FIGURE 17.19 High pressure packing with discharge pressure surrounding the case.

that might raise stress beyond acceptable levels. Compounding of cups, autofrettaging, or pressure loading the outside of the case are typical ways to ensure long life of the case parts. Typical high pressure packing is shown in Figs. 17.18 and 17.19.

17.23 COMPRESSOR PISTON RINGS

Although sealing principles for rings on the piston are the same as those in the packing, their construction is somewhat different. In normal compressors, the requirement for sealing at the piston is not as stringent as it is in the packing. In fact, there is some reduction in wear if piston rings do leak slightly and thus distribute the pressure drop over more than one ring. The predominant type of joint for double-acting cylinders is the angle, or butt cut as in Fig. 17.20. For single-acting cylinders where leakage is more important, a seal joint is sometimes used. Regardless of the joint style, a large percentage of compressors use segmental rings—either two- or three-piece. The segmental type allows a ring with more radial thickness, which exerts less load against the cylinder wall than a radially thin one-piece ring.

Choosing the number of piston rings to use is, to some degree, an art. The quantity, no doubt, influences one thing of primary importance—ring life, and attempts have been and are being made to put this in to a good relation. However, at the moment, the number of rings selected for most applications is based essentially on experience. A guide for number of sealing rings generally used is included with Fig. 17.21.

17.24 COMPRESSOR RIDER RINGS

In some lubricated applications, and all nonlube ones, it is necessary to use a separate bearing, or rider ring(s), on the piston. This can be metal or plastic and serves only to keep the piston from contacting the cylinder. The rider must be made wide enough to keep bearing pressure between rider and cylinder very light, since

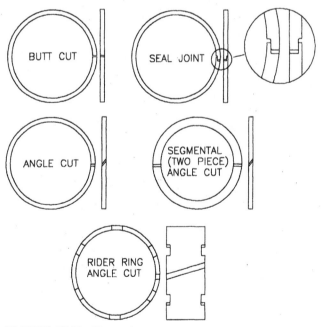

FIGURE 17.20 Piston ring types.

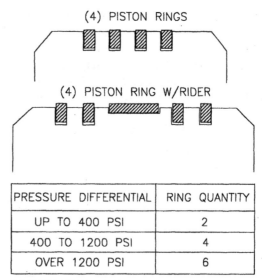

FIGURE 17.21 Typical piston ring arrangements and number of rings required versus pressure differential.

tolerance for wear will be less than for piston rings. This is because even with a rider, relatively little clearance separates the piston and the cylinder, and no more than this can be worn from the rider before piston and cylinder come into contact.

One major problem with riders is preventing them from pressure actuating like the sealing rings. They are usually notched on the sides or across the face and, in some instances, grooved or drilled in such a manner that they will not trap gas and thus seal like piston rings. They can be made either with a cut, as shown in the illustration, or uncut. Both of these types have advantages and disadvantages. The uncut ring is more difficult to install, will not tolerate even moderate temperature increases, but is slightly less prone to act as a seal as long as it remains tight against the groove bottom. A cut ring is easily installed, has room for expansion circumferentially, and has the advantage of large end clearance, through which gas can readily flow.

The rider supports piston weight plus one-half rod weight. This load is considered to be carried by the projected contact area of a 120° arc. Loading is usually acceptable if kept below 5 psi for nonlubricated cylinders. For lubricated service, American Petroleum Institute Standard 618 limits rider loading to 10 psi, but this has been extended to 50 psi successfully in a number of applications.

17.25 PISTON RING LEAKAGE

The average compressor has from two to six piston rings. The most common ring joint is an "open type" butt, or angle cut as in Fig. 17.20. Leakage wise, these are about the same. There is some slight advantage to the angle cut, but this is often overshadowed by the other factors affecting leakage.

Nearly all the leakage occurs through the joint since this is the only point in the ring where there is a definite opening or orifice. The opening is a rectangular passage with one dimension equal to the ring gap and the other to the piston clearance. This path is subject to wide variation—it will be almost zero when positioned at bottom of the piston, changing to maximum when at the top. It also constantly increases as the ring wears. Both of the leak path dimensions are a function of cylinder diameter, so in general, leakage can also be related to cylinder diameter.

Pressure distribution across the rings has been analyzed and measured and, for a two-ring piston, would look as illustrated in Fig. 17.22. For additional rings, this becomes more complicated, but in essence pressure within the ring pack cycles through a range somewhere between suction and discharge, resulting in a differential across the rings, first in one direction and then in the other.

Leakage through the rings then is not a result of steady pressure drop, but changes constantly. Leakage can be expressed however, as an average flow of gas during any particular compression cycle. A representation of approximate leak rates is pictured in Fig. 17.23. Quantity wise, there can be large variations. For example, .03 scfm in a small cylinder all the way up to 40 in a very large one.

In a double-acting cylinder this is not actually leakage, as gas is not lost. It simply passes from one side of the piston to the other. It is really a loss only from compressor discharge capacity. So, a better way to look at this is as a percentage change leakage causes to cylinder volumetric efficiency. For new rings in lubricated applications, loss of V.E. with open joint rings will be about .5% up to approximately 3%.

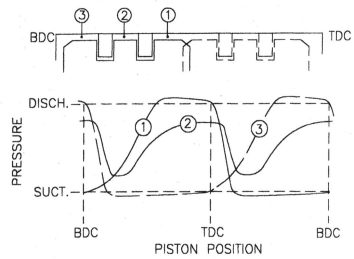

FIGURE 17.22 Instantaneous pressure between piston rings.

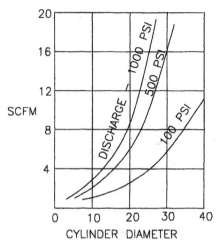

FIGURE 17.23 Average leakage by piston rings (for light gases in a double acting cylinder at ratio of 4).

Rings in nonlube cylinders suffer in two respects. First, no oil is present to reduce leak paths, and second, piston clearances are larger. Both of these have the effect of increasing leakage by roughly three times. To recover the loss of capacity, seal joint rings are often used as in Fig. 17.20. These rings have no theoretical leakage paths, plus leakage remains, essentially constant during the life of the ring, that is, until the gaps open. Based on tests, it is a good assumption these styles of rings will reduce leakage, when compared to open gap rings, by as much as 90%.

17.26 COMPRESSOR RING MATERIALS

The trend to plastics has not completely left metals behind. For lubricated service, time-proven bronze and cast iron are still commonly used materials. These are good simply because they are excellent bearing materials. They have the ability to carry and hold lubricant because of their porosity, the chemistry or structure to supply their own lubrication when oil is lacking, and heat transfer properties to quickly carry frictional heat away from the rubbing surface.

To replace these metals with plastics with equally good properties requires selection from an almost infinite number of plastic-filler or plastic-plastic composites. The first plastics that made successful rings were the phenolic and cloth laminates. These are resistant to many chemicals, will work under marginal lubrication, and are relatively inexpensive. The other important group has been the low-friction, but weaker, plastics blended with a strengthening filler or another stronger plastic. In this last group, there have been only a few with frictional properties good enough to run without lubrication.

It is difficult to put plastics in categories, but the most useful as related to compressors might be described as thus:

*Higher friction plastics needing at least some lubrication:

Polyimide (PI)

Poly(amide-imide) (PAI)

Polyetheretherketone (PEEK)

Polyphynylene Sulfide (PPS)

Polyamide (Nylon)

Phenolic-Cloth laminates

Low-friction materials capable of running without lubrication:

TFE plus strengthening and wear reducing fillers

PI plus friction reducing fillers

PEEK with friction reducing fillers

This is just a general grouping of the materials, but indications are that from these come most of the best, or most common, seal ring materials.

The properties of these materials influence the types of piston and packing rings used. For example, the relatively low yield strength of TFE blends has dictated the use of Type TR rings; the high elongation and "plastic memory" of TFE allows its use in stretch-on riders, while the strength and stiffness of some newer plastics make them useful for BT or M rings or as anti-extrusion rings. This, plus the fact that compressors face such a variety of conditions, is the reason there may never be universal ring "standards" in material or configuration. As new materials come along, the rings applied to compressors are designed around material properties, as well as operating conditions.

17.27 SEAL RING FRICTION

More than any other characteristic, friction serves as an indicator of how well compressor seals are performing. Like wear, this is very dependent upon lubrication. A certain amount of power must be put into a compressor to overcome seal ring friction, but as indicated in Fig. 17.24, this is relatively low compared to the power needed for gas compression. These curves are based on normal size piston and rod packing rings.

Power needed to overcome ring friction will usually be only about .5% to 2% of the compression HP. Essentially, all the frictional horsepower changes to heat

---

* All these materials have friction reducing fillers.

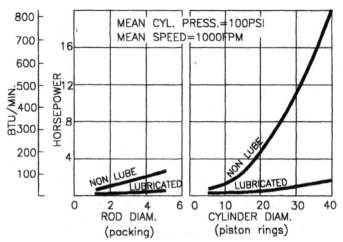

FIGURE 17.24 Power required (heat generated) to overcome ring friction versus ring diameter.

and, if not conducted away from the packing or piston rings, can cause wear and leakage. To illustrate the affect of this heat, the approximate 1.6 HP (68 BTU/Min.) as shown, which might be generated on a 3 inch rod, will cause approximately 100°F rise in four minutes if not conducted away. The principle ways to reduce friction are proper lubrication, low friction materials, and narrow rings.

17.28 COOLING RECIPROCATING COMPRESSOR PACKING

One of the critical, if not most critical, factors in obtaining good service from compressor packing is proper cooling. A primary source of heat is from the work required to overcome frictional resistance of the seal rings.

This is influenced by material selection, ring dimensions, characteristics of the compressor, and operating conditions. The relation between cooling requirements and the various influencing factors is not known precisely. What follows is intended to serve as a guide, indicating when special cooling is required and to help in sizing the equipment needed to provide the cooling.

Low friction materials, such as TFE blends, or carbon graphite, have made it possible for packing and piston rings to operate without lubrication. Frictional characteristics of these materials are good, but not nearly as good as when lubrication is used. Configuration of the seal rings affects this somewhat, but with most designs considerable frictional heat is generated.

The primary purpose of cooling packing is to remove heat generated due to friction between seal rings and the rod. Nearly all the work done to overcome friction converts to heat at the ring and rod mating surface. This heat is transferred to the case, gas passing through the cylinder, distance piece, and the crankcase.

Examples of methods to cool cases are shown in Fig. 17.25. Coolants in successful applications range from oil, circulated only by convection, to special fluids chilled and pumped through the case. In some instances, gas is blown through the case or over the rod for cooling.

Currently, the best available method to affect cooling is to use a case with internal channels through which water or a water anti-freeze mixture is circulated. Oil is not often used because it is ineffective at removing heat compared to water, or water with anti-freeze.

Factors affecting generation of heat and heat flow vary from compressor to compressor, making accurate predictions of the quantities involved very difficult. A general method of calculation, coupled with certain assumptions, is a starting point that can be modified by empirical data gathered in actual field installations.

This will provide reasonably accurate results for most applications. Estimating the coefficient of friction is difficult in any event, but especially difficult for applications with less than full lubrication.

When a compressor is lubricated and pressures are relatively low, friction loads can be estimated fairly accurately. However, at low pressures, cooling is frequently not required. At higher pressures, the lubricant film separating the ring and rod surfaces is, at best, partially effective and the coefficient of friction is more difficult to determine without actual operating experience and empirical data.

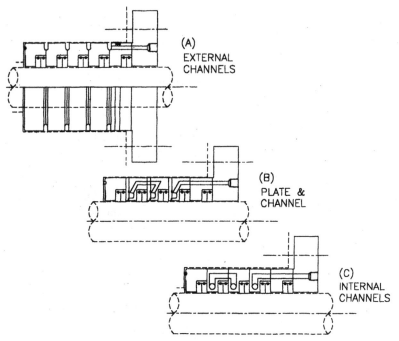

FIGURE 17.25 The most common methods for cooling packing cases.

17.28.1 Heat Generation

In a packing case containing several rings, most of the pressure drop will be across one ring. For calculation purposes, the heat generated by friction can be considered to be totally from one ring carrying the entire pressure drop. A pressure drop distributed in any manner over a complete set of rings generates essentially the same heat.

For purposes of simplifying the calculations, it can be assumed that the coefficient of friction is independent of load and contact area. The work required to overcome friction and the heat generated is:

$$\frac{\text{Work}}{\text{Revolution}} = (F)(2)(S) \quad (17.1)$$

$$\text{BTU/MIN} = \frac{(F)(FPM)}{777} \quad (17.2)$$

where: F = Force required to move rod against friction (lb.)
S = Compressor stroke length (In.)
FPM = Average rod speed (Ft/Min.)

Velocity of the rod is not constant throughout the stroke, but again, the coefficient of friction, f, can be estimated on the assumption that it is independent of velocity.

Friction force, F, is dependent upon the coefficient of friction, ring dimensions and average pressure, Pa, acting on the ring. Reasonably accurate results can be obtained using mean pressure between suction and discharge. A more accurate calculation of friction force can be made using an average pressure, Pa, as follows:

$$Pa = \frac{\left[(Pd)(n)\left(\frac{Ps}{Pd}\right)^{1/n}\right] - (Ps)(2 - n)}{2(n - 1)} \quad (17.3)$$

where: Pd = Compressor discharge pressure (psia)
Ps = Compressor suction pressure (psia)
n = Gas constant

The various ring configurations found in packing cases can be broken down into four groups based on the load exerted against the rod. At a given pressure and width, W, each group exerts a different load due to the type of cut between segments. The friction force for various ring types is shown in Fig. 17.26, as related to pressure and ring dimensions.

The coefficient of friction can vary over a broad range as illustrated in Fig. 17.27. The lower figures correspond to well-lubricated surfaces, while higher apply where dirt and/or abrasives are present. When f exceeds approximately 0.3, accompanied by high pressure, the packing is usually unable to function and can be totally destroyed. At very high frictional resistance, it becomes nearly impossible to get rid of the heat generated or to maintain a reasonable seal.

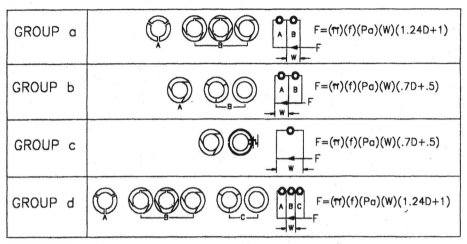

FIGURE 17.26 Friction loading for various packing ring types.

OPERATING CONDITIONS	RANGE OF f	SUGGESTED f TO USE WHEN ESTIMATING BTU/MIN
LUBRICATED		
METAL RINGS Pd<2000 PSI	.005 – .02	.02
Pd>2000 PSI	.01 – .05	
NONMETALLIC RINGS	.01 – .1	.05
POORLY LUBRICATED		
METAL RINGS	.1 – .2	.2
NONMETALLIC RINGS	.12 – .2	
DIRT OR SOLIDS PRESENT	.12 – .7	.3
NON LUBE		
CARBON OR TFE RINGS	.15 – .25	.25
DIRT OR SOLIDS PRESENT	.15 – .7	.3
DRY INERT GASES		
CARBON GRAPHITE RINGS	.2 – .3	.3
TFE RINGS	.15 – .3	.25

FIGURE 17.27 Coefficient of friction for various materials and levels of lubrication.

The value for f, along with ring dimensions, operating pressures and rod speed can be used to calculate BTU/Min. generated by the seal rings. (From Eq. 17.2).)

17.28.2 Heat Transfer to Gas

The heat generated is dissipated through several means. For most compressors, the two major paths are:

1. Through the case or case coolant
2. Through the gas flowing in the cylinder

Heat is lost to gas passing through the cylinder when the rod, warmed by friction, moves into the cylinder and releases some of its heat to the inlet gas at suction temperatures, and also during a portion of the discharge stroke.

There are several formulas, or empirical relationships, used to describe flow of heat from the rod into the gas. One which might be used is Eq. (17.4), for calculating the surface coefficient for gas passing over a smooth surface.

$$hc = .05 \, k \left(\frac{Vd}{u}\right)^{.75} \tag{17.4}$$

where: hc = Film coefficient (BTU/Hr-Ft²-°F)
k = Thermal conductivity (BTU-Ft/Hr-Ft²-°F)
V = Gas velocity (Ft/Hr)
u = Gas viscosity (Lb/Hr-Ft)
d = Gas density (Lb/Ft³)

The total heat flow, Q, into the gas, may be expressed as

$$Q = \frac{(hc)(\Delta T)(D)(S)}{5500} \tag{17.5}$$

where: Q = Heat Flow (BTU/Min.)
ΔT = Temperature difference between rod and gas (°F)
D = Rod diameter (In.)
S = Stroke length (In.)

OR

$$Q = \frac{(k)(D)(S)}{5000}\left(\frac{(d)(FPM)}{u}\right)^{.75}(Tr - Ts) \tag{17.6}$$

where: Tr = Rod temperature (F)
Ts = Suction gas temperature (F)

The rod is not a flat surface with gas flowing exactly parallel to it, gas velocity and direction of flow vary widely from point to point on the surface of the rod, and rod temperature and area of exposure are constantly changing as well. Essentially then, the calculation of heat flow is an approximation.

In making this approximation, values can readily be assigned to everything except the rod temperature, Tr. Rod temperature is one of the conditions to be controlled with cooling. Using the maximum value of rod temperature will give the maximum heat flow, both into the gas and the case.

If the value is lower, the heat transfer is lower, and the rod temperature will tend to rise. Therefore, it is logical to base heat flow predictions on maximum allowable rod temperature. In general, nonlubricated machines can run with rod temperatures as high as 250°F. In lubricated machines, the limitations is approximately 150°F.

17.28.3 Coolant Requirements

When heat flow into the gas exceeds heat generated, no separate liquid coolant is required. Experience indicates that, unless total heat to be removed from the case exceeds 20 BTU/min. per inch of rod diameter, it is not necessary to provide coolant.

Once it is established that coolant is required, it is necessary to determine required coolant temperature and amount of coolant flow. Usually, one gallon per minute per inch of rod diameter provides sufficient velocity through most cases to ensure good heat transfer. The increased flow for larger diameter rods absorbs the increased heat generated and also compensates for larger coolant passages. Larger rod diameters generally have larger cases and thus more room for coolant passages.

It is difficult to find a good correlation between calculated thermal resistance of a packing case and observed heat rejection rates. Because of this, coolant temperatures are determined by first setting the temperature of the coolant leaving the case at about 90°F and then calculating the inlet temperature.

For example: If 200 BTU's per minute are to be removed from a particular packing and circulation is two gallons (16.6 pounds) per minute, there are 16.6 pounds of water available to absorb the heat. Dividing 200 by 16.6 yields a temperature rise of 12°F. Subtracting this from 90°F exit temperature gives a maximum allowable inlet temperature to achieve this of 78°F. A definite temperature difference between coolant and rod is required for any given amount of heat to be conducted from the case. Using the method previously outlined, it is apparent that there are instances where rod temperatures will vary from the 250°F or 150°F level for un-lubricated and lubricated service.

17.28.4 Materials

The influence of materials on heat generation is illustrated in a general way in Fig. 17.27. In addition to frictional properties, heat transfer characteristics also affect temperature control. These two parameters are not the only basis for choosing a material to be used for packing, as strength, resistance to the medium, cost, and wear resistance are also important.

At the extremes of lubrication, choice of material is limited. With full lubrication, metals such as bronze or cast iron are best. Plastics such as phenolic, nylon or TFE may be used due to conditions other than heat transfer characteristics. For nonlube service, filled TFE is usually the first choice. Filled polyimides are also excellent but costly, while some of the less expensive plastics do not have the frictional properties to allow them to be effective for nonlube service.

Compressor and Gas Data		
	Example 1	Example 2
Rod Diameter, D	2	4
Pd	2000	6000
Ps	1000	3000
Td	200	220
Ts	100	100
RPM	500	300
Stroke, S	9	12
Service	nonlube	lubricated
Gas	Hydrogen	Nitrogen
Gas Properties		
n	1.4	1.4
k	.11	.015
p	.33	14.1
u	.022	.044
Ring Type	Group b	Group a
Ring Width, W	.18	.31
Calculated Values for Cooling Requirements		
Pa	1380	4200
FPM	750	600
Assumed f (fig. 27)	.25	.02
BTU/Min Generated by Ring (Equ. (2))	370	379
BTU/Min into Gas (Equ. (6)), Q	64	66
BTU/Min to be Transfered into Coolant	304	313
Coolant Inlet Temp. (water out @ 90F and 1 GPM/in. rod diameter)	72	81

FIGURE 17.28 Examples of coolant calculations.

Between these two extremes, compressors operate in mini- or semi-lube service or, as shown in Fig. 17.27, "poorly lubricated." In this type of service, it is more difficult to select optimum material to provide the lowest operating temperature due to the overlapping performance of metals and non-metals. For example, at a certain level of lubrication, metal rings will operate at a satisfactory temperature level. However, with a slight change of conditions, nonmetallic rings may perform better. Metals transfer heat faster and can run in conditions where the coefficient of friction is a bit higher, whereas nonmetallics require low friction for optimum results. It is sometimes possible to realize best properties of both materials by combining nonmetallic seal rings with a metallic backup ring, which not only acts as an anti-extrusion ring, but aids the nonmetallic ring, by using its contact with the rod to conduct heat away from the ring-rod interface.

Figure 17.28 contains two examples of calculated coolant requirements. Due to generalizations and assumptions made, the results are approximations, as stated previously. However, designs often are put into practice in the field without benefit of even these rough calculations. Since many problems experienced with compressor packing stem from inadequate cooling, the method outlined here should help eliminate those problems.

Many machines operate under conditions that do not exactly match the assumptions or descriptions used. They have features, or use materials, that could change to some degree the calculated values of either heat generation or heat flow. For example, in a compressor with large clearance volume, the temperature in the cylinder may have an effect not allowed for in the calculations. Units which have short strokes will have a very limited amount of heat flow into the gas and a more concentrated input of heat to the rod than normal. (A good approximation of this is to ignore the heat calculated from formula (6), and plan on removing all the heat through the case.) There are also materials and material-lubrication combinations which provide a different coefficient of friction than found in the coefficient of friction chart. The area designated as "poorly lubricated" is actually a broad range, and to assign one value for the coefficient may be an oversimplification.

CHAPTER 18
COMPRESSOR LUBRICATION

Glen Majors, P.E.
C.E.S. Associates, Inc.

Lubrication of compressors must accomplish one or more of the following:

1. Reduce friction between moving parts
2. Carry heat away from bearing surfaces
3. Prevent corrosion both during operation and when compressor is stopped
4. Reduce gas leakage between seal faces and close clearances

In rotary screw compressors, oil is used to remove the heat of compression of the gas, seal the rotors, and lubricate the bearings.

In the frames of reciprocating compressors, the crankcase oil lubricates the bearings, carries away bearing heat, reduces friction, and prevents corrosion.

In reciprocating compressor cylinders, the oil is a once through operation designed to reduce friction and wear as well as preventing corrosion.

18.1 ROTARY SCREW COMPRESSORS

Figure 18.1 shows a typical piping flow diagram for an oil-flooded screw compressor. Oil is injected directly into the compressor intake at a rate of 0.25 to 0.50 gpm/bhp. The discharge gas stream is a mixture of oil and gas at 185 to 200 °F flowing into a series of separators and filters which reduces the oil content to around 2 ppm. The discharge air pressure on the oil reservoir permits oil flow in this operation without use of a pump.

Screw compressor manufacturers have their own unique temperature controls for preventing water condensation. For this reason, they have proprietary oil specifications for best operation. They use different synthetic fluids or combination of synthetic fluids for controlling oxidation, oil emulsions, water separation, and corrosion.

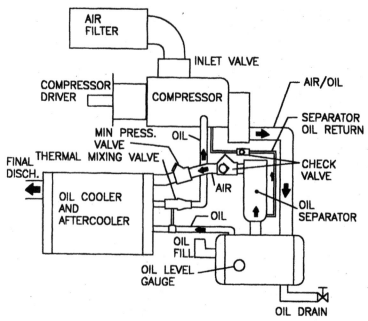

FIGURE 18.1 Flow diagram—air/oil systems rotary screw compressor.

18.2 RECIPROCATING COMPRESSOR CRANKCASE

In reciprocating compressor crankcases (Fig. 18.2), the oil pump delivers a continuous flow of 40-45 psi oil to the main and connecting rod bearings in order to reduce friction, and carry away heat. Oil sump temperature is usually maintained at 135 to 160 °F to prevent moisture condensation. The pump picks up oil from the crankcase, passes it through an oil filter and thermostatically controlled cooler and back into the main bearing header.

Compressor manufacturers generally recommend for the crankcase an SAE 30 to 40 lube oil with rust and oxidation (R&O) inhibitors. If a high viscosity oil is used, it will reduce the oil flow to the bearings causing hotter bearing surfaces. Low viscosity oils may be inadequate to lubricate the compressor cylinders or packing in those units using crankcase oil for the dual purpose. Ⓐ on Figure 18.3 shows the operating temperature range for crankcase oils, while Ⓑ on Fig. 18.3 shows the oil pump cavitation region for those oils on a cold start.

18.3 COMPRESSOR CYLINDERS

Compressor cylinder lubrication is completely different in that oil passes "once through" the cylinder with no recycling. Successful operation depends upon uninterrupted, continuous, metered flow to a cylinder bore and piston rod.

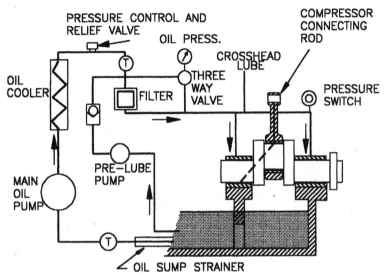

FIGURE 18.2 Oil flow diagram—compressor crankcase.

The gases being pumped range from air, sweet gases, sour gases, refrigeration gases, entrained hydrocarbon-liquid gases, and liquid water-entrained gases. These gas properties affect the lubrication by oxidation, corrosion, chemical reaction, water washing, dilution, and gas absorption.

The pressures range from vacuum to as high as 60,000 psi. Temperatures range from as low as −60 °F to as high as 400 °F. The piston rings, packing rings, and valves may be either metallic or non-metallic.

18.4 LUBE OIL SELECTION

All of the above conditions require consideration before selecting a compressor lubricating oil. With proper attention to the selection, the life of the wearing parts can be extended several years. With ineffective lubrication, the life may be only minutes.

18.4.1 Oil Viscosity

Any oil selected must have sufficient viscosity at operating temperatures to keep the moving parts from coming into contact with each other and to minimize wear if abrasive particles pass through the system. There are two different temperatures that must be considered for oil selection: a) "cold flow;" and b) cylinder discharge temperature. Cold flow temperature should not be confused with the "pour point" in oil properties. Cold flow is that temperature at which oil pumps cavitate or plungers fail to consistently fill on the suction stroke. Regardless of the oil selected,

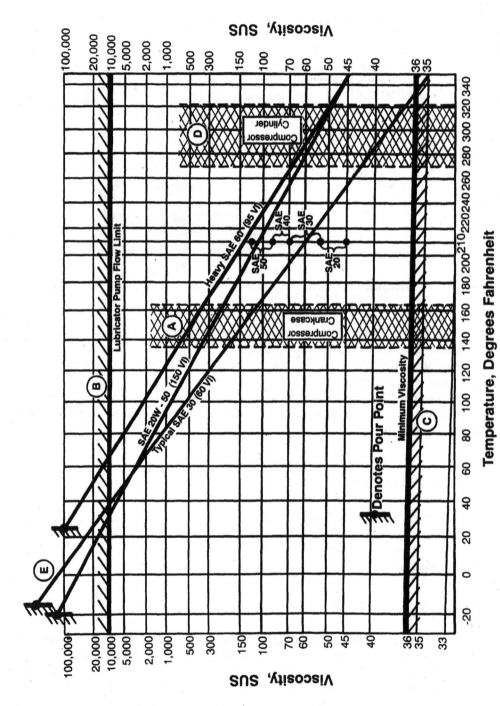

FIGURE 18.3 Flow diagram—air/oil systems rotary screw compressor.

the cold flow limit occurs at the 6,000 to 10,000 SUS viscosity range (Fig. 18.3⒝).

The oil viscosity selection is always made on the basis of operating temperature or on the maximum cylinder discharge temperature. If that oil then has a cold flow problem in cold weather, heaters and insulation must be added to the oil reservoir and meter pumps.

18.4.2 Minimum Oil Viscosity

When lubricating oil reaches the viscosity equivalent to water, the oil film no longer supports dynamic loads resulting in rapid failure. This minimum viscosity is recognized as about 36 SUS (Fig. 18.3©).

18.4.3 Gas Absorption

All petroleum base compressor oils will absorb gases. The higher the gas pressure, the more gases will be absorbed into the oil. The oil then becomes less viscous in the compressor cylinder. This gas dilution effect is hard to accurately measure and/or predict without time-consuming laboratory tests using the actual gas stream components elevated to the operating cylinder pressure and temperature. A laboratory test using natural gas at 980 psi pressure showed that one gallon of oil absorbed 0.75 gallons of gas when de-pressurized. A somewhat reasonable and practical way to offset this gas dilution effect is to select an oil having 5 to 10 SUS higher viscosity at operating temperature (See Fig. 18.3Ⓓ). The oil supplier has temperature viscosity curves (as in Fig. 18.3) for all compressor oils under consideration. The suggested upgrade in viscosity of 5 to 10 SUS around cylinder operating temperatures generally requires the selection of the next higher SAE grade of oil.

18.4.4 Liquid Hydrocarbon Dilution

Many compressors and particularly field gas gathering units encounter hydrocarbon liquids in the gas stream. Petroleum base lube oils are also hydrocarbons and will be diluted or washed away by gas stream liquids. The magnitude of this dilution will vary as liquid carryover usually occurs in slugs. There are two choices: a) remove the liquid hydrocarbons from the gas stream; or b) select the highest viscosity oil.

18.5 OIL ADDITIVES

No amount of flood lubrication will solve an oil quality problem. Various oil additives are often required to overcome unusual operating conditions.

18.5.1 Water Displacing and Metal Wetting Additive

The majority of compressor cylinder wear problems result from water carryover in the gas stream. When water reaches the compressor cylinder or is chemically formed inside the cylinder, "water washing" causes the oil to float away leading to drastic wear. A synthetic polar-type additive (not animal fatty oil) that has metal wetting and water displacing properties may be used to minimize the effects of water.

18.5.2 Corrosion Inhibitors

Small amounts of CO_2, H_2S, chlorides, and other potentially corrosive gases can be successfully handled with "standard" compressor components, providing the gas stream is absolutely dry. If any moisture is present, then either: a) corrosion and chemical resistant materials in all components touched by the gas; or b) a fortified corrosion inhibited compressor oil may be used. The corrosion inhibitors should be combined with the water displacing and metal wetting additives discussed above. These additive combinations promote an oil film that tightly adheres to all metal surfaces even in the presence of water. The object is to let the oil be the sacrificial agent to neutralize the acids and protect the critical metal parts.

18.5.3 Oxidation Inhibitor

Oxygen reacts with the hydrocarbon molecules of lube oil to form a brownish crystalline volcanic-ash type deposit that cannot be dissolved with petroleum solvents or cleaners. Oxidation rates double for every 18 °F increase in temperature. Normal R&O inhibitors are adequate for compressor crankcase oil operating at 140 °F, but totally inadequate for air compressor cylinders operating over 300 °F. As little as 2% excess O_2 in a gas stream will cause serious ash type build up in a matter of weeks. In order to handle this problem, the compressor oil must be fortified with a high-temperature, anti-oxidation inhibitor.

18.5.4 Anti-Foam Additive

Lube oil leaving a compressor cylinder may be highly agitated in a foamy "mayonnaise" state that will pass through liquid knock out traps. If the oil has to be removed from the gas stream, the oil must be fortified with a 3 to 5 ppm active anti-foam inhibitor to quickly break down the gas bubbles.

18.5.5 Anti-Emulsion Additive

At high discharge gas temperatures, the aerated oil combined with moisture forms a black "soap" deposit on the inside of the pipes and inside the cooler tubes. To overcome this problem, an anti-emulsion additive should be used.

18.5.6 Viscosity Index (VI) Improver

All engine-type motor oils have VI improver additives for cold starts and cold weather operation. Figure 18.3 shows how oil viscosity changes with temperature and how 60 VI to 150 VI oils might be compared. These VI improvers serve practically no useful purpose in compressor cylinders. They do, however, make it possible for lubricator pumps to operate at considerably colder temperature without heaters (Fig. 18.3Ⓑ).

18.6 OPTIMUM LUBRICATION

Figure 18.4 gives empirical guidelines for optimum quantity of oil for various compressor cylinders operating in different gas streams, with different ring materials, under low-to-high pressures, and in a wide range of speeds.

"Optimum" lubrication gives years of compressor life, while "starved" lubrication produces rapid wear and short life. "Over lubrication" gains little in operating life and requires more oil. Figure 18.5 illustrates the compressor component life with various lubrication rates.

There are various lube oil rate formulas or guidelines proposed by compressor manufacturers, oil suppliers and seal manufacturers. They provide an estimate for quantity of oil to lubricate gas transmission type compressor cylinders. These formulas are similar in that they are based on total swept surface area to be lubricated.

No formula or graph can cover all possible conditions, pressures, speeds, gases, and ring materials. Figure 18.4 graphically covers a broad range in cylinder sizes, compressor speeds, pressures, and ring materials. The oil usage for optimum lubrication is given in pints per day per cylinder for PTFE-equipped cylinders. For other cylinders with different rings and different gas streams, the appropriate multiplier is listed on the graph.

18.7 OIL REMOVAL

When sensitive downstream catalyst is involved, oil carryover from compressor cylinders may be objectionable. To overcome this problem: a) the compressor should be converted to non-lubricated construction; or b) adequate oil removal equipment should be installed. Field experience with the following devices will permit inexpensive removal of lube oil from gas streams:

Percentage Of Oil Removed From Stream
1. Regular lube oil with **no** anti-foam additive with KO (knock out) Traps — 16%
2. Oil **with** anti-foam additives + KO trap — 40%
3. Oil with anti-foam additive + aftercooler + KO trap — 84%
4. Oil with anti-foam additive + aftercooler + KO trap + coalescing filter — 96%
5. Oil with anti-foam additives + aftercooler + KO trap + coalescing filter + molecular sieve — 99.6%

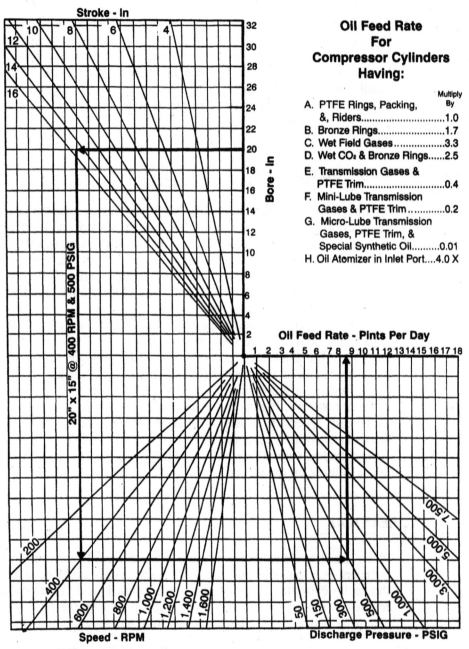

FIGURE 18.4

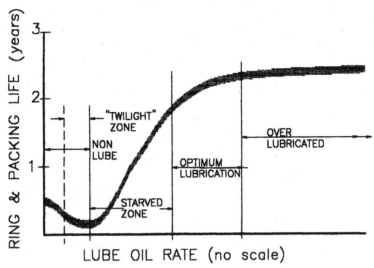

FIGURE 18.5 Compressor seal life vs. oil usage.

18.8 NON-LUBE (NL) COMPRESSORS

There are certain compressor applications where no oil can be tolerated in the gas stream. To be most effective, these compressors require PTFE piston rings, packing rings, wear bands, and nonmetallic compressor valve parts, plus a discharge gas temperature under 275 °F. They also require double distance pieces and a slinger ring on the rod between the cylinder and the frame in order to prevent crankcase oil migration along the piston rod. If a very small quantity of oil gets into the NL compressor cylinder, "twilight zone" operation occurs in that area between non-lube and minimum lube where the ring and the packing life may be only a few weeks. Figure 18.5 is a representation of various degrees of lubrication up to "over" lubrication.

18.9 SYNTHETIC LUBRICANTS

A relatively small percentage of compressors are lubricated with synthetic lubricants. They are more expensive, have special properties not found in petroleum oils, and some have excellent fire resistant properties while others enter into the reaction of the gas process. Sometimes the synthetics attack paints, gaskets, o-rings and form corrosive acids in the presence of water.

Table 18.1 compares the properties of the most common synthetic lubricants with mineral oil and their compatibility with compressor components.

TABLE 18.1 Comparison of Synthetic Lube Oils

	Mineral Oil	Polyisobutenes	Polyalphaolefins	Alkylated Aromatics	Polyalkylenegycols	Perfluoroalkylethers	Polyphenylethers	Dicarbonxylic Acid Esters	Neopentyl Polyesters	Triaryl Phosphate Esters	Trialkyl Phosphate Esters	Silicone Oils
Viscosity Temp. Behavior (VI)	M	P	V	M	V	M	P	V	V	P	E	E
Low Temp. Behavior (Pourpoint)	P	M	E	G	G	G	P	E	V	M	E	E
Liquid Range	M	P	V	G	G	E	P	V	V	V	G	E
Oxidation Stability (Aging)	M	M	V	M	G	E	V	V	V	V	M	V
Thermal Stability	M	M	M	M	G	E	E	G	V	V	G	V
Evaporation Loss, Volatility	M	M	V	G	G	E	G	E	E	V	V	V
Fire Resistance, Flash Temp.	P	P	P	P	M	E	M	M	M	V	V	G
Hydrolytic Stability	E	E	E	E	G	E	E	M	M	M	G	G
Corrosion Protection Properties	E	E	E	E	G	P	M	M	M	M	M	G
Seal Material Compatibility	G	G	V	G	G	E	G	M	M	P	P	G
Paint and Lacquer Compatibility	E	E	E	E	M	V	M	M	M	P	P	G
Miscibility with Mineral Oil	E	E	E	E	P	P	G	V	V	M	M	P
Solubility of Additives	E	E	V	E	M	P	V	V	V	E	E	P
Lubricating Properties	G	G	G	G	V	E	E	V	V	E	M	P
Toxicity	G	E	E	P	G	E	G	G	G	V	M	E
Biodegradability	M	P	G	P	E	P	P	E	E	V	V	P
Price Relation Against Mineral Oil	1	4	4	4	8	500	400	8	8	8	8	60

E — Excellent
V — Very Good
G — Good
M — Moderate
P — Poor

18.10 COMPRESSOR LUBRICATION EQUIPMENT

Reciprocating compressors require equipment that can reliably and consistently inject small quantities of oil under pressure to different locations on the cylinder and packing. There are two basic systems in general use: a) pump-to-point, and b) divider block.

18.10.1 Pump-To-Point

These metering pumps are driven from either the main compressor or from an electric motor. Figure 18.6 shows a cut-away view of an individual lubricator pump that goes into a 5 to 18 compartment unit. Each compressor cylinder may have 2 to 6 oil injection points. The no-flow/shut-down device is to prevent compressor operation when there is no oil flowing. These pumps come in different sizes, different pressure ratings, and have adjustable outputs.

18.10.2 Divider Block System

This system (Fig. 18.7) uses one adjustable output pump and various sized divider blocks to meter a specific amount of oil to multiple cylinder locations. Like any

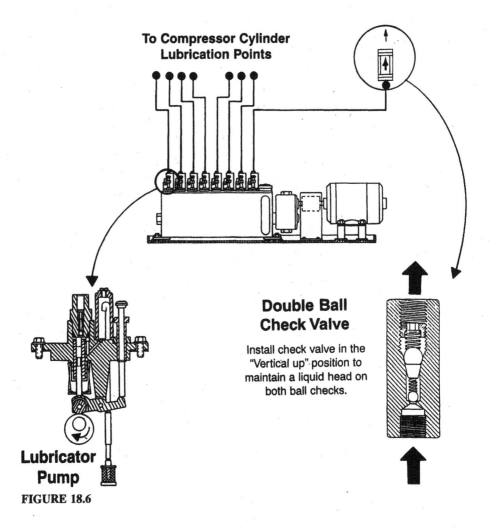

FIGURE 18.6

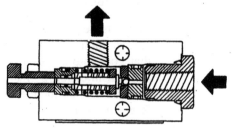

Balancing Valve

Balancing valves assists divider block valves to accurately proportion lubricants to different pressure points. Set each balancing valve at slightly higher pressure than lubrication points and with the outlet in the vertical up position.

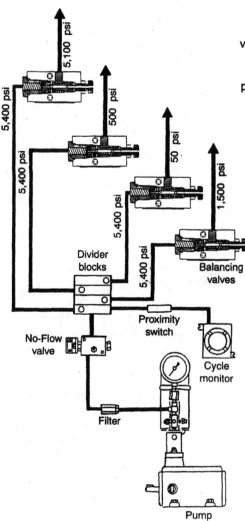

FIGURE 18.7

other lubricator system, it must have a no-flow shut-down device to prevent compressor operation when no oil is flowing. Reliable operation can only be achieved when "balancing valves" (Fig. 18.7) are in each divider block discharge line and manually adjusted for higher than the highest cylinder pressure.

18.10.3 Cylinder Check Valves

Each oil injection point on compressor cylinders requires a check valve (Fig. 18.6) to be installed close to the cylinder to prevent gas back-flow into the oil system. The check valve should be installed in the vertical **up** position so as to have a liquid "oil leg" covering the ball check to prevent gas leakage and air binding of the metering system.

18.10.4 Balancing Valves

Figure 18.7 shows a cut-away view of an adjustable balancing valve used in divider block systems. All are set slightly higher than the highest cylinder pressure. Balancing valves should be installed with the outlet **up**; otherwise, they will trap gases and cause "soft pump" operation.

18.10.5 Air Binding

The biggest operational problem with divider block systems is "air binding." Air and gases may get entrained in the oil supply, do get injected into it during lubricator reservoir filling, and do flow back through improperly installed cylinder check valves. These troubles can be overcome by making the lubricator system self-venting. Gases seek the highest point; therefore the piping and devices should be installed so that oil always flows **up** with no downward tubing loops.

CHAPTER 19
PRINCIPLES OF BEARING DESIGN

Hooshang Heshmat, Ph.D.
H. Ming Chen, Ph.D., P.E.
Mohawk Innovative Technology, Inc.

19.1 NOMENCLATURE

Unless noted otherwise, the following symbols are used throughout the text.

		Units
A	Area	(in.^{2})
AMB	Active Magnetic Bearing	
B	Damping coefficient	(lbf-s)/in
$\overline{B}$	$(\omega B)/2\mu ND(L/D)(R/C)^3 = (\pi B)/\mu L(R/C)^3$	
C	Radial clearance	(in.)
C_m	Smallest clearance for $\epsilon = 0$	(in.)
C_M	Largest clearance for $\epsilon = 0$	(in.)
CSB	Compliant Surface Bearing	
D	Diameter of bearing for journal	(in.)
E	Elastic modulus	(psi)
F_τ	Frictional force	(lb)
G	Mass flow rate	(lbm/s)
$\overline{G}$	$G/1/2\, \rho_a$ ULC	(—)
G_x	Turbulence coefficient in x or θ	(—)
G_z	Turbulence coefficient in z or r direction	(—)
G_τ	Turbulence coefficient for viscous shear	(—)
H	Power loss	(hp)
K	Spring coefficient	(lb/in.)
$\overline{K}$	$K/2\mu\, NL\, (R/C)^3$	(—)
K_B	Structural stiffness of CSB	(lb/in./in.)
L	Width of bearing	(in.)
	— in the z direction in journal bearings	
	— in the r direction in thrust bearings	
M_{CR}	Critical mass for journal bearing stability	(lbm)
$\overline{M}_{CR}$	$(M_{CR}N)/2\mu\, L(R/C)^3$	
N	Revolutions per unit time (rpm)	(—)
P	Unit load	(psi)
	— (W/LD) in journal bearing	(lbf/in.^{2})
	— (W/A) in thrust bearings	(lbf/in.^{2})

		Units
Q	Volumetric lubricant flow	(in.³/s)
$\hat{Q}_z =$	(Q_z/Q_{zF}) Level of starvation	(−)
Q_z, Q_r	Side leakage of lubricant (hydrodynamic)	(in.³/s)
Q_1, Q_{in}	Flow in at leading edge	(in.³/s)
Q_2	Flow out at trailing edge	(in.³/s)
Q_{2P}	Flow out at trailing edge due to pressure gradient	(in.³/s)
R	Radius of bearing or journal	(in.)
R_1	Inner radius	(in.)
R_2	Outer radius	(in.)
$\mathcal{R}$	Perfect gas constant (ft-lbf/lbm-°R)	(−)
Re	Reynolds number	(−)
	− $(\rho R\omega h/\mu)$ in journal bearings	
	− $(\rho r\omega h/\mu)$ in thrust bearings	
REB	Rolling Element Bearings	
S	Sommerfeld number $(LD\mu N/W)(R/C)^2 = 1/\overline{W}$	(−)
T	Temperature	(°R or °F)
ΔT	Temperature rise $(T - T_i)$	(°R or °F)
U	Linear velocity	(in./s)
W	Load	(lb)
W_{CR}	Critical load for journal bearing stability	(lb)
$\overline{W}_{CR}$	$(W_{CR}/W)(C\omega^2/g)$	(−)
$\overline{W}$	− $(W/LD\mu N)(C/R)^2$ in journal bearings	(−)
	− $(W/R_2^2\mu\omega)(h_2/L)^2$ in thrust bearings	
b	Tapered fraction of thrust bearing	(−)
c	Specific heat of lubricant	(BTU/°F or °R)
d	Amount of center offset or preload	(in.)
e	Eccentricity	(in.)
f	Friction coefficient $(F\tau/W)$	(−)
g	Gravitational constant	(32.174 ft/s²)
h	Film thickness	(in.)
h_{min}	Minimum film thickness	(in.)
h_{11}	Value of $(h_1 - h_2)$ at $R_1, 0$)	(in.)
h_p	Film thickness over pivot	(in.)
h_2	Film thickness at end of fluid film	(in.)
$\overline{h}$	Dimensionless film thickness	(−)
	− (h/C) for journal bearings	
	− (h/h_2) for thrust bearings	
h_N	Nominal film thickness	(in.)
h^*	(h/δ)	(−)
h_N^*	(H_n/δ)	(−)
h_1^*	(h_1/δ)	(−)
h_2^*	(h_2/δ)	(−)
l_o	Half length of bump in angular direction	(in.)

		Units
m	Preload (d/C)	(—)
n	Number of pads in bearing	(—)
p	Pressure	(psi)
p_a	Ambient pressure	(psia)
p_s	Supply pressure	(psi)
r	Radial coordinate	(—)
$\bar{r}$	(r/R_2)	(—)
s	Distance between bumps (pitch)	(in.)
t	Time, foil thickness	(s, in.)
w	Specific weight	(—)
x, y	Rectangular coordinates	(—)
z	Axial coordinate	(—)
α	Bearing compliance:	

$$\frac{2p_a s}{CE}\left(\frac{l_o}{t}\right)^3 (1 - \nu^2); \quad \frac{2p_a s}{h_2 E}\left(\frac{l_o}{t}\right)^3 (1 - \nu^2)$$

β	Angular extent of bearing pad (rad or deg.)	(deg. or rad)
β_p	Angular extent of pad from start to pivot	(deg. or rad)
β_s	Fluid film arc $(\theta_2 - \theta_1)$	(deg. or rad)
$\bar{\beta}_p$	(β_p/β)	(—)
δ	Taper	(in.)
δ_θ	Circumferential taper $(h_1 - h_2)$	(in.)
δ_r	Radial taper at $\theta = 0$, $(h_{11} - h_{12})$	(in.)
ϵ	Eccentricity ratio (e/C)	(—)
ϵ_m	(e/C_m)	(—)
θ	Angular coordinate	(deg. or rad)
θ_S	Start of bearing pad	(deg. or rad)
θ_p	Location of pivot	(deg. or rad)
θ_E	End of bearing pad	(deg. or rad)
θ_2	End of hydrodynamic film	(deg. or rad)
θ_{min}	Location of h_{min}	(deg. or rad)
μ	Lubricant viscosity	(lbf-s/in.^{2})
ν	Poisson's ratio	(—)
ρ	Density of lubricant	(lbm/in.^{3})
ϕ	Attitude angle $(\theta_{min} - \pi)$	(rad)
ϕ_L	Load angle	(deg. or rad)
ω	Angular velocity	(rad/s)
ω_i	Threshold instability frequency	(rad/s)
ω_n	Natural frequency of system	(rad/s)
$\bar{\omega}_i$	(ω_i/ω)	(—)
Λ	$12\pi (\mu N/p_a)(R/C)^2$	(—)

Units

Subscripts

a	Ambient
E	End
r	Radial
F	Full film
R	Ring
S	Start
xx	Force in the x direction due to a displacement in the x direction
xy	Force in the x direction due to a displacement in the y direction
yx	Force in the y direction due to a displacement in the x direction
yy	Force in the y direction due to a displacement in the y direction
τ	Friction
1, 2, 3	Lobe 1, 2 or 3

19.2 COMPRESSORS AND THEIR BEARINGS

From the standpoint of speed and pressure heads one can, in a general fashion, classify compressors as follows:

- Reciprocating—low speeds generating both moderate and high pressures
- Centrifugal—high speeds and moderate pressures
- Axial flow—very high speeds and low pressure

This picture of compressor operation is portrayed in Fig. 19.1. The temperatures of the compressed fluid are usually kept below 600°F and for this intercoolers are often required. The drives employed are shown in Fig. 19.2, along with compressor mass flows; the flows are high at low speeds and low at the higher speeds. Thus bearings in compressors operate at speeds up to 30,000 rpm and, in some unusual applications such as in cyrogenic pumps, may reach speeds of 100,000 rpm.

19.2.1 Bearings in Reciprocating Compressors

A generic picture of the use of bearings in reciprocating compressors is shown in Fig. 19.3. The unit contains four sets of bearings, each subject to different operating conditions. The main shaft bearing is the most conventional, its position and size permitting a proper supply of lubricant. It runs typically in the range of 125 to 500 rpm, driven usually by a steam turbine. Next comes the crankshaft bearing which is the most heavily loaded, this being due to the variable forces exerted on it during the power stroke; its speed, too, is variable during each cycle. Then there is the wrist pin bearing at the end of the connecting rod undergoing an oscillatory motion.

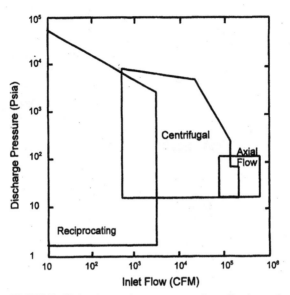

FIGURE 19.1 Approximate range of application of various compressors.

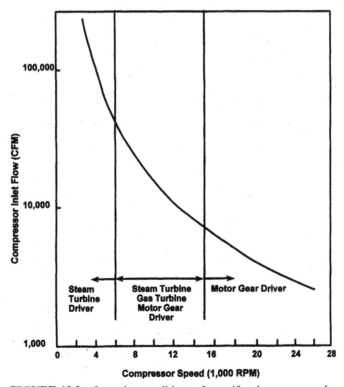

FIGURE 19.2 Operating conditions of centrifugal compressors.[6]

19.6 CHAPTER NINETEEN

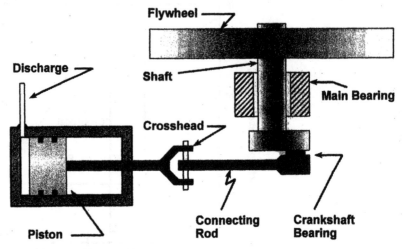

FIGURE 19.3 Schematic of bearing locations in reciprocating compressors.

The crosshead usually has a bronze bushing operating in the boundary lubrication regime.

19.2.2 Bearings in Centrifugal Compressors

By far the most commonly used compressors are of the centrifugal type, ranging from simple fans with pressures of no more than a few psi to multistage units employing intercoolers and often gear trains linked to electric motors or gas turbines. A typical relation between volume flow and speed would be as follows:

Volume, cfm	Speed, rpm
3,000	10,000
12,000	6,000
40,000	3,600
90,000	2,700

Since high cfm also implies large shaft sizes the actual bearing linear speeds do not vary much for the different conditions listed above.

From the standpoint of bearings, centrifugal compressors can be subdivided into the following groups:

Single Stage with Overhung Impellers. These are shown in Figs. 19.4a and 19.4b and although different in construction they produce similar effects on the journal and thrust bearings, namely misalignment. Naturally, the one with no bearing outside the drive would be more severely misaligned. Due to its overhung construction, the unit would also be more prone to vibration and instability. The preferred construction is that shown in Fig. 19.4c which has an additional bearing at the outboard end of the shaft.

Multistage Compressors. Typical bearing locations in multistage compressors are shown in Fig. 19.5. Here there is no overhung mass and unless the shaft is severely bowed there should be no misalignment. In the high speed units, couplings, electric motors and gear sets are used introducing external excitation forces in addition to possible unbalance forces. These come from the power line frequencies, the gear

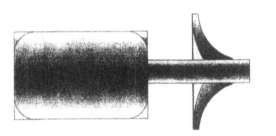

a. Unsupported Overhung Impeller

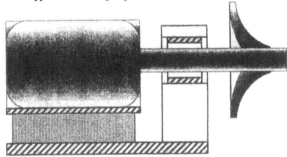

b. Single Bearing Overhung Impeller

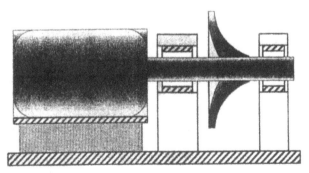

c. Two Bearing Impeller Shaft

FIGURE 19.4 Single stage centrifugal compressors.

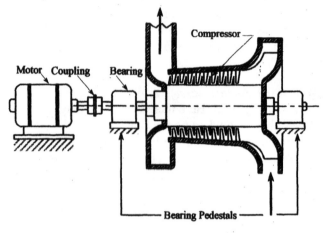

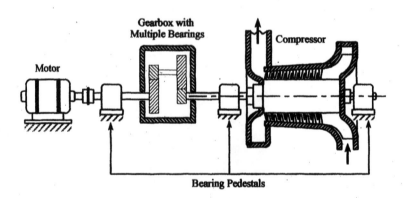

FIGURE 19.5 Multistage centrifugal compressors.

teeth, from misalignment problems with the coupling, and occasionally even from pedestal vibrations.

The greatest complexity arises in accommodating the generated axial forces. Where large thrust loads are present, an attempt is usually made to reduce them by the use of a balancing piston. A schematic of such an arrangement is shown in Fig. 19.6. Still there is always a net thrust load present which must be carried by a properly designed thrust bearing. Often there are two such bearings, one active and one inactive; the first is designed to carry the steady load while the other is meant to accommodate any dynamic loads. This is made possible by having the shaft float axially some 10 to 20 mils, thereby transferring the thrust load from the active to the inactive bearing. In some cases the thrust load may actually reverse

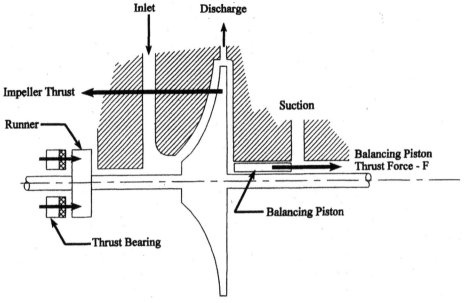

FIGURE 19.6 Thrust loads in centrifugal compressor.

itself, in which case two thrust bearings are needed for supporting a thrust in either direction.

19.3 GENERAL BEARING PRINCIPLES

Five kinds of bearing can be considered candidates for use in compressors, each preferable for a particular set of operating conditions. From a tribological standpoint, these can be grouped as follows:

Fluid Film Bearings. These bearings carry the imposed load on a fully developed fluid film, be it a liquid or gas. To this group belong the various hydrodynamic journal and thrust bearings which generate support pressures by virtue of the relative motions between the surfaces; and hydrostatic bearings whose load capacity is provided by an externally supplied high-pressure fluid.

Elastohydrodynamic Bearings. These bearings operate on a combination of hydrodynamic pressures and forces generated by deflections of the elastic bearing surfaces. Bearings in this category are the compliant surface (or foil bearings), and rolling element bearings.

Magnetic Bearings. The third kind considered here is a unique support system, known as the magnetic bearing. In this device the load is carried on forces generated by an electromagnetic field in the bearing structure.

The basic elements of these various fluid film bearing designs are sketched in Fig. 19.7. In the following subsections, the theoretical principles of bearing operation will be reviewed briefly. Sections 19.3.0 through 8.0 will provide the practical aspects of bearing design and application.

19.3.1 Hydrodynamic Bearings

Figure 19.8 shows a generalized sketch of a bearing consisting of a moving and a stationary surface. The two surfaces are separated by a fluid film of variable thickness. As the lubricant is sheared through the clearance from inlet to outlet, its velocities, pressures, temperatures and viscosity undergo considerable variations which bear directly on bearing performance.

Incompressible Lubrication. For bearings using incompressible lubricants, the basic mathematical expression that relates performance to the bearing's geometrical and operational parameters is the Reynolds differential equation given by

$$\frac{\partial}{\partial x}\left[\frac{h^3}{\mu}\left(\frac{\partial p}{\partial x}\right)\right] + \frac{\partial}{\partial z}\left[\frac{h^3}{\mu}\left(\frac{\partial p}{\partial z}\right)\right] = 6\frac{\partial h}{\partial x} \qquad (19.1)$$

When solved, the Reynolds equation yields the pressure field $p(x,z)$, as well as the

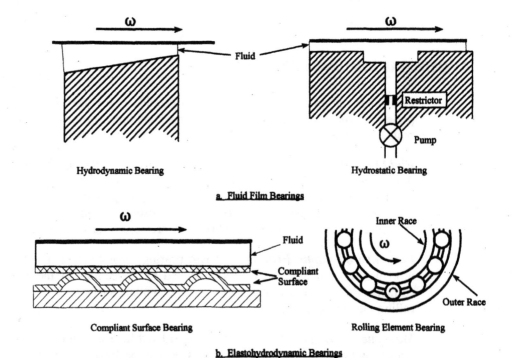

FIGURE 19.7 Four generic kinds of bearings.

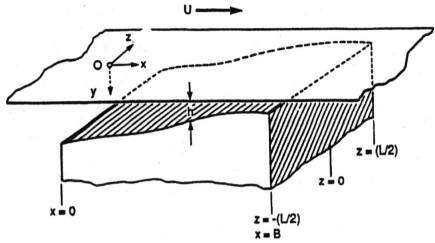

FIGURE 19.8 Schematic of a hydrostatic bearing.

pressure gradients $(\partial p/\partial x)$ and $(\partial p/\partial z)$ in the bearing. By integration one then obtains

- Load capacity

$$W = \int_0^L \int_0^B p(x, z) dx dz \tag{19.2}$$

- Frictional force and power loss

$$F_\tau = \int_0^L \int_0^B \left[\pm \frac{h}{2} \left(\frac{\partial p}{\partial x} \right) + \frac{\mu U}{h} \right] dx dz; \; H = F_\tau U \tag{19.3}$$

- Component flows in the x and z directions

$$Q_x = \int_0^L \left[\frac{h^3}{12\mu} \left(\frac{\partial p}{\partial x} \right) + \frac{hU}{2} \right] dz \tag{19.4a}$$

$$Q_z = \int_0^B \frac{h^3}{12\mu} \left(\frac{\partial p}{\partial z} \right) dx \tag{19.4b}$$

Compressible Lubrication. For a bearing lubricated by gas, the Reynolds equation is similar but because of gas compressibility a density parameter now makes an appearance, namely

$$\frac{\partial}{\partial x}\left[\frac{\rho h^3}{\mu} \left(\frac{\partial p}{\partial x} \right) \right] + \frac{\partial}{\partial z}\left[\frac{\rho h^3}{\mu} \left(\frac{\partial P}{\partial z} \right) \right] = 6U \frac{\partial(\rho h)}{\partial x} \tag{19.5}$$

The basic differences between gas and liquid lubricated bearings are:

- The load capacity depends on the prevailing ambient pressure, whereas it does not with incompressible fluids.
- In liquid lubricated bearings, there is often cavitation at the end of the fluid film; there is no cavitation in gas films.
- Opposite to that of liquids, gas viscosity rises with a rise in temperature.
- Film thicknesses in gas bearings are one or two orders of magnitude smaller than with liquid-lubricated bearings; this is due to the much lower viscosity of gases, (see Fig. 19.9). For the same reason, frictional losses in gas bearings are considerably lower.
- No special arrangements for lubricant supply are required.

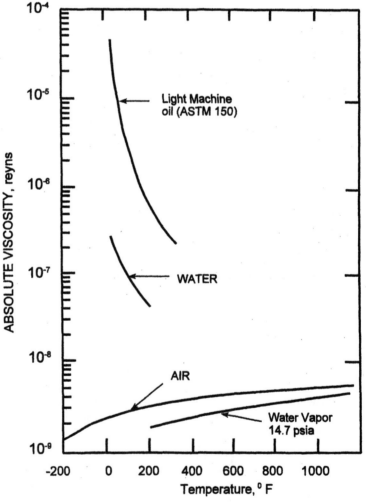

FIGURE 19.9 Comparative viscosity of various fluids.

Effects of Turbulence. The above equations are all for laminar flow, i.e., when the lubricant velocity is below a Reynolds number of about 1,500, where the Reynolds number is given by

$$Re = \rho R \omega h / \mu \qquad (19.6)$$

Should the flow in any region of the film exceed the above value, then lubricant flow would be reduced and both power loss and temperatures would rise. The Reynolds equation that defines turbulent bearing operation is given by:

$$\frac{\partial}{\partial x}\left[G_x \frac{h^3}{\mu}\left(\frac{\partial p}{\partial x}\right)\right] + \frac{\partial}{\partial z}\left[G_z \frac{h^3}{\mu}\left(\frac{\partial p}{\partial z}\right)\right] = 6U\left(\frac{dh}{dx}\right) \qquad (19.7)$$

where the functions G_x, G_z and G_τ (G factor) are given as a function of Re number in Fig. 19.10. Thus the G factors are a function of the turbulence level (i.e., G factors are a function of the Reynolds number). The function G_τ is a multiplying factor of the power loss given by Eq. (19.3) (i.e., $G_\tau F_\tau$).

Stability. Rotordynamics plays a crucial part in determining the critical speeds, vibrational amplitudes and possible instabilities in rotating machinery. Since the fluid film possesses both stiffness and damping properties, it affects considerably the overall characteristics of such a system. Moreover, each bearing possesses four

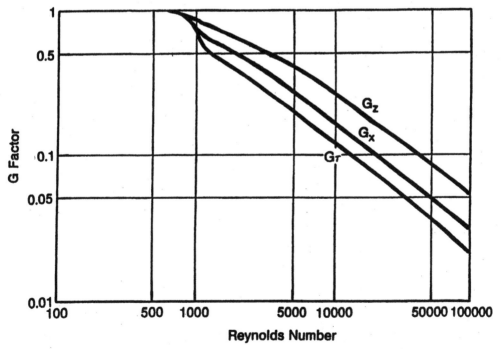

FIGURE 19.10 Values of turbulence coefficients.

spring and four damping coefficients which must be known to determine the effect of bearings on a rotordynamic system. The origin of these eight coefficients derives from the fact that a load imparted either along x or y produces a journal displacement in two perpendicular directions; conversely a journal motion either along x or y generates incremental forces in two perpendicular directions. The eight stiffness coefficients thus generated are defined by

$$K_{xx} = -\frac{\partial F_x}{\partial e_x} \qquad B_{xx} = -\frac{\partial F_x}{\partial \dot{e}_x}$$

$$K_{xy} = -\frac{\partial F_x}{\partial e_y} \qquad B_{xy} = \frac{\partial F_x}{\partial \dot{e}_y}$$

$$K_{yx} = -\frac{\partial F_y}{\partial e_x} \qquad B_{yx} = -\frac{\partial F_y}{\partial \dot{e}_x}$$

$$K_{yy} = -\frac{\partial F_y}{\partial e_y} \qquad B_{yy} = -\frac{\partial F_y}{\partial \dot{e}_x}$$

where the e's represent displacement and $\dot{e}$ velocities. The terms with subscripts xx and yy are called colinear and those with xy and yx cross-coupling coefficients. These eight springs and dashpots are schematically represented in Fig. 19.11.

The dynamics for a rotor bearing system is then determined by the set of two dynamic equations

$$K_{xx}x + K_{xy}y + B_{xx}\dot{x} + B_{xy}\dot{y} + M\ddot{x} = 0 \qquad (19.8a)$$

$$K_{yx}x + K_{yy}y + B_{yx}\dot{x} + B_{yy}\dot{y} + M\ddot{y} = 0 \qquad (19.8b)$$

where the dotted quantities denote velocities and the double dot are accelerations.

The solution of this set of equations yields two important quantities with regard to system stability

- Threshold instability frequency ω_i
- Critical mass, M_{CR}

The highest rotor mass M that a bearing will support before becoming unstable is then given

$M < M_{CR}$ system stable

$M > M_{CR}$ system unstable

The threshold instability frequency ω_i gives the vibrational frequency at the onset of stability. Given ω_n as the natural frequency of the system, we then have

If $\omega_i < \omega_n$ system is stable

If $\omega_i > \omega_n$ system is unstable

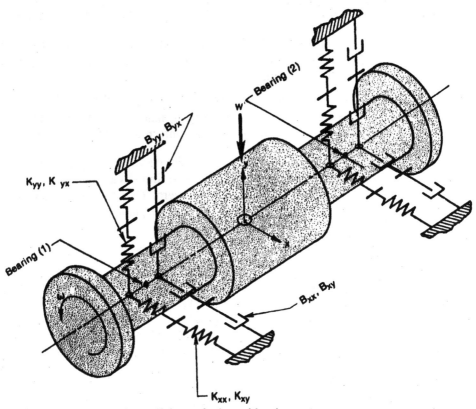

FIGURE 19.11 Dynamic coefficients of a journal bearing.

The two quantities M_{CR} and ω_i are determined from the eight coefficients as follows:

$$\overline{M}_{CR} = (\overline{Z}/\overline{\omega}_i^2) = \overline{Z}/4\pi^2\omega_i^2 \tag{19.9a}$$

$$\overline{Z} = \frac{\overline{K}_{xx}\overline{B}_{yy} + \overline{K}_{yy}\overline{B}_{xx} - \overline{K}_{xy}\overline{B}_{yx} - \overline{K}_{yx}\overline{B}_{xy}}{\overline{B}_{xx} + \overline{B}_{yy}} \tag{19.9b}$$

$$\overline{\omega}_i = (\omega_i/\omega) = \left[\frac{\overline{K}_{xx}\overline{K}_{yy} - \overline{Z}(\overline{K}_{xx} + \overline{K}_{yy}) - \overline{K}_{xy}\overline{K}_{yx} + \overline{Z}}{\overline{B}_{xx}\overline{B}_{yy} - \overline{B}_{xy}\overline{B}_{yx}}\right]^{1/2} \tag{19.9c}$$

19.3.2 Hydrostatic Bearings

There are several distinct advantages to the use of hydrostatic bearings, namely

- They can operate at low or zero speeds without affecting the load capacity. They can therefore be easily started under load.
- Due to the depth of the fluid pocket which is one or more orders of magnitude higher than in hydrodynamic films, they have very low drag losses.

The above merits are counteracted by some inherent disadvantages:

- They need an external pump or some other source to supply the pressurized fluid; thus even though the bearing power loss is low, the pumping power is part of the energy expenditure.
- Due to the depth of the pocket, they are prone to turbulence in the fluid.
- They require a compensation system, that is, a regulatory mechanism to adjust the fluid pressure in the pocket in conformity with changes in the imposed load.

Since the pockets containing the pressurized fluid are very deep, there is usually no significant hydrodynamic effect. The film thickness can be considered constant and so $(\partial h/\partial x)$ in Eq. (19.1) becomes zero. Thus the Reynolds equation for hydrostatic bearings reduces itself to a Laplace equation

$$\frac{\partial^2 p}{\partial x^2} + \frac{\partial^2 p}{\partial z^2} = 0 \tag{19.9}$$

The proper boundary conditions are determined by the supply pressure and the form of compensation used in the system.

The most common restrictor used for compensation is either an orifice or a capillary. Referring to Fig. 19.12, the relation between the supply pressure p_s, the pocket pressure p_o, and the ambient pressure p_a is for an incompressible fluid given by

a. For orifice compensation

$$\frac{C_D \pi a^2}{2} \left[\frac{p_s - p_a}{2\rho}\right]^{1/2} = \frac{\pi h^3 (p_o - p_a)}{6\mu \ln(R_2/R_1)} \tag{19.10}$$

b. For capillary compensation

$$\frac{(p_s - p_o) r_0^4}{4l} = \frac{(p_o - p_a) h^3}{3 \ln(R_2/R_1)} \tag{19.11}$$

Using p_o obtained from the above relations, one can subsequently calculate load, film thickness and the flow in the bearing.

19.3.3 Compliant Surface Bearings

The compliant bearing combines features of fluid dynamics and elastic response which make for a number of unique features of this device. It also accounts for its greater analytical complexity since, in addition to the Reynolds equation, one must consider the corresponding mechanical response of its surfaces. Given the large number of possible configurations, no single deflection equation can be written for them all; rather each family of compliant surface bearings will have its own set of

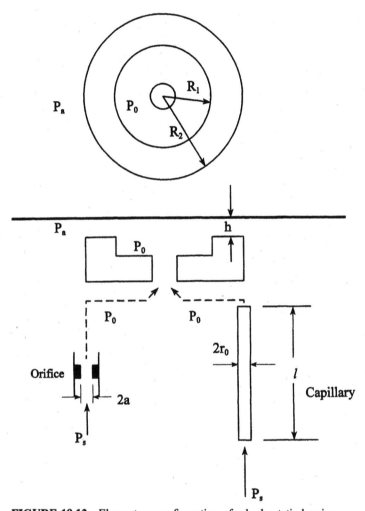

FIGURE 19.12 Elementary configuration of a hydrostatic bearing.

characteristic expressions, which together with the Reynolds equation will provide the necessary solution.

There are multiple advantages to the use of compliant bearings.

- Due to their ability to yield under load, they perform well at high speeds and at high temperatures.
- By introducing friction between the compliant and rigid surfaces, Coulomb damping can be introduced.
- Compared to either fluid film or rolling element bearings, they have very low power losses.

- They can operate with either liquid or gas lubrication. By virtue of their ability to deflect, the difference between performance with a liquid or a gas is not as pronounced as in rigid bearings.

One of the major disadvantages is the high starting torque due to rubbing. This problem can be alleviated by applying coatings to the mating surfaces.

The two major groups of compliant bearings are those in which the dominant mode of deformation is extension and those where bending is the main mode. Quite often both modes are present. A model of the tension dominated configuration is given in Fig. 19.13 where a highly flexible foil is stretched between fixed fulcrums. Here the imposed bearing load is supported by the radial component of the tension forces; the load capacity can be further increased by imparting to the foil a preload tension. Figure 19.14 shows a bearing belonging to the second category. The figure part (b) shows one segment of the surface consisting of a strip of foil supported on a compliant subsurface. The corrugated subsurface rests on the rigid bearing shell.

19.3.4 Rolling Element Bearings

Most of the literature, not to speak of handbooks, on ball and roller bearings is based on the premise that there is metal-to-metal contact between the races and the rolling elements. Under load, the ball or roller is compressed to form a circular, elliptical or rectangular contact area, generating conventional Hertzian stresses. In reality, however, the contact area is covered with a lubricant film, albeit several orders of magnitude smaller than in hydrodynamic bearings. Under these conditions, as was asserted in the opening paragraph of this chapter, there is superimposed onto the Hertzian stresses a hydrodynamic pressure field. These combined effects produce what is called an elastohydrodynamic pressure distribution along with a finite film thickness, whose basic shapes are given in Fig. 19.15. The calculation of these pressures, film thicknesses, power loss and so on, must, as in the

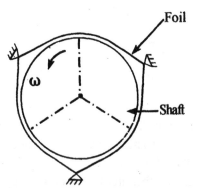

FIGURE 19.13 Extension dominated compliant surface bearing.

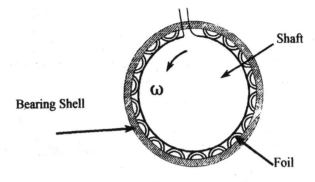

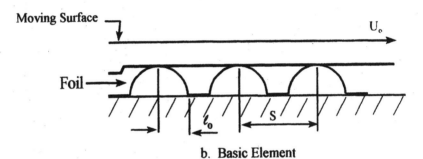

FIGURE 19.14 Bending dominated compliant surface bearing.

case of compliant bearings, be based on the use of the Reynolds equation in conjunction with the elasticity equations as they pertain to the particular geometry of the bearing.

The more significant attributes of rolling element bearings are as follows:

- They can start under load and operate well at lower, including zero, speed.
- When on the point of failure, they will continue to operate for some time providing an opportunity to shut down the machine.
- They can support both radial and axial loads.

The main shortcoming of rolling element bearings is that, depending on speed and load, they have a finite life span. They must, therefore, be periodically replaced, unlike hydrodynamic bearings which can operate indefinitely. They also require more installation space and are noisy.

Given the complexity of the elastohydrodynamic solutions, design and performance calculations of REB's are based on empirical relations, related primarily to the fatigue properties of the structures involved. Since fatigue failure is a function of level of stress and number of cycles, the viability of an REB can thus be related to rotational speed of the machine, shaft size and bearing loading. The relation between the first two parameters, known as the DN curve, is shown in Fig. 19.16.

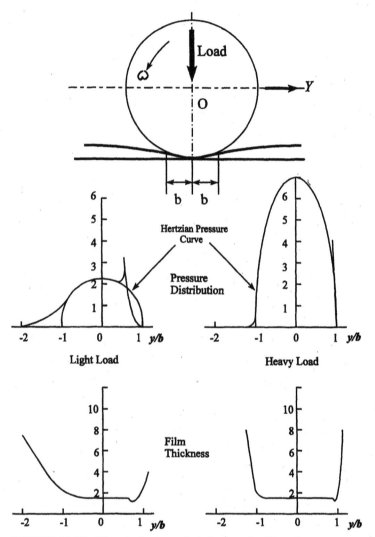

FIGURE 19.15 Elastohydrodynamic behavior of rolling elements.

This limit is imposed by the dynamics of the spinning balls or rollers and the rotating races and cages. Thus, regardless of load this DN limit ought not be exceeded. The longevity of a given bearing when operated under the limitations of Fig. 19.16 does depend on the imposed load. This dependence is given by

$$L_{10} = \frac{1.67 \cdot 10^3}{N} \left(\frac{W_D}{W}\right)^\gamma \qquad (19.12)$$

where

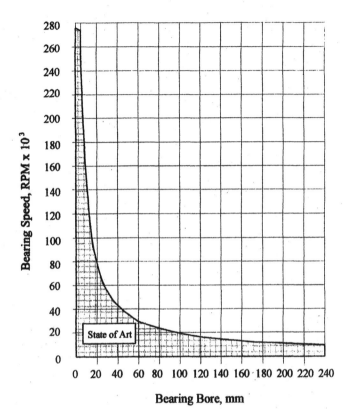

FIGURE 19.16 Limits in ball bearing DN values.

L_{10} = is the predicted life of the bearing carrying a 90% probability of survival
N = shaft speed, rpm
W = load on the bearing, (lbf)
W_D = the dynamic load, (lbf) a bearing will endure for one million revolutions

and

$$\gamma \begin{cases} = 3 \text{ for ball bearings} \\ = 3.3 \text{ to } 4.0 \text{ for roller bearings} \end{cases}$$

The value of L_{10} is influenced by a host of additional factors such as size and number of rolling elements, kind of materials used, the bearing's angle of contact, lubrication method, surface finish, and others.

19.3.5 Magnetic Bearings

Figure 19.17 shows basic designs of radial and thrust AMB's. These contain a stator wound with coils to create the magnetic field and a rotor which has ferro-

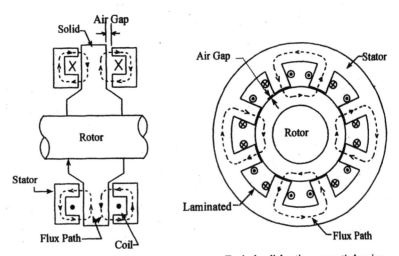

FIGURE 19.17 Basic designs of magnetic bearings.

magnetic laminations or solid materials to interact with the magnetic field. In a radial bearing, it is common to have two sets of opposing coils with a sensor to provide feedback of rotor position along two axes. In a thrust bearing, position is maintained by a set of stationary coils on either side of the runner. Since each pair of poles acts independently, it is sufficient to illustrate the action of one set to make clear the operation of the bearing as a whole. The attractive force on a rotor, as shown in Fig. 19.18, is given by

$$F \propto (I/C)^2$$

where F is the magnetic force and I the current. With attractive poles, as the rotor moves toward the poles, the force on this face increases while on the opposite poles it decreases; thus without control the system is unstable leading to metal-to-metal contact. With a sensor and feedback system, current is decreased on the approaching side and increased on the opposite to force the rotor back toward the center. Direct feedback control is non-linear and not responsive enough; this is remedied by the introduction of a large bias current compared to which the regulating current is small.

The above outlined arrangement of displacement feedback is, however, not sufficient to control resonances. For this, the system must account not only for displacement but also for rotor velocities. Thus a feedback for velocity sensor is also required, or a provision to convert the recorded displacements into velocities.

A major advantage of AMB's is that stiffness and damping can be varied, unlike in any other bearings where they are fixed for a given design. With such control, optimum performance can be obtained over a wide range of operating conditions. Moreover, this flexibility offers means for virtual rotor balancing to counteract

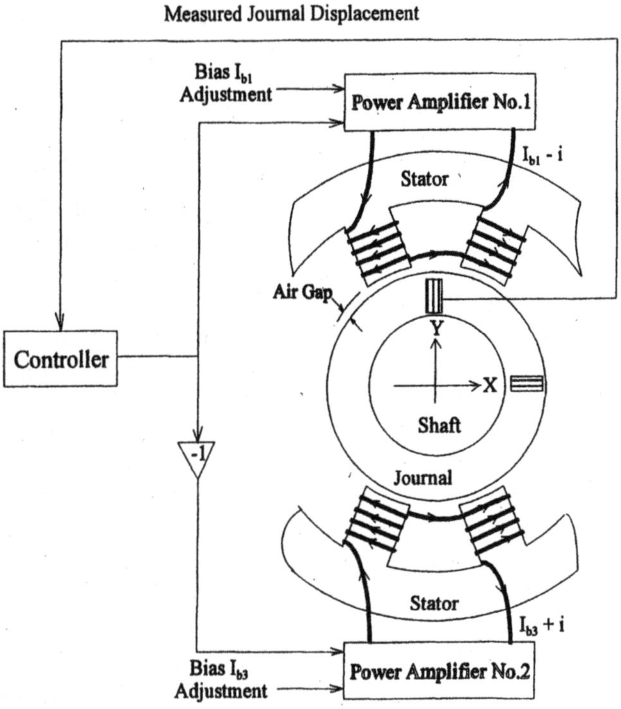

FIGURE 19.18 Elements of AMB's servo control.

residual unbalance in the system. The system can even account for pedestal vibrations once they are measured and fed into the control circuit of the bearing.

All the varied bearing designs are combinations of three principles of magnetic operation

- Attraction by the use of electromagnets
- Repulsion by the use of permanent magnets
- Reluctance using permanent magnets

Figure 19.19 illustrates these principles. Part A) shows the attraction system which has been applied in a number of machine designs employing five active servo controls. The repulsion principle illustrated in part B) yields a radially stable bearing by means of passive magnets, but is axially unstable and requires at least one active control. In the reluctance design shown in part C), the magnetic circuit always seeks a geometry so as to align itself with the edges of the rectangular salient pole faces. In the configuration shown, axial stability is achieved at the expense of radial instability that must be overcome by an active control. Part D) is shown not because it represents another principle, but because using permanent magnets to energize the air gaps and electromagnets for control yields a metastable system that can operate at almost zero control power when disturbance loads are absent; and it produces a linear force versus current which simplifies the servo control design.

Attraction Electromagnets. These are by far the most widely used. A straightforward approach is to use two sets of orthogonally disposed magnet pairs to control the rotating shaft, as shown in Fig. 19.19. This approach controls x and y as well as the two angular directions. The axial direction must be controlled by either an active or passive bearing. With today's materials these electromagnets can be quite compact. Unit loadings of 58 lb/in.^{2} per electromagnet pole face can be realized with silicon steel; and 116 lb/in.^{2} with iron-cobalt-vanadium alloys. As an example, to provide an active radial magnetic bearing capable of supporting a 100 lb. Static load with a stiffness of 100,000 lb/in. on a 2 inch shaft diameter would call for a 2 inch long bearing.

Radial Repulsion Design. One radially passive design is shown in Fig. 19.20. Here multiples of radially polarized permanent magnets of alternating polarization are stacked axially with matching sets so that the disks are in repulsion with the housing. This provides a radially centering force. However, the bearing is axially unstable and requires a servo-controlled axial magnet. The bearing has limited stiffness; Fig. 19.20 shows that for a stiffness of 100,000 lb/in. such a bearing would be 20 inches long. Furthermore, the bearing does not by itself develop much damping and the introduction of eddy currents is required to provide radial damping.

Reluctance Designs. A radially passive bearing using multiple concentric rings on a stator and rotor is shown in Fig. 19.21. The design uses permanent magnets to energize the radially passive gaps but electromagnetic power can be used as well. Here the ferromagnetic circuit will seek a geometric arrangement so as to minimize total magnetic "reluctance." This design is radially stable and it will maintain a uniform centering force, provided the axial magnetic gaps have an active servo control. Its greatest disadvantage is that the maximum radial stiffness is low compared to the axially destabilizing stiffness. Consequently, to achieve good radial stiffness, large axial electromagnets are required.

19.4 CONVENTIONAL BEARINGS

The bearings described in the present chapter are the most commonly used bearings in rotating machinery. These are oil lubricated, mostly babbitted journal and thrust bearings which, when properly designed and maintained, will last throughout the life of the machine. The parameters that determine the adequacy of a bearing design are described first, following which equations, tables and charts supply the numerical values of these parameters for various bearing configurations and various sets of operating conditions.

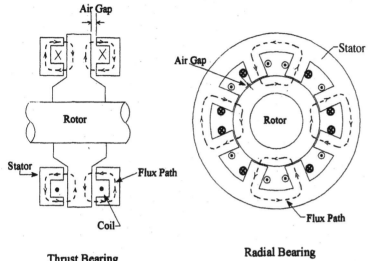
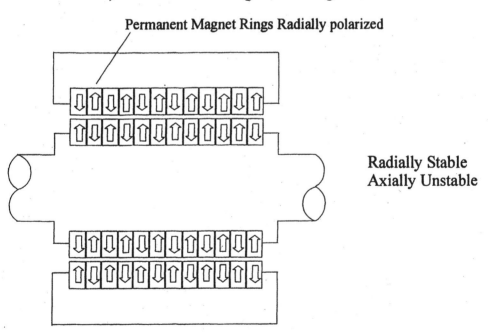

FIGURE 19.19 The direction stabilities of magnetic bearings

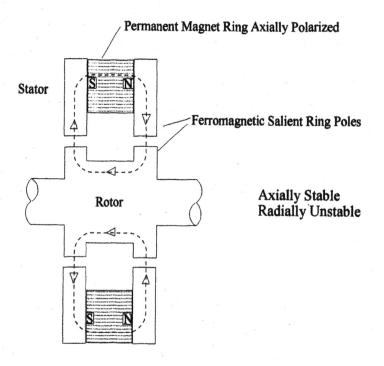

C) Reluctance Centering Thrust Bearing

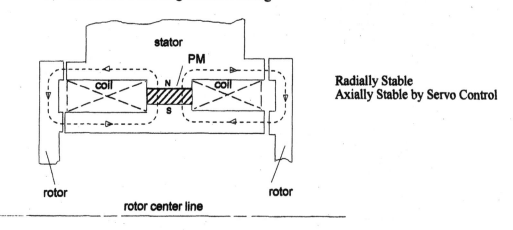

D) Reluctance Centering Radial Bearings
and Actively Controlled Thrust Bearing with Permanent Magnet Bias

FIGURE 19.19 (*Continued*)

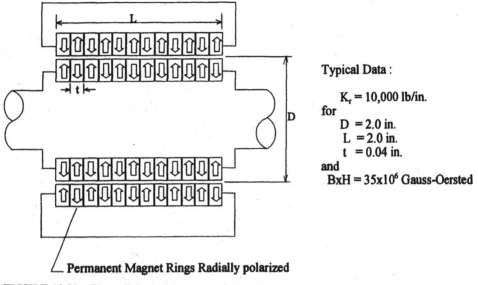

FIGURE 19.20 The radial repulsive magnetic bearing.

Typical Data:

$K_r = 10,000$ lb/in.

for
$D = 2.0$ in.
$L = 2.0$ in.
$t = 0.04$ in.

and
$B \times H = 35 \times 10^6$ Gauss-Oersted

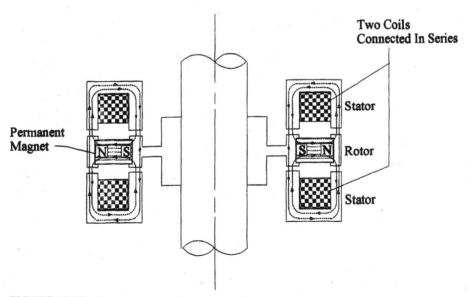

FIGURE 19.21 The "reluctance" type magnetic bearing.

19.4.1 Bearing Parameters

The Minimum Film Thickness. All discussion of bearing performance includes the term "load capacity" by which is usually meant the load a bearing can support at a given minimum film thickness. The minimum film thickness is important because it gives an indication of the following:

- The likelihood of physical contact between the mating surfaces which may lead to failure
- The intensities of the peak pressures and temperatures which tend to rise steeply with a decrease in h_{min}
- The reserves available in the bearing to accommodate any unexpected increase in load or sudden shaft excursions

There is no fixed value for a satisfactory h_{min}. It depends upon a number of factors including size of bearing, nature of application, operating conditions, degree of reliability required, and others. Naturally, in no case should h_{min} be smaller than the sum of the asperities of the two mating surfaces.

The Maximum Temperature. The importance of knowing the value of the film maximum temperature, T_{max}, is often on a par with that of h_{min}. While too small, an h_{min} can cause damage by physical contact, excessive temperatures cause failure either by softening or melting the bearing surface, and this can occur even when there is an ample film thickness. T_{max} usually occurs near the trailing edge of the bearing pad. In aligned journal bearings, this T_{max} occurs on the axial centerline; in thrust bearings, it occurs near the outer radius where both linear speed and the circumferential path of the lubricant are largest.

Temperature Rise. It is common practice in industry to use ΔT as a criterion of bearing performance, this quantity being the difference between the bulk temperature of the oil discharging from the bearing and the oil supply temperature.

While monitoring ΔT may be helpful in spotting sudden changes in bearing behavior, it is a poor indicator of the magnitude of T_{max}. Two bearings may have the same ΔT, yet two radically different values of T_{max}, and if a bearing is to fail because of excessive temperature, it will be T_{max} that would initiate the failure.

Thus, while ΔT remains a useful overall indicator of the amount of total heat generation and of any untoward changes in bearing behavior, it cannot represent or replace the crucial quantity T_{max}.

19.4.2 Journal Bearing Performance

For a journal bearing with the nomenclature as sketched in Fig. 19.22, the performance quantities of interest can be calculated as follows:

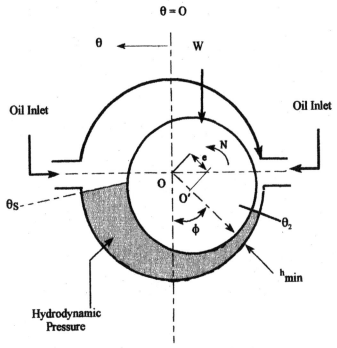

FIGURE 19.22 Conventional 2-axial-groove bearing.

- *Film thickness.* For an aligned journal it is

$$\overline{h} = (h/C) = 1 + \epsilon \cos(\theta - \phi) \qquad (19.13)$$

- *Sommerfeld number and load parameter.* The Sommerfeld number, given by

$$S = (\mu N/P)(R/C)^2 \qquad (19.14a)$$

has traditionally been the most important parameter. However, a more convenient quantity is the inverse of S, here called load parameter, given by

$$\overline{W} = [P/(\mu N)](C/R)^2 = [W/(LD\mu N)](C/R)^2 \qquad (19.14b)$$

where $P = (W/LD)$ is the unit loading. What this parameter says is that any combination of P, μ, N, C and R, such as to leave the value of W unchanged, would result in the same bearing eccentricity ratio ϵ and attitude angle ϕ.

- *Minimum film thickness.* This is the smallest distance between the journal and bearing surfaces and is given by:

$$\overline{h}_{min} = \frac{h_{min}}{C} = (1 - \epsilon) \qquad (19.15)$$

What is normally referred to as load capacity related to the load W which this h_{min} can support.

- *Friction coefficient.* This is the ratio between the frictional force and bearing load (W). It is normally expressed in the form of:

$$f = \frac{F_\tau}{W} \tag{19.16}$$

- *Power loss.* This, of course, can be obtained from the value of F_τ, namely:

$$H = F_\tau R\omega = fWR\omega \tag{19.17}$$

However, the power loss is often given directly in the form of

$$\overline{H} = \frac{HC}{\pi^3 \mu N^2 L D^3} \tag{19.18}$$

In Eq. (19.18), the quantity by which H is normalized represents the power loss in an unloaded concentric journal bearing, i.e, one in which $\epsilon = 0$. It is known as the Petroff equation.

- *Flow.* In a bearing with a single or a multi-pad arrangement, flow of lubricant Q_1 enters at the leading edge; an amount of lubricant Q_z is lost from sides of the bearing; and flow Q_2 leaves at the trailing end of the pad and adheres to the moving surface in most cases. The adhered flow, Q_2, is carried to the next pad and/or the bearing convergent section. Thus, the net amount of lubricant to be replenished is Q_z. The latter is referred to as side leakage. Clearly we must always have:

$$Q_1 = Q_2 + Q_z$$

All of these flows are given in dimensionless form as:

$$\overline{Q} = Q/(\pi NDLC/2) \tag{19.19}$$

the denominator representing the flow in an unloaded, concentric bearing, i.e., at $\epsilon = 0$ (for which case $Q_z = 0$ and $Q_1 = Q_2$).

- *Temperatures.* For isothermal conditions, a bulk temperature rise can be estimated from the values of power loss and side leakage, namely:

$$\Delta T = (T_{av} - T_1) = \frac{H}{cwQ_z} \tag{19.20}$$

The parameters ϵ, f, $\overline{Q}_1$, $\overline{Q}_z$ and ΔT_{max}, which serve to evaluate the various bearing performance items, are to be obtained from a solution of aforementioned governing equations for the specific geometry and specific operating regimes of the bearing under consideration.

- *The problem of viscosity.* In most of the expressions, μ appears either as a variable, or as one of the normalizing quantities. Since in practice viscosity varies throughout the fluid film, the question arises as to what value to assign to μ when quantitatively evaluating bearing performance. Without going into much detail, several approaches are possible.

a) *Inlet viscosity:* $\mu = \mu_1 = const$. This is the approach most widely used because it is the simplest; it also introduces the largest errors. In this approach, a viscosity, μ_1, corresponding to the known inlet temperature, T_1, is used for all calculations. This may be acceptable for cases where the ΔT is expected to be low. Even then, it should be kept in mind that in most applications, a given variation in T produces a much more pronounced variation in μ so that a small ΔT may still produce appreciable variations in viscosity.

b) *Average viscosity:* $\Delta\mu = \Delta\mu_{av} = const$. The overall aim in this approach is to assume an average temperature, T_{av}, which yields levels of power loss and side leakage such that when used in Eq. (19.20), the calculated T_{av}, will be the same as the assumed one. The steps involved here are as follows:

1. Assume a $T_{av} > T_1$ which then specifies μ_{av}.
2. Using the above μ_{av}, calculate H and Q_z.
3. Determine T_{av} from Eq. (19.20).
4. If the assumed T_{av} from step 1 differs from the calculated T_{av} in step 3, assume a different T_{av} and repeat procedure.
5. Continue until the assumed T_{av} equals the T_{av} calculated from Eq. (19.20).

Figure 19.23 shows the procedure for arriving at a correct T_{av}. Plotting the assumed T_{av} versus the calculated T_{av} from Eq. (19.20), one can, after several trials, arrive at the correct result which is a point lying on the 45° line. While a correct convergence may require four or five trials, Fig. 19.23 shows that drawing a straight line through the first two guesses may yield an approximate T_{av} sufficient for many applications.

The use of μ_{av} based on the above approach is a very useful and efficacious means for calculating bearing performance, and it provides results of good accuracy.

Types of Bearing Used. Compressors employ at least half a dozen types of journal bearings. Essentially all of these designs consist of partial arc pads having a circular geometry. The differences are mainly in the number and arrangement of the pads and in whether or not the centers of curvature of the pads coincide with the geometric center of the assembled bearing. This will become clear during discussion of the individual bearing types.

Circular Bearings. The most common journal bearing is what is often referred to as a "2-axial-groove bearing," portrayed in Fig. 19.24. Its bore is circular and each of its two identical pads may span anywhere from 120° to 160° in angular extent. A variation of this bearing, shown in Fig. 19.24b, is called the "overshot groove bearing." The bottom pad is identical to that of part (a); however, the top pad is cut circumferentially by a deep channel.

The rationale for the overshot groove is to increase the amount of oil flow. This it does. However, by flooding the upper half, it also increased the power losses. In

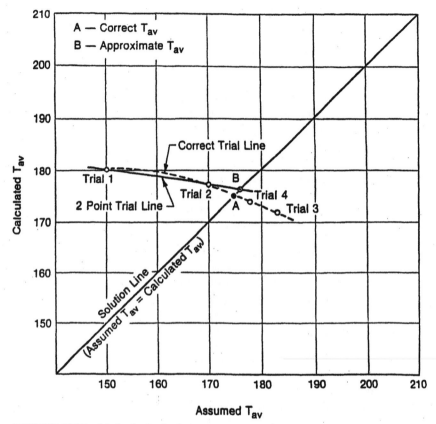

FIGURE 19.23 Method of calculating temperature rise.

general, since the load is carried primarily by the bottom pad, the overshot groove in circular bearings does not affect h_{min} or T_{max} one way or another.

Circular bearings, made up of more than two pads, are called 3, 4, or 5 axial-groove bearings. Similar to the 2 axial-groove bearings, the individual pads in these designs are separated by oil grooves with the load passing midway through the bottom pad.

Circular bearings are used extensively since they are easy to manufacture, to install, and to repair. Their performance is good as long as stability is not a problem. The angular extent of the pads can be reduced down to 120° without penalty, the load capacity being about the same for all $360° < \beta < 120°$. If the bottom pad arc is reduced below 120°, a deterioration in load capacity sets in which accelerates rapidly when arcs of less than 80° are used.

As is true for most bearings, the smaller the (L/D) ratio, the lower the load capacity, through a low L/D, by increasing eccentricity, improves a bearing from the standpoint of stability.

Table 19.1 gives the characteristics of 2, 3 and 4 axial-groove bearings.

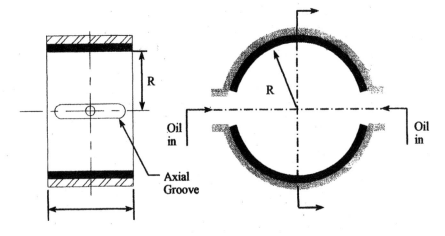

a) Two-Axial Groove Bearing

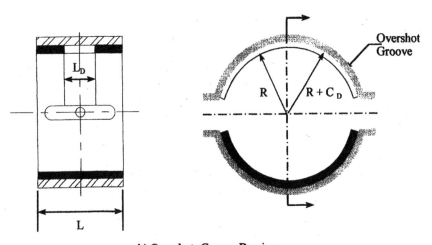

b) Overshot- Groove Bearing

FIGURE 19.24 The "overshot" journal bearing.

Elliptical Design. A journal bearing which has an increased capability to suppress instability is the elliptical bearing shown in Fig. 19.25. This bearing looks much like the 2 axial groove bearing, but the two lobes are assembled so that their centers of curvature are not coincident. Each lobe has been displaced inward, this displacement, expressed as a percentage of the machined radial clearance, being denoted as ellipticity, or "preload." In effect, then, even for operating at the geometric bearing center, both lobes have eccentricities greater than zero. Larger amounts of ellipticity increase the pad eccentricities and thus provide more stable operation. The design is penalized by higher friction losses, higher flow rates, and reduced

TABLE 19.1 Performance of Axial Groove Bearings[21]

L/D	ϵ	2-Groove ϕ	2-Groove S	2-Groove $\overline{Q}_1$	3-Groove ϕ	3-Groove S	3-Groove $\overline{Q}_1$	4-Groove ϕ	4-Groove S	4-Groove $\overline{Q}_1$
1-1/2	0.2	62	0.435	0.165	67	0.99	0.11	80	1.9	0.081
	0.4	51	0.179	0.290	49	0.37	0.205	72	0.71	0.16
	0.6	42	0.0943	0.350	36	0.15	0.29	49	0.245	0.29
	0.8	29	0.037	0.320	27	0.05	0.39	25	0.259	0.38
	0.9	23	0.0167	0.290	20	0.019	0.40	18	0.0222	0.37
1	0.2	64	0.714	0.27	69	1.33	0.16	79	2.04	0.12
	0.4	53	0.275	0.43	50	0.48	0.30	72	0.87	
	0.6	45	0.125	0.56	37	0.187	0.435	49	0.30	0.39
	0.8	28	0.041	0.46	26	0.059	0.059	25	0.07	0.51
	0.9	22	0.019	0.43	20	0.022	0.022	19	0.025	0.53
1/2	0.2	64	2.10	0.32	72	3.00	0.245	80	4.23	0.20
	0.4	53	0.80	0.56	54	1.06	0.47	72	1.7	0.42
	0.6	42	0.31	0.715	39	0.385	0.68	49	0.55	0.67
	0.8	24	0.082	0.725	27	0.103	0.92	25	0.11	0.84
	0.9	22	0.031	0.695	20	0.033	1.04	19	0.035	0.90
1/4	0.2	70	7.94	0.36	74	9.61	0.30	79	11.6	0.272
	0.4	57	2.86	0.67	56	3.30	0.625	72	4.5	0.56
	0.6	43	1.06	0.86	41	1.17	0.90	49	1.3	0.89
	0.8	27	0.256	0.93	28	0.27	1.17	25	0.23	1.15
	0.9	20	0.074	0.87	20	0.0735	1.29	19	0.73	1.22

load capacity at low eccentricities. At high eccentricities, the behavior of the elliptical bearing approaches that of the circular design. The bearing is commonly used because it is effective and is easy to manufacture; with shims placed at the horizontal split, the bearing is given a circular bore; then the shims are removed yielding the desired "elliptical" geometry.

In order to properly understand and handle the elliptical bearing, the following central fact must be stressed. While, as is customary, journal positions are referred to the geometric center of the bearing (ϵ and ϕ), the quantities that count, i.e., the parameters that determine load capacity, h_{min}, etc., are the eccentricities and attitude angles with respect to the lobe centers 0_1 and 0_2.

The relation between the bearing parameters and the two lobe parameters are, referring to part (b) of Fig. 19.25, as follows:

$$\epsilon_1 = (\epsilon^2 + m^2 + 2\epsilon \cos \phi)^{0.5} \tag{19.21a}$$

$$\phi_1 = \sin^{-1} \left[\frac{\epsilon \sin \phi}{e_1} \right] \tag{19.21b}$$

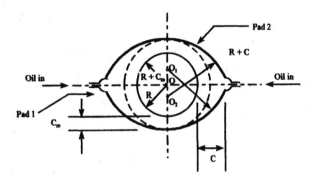

a.) Concentric Journal Position

$m = (d/C)$

$\epsilon_{1,2} = (e_{1,2}/C)$

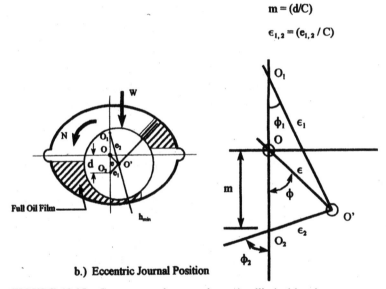

b.) Eccentric Journal Position

FIGURE 19.25 Geometry and nomenclature in elliptical bearings.

$$\epsilon_2 = [\epsilon^2 + m^2 - 2\epsilon m \cos \phi]^{0.5} \tag{19.22a}$$

$$\phi_1 = \pi - \sin^{-1}\left[\frac{\epsilon \sin \phi}{\epsilon_2}\right] \tag{19.22b}$$

where:

$$\epsilon = \frac{e}{C}, \quad \epsilon_1 = \frac{e_1}{C}, \quad \epsilon_2 = \frac{e_2}{C}, \quad m = \frac{d}{C}$$

Thus, for a bearing with a certain ellipticity ratio m, when bearing eccentricity and attitude angle (ϵ and ϕ) are given, all other quantities can be determined from the foregoing relationships.

It should be noted here that in all the normalized quantities, the machined bearing clearance C is the relevant parameter. Also, the value of h_{min}, by virtue of what was said above, is determined not by ϵ and ϕ, but by the eccentricity of either the upper or lower lobe. If the shaft center is above the horizontal centerline (and, as will be seen later, this is possible) then h_{min} is located in lobe No. 2 and h_{min} is determined by:

$$h_{min} = C(1 - \epsilon_2) \tag{19.23a}$$

$$\theta_{min} = 3\pi/2 + \phi_2 \tag{19.23b}$$

If, as happens in the majority of cases, the shaft center is below the horizontal centerline, then the relevant equations are

$$h_{min} = C(1 - \epsilon_1) \tag{19.24a}$$

$$\theta_{min} = \pi + \phi_2 \tag{19.24b}$$

A convenient quantity in dealing with elliptical bearings is to introduce a bearing eccentricity ratio based not on C, but on the minor clearance C_m. Thus, since

$$\overline{C}_m = (C_m/C) = 1 - m \tag{19.25}$$

we have for this new eccentricity ratio:

$$\epsilon_m = (e/C_m) = \epsilon/(1 - m) \tag{19.26}$$

The usefulness of the above expression lies in the fact that as in circular bearings, the journal cannot travel in the vertical direction beyond $\epsilon_m = 1$. (It can do so in other directions.) It is thus possible to plot the loci of journal center for all values of m against a common vertical scale, as shown in Fig. 19.26. As seen, at low eccentricities, the shaft center is often above the horizontal centerline locating the bearing's h_{min} in the upper lobe. The plot includes, for comparative purposes, $m = 0$, which corresponds to a circular bearing.

The general performance characteristics of the elliptical bearing are given in Table 19.2 and Fig. 19.27. These are given for the entire practical range of (L/D) ratios and for ellipticity ratios from 0.25 to 0.75. Since $m = 0$ represents a circular bearing and $m = 1$ is the limit for possible ellipticity ratios (at $m = 1$, there is contact between the surfaces), the plots cover the entire spectrum of elliptical bearing design.

Unlike the case of circular bearings, the introduction of an overshot groove in elliptical designs has a telling effect on performance. In essence, an extensive overshot groove destroys the effectiveness of the bearing as an elliptical design, approximating it to a circular bearing of clearance C.

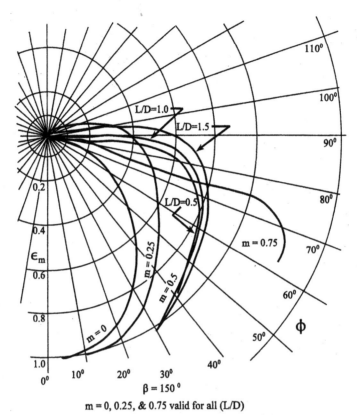

m = 0, 0.25, & 0.75 valid for all (L/D)

FIGURE 19.26 Locus of shaft center for elliptical bearings.[23]

Three-Lobe Design. The 3-lobe bearing represents a further accentuation of the features that characterize the elliptical design. It is more stable than the circular and elliptical varieties, but due to its lower average clearance and smaller arc of bottom lobe, it shows higher losses and lower load capacities than the two other designs.

As the name implies, the conventional 3-lobe design consists of three pads each of about 80° to 120° in angular extent, with the bottom lobe placed symmetrically about the vertical load. The bearing is shown schematically in Fig. 19.28. All remarks made in connection with the elliptical bearing regarding the relationships, usage, and importance of the bearing and lobe parameters hold here too, though clearly the quantitative relations will be somewhat different. Thus, the relation between the parameters of the geometric center (ϵ, ϕ) and of the individual lobes are here as follows:

$$\epsilon_1 = [\epsilon^2 + m^2 + 2\epsilon m \cos \phi]^{0.5} \qquad (19.27a)$$

$$\phi_1 = \sin^{-1}\left[\frac{\epsilon \sin \phi}{\epsilon_1}\right] \qquad (19.27b)$$

TABLE 19.2 Performance of Elliptical Bearings[23]

L/D	ϵ_m	$m = 1/4$				$m = 1/2$				$m = 3/4$			
		ϕ	$\epsilon_{1,2}$	S	$\bar{Q}_z$	ϕ	$\epsilon_{1,2}$	S	$\bar{Q}_z$	ϕ	$\epsilon_{1,2}$	S	$\bar{Q}_z$
1-1/2	0.2	110	0.33	0.572	0.17	98	0.52	0.625	0.23	87	0.75	0.25	0.27
	0.4	90	0.39	0.185	0.29	98	0.56	0.313	0.25	75	0.178	0.11	0.275
	0.6	62	0.61	0.090	0.35	92	0.59	0.156	0.33	75	0.82	0.045	0.295
	0.8	38	0.81	0.0333	0.305	70	0.74	0.051	0.333	75	0.835	0.034	0.30
	0.9	32	0.90	0.0107	0.29	53	0.85	0.024	0.30	—	0.84	0.03	0.30
	1.0	—	—	—	—	—	—	—	—	75	0.85	0.029	0.30
	1.2	—	—	—	—	—	—	—	—	70	0.895	0.019	0.30
1	0.2	105	0.32	0.834	0.225	95	0.52	0.415	0.322	85	0.75	0.285	0.41
	0.4	90	0.39	0.305	0.412	90	0.54	0.274	0.37	80	0.77	0.143	0.41
	0.6	68	0.59	0.120	0.546	87	0.60	0.155	0.44	81	0.79	0.087	0.41
	0.8	35	0.81	0.040	0.436	67	0.75	0.056	0.47	79	0.81	0.0607	0.44
	0.9	30	0.90	0.019	0.430	48	0.80	0.024	0.43	78	0.94	0.0155	0.44
	1.2	—	—	—	—	—	—	—	—	65	0.90	0.0202	0.45
1/2	0.2	100	0.31	2.0	0.275	90	0.51	1.3	0.52	80	0.76	0.53	0.68
	0.4	90	0.39	0.57	0.53	85	0.54	0.67	0.57	80	0.78	0.26	0.69
	0.6	82	0.51	0.29	0.72	83	0.61	0.32	0.67	75	0.82	0.085	0.72
	0.8	38	0.81	0.071	0.72	65	0.76	0.11	0.72	75	0.83	0.067	0.73
	0.9	30	0.90	0.0305	0.69	48	0.86	0.040	0.675	75	0.84	0.058	0.73
	1.0	—	—	—	—	—	—	—	—	75	0.85	0.053	0.74
	1.2	—	—	—	—	—	—	—	—	67	0.90	0.030	0.76
1/4	0.2	100	0.31	7.15	0.42	90	0.51	5.55	0.56	75	0.76	1.07	0.93
	0.4	95	0.41	3.33	0.59	85	0.545	2.32	0.72	75	0.78	0.715	0.96
	0.6	62	0.61	0.96	0.94	80	0.62	0.945	0.825	75	0.80	0.488	0.97
	0.8	30	0.81	0.24	0.93	70	0.74	0.377	0.91	75	0.83	0.260	0.98
	0.9	28	0.90	0.074	0.93	45	0.875	0.080	0.875	75	0.84	0.170	0.98
	1.0	—	—	—	—	—	—	—	—	75	0.85	0.130	1.00
	1.2	—	—	—	—	—	—	—	—	68	0.88	0.091	1.00

Note: Wherever $\phi > 90$, $\epsilon_{1,2}$ denoted ϵ_2; otherwise it denotes ϵ_1.

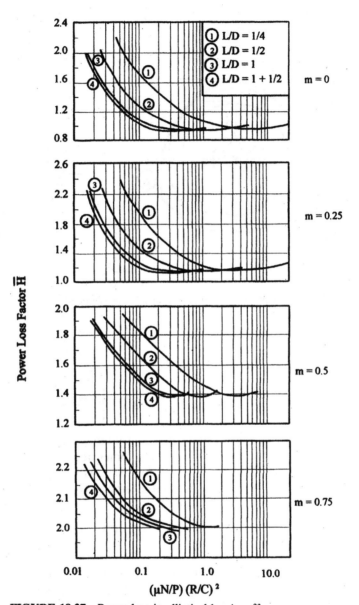

FIGURE 19.27 Power loss in elliptical bearings.[23]

$$\epsilon_2 = \left[\epsilon^2 + m^2 - 2\epsilon m \cos\left(\frac{\pi}{3} + \phi\right) \right]^{0.5} \quad (19.28a)$$

$$\phi_2 = \frac{2\pi}{3} - \sin^{-1}\left[\frac{\epsilon \sin(\pi/3 + \phi)}{\epsilon_2}\right] \quad (19.28b)$$

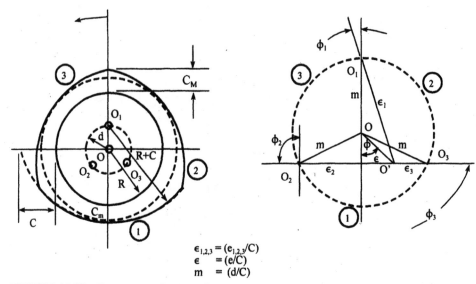

$$\epsilon_{1,2,3} = (e_{1,2,3}/C)$$
$$\epsilon = (e/C)$$
$$m = (d/C)$$

FIGURE 19.28 Geometry and nomenclature of three-lobe bearings.

$$\epsilon_3 = \left[\epsilon^2 + m^2 - 2\epsilon m \cos\left(\frac{\pi}{3} + \phi\right)\right]^{0.5} \qquad (19.29a)$$

$$\phi_3 = \frac{4\pi}{3} + \sin^{-1}\left[\frac{\epsilon \sin(\pi/3 - \phi)}{\epsilon_3}\right] \qquad (19.29b)$$

Another detail to be noted is that although C is the machined clearance in all three lobes; when assembled, this dimension does not physically appear in the bearing. The largest concentric clearance, which occurs at the junction of the three lobes, is less than the machined clearance C, and will here be denoted by C_M. It is given by

$$\overline{C}_M = (C_M/C) = (1 - m/2) \qquad (19.30)$$

whereas $\overline{C}_m$, as before, is given by $(1 - m)$.

Figure 19.29 shows the locus of shaft center in a 3-lobe bearing of 100° arc extent. Since, as seen here, the shaft never rises above the horizontal centerline, the minimum film thickness will always be located in the bottom pad, and is given by

$$\overline{h}_{min} = (1 - \epsilon_1) \qquad (19.31)$$

Table 19.3 and Figs. 19.29 and 19.30 give the performance characteristics of the 3-lobe bearing for three (L/D) ratios and two ellipticity ratios. Together with the 3-groove circular bearing given in Table 19.1, which represents the case $m = 0$, a 3-point variation in m is provided from which performance for intermediate values of m can be obtained by cross plotting.

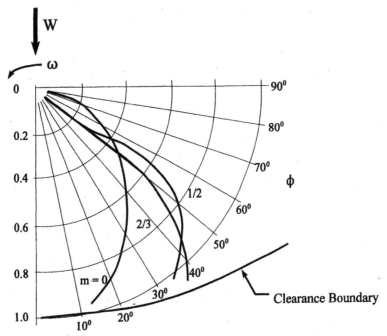

FIGURE 19.29 Locus of shaft center for three-lobe bearings.[24]

It was said previously that the 3-lobe bearing is usually chosen for its superior stability characteristics and following is a brief illustration of its features vis-à-vis the elliptical bearing. Figure 19.31 shows the stable and unstable regimes plotted against Sommerfeld number for the two bearing types. Each of the constant lines gives the locus of the operation of a bearing with a fixed geometry as its speed is varied. The bearing parameter, η, is given by

$$\eta = \frac{\mu LD}{2\pi w} \left[\frac{R}{C}\right]^2 \left[\frac{W}{CM}\right]^{0.5}$$

which is independent of rotor speed and describes a certain bearing geometry. As speed is increased, the bearing will, via S, proceed along a line of constant η and eventually enter the unstable region. The parameter η is most sensitive to bearing diameter and lobe clearance and less so to the length, L, and viscosity, μ. A move to a higher value of η, i.e., to a more stable region is accomplished by either increasing μ, increasing L, increasing D, or decreasing C. Because the ordinate parameter $\mu(MC/W)^{0.5}$ also depends upon the clearance, changes in C will follow the slightly inclined dashed lines—toward the left if C is increased, toward the right if C is reduced. At light loads and small clearances, the 3-lobe bearing is better than the elliptical bearing. Under heavier loads, the elliptical design is the more stable bearing. It should, however, be kept in mind that it is precisely the low load range that is the troublesome region of bearing stability.

TABLE 19.3 Performance of Three-lobe Bearings*[24]

			$m = 1/2$				$m = 2/3$		
L/D	ϵ_m	ϕ	ϵ_1	S	$\overline{Q}$	ϕ	ϵ_1	S	$\overline{Q}$
1	0.2	42	0.58	0.45	0.125	50	0.71	0.21	0.18
	0.4	53	0.63	0.18	0.185	50	0.75	0.12	0.185
	0.6	55	0.71	0.10	0.20	50	0.81	0.071	0.20
	0.8	50	0.815	0.048	0.235	45	0.85	0.039	0.21
	1.0	30	0.965	0.0063	0.28	40	0.945	0.0095	0.23
1/2	0.2	45	0.57	0.84	0.265	50	0.71	0.43	0.355
	0.4	55	0.63	0.40	0.31	52	0.76	0.21	0.35
	0.6	55	0.71	0.20	0.335	50	0.81	0.11	0.38
	0.8	50	0.815	0.084	0.425	45	0.86	0.054	0.41
	1.0	30	0.905	0.011	0.50	40	0.945	0.0125	0.465
1	0.2	45	0.575	2.5	0.37	62	0.70	1.16	0.53
	0.4	45	0.65	1.0	0.51	55	0.75	0.59	0.53
	0.6	45	0.745	0.41	0.54	58	0.80	0.33	0.53
	0.8	40	0.845	0.13	0.61	52	0.86	0.12	0.574
	1.0	30	0.965	0.021	0.75	44	0.945	0.025	0.58

*For $m = 0$ see Table 19.1.

When a bearing is absolutely stable, the whirl ratio approaches zero and Fig. 19.31b shows the variation of the whirl ratio at the threshold of instability. The whirl ratios with the 3-lobe and the elliptical bearings share similar characteristics. They both rise sharply at $S = 0.1 - 0.2$ and remain fairly constant thereafter. The elliptical bearing which is the less desirable for lightly loaded applications is seen to have a whirl ratio in excess of 0.5, thus a worse ratio than the 3-lobe bearing. The whirl ratio is an important parameter in determining the stability threshold for a flexible rotor which is always lower than that for a rigid rotor. Thus, Fig. 19.31b can be viewed as the highest possible stability that can be achieved with these bearings.

Tilting Pad Bearings. Unlike the previously considered designs, the tilting pad bearing is a generic name which covers many permutations. Its primary characteristic is that the individual pads are not fixed in position, but are pivot-supported so that during operation not only does the shaft move in response to operational conditions, but so do the pads, and each pad in a different fashion. A general picture of a pivoted shoe bearing is shown in Fig. 19.32. The complexity of the design is partly evidenced in the configuration of a single pad given in Fig. 19.32b.

Several things ought to be noticed. In the first place, the criterion of h_{min} as a measure of load capacity somewhat loses its meaning here since this h_{min} is not a

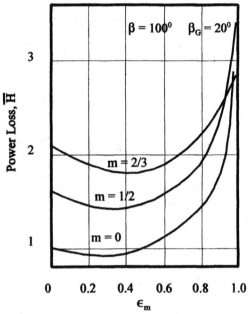

FIGURE 19.30 Power loss in three-lobe bearings.[24]

fixed distance; rather the film thickness over the pivot, h_p, is a geometrically fixed point. Under excessive load, given that the pad at h_{min} can yield whereas at h_p it cannot, failure is more likely to occur at h_p. Thus, the critical quantity here is perhaps h_p rather than h_{min}. The next thing to realize is that the center of curvature of the pad is not fixed in space; when the pad rocks above the pivot, its center of curvature moves either in a positive or negative angular direction, shown in Fig. 19.32b by $\pm\gamma$. Next, should the preload be too low, some of the pads on the top of the bearing may become unloaded, in which case, as shown in Fig. 19.33, the fluid film frictional moment about the pivot will make the leading edge of the pad scrape against the journal and cause "flutter," obviously an undesirable contingency. The condition that the top pads not be unloaded is dictated by the amount of preload and shaft position; the lower the preload and the higher ϵ, the more likely that some of the pads will become unloaded. Figure 19.34 gives a sample graph, in terms of ϵ_m and m, when a pad is likely to be unloaded. In this respect, loading between the pads, when the journal may reach $\epsilon_m > 1$, is a more undesirable mode of operation.

The number of possible design parameters and operating modes in a tilting pad bearing is large. Some of these design options are:

- Number of pads, $3 < n < 8$
- Angular extent of pads, including the option of a variation of β for various pads

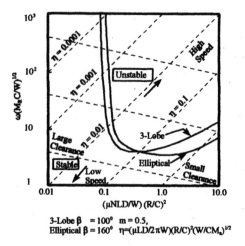

a. Stability Regimes

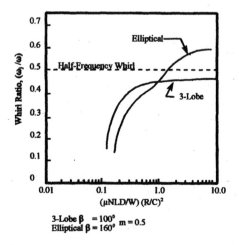

b. Instability Threshold Frequency

FIGURE 19.31 Stability characteristics of elliptical and 3-lobe bearings.[1]

- Load vector passing between pads or over a pad
- Central or eccentric pivot location, i.e., a choice of value for (β_p/β)
- (L/D) ratio
- Preload, $m = 1 - (C_m/C)$
- Pad inertia, which often determines the ability of the pad to follow or track journal motion.

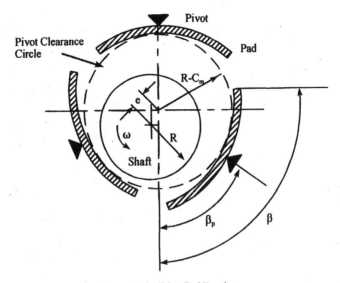
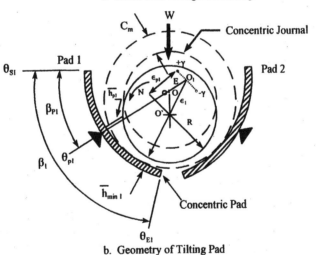

FIGURE 19.32 Tilting pad journal bearing.

19.4.3 Thrust Bearings

Much of what has been said previously about journal bearings applies also to the behavior of thrust bearings. It thus remains only to point out the differences that arise due to the different geometry of a thrust bearing shown in its generic elements in Fig. 19.35.

Thrust bearings are simpler to handle in that no cavitation occurs and in that it is sufficient to solve only for one pad (for a parallel runner, all pads are identical). The geometry of the film, on the other hand is more varied. The film shape in

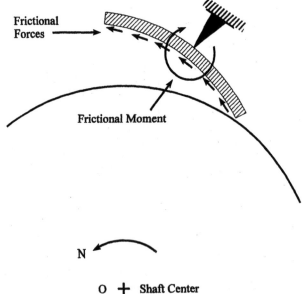

FIGURE 19.33 Unloaded tilting pad.

journal bearings is more or less universal, namely that prevailing between two eccentric circular cylinders. In thrust bearings—it can be anything—shapes with one or two directional tapers, with or without flats, crowned profiles, pocket bearings, and finally tilting pad designs. Another simplification with thrust bearings is that no instability problems, such as occur with journal bearings, arise in their operation. There is, thus, no need to evaluate stiffness and damping.

The Reynolds equation for thrust bearings has to be written in polar instead of rectangular coordinates. In parallel to journal bearings, turbulence is accounted for on a point-by-point basis, here a function of r as well as θ

$$\frac{\partial}{\partial \bar{r}}\left[\frac{\bar{r}\bar{h}^3 G_z}{\mu}\left(\frac{\partial \bar{p}}{\partial \bar{r}}\right)\right] + \frac{1}{\bar{r}}\frac{\partial}{\partial \theta}\left[\frac{\bar{h}^3 G_x}{\mu}\left(\frac{\partial \bar{p}}{\partial \theta}\right)\right] = 6\frac{\bar{r}}{(L/R_2)^2} \quad (19.32)$$

where G_x, G_z are the turbulence coefficients in the θ and r directions respectively, both functions of the Reynolds number given by:

$$Re = \rho r \omega h/\mu = f(r, \theta)$$

The expressions above differ from those for the journal bearing in that they have a dependence on r. This is due to the variation of the velocity profile and flow with r and to the film thickness being a function of both coordinates, r and θ.

Tapered Land Bearings. The simplest tapered land bearing is one which has a constant angular taper, or:

$$h(\theta) = h_2 + \delta_\theta(1 - \theta/\beta); \quad \delta_\theta = (h_2 - h_1) \quad (19.33)$$

with its geometry as shown in Fig. 19.35. This equation, independent of r, is valid

PRINCIPLES OF BEARING DESIGN   19.47

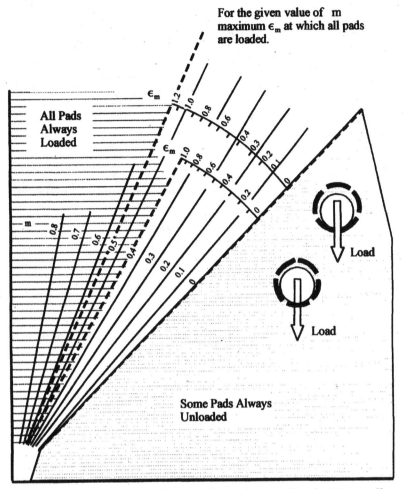

FIGURE 19.34 Regime of unloaded pads in a five-pad tilting pad bearing.[18]

for a bearing surface with a circumferential taper alone. As will be shown later, the exact shape of the fluid film between fixed values of h_1 and h_2 does not affect the results appreciably. Thus, by their simplicity, the one-dimensional taper solutions provide a useful key for evaluating the performance of thrust bearings in general.

The several crucial parameters in journal bearings are β, (L/D) and (e/C). Parallel quantities appear in thrust bearings, namely, β, the angular extent of the pad; (L/R_2); and (h_2/δ_θ) with δ_θ (like C) being a geometric quantity and h_2 being the trailing film thickness at which the bearing is run. It should be also noted that here

$$h_{\min} = h_2 \qquad (19.34)$$

Solutions for the tapered land bearing are given in Table 19.4, where:

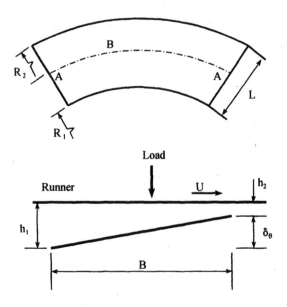

Cut A-A

FIGURE 19.35 Elements of tapered land thrust bearing.

$$\overline{Q}_r = (Q_r / \pi R_2 N L \delta_\theta) \tag{19.35}$$

is the side leakage, the index R_1 indicating the leakage along the inner radius, and R_2 indicating the leakage along the outer radius. The total side leakage is then

$$Q_r = [\overline{Q}_r |_{R_1} + \overline{Q}_r |_{R_2}] \pi R_2 N L \delta_\theta$$

The leakage out the end of the pad, Q_2, is given by:

$$Q_2 = 0.5 \pi N L h_2 (R_1 + R_2) + \overline{Q}_{2P} \pi R_2 N \delta_\theta \tag{19.36}$$

where the first right-hand term is the shear flow and does not involve any computer obtained coefficients. The value of $\overline{Q}_{2P}$ can be obtained from Table 19.4 by subtracting $\overline{Q}_r$ from $(\overline{Q}_r + \overline{Q}_{2P})$.

Table 19.5 shows the relative load capacities and friction of three different thrust bearing configurations. One is a plane slider, i.e, an inclined rectangular block; the second, a slider with an exponential film profile; and the third is the tapered land geometry of Eq. (19.33). As seen, the results for a given value of (h_1/h_2) are nearly identical, confirming the assertion that once h_1 and h_2 are fixed, the exact variation in h between these values is not of great importance.

In all of the above results, it should be noted that P is the unit pressure given by:

TABLE 19.4 Solutions for Tapered Land Thrust Bearings[25]

All values are for single-pad

$\dfrac{L}{R_2}$	$\dfrac{h_1}{\delta_\theta}$	β (deg)	$\dfrac{\mu N}{P}\left(\dfrac{L}{\delta_0}\right)^2$	$\overline{Q}_r$ at R_1	$\overline{Q}_r$ at R_2	$\dfrac{\overline{Q}_r}{\overline{Q}_{2p}}$	Center of pressure $\overline{\theta}$	Center of pressure $\overline{r}$	$\dfrac{H\delta_\theta}{\pi\mu N^2 R_2^4}$
1/3	1	80	1.423	0.34	0.40	0.87	0.64	0.37	2.44
		55	1.180	0.32	0.44	0.84	0.025	0.45	1.685
		40	0.947	0.28	0.81	0.81	0.61	0.49	1.20
		30	0.870	0.235	0.75	0.75	0.605	0.51	0.95
	1/2	80	0.321	0.35	0.47	0.87	0.71	0.37	3.94
		55	0.257	0.32	0.44	0.84	0.69	0.47	2.70
		40	0.225	0.29	0.40	0.79	0.67	0.50	2.00
		30	0.211	0.245	0.36	0.74	0.66	0.51	1.57
	1/4	80	0.0855	0.35	0.47	0.87	0.78	0.41	5.96
		55	0.714	0.32	0.44	0.83	0.76	0.45	4.25
		40	0.0652	0.29	0.41	0.78	0.74	0.505	3.23
		30	0.0635	0.235	0.36	0.70	0.73	0.52	2.54
	1/8	80	0.0278	0.36	0.48	0.85	0.83	0.465	8.51
		55	0.0247	0.33	0.45	0.81	0.815	0.50	6.23
		40	0.0238	0.29	0.41	0.75	0.795	0.51	4.88
		30	0.0242	0.25	0.37	0.67	0.78	0.565	3.91

$r = (r - R_1)/L$
$\theta = \theta/\beta$

TABLE 19.4 Solutions for Tapered Land Thrust Bearings[25] (*Continued*)

All values are for single-pad

$\dfrac{L}{R_2}$	$\dfrac{h_1}{\delta_\theta}$	β (deg)	$\dfrac{\mu N}{P}\left(\dfrac{L}{\delta_0}\right)^2$	$\overline{Q}_r$ at R_1	$\overline{Q}_r$ at R_2	$\overline{Q}_r + \overline{Q}_{2p}$	Center of pressure $\overline{\theta}$ $r = (r - R_1)/L$ $\theta = \theta/\beta$	Center of pressure $\overline{r}$	$\dfrac{H\delta_\theta}{\pi\mu N^2 R_2^4}$
	1	80	1.72	0.23	0.405	0.75	0.62	0.48	2.90
		55	1.494	0.19	0.36	0.69	0.61	0.51	1.96
		40	1.435	0.145	0.31	0.61	0.60	0.53	1.47
		30	1.489	0.11	0.20	0.57	0.59	0.55	1.13
	1/2	80	0.402	0.23	0.41	0.74	0.685	0.46	4.72
		55	0.3585	0.19	0.33	0.61	0.67	0.52	3.33
		40	0.352	0.15	0.31	0.60	0.655	0.53	2.49
1/2		30	0.370	0.11	0.26	0.53	0.65	0.55	1.92
	1/4	80	0.1138	0.24	0.42	0.72	0.755	0.48	7.32
		55	0.1062	0.20	0.27	0.65	0.735	0.52	5.29
		40	0.1080	0.15	0.32	0.56	0.72	0.54	4.065
		30	0.1103	0.11	0.27	0.49	0.71	0.56	3.18
	3/8	80	0.0402	0.25	0.42	0.70	0.81	0.50	10.81
		55	0.0399	0.20	0.28	0.62	0.78	0.53	8.06
		40	0.0423	0.16	0.32	0.53	0.77	0.55	6.30
		30	0.0470	0.11	0.27	0.44	0.765	0.57	5.01

1	80	2.240	0.12	0.35	0.60	0.61	0.50	3.06
	55	2.185	0.082	0.295	0.53	0.60	0.55	2.12
	40	2.320	0.052	0.245	0.48	0.59	0.58	1.57
	30	2.590	0.033	0.200	0.44	0.59	0.61	1.20
1/2	80	0.538	0.13	0.35	0.58	0.67	0.51	5.07
	55	0.537	0.084	0.30	0.51	0.66	0.56	3.59
	40	0.578	0.053	0.25	0.45	0.65	0.59	2.70
	30	0.653	0.034	0.20	0.40	0.645	0.61	2.07
1/4	80	0.1598	0.13	0.36	0.56	0.735	0.53	8.00
	55	0.1655	0.087	0.30	0.46	0.72	0.57	5.70
	40	0.1820	0.055	0.25	0.40	0.71	0.60	4.43
	30	0.2085	0.035	0.21	0.38	0.705	0.62	3.46
1/8	80	0.0599	0.14	0.365	0.53	0.79	0.55	12.07
	55	0.0649	0.09	0.31	0.44	0.78	0.58	8.98
	40	0.0737	0.056	0.25	0.35	0.765	0.61	6.94
2/3	30	0.0861	0.036	0.21	0.29	0.75	0.63	5.47

TABLE 19.5 Performance of Thrust Bearings with Various Film Configurations[25]

α	Plane slider*	Exponential slider**	Sector pad***
		$\overline{P} = \dfrac{PL^2 h_2^2}{\mu \omega R_2^4}$	
2.00	0.0810	0.0819	0.0826
2.50	0.113	0.1137	0.106
2.85	0.135	0.135	0.125
		$\overline{F} = \dfrac{Fh_2}{\mu \omega R_2^4}$	
2.00	0.66	0.81	0.78
2.50	0.74	0.875	0.825
3.04	0.84	0.95	0.88

*$h = \alpha x$
**$h = k_1 e^{k_2}$
***$h = h_2 + \delta(1 - \theta/\beta)$

$$P = W/Area = 360 W_T / [n\beta\pi L(R_2 + R_2)]$$

where β is in degrees, W_T is the total load on the thrust bearing and n the number of pads. Also it should be noted that the data for flow and power loss in Table 19.4 are for a single pad so that the total flow and losses are

$$Q_T = nQ_{pad} \qquad H_T = nH_{pad}$$

Composite Tapered Land Bearings. A more practical and preferred thrust bearing geometry is a tapered land bearing having tapers in both the circumferential and radial directions with a flat portion at the end of the film. Its advantages are: (1) it has higher load capacity; (2) has lower side leakage and, (3) at low speed and during starts and stops it provides a flat surface for supporting the load, thus minimizing wear. The geometry of such a bearing is shown in Fig. 19.36. Its film thickness is given by:

$$h = h_{11} - [(h_{11} - h_{12})/L](r - R_1)$$
$$- \left[\frac{h_{11} - [(h_{11} - h_{12})/L \div (r - R_1) - h_2}{b\beta} \right] \theta \qquad (19.37)$$

$$\text{for } 0 \leq \theta \leq b\beta$$
$$h = h_2 \text{ constant} \qquad \text{for } b\beta \leq \theta \leq \beta$$

Normalizing all h's by h_2 and all radii by R_2 we have:

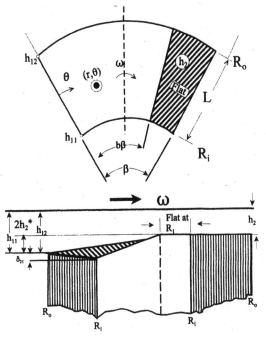

FIGURE 19.36 Composite tapered land bearing.

$$\bar{h} = \bar{h}_{11} - (\bar{h}_{11} - 1)\frac{\theta}{b\beta} - \bar{\delta}_r \left[\frac{\bar{r} = (L/R_2) - l}{(L/R_2)}\right]\left(1 - \frac{\theta}{b\beta}\right) \quad (19.38)$$

$$\text{for } 0 \leq \theta \leq b\beta$$
$$\bar{h} = 1 \quad \text{for } b\beta \leq \theta \leq \beta$$

The expression for $\bar{h}$ has, as seen, three arbitrary parameters:

- $\bar{h}_{11} = (h_{11}/h_2)$ – the dimensionless maximum film thickness at the lower left corner
- $\bar{\delta}_r = (h_{11} - h_{12})/h_2$ – the radial taper along the leading edge $\theta = 0$
- b = the friction of β tapered

In an optimization study in which both load capacity and lower power losses were considered, the following desirable proportions for the above three parameters were arrived at:

$$\bar{h}_{11} = 3.0 \quad \bar{\delta}_r = 0.5 \quad b = 0.8$$

Physically, the above numbers imply a maximum film thickness at $(R_1, 0)$ of three times the one over the flat; an outward decrease in film thickness at the leading edge half that of h_2; and a flat portion equal to 20% of the pad's angular extent. The performance of such a bearing for the case of a 40° bearing pad an (*OD/ID*) ratio of 2 is given in Table 19.6. The table provides data for both turbulent and laminar operation.

The following comments will, perhaps, be useful:

- The values of the Reynolds number

$$Re_1 = \rho R_1 \omega h_2 / \mu_1$$

- The losses as represented by H_C in column 4 include the losses over a 10° oil groove; the losses H over the pad only can be obtained from the last column in Table 19.6.
- The flow Q_{IN} represents the inflow at $\theta = 0$. The outflow will, be given by:

$$Q_2 = Q_{IN} - (Q_{R_1} - Q_{R_2})$$

- The lowest value of Re_1 given is 500. This value is close to laminar operation.
- For the total bearing, the values of W, H_C, Q and H should all be multiplied by the number of pads.

Tilting Pad Bearings The comments made about the tilting pad journal bearing regarding its complexity and large number of parameters apply equally well to the thrust bearing. However, in the case of a pivoted thrust pad such as the one shown

TABLE 19.6 Composite Tapered Land Thrust Bearings[22s] $(R_2/R_1) = 2$; $\beta = 40°$, $h_{11} = 3$, $\delta_r = 0.5$; $b = 0.8$

Re_1	$\dfrac{Wh_2^2}{\mu R_1^4 \alpha}$	$\dfrac{H_c * h_2}{R_1^4 \omega^2}$	$Q/R_1^2 h_2 \omega$ at $\theta = 0$	at R_1	at R_2	$\dfrac{H}{W\omega h_2}$
500	0.192	3.92	1.58	0.296	0.437	17.6
	0.182	3.88	1.58	0.294	0.445	18.3
	0.175	3.78	1.59	0.293	0.451	18.5
	.0168	3.68	1.59	0.292	0.456	18.7
	.0151	3.42	1.60	0.290	0.469	19.1
1500	0.337	8.45	1.61	0.324	0.455	21.5
	0.307	7.93	1.62	0.321	0.465	21.6
	0.289	7.59	1.62	0.319	0.471	21.8
	0.275	7.31	1.63	0.318	0.476	21.9
	0.240	6.63	1.64	0.314	0.490	22.3
3500	0.567	15.6	1.63	0.339	0.462	23.2
	0.499	14.0	1.64	0.334	0.475	23.2
	0.463	13.2	1.64	0.331	0.482	23.3
	0.435	1.25	1.65	0.329	0.488	23.4
	0.369	11.2	1.66	0.325	0.504	23.7

*H_c includes losses over a 10° oil groove. All results are per individual pad.

in Fig. 19.37, an additional complication overshadows the other difficulties; there is theoretically no solution to a planar centrally pivoted sector. This can be deduced from the pressure profile sketched in Fig. 19.37a. Such a profile must always be asymmetrical with respect to the center of the pad; an asymmetrical pressure profile would impose a moment about the pivot tending to align the pad parallel to the runner. However, a parallel pad produces no hydrodynamic pressures, thus making the working of such an arrangement impossible. Yet such centrally pivoted, planar surface thrust bearings are widely used and they perform exceedingly well.

Various theories have been advanced and stratagems employed to explain the workings of these bearings and obtain a solution to the problems. Among these are:

- *Thermal or density wedge*—The variation in viscosity or density of the oil is often credited with generating hydrodynamic forces in the parallel film. At best, such effects produce forces which come nowhere near the heavy loading supported by such bearings.

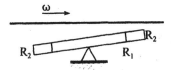

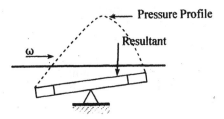

a. Zero load capacity of centrally pivoted slider

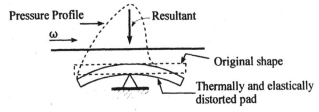

b. Generation of a hydrodynamic film due to thermal and elastic distortions

FIGURE 19.37 The hydrodynamics of a tilting pad thrust bearing.

- *Thermal and elastic distortion of the pad*—As shown in Fig. 19.37b, thermal and elastic stresses may crown a pad, so that in essence it produces a convergent-divergent film. In that case, it is possible for the resultant load to pass through the pivot and the pad can support a load. However, such bending can occur only with very thin pads or extremely high temperature gradients. Yet such bearings perform satisfactorily even with very thick pads and under conditions of minimal heat generation.
- *Incidental effects*—There are a number of incidental features which may play a more important role than the above theoretical explanations. Among these are:
  - Machining inaccuracies on the faces of both runner and bearing and rounded off edge at entrance to the pad, which in effect constitute a built-in taper
  - Misalignment between runner and pads during assembly or during operation
  - Pivot location not exactly at 50% of pad angular extent

These factors would combine to generate hydrodynamic forces and they are perhaps the most likely explanation for the satisfactory working of tilting pad thrust bearing.

19.5 LOW-SPEED BEARINGS

One of the requirements in the bearing described in the previous section is a proper lubrication system. This includes a pump delivering oil at 10 to 50 psi supply pressure with all the accompanying equipment such as oil tank, filters, piping, sump and cooling arrangements. When the bearings run at relatively low speed involving low power dissipation and therefore low bearing temperatures—as in fans, blowers and some compressors—one can simplify the system by employing oil-ring lubrication. This consists of a self-contained oil delivery package placed adjacent to the bearing which dispenses with all the auxiliary equipment required for a more demanding operation.

Figure 19.38 shows the components of an oil-ring lubrication setup. The ring, riding on the top of the exposed shaft, is a sort of viscous drag device that lifts oil from the sump and deposits it on the shaft. It is clear that in comparison to a pressurized supply system where the oil is distributed along an axial groove, here the amount of oil lifted is not sufficient to provide the bearing with a complete oil film, and therefore an important parameter in oil ring operation is the amount of oil the bearing receives relative to what it needs for a full film. This is called the starvation ratio and is given by

$$\hat{Q}_z = Q_z/Q_{zF}$$

where Q_z is the side leakage under starved conditions and Q_{zF} is the side leakage for a full film.

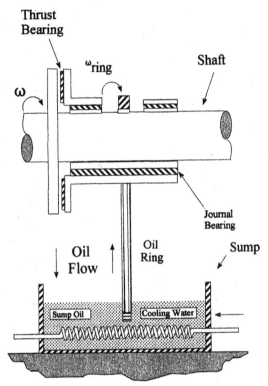

FIGURE 19.38 Oil-ring lubrication system.

19.5.1 Regimes of Operation

The value of starvation ratio depends on shaft and ring speeds. Due to the centrifugal effects of the rotating ring and attached oil one can distinguish four regimes of ring behavior, as shown generically in Fig. 19.39. The characteristics of these regimes are as follows:

Regime I. At the low end of journal rotation, there is contact between ring inner surface and the journal, and the linear speeds of the two mating surfaces are about the same. There is thus a linear rise in ring rpm with the rpm of the journal. The oil delivered by the ring rises throughout this regime. At the upper end of Regime I, ring speed and oil delivery reach a local maximum.

Regime II. At the beginning of this regime, direct frictional drag yields to a state of boundary lubrication between ring and journal. Due to this, slippage occurs and ring speed drops. Since the speed has decreased, so too does the amount of oil delivered by the ring. However, with further rise in journal speed, a full hydrodynamic film is established between journal and ring. The reduced viscous friction (the friction coefficient may drop from 0.1 to 0.01) and the larger film between the

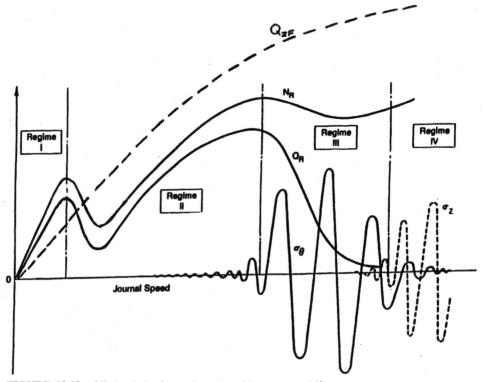

FIGURE 19.39 Oil ring behavior as function of journal speed.[15]

mating surfaces bring about a rapid increase in both ring speed and oil delivery. Once again, at the upper end, a local maximum in ring speed is achieved. Oil flow, however, at the end of this regime is an absolute maximum and represents the highest possible oil delivery by the ring.

Regime III. The drop in ring speed and oil delivery following Regime II is associated with the onset of ring oscillations in the plane of rotation. While small oscillations already appear during the trailing portion of Regime II, the values of σ_θ become, within a short span, very large, bringing about a drastic reduction in oil delivery. While, due to these planar oscillations, the ring speed drops only slightly, the oil delivery is affected to such a point that at the end of this regime it approaches asymptotically zero. The angular swing of the ring at its maximum values can be of the order of $\pm 5°$ to $\pm 10°$ with an oscillatory frequency equal to that of ring rotational frequency.

Regime IV. This regime, essentially beyond our interest, is characterized by conical and translatory vibrations of the ring. While the oscillatory motion abates and the ring speed once again tends to increase with journal speed, oil delivery is essentially zero. The frequency of both the conical and translatory (or axial) vibrations is that of ring rotational frequency. Starting with the oscillatory vibrations and proceeding through the two other modes of instability, the violent motion of the ring causes splash and a throw-off of oil from the surface of the ring, and partly also from the journal, so that little oil reaches the bearing.

Figure 19.40 presents the hydrodynamics of a starved journal bearing vis-à-vis a full film or flooded condition. For a fixed load, speed and supply oil temperature, the deviations of a starved bearing from one operating with a full fluid film are as follows:

- The film starts later and terminates earlier, that is $\theta_1 > \theta_{IF}$ and $\theta_2 < \theta_{IF}$, producing in essence something similar to a partial bearing, though the upstream, and sometimes also the downstream, boundary conditions are different.
- The eccentricity increases, producing smaller values of h_{min}.
- The attitude angle decreases, yielding a more vertical locus of shaft center.
- Since oil supply is equivalent to side leakage, there is a reduction of bearing side leakage and consequently an increase in fluid film temperatures. This is somewhat mitigated by a reduction in power loss due to a shorter extent of the fluid film.

Table 19.7 gives a set of solutions for a wide range of loads and levels of starvation. It is seen that the effects of starvation are much more pronounced at light and moderate loads, which are due to the fact that, at heavy loads, the full film extends over a narrower arc and the pressure gradients at Q_1 being higher, a lower amount of Q_1 is required to form a film. The loci of shaft center, both for constant oil supply values $\overline{Q}_1$ and for constant loads, W, are given in Fig. 19.41. It

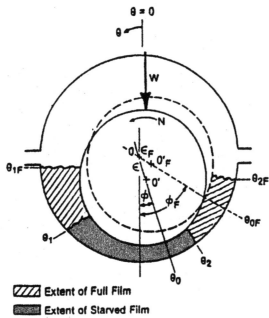

FIGURE 19.40 Full and starved fluid films.

TABLE 19.7 Theoretical Performance of Starved Journal Bearings[16]

μ = const; β = 150°; (L/D) = 0.93

$\overline{W}$	$\hat{Q}_z\%$	ϵ	ϕ, degs	$\overline{Q}_1$	$\overline{Q}_z$	θ_1, deg	β_z, degs
0.491	0	1.00	0	0	0.	180	0
	0.9	.903	3	.100	$.899 \times 10^{-3}$	175	13
	3.1	.808	6	.200	$.309 \times 10^{-2}$	171	22
	12.0	.625	11	.400	.0122	163	40
	29.0	.446	18	.600	.0289	155	60
	53.0	.272	28	.800	.0534	144	86
	100*	.081	78	1.03	.100	105	150
1.965	0.	1.00	0	0	0	180	0
	0.7	.906	5	.100	$.208 \times 10^{-2}$	172	16
	2.8	.818	9	.200	.0817	167	29
	12.4	.657	17	.40	.0357	155	57
	30.3	.516	26	.600	.0875	144	82
	50.2	.395	37	.800	.162	129	101
	100*	.285	57	1.03	.289	105	150
9.825	0.	1.00	0	0.	0.	100	0
	1.5	.914	8	.100	$.659 \times 10^{-2}$	168	25
	6.4	.846	14	.200	.0291	158	46
	15.6	.793	19	.300	.0697	148	69
	28.0	.753	24	.400	.125	139	82
	60.0	.702	31	.600	.267	122	108
	100*	.672	37	.809	.445	105	132
34.39	0.	1.00	0	0.	0	180	0
	4.8	.929	11	.100	.0192	160	38
	19.4	.896	17	.200	.0802	144	73
	38.7	.881	21	.300	.162	131	79
	61.3	.073	23	.400	.253	120	86
	100*	.866	26	.560	.413	105	110
98.25	0	1.0	0	0.	0	180	0
	12.9	.954	12	.100	.0463	149	50
	38.7	.947	16	.200	.138	131	70
	66.1	.946	17	.300	.235	118	87
	93.8	.945	18	.400	.335	107	100
	100*	.944	18	.423	.357	105	101

*Full fluid film.

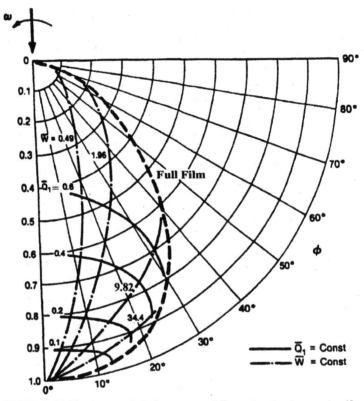

FIGURE 19.41 Locus of shaft center at different levels of starvation.[15] $L/D = 0.93; \beta = 150°$

is seen that starvation displaces the locus of shaft center inward of the full film line, i.e., towards higher eccentricities and lower values of attitude angle ϕ.

From the above and the parametric studies described in Section 19.9 the following conclusions can be drawn:

- Except at very low speeds, most oil ring bearings operate under starved conditions.
- The load capacity of oil ring bearings, first increases then decreases with rising shaft speed.
- The locus of shaft center of oil ring bearings is much closer to the vertical axis than in full film bearings.
- An optimum in the (L/D) ratio exists in oil ring bearings which ranges from 0.6 to 0.8.
- The effects of starvation are much more pronounced at low and intermediate loads than at high loadings.

19.6 HIGH-SPEED AND HIGH-TEMPERATURE BEARINGS

This chapter will consider bearings suitable for operation at extreme speed and temperature ranges. These two parameters may occur either together or independently, that is, although usually high speed implies high temperatures, the reverse is not always so. One can have lower or moderate velocities, but the environment may be such that the bearing will be exposed to high temperatures and the fluid film, the lubricant and bearing materials must be able to cope with it. The other important consideration in high-speed bearings is that of stability. As will be seen below, the likelihood of bearing instability, known as *half-frequency whirl*, rises with rotational speed. This becomes particularly intense when speeds twice the natural frequency of the system are reached.

The kinds of bearings that are possible candidates for such applications are gas bearings, either hydrodynamic or hydrostatic; compliant surface geometries; and magnetic bearings.

19.6.1 Gas Bearings

The differential equation governing the behavior of gas bearings is that given by Eq. (19.39). For liquids, $\rho =$ constant and the density terms fall out of the equation. In gas bearings, it varies with both pressure and temperature. In most cases, the perfect gas equation is applicable, or

$$\frac{P}{\rho} = \mathcal{R}T \tag{19.39}$$

Two factors make the gas film in bearings isothermal; one is the low heat generation; the other is the high thermal capacity of the bearing shell as compared to the tiny volume of gas in the film. We therefore have

$$\rho = \mathcal{R}T \cdot P = const \cdot P \tag{19.40}$$

and Eq. (19.39) becomes

$$\frac{\partial}{\partial x}\left[\frac{ph^3}{\mu}\left(\frac{\partial P}{\partial x}\right)\right] + \frac{\partial}{\partial z}\left[\frac{ph^3}{\mu}\left(\frac{\partial p}{\partial z}\right)\right] = 6U\frac{\partial(ph)}{\partial x} \tag{19.41}$$

All the solutions given subsequently are based on this expression with the proper boundary conditions applied to each specific geometry.

As was pointed out in Section 19.3, unlike with liquids, gas bearing behavior depends on the ambient pressure. Thus, for example load capacity increases with a rise in p_a. A new dimensionless parameter now makes an appearance which governs gas bearing behavior given by:

$$\Lambda = \frac{6\mu\omega}{P_a}\left(\frac{R}{C}\right)^2 \qquad (19.42)$$

called the Bearing number. Thus, along with geometry and such variables as (L/D) ratio, load, speed, etc., the value of p, or, in dimensionless form, Λ, constitutes now an additional input.

Full Circular Bearings. These bearings are the most commonly used if for no other reason that they are easy to manufacture, requiring no grooves or holes for lubricant supply. In obtaining a solution it is only required that at the sides of the bearing the hydrodynamic pressures fall to ambient pressure p, in most cases the atmosphere, P_a. Figures 19.42 and 19.43 give the load capacities and frictional losses for full (360°) gas journal bearings for a range of (L/D) ratios from 1/2 to 2 and for the entire spectrum of possible Λ values. It can be shown that when $\Lambda \to 0$, the gas bearing solution approaches that of a liquid; thus the solutions in the figures comprise cases from liquid lubricants to gases of very high compressibility. The reason that the (L/D) ratios range as high as 2 is to compensate for the inherently low load capacity for gas bearings.

Non-Circular Geometries. As was pointed out in Section 9.4, elliptical and 3-lobe geometries are often resorted to because of their higher stability characteristics. This is particularly desirable with gas bearings which tend to become unstable at high speeds. The solutions given in Section 19.4.2 for these bearings are for centrally loaded cases, that is when the load vector passes midway through the bottom lobe. However, this is not the optimum mode of loading; better results can be obtained when the bearing is so positioned in the housing that the load is made to pass through the bottom lobe at an angle ϕ_L, see Fig. 19.44, called the load angle.

Load Capacity. There is a rather large number of independent parameters when dealing with noncircular gas bearings. Assuming even, as was done here, that the space or slots between the individual lobes occupy a negligible portion of the arc, i.e., assigning to the elliptical bearing two arcs of 180° span each, and to the 3-lobe bearing, three symmetrical (they do not have to be equal) arcs of 120° each, we are still left with five independent parameters, namely

$$p = p[(L/D), m, \epsilon_B, \alpha_B, \Lambda]$$

A solution for any set of these five parameters will yield the load capacity in the form of the Sommerfeld number S, and the line of action of the resultant force, or load angle ϕ_L with respect to the geometry of the bearing.

Tables 19.8 to 19.12 give the results with regard to load capacity and some of the other bearing performance characteristics. It should be kept in mind that since load capacity means the relation between Sommerfeld number and minimum film thickness, this $h_{\min}$ is provided not by the bearing eccentricity which is unrelated to the surface curvature, but by the value of one of the lobes, i.e.

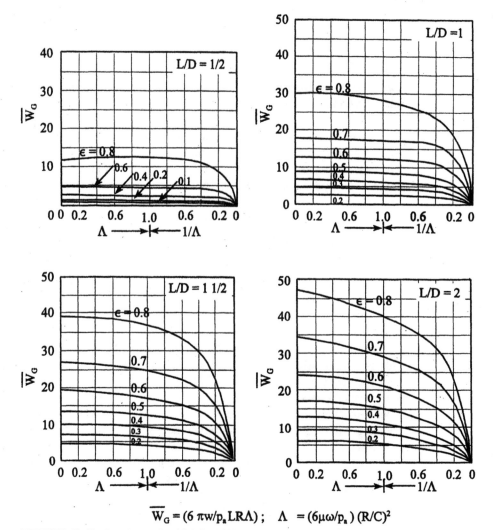

$$\overline{W}_G = (6\pi w/p_a LR\Lambda); \quad \Lambda = (6\mu\omega/p_a)(R/C)^2$$

FIGURE 19.42 Load capacity of full (360°) gas journal bearing.[7]

$$h_{min} = C(1 - \epsilon) \tag{19.43}$$

In order to obtain this h_{min}, we must search from among the two or three lobes the maximum lobe eccentricity ratio. The values of ϵ listed in the tables and figures are these maximum eccentricity ratios. These most often occur in the bottom lobe. In the few cases when the maximum ϵ occurs in Lobe No. 2, this can be identified in the tables from the fact that for the elliptical bearing this would require $\alpha_\beta > 90°$; and for the 3-lobe bearing $\alpha_\beta > 60°$.

While in a circular ungrooved bearing the direction of load application is immaterial, this is not so in the case of non-circular designs. The most common mode of bearing operation is with the load vector parallel to the vertical line of symmetry.

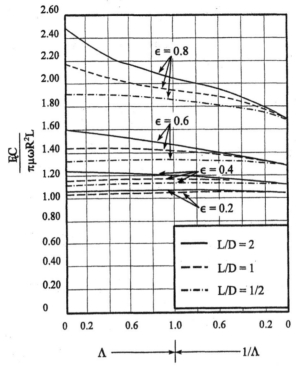

FIGURE 19.43 Friction in full (360°) gas journal bearings.[7]

This is the natural way of mounting the bearing and it is also useful in that it enables two-directional rotation. However, this is not necessarily the optimum arrangement. Applying the load at various angles to the vertical centerline would yield different values of bearing performance. Somewhere an optimum angle exists for the direction of load application and these are shown in Tables 19.9 and 19.11. In practice it means that depending on the parameters S and Λ, the bearing should be rotated with respect to the load vector anywhere from a few degrees to as much as 25° and nearly always in the clockwise direction, in order to obtain the maximum load capacity.

Figure 19.45 summarizes graphically some of the data contained in the tables. Table 19.12 is a practical summary of the various implications contained in the previously discussed results. In practice, one is usually confronted with the given requirements of speed, load, ambient conditions, etc. In other words, S and Λ are fixed. Given these parameters, Table 19.12 shows what eccentricities one can obtain by using a circular or non-circular design.

Stiffness Characteristics. As was done with load capacity and friction, the stiffnesses of the various bearings will be considered at identical Sommerfeld numbers for all the designs. This means that they will be evaluated at different values of S. This is pertinent from a practical viewpoint since the designer wants to know what the stiffness of the bearing will be under the given conditions of load, speed, etc.

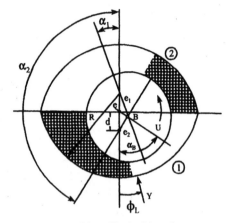

a. The elliptical bearing

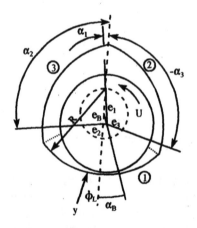

b. Three-lobe bearing

FIGURE 19.44 Nomenclature for non-circular bearings.

regardless of where the journal positions itself in the bearing clearance as a consequence of the imposed operating conditions. Since shaft displacement in different directions produces different responses, here the displacement will be considered in the direction of the load vector. Thus here the definition of the spring constant is given by $K = (dF/de)$, where F is the response force to a displacement along the load line. Table 19.13 and Fig. 19.46 give the results. We see immediately the profound improvement in stiffness in the non-circular over the circular design. In the region of low eccentricities where instability usually occurs (low load, high speed), the value of the spring constants for the elliptical and 3-lobe designs are nearly an order of magnitude higher.

TABLE 19.8 Centrally Located Elliptical Gas Bearings[26]

	$L/D = 1 \quad m = 1/2$								
	α_B at $\Lambda =$			ϵ at $\Lambda =$			α at $\Lambda =$		
ϵ_B	1/2	1	3	1/2	1	3	1/2	1	3
0.1	75	60	30	0.53	0.56	0.59	10	9	5.
0.2	75	60	27	0.58	0.62	0.68	19	16	7.5
0.3	70	50	20	0.67	0.73	0.79	25	18	7.5
0.4	50	30	—	0.82	0.87	—	22	13	—
0.45	30	—	—	0.92	—	—	14	—	—
	S at Λ			$\overline{G}$ at Λ			$\overline{E} = \Lambda$		
ϵ_B	1/2	1	3	1/2	1	3	1/2	1	3
0.1	0.283	0.250	0.282	0.079	0.127	0.122	3.28	3.28	3.21
0.2	0.135	0.117	0.128	0.0703	0.125	0.101	3.41	3.40	3.37
0.3	0.0772	0.0625	0.700	0.0455	0.0620	0.0717	3.70	3.76	3.86
0.4	0.0341	0.0270	—	−0.0204	0.0444	—	4.54	4.90	—
0.45	0.0147	—	—	0.0412	—	—	6.27	—	—

Special Design. Several unorthodox configurations which have in the past been used on high-speed equipment, including automotive gas turbines, are bearings with grooved surfaces and foil bearings. Figures 19.47a and 19.47b show a herringbone grooved journal bearing and two versions of spirally grooved thrust bearings. In both designs, the bearing or runner surface consists of a lattice of grooves and ridges. From a hydrodynamic point of view, the geometry essentially consists of a series of step bearings though, unlike with conventional steps, these are at an angle to the direction of motion. One of the achievements of such a design is that the fluid is being driven away from the edges of the bearing, minimizing side leakage and raising load capacity. By a proper orientation of the grooves, the fluid can be pumped away from either the inner or outer periphery or from both edges. The advantages of these bearings also lie in the fact that, whereas an ordinary geometry has poor stability characteristics for a concentric shaft position ($\epsilon = 0$), the herringbone bearing is superior to a conventional bearing at low eccentricities. These designs can, of course, be used also with liquid lubricants.

Hydrostatic Bearings

Thrust Bearings. Hydrostatic gas bearings are subject to an instability called pneumatic hammer. It is therefore necessary that their recess volume be kept to a minimum. Referring to Fig. 19.48, this means that r_1 and δ are small. The entrance

TABLE 19.9 Optimally Loaded Elliptical Gas Bearings[26]

			$L/D = 1$	$m = 1/2$			
Λ	β_B	ϵ_B	ϵ	α	S	ϕ_L	$\overline{G}$
1/2	0.1	80	0.53	11	0.297	−3	0.0798
	0.2	80	0.57	20	0.143	−4	0.0735
	0.3	80	0.63	28	0.0879	−7	0.0608
	0.4	80	0.69	35	0.0581	−10	0.0408
	0.45	80	0.73	37.5	0.0474	−13	0.0279
1	0.85	80	0.52	9	0.362	−12	0.137
	0.10	80	0.53	11	0.308	−12	0.131
	0.175	75	0.57	17	0.160	−9	0.124
	0.20	75	0.575	20	0.149	−13	0.120
	0.29	75	0.64	26	0.0888	−12	0.0995
	0.40	80	0.69	35	0.0619	−18	0.0753
	0.475	75	0.77	36	0.0411	−19	0.0204
3	0.09	80	0.52	10	0.449	−23	0.198
	0.11	80	0.53	12	0.336	−23	0.196
	0.22	75	0.60	21	0.163	−21	0.168

flow area $2\pi r_1 h$ from the recess into the clearance will then become more restrictive than the orifice area ($\pi d_s^2/4$) in which case the bearing is said to be inherently compensated. With an inherently compensated bearing the recess pressure p_o will equal the supply pressure p_s and the drop ($p_s - p_1$) across the bearing film is governed by the equation:

$$m = C_D A_o p_s \left\{ \frac{2\gamma}{(\gamma - 1)\mathcal{R}T} \left[\left(\frac{p_1}{p_s}\right)^{2/\gamma} - \left(\frac{p_1}{p_s}\right)^{\gamma+1/\gamma} \right] \right\} \qquad (19.44)$$

where m is the mass flow of the gas in lbm/s; A_o is the entrance throat area $2\pi r_1 h$ in in.²; T is the gas temperature in °R; and γ is the ratio of specific heats.

To avoid pneumatic hammer gas thrust, bearings must be designed with inherent compensation. For a design with an (r_1/r_2) ratio of 0.1. Figures 19.49 and 19.50 provide appropriate design charts, given in terms of a parameter B defined as

$$B = \frac{12\mu A_o C_D}{h^3 p_a} \left[\left(\frac{2\gamma \mathcal{R}T}{\gamma - 1}\right)^{1/2} \frac{\ln(r_1/r_2)}{\pi} \right] \qquad (19.45)$$

The chart in Fig. 19.49 presents the dimensionless flow, m', for various values of $\bar{p} = (p_s/p_a)$ while Fig. 19.50 gives load capacity. Chart 5-10 offers values of bearing stiffness K.

TABLE 19.10 Centrally Located Three-lobe Gas Bearings

$L/D = 1 \quad M = 1/2$

ϵ_B	α_B at $\Lambda =$			ϵ at $\Lambda =$			α at $\Lambda =$		
	1/2	1	3	1/2	1	3	1/2	1	3
0.1	70	44	0.57(2)*	0.56	0.58	112(2)	8.4	7.1	
0.2	67	42	0.64(2)	0.64	0.66	106(2)	14.9	11.7	
0.3	63	37	0.71(2)	0.73	0.76	99(2)	18.3	13.5	
0.4	53	—	0.81	0.85	—	23	17.2		

ϵ_B	S $\Lambda =$			$\overline{G}$ at $\Lambda =$			$\overline{F}$ at $\Lambda =$		
	1/2	1	3	1/2	1	3	1/2	1	3
0.1	0.293	0.301	0.437	0.0358	0.0902	0.0736	3.62	3.60	3.55
0.2	0.134	0.135	0.194	0.0368	0.0860	0.0675	3.86	3.82	3.76
0.3	0.0734	0.0734	0.106	0.0370	0.0773	0.0575	4.30	4.25	4.13
0.4	0.0370	0.0352	—	0.0314	0.0550	—	—	—	—

*(2) indicates that minimum film thickness occurs in right-hand lobe.

TABLE 19.11 Optimally Loaded Three-lobe Gas Bearings[26]

Λ	ϵ_B	α_B	ϵ	α	S	ϕ_L	$\overline{G}$
1/2	0.1	60	0.56	9	0.284	9	0.0358
	0.2	60	0.62	16	0.128	7	0.0364
	0.3	60	0.70	22	0.0720	3	0.0362
	0.4	60	0.78	26	0.0418	−5	0.0336
1	0.085	55	0.55	7	0.385	1.5	0.0580
	0.10	55	0.56	8	0.300	1	0.0902
	0.175	60	0.61	14.5	0.165	−4	0.0564
	0.20	60	0.62	16	0.140	−5	0.0862
	0.28	60	0.68	21	0.0893	−6.5	0.0521
	0.30	60	0.70	22	0.0810	−8	0.0790
	0.40	60	0.78	26	0.0477	−13	0.0425
	0.455	60	0.83	28	0.0347	−17	0.0359
3	0.10	50	0.57	8	0.444	−6	0.117
	0.225	50	0.67	15	0.171	−7	0.0655
	0.30	60	0.70	22	0.121	−17	0.0578

TABLE 19.12 Comparison of Load Capacity of Various Gas Bearings; $(L/D) = 1$ Values of ϵ for Given Λ and S

Λ	s	$m = 0$ Circular	Elliptical[a] $m = 1/2$		3-lobe[a] $m = 1/2$	
			$\lambda_{L=\theta}$	Optimum	$\phi_L = 0$	Optimum
1	0.365	0.2	0.54 (−170)	0.525 (−162)	0.55 (−175)	0.545 (−172)
	0.162	0.4	0.59 (−48)	0.56 (−40)	0.615 (−54)	0.605 (−51)
	0.866	0.6	0.67 (−12)	0.635 (−6)	0.70 (−17)	0.685 (−14)
	0.0381	0.8	0.815 (−2)	0.77 (+4)	0.84 (−5)	0.815 (−2)
3	0.365	0.25	0.57 (−132)	0.535 (−114)	0.595 (−38)	0.57 (−128)
	0.162	0.475	0.65 (−37)	0.59 (−24)	0.68 (−43)	0.645 (−36)
	0.0860	0.655	0.75 (−14)	0.685 (−4)	0.80 (−22)	0.755 (−15)

[a] Numbers in parentheses refer to percentage reduction in load capacity from a circular design.

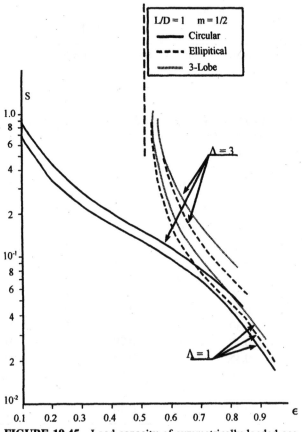

FIGURE 19.45 Load capacity of symmetrically loaded gas bearings.[26]

TABLE 19.13 Values of $\overline{K} = K/2\mu LN\,(C/R)^3$ [26]

| | | | \multicolumn{4}{c}{$m = 1/2$} | |
| | | | Elliptical | | 3-Lobe | |
Λ	S	$m = 0$ Circular	$\phi_L = 0$	Optimum, ϕ_L	$\phi_L = 0$	Optimum, ϕ_L
\multicolumn{7}{c}{For $L/D = \Lambda = 1$}
1	0.365	2.8	26.2	10.2	22.4	—
	0.162	5.4	29.4	—	28.6	20.0
	0.0866	12.8	42.4	26.4	37.4	26.4
	0.0381	87.2	96.0	71.6	115.6	152.6
3	0.365	5.5	20.0	26.4	16.4	—
	0.162	6.8	25.8	31.2	46.2	40.4

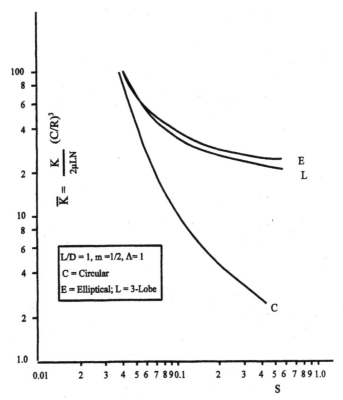

FIGURE 19.46 Spring constants for symmetrically loaded gas bearings.[26]

19.72 CHAPTER NINETEEN

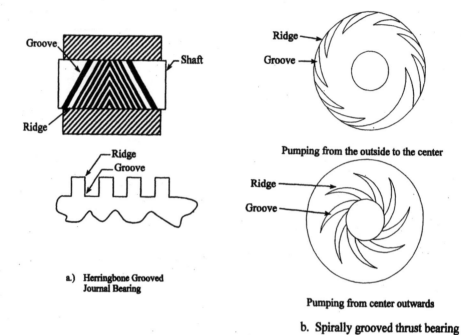

a.) Herringbone Grooved Journal Bearing

b. Spirally grooved thrust bearing

FIGURE 19.47 Special bearing designs.

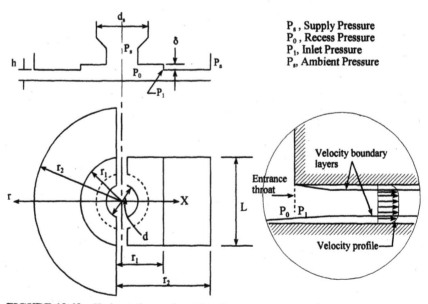

P_s, Supply Pressure
P_0, Recess Pressure
P_1, Inlet Pressure
P_a, Ambient Pressure

FIGURE 19.48 Hydrostatic gas thrust bearing.

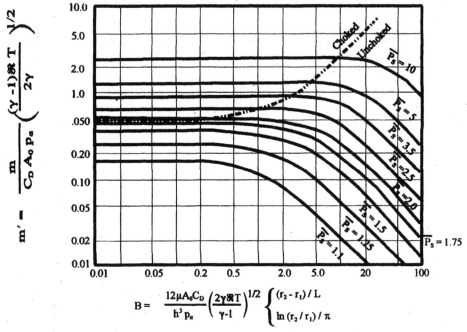

FIGURE 19.49 Mass flow rate in hydrostatic thrust bearing.[34]

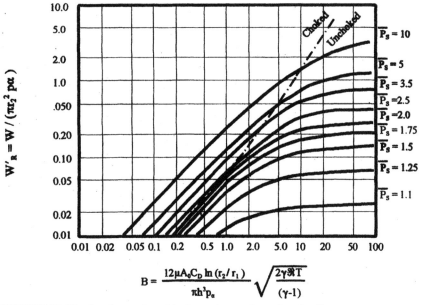

FIGURE 19.50 Load capacity of hydrostatic gas thrust bearing.[34]

To use the charts for design purposes one first needs to know value of p_s to determine B for the maximum obtainable stiffness from Fig. 19.51. From Fig. 19.50 one then determines the r_2 that will carry the required load W. The value of (h/r_2) is set within the limits of $0.5 \cdot 10^{-3} < (h/r_2) < 2 \cdot 10^{-3}$. With r_2 and h established, the actual dimensional values of stiffness k and flow rates can be calculated from Figs. 19.49 and 19.50. Refinements are possible by a few more iterations.

Journal Bearings. Typical configurations for journal bearings are shown in Fig. 19.52a and b. The former is an inherently compensated design; the latter has an orifice restrictor. Here, too, the recess must be small, of the order of 10% of an incompressible fluid pocket. Capillary restrictors are not used because they cause pneumatic hammer.

The analysis of these bearings is very complex and possible only with numerical methods. A typical set of performance curves for an $L/D = 1/2$ is shown in Figs. 19.53 and 19.54. The first shows stiffness as a function of a restrictor coefficient Λ_s for various ratios of (p_s/p_a). This Λ_s is similar to the B parameter in thrust bearings for it represents the ratio of fluid film resistance to the resistance of the restrictor. The parameter δ appearing in the coordinates is defined as

$$\delta = (d^2/4hr_1) \qquad (19.46)$$

which gives the ratio of the throat area for an orifice restrictor to the throat area

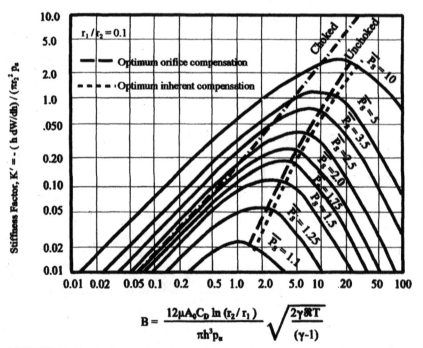

FIGURE 19.51 Stiffness of orifice and inherently-compensated hydrostatic thrust bearing.[34]

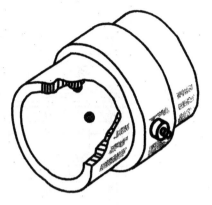

a. Inherently compensated

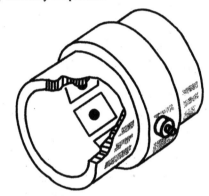

b. Orifice compensated

FIGURE 19.52 Two bearing geometries of hydrostatic gas journal bearings.

represented by the restriction of the bearing film. Its inclusion in the figures permits one to use these charts for both modes of restriction.

The K values presented are the center stiffness of the bearing ($\epsilon = 0$). Since the stiffness remains essentially constant up to $\epsilon = 1/2$, load capacity of the bearing can be calculated from $W = \epsilon CK$; or since it is not recommended that the bearings operate at higher eccentricities, the load capacity is given by $W = 0.5\ CK$ where K is obtained from Fig. 19.53.

19.6.2 Complaint Surface Bearings (CSB)

In high-speed, high-temperature applications the CSB's have the advantage of developing higher load capacities than with fixed geometry gas bearings. Since CSB's have a complex structure consisting of a spring-like substructure with an overlying flexible surface, there exists a great variety of permutations on any given design. The presentation here will start with two basis models of a journal and a thrust

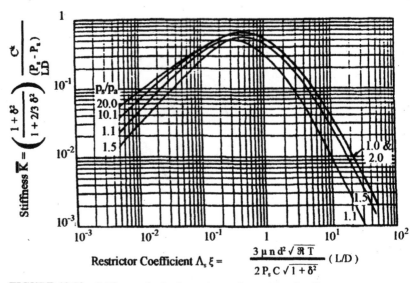

FIGURE 19.53 Stiffness of a hydrostatic gas journal bearing.[34]

bearing, to be followed with some more elaborate geometries. In all cases the lubricant will be that of air.

Foil Journal Bearing. The solution of this bearing is based on Eq. (19.39), coupled with additional expressions accounting for the elastic behavior of the top and bottom surfaces. The journal bearing is portrayed in Fig. 19.55 and its solution is based on the following postulates.

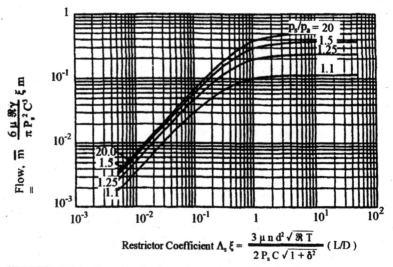

FIGURE 19.54 Flow in a hydrostatic gas journal bearing.[34]

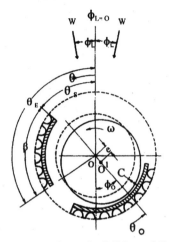

a. Nomenclature for Foil Journal Bearing

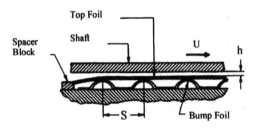

b. Basic Elements of Bearing

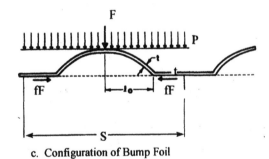

c. Configuration of Bump Foil

FIGURE 19.55 Configuration of a foil journal bearing.

- The stiffness of the foil is uniformly distributed around the circumference and is linear with the amount of deflection.
- The foil is assumed not to "sag" between bumps but to follow the deflection of the bumps.
- In response to the hydrodynamic pressures, the deflections are local, i.e., they depend only on the force acting directly over a particular point.

Under the above conditions the variation in h is due to the eccentricity e and the deflection of the foils. We then have:

$$h = C + \theta \cos(\theta - \phi_o) + K_1(p - p_a) \qquad (19.47a)$$

where K_1 is a constant reflecting the structural rigidity of the bumps, given by:

$$K_1 = \frac{\alpha C}{p_a}; \quad \alpha = \frac{2p_a s}{CE}\left(\frac{l_o}{t}\right)^3 (1 - \nu^2) \qquad (19.47b)$$

K_1 is then the compliance of the bearing with the quantities, s, l_o and t portrayed in Fig. 19.55c.

The Nominal Film Thickness. In rigid journal bearings, the minimum film thickness is a clear and fixed quantity. It occurs at the line of centers and its value is constant across the axial width of the bearing. Also, generally, the film thickness anywhere is constant in the z direction. Since in our case pressures cause proportional deflections of the bearing surface, the film thickness in the interior of the bearing, where pressures are highest, will be larger than at the edges ($z = \pm L/2$); also since the maximum pressures occur near the line of center, the film thickness in the interior at $\theta = \theta_0$, are larger than at angular position $\theta = \theta_N$. Figure 19.57 shows a three-dimensional film thickness plot for a 120° pad in which, while film thickness at the edge ($z = L/2$) is small over most of the pad area, the surface has been deflected into much larger values of h.

For these reasons, a nominal film thickness h_N will be defined as the minimum film thickness that occurs along the bearing centerline, i.e, at $z = 0$ at various values of α as shown in Fig. 19.58. While h_{min} for the rigid case occurs at $\theta = 180°$, with increasing values of α, the value of this h_{min}, or our h_N, shifts downstream and increases in value; at $\alpha = 5$, it is twice the value of the rigid case and has shifted downstream by nearly 100°. This should be kept in mind later on, when load capacity, i.e., the $W - h_N$ relation is plotted; an increase in load while increasing ϵ may also produce an increase in the nominal film thickness.

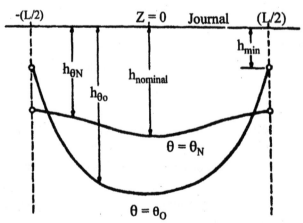

FIGURE 19.56 Minimum and nominal film thicknesses.

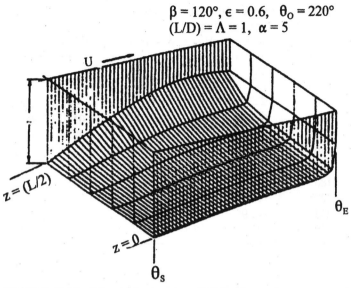

FIGURE 19.57 Film thicknesses in a 120° bearing pad.

Active and Effective Bearing Arc. Compliant foil bearings suffer a penalty in their ability to generate hydrodynamic pressures whenever the pad arc commences in a diverging region. The effect can be seen in Table 19.14 which show that by shifting in a 360° bearing the line of centers from 180 to 270°, there was a loss in load capacity of nearly 30% as well as a reduction in h_N.

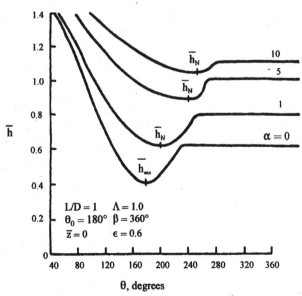

FIGURE 19.58 Location of nominal film thickness.[9]

TABLE 19.14 Effect of Load Angle in 360° Bearings[9]

$\epsilon = 0.6;\ (l/d) = \Lambda = \alpha = 1;\ \theta_1 = 0,\ \theta_E = 360°$

θ_o	ϕ_L	ϕ	θ_2	$\overline{P}_{max}$	h_N	$W \times 10^2$	$T \times 10$
0	—	—	—	0	0.40	0	—
20°	163.1°	4°	32°	1.018	0.41	1.04	42.9
45°	141.6°	11.6°	74°	1.087	0.48	11.5	38.1
60°	129.8°	20.2°	95°	1.133	0.52	21.4	33.9
90°	107.2°	27.2°	142°	1.201	0.58	40.0	27.2
180°	33.1°	33.1°	240°	1.253	0.62	56.8	24.6
220°	−11.9°	28.1°	272°	1.245	0.62	54.9	23.7
245°	−41.9°	23.1°	300°	1.228	0.61	50.0	23.7
270°	−72.9°	17.1°	325°	1.201	0.58	40.1	25.5

In designing a foil bearing, if the eccentricity is fixed for the particular application, it is best to start the bearing at $\theta_S = \phi$ (for a vertical load); if the eccentricities are liable to vary, some compromise value of $\theta_S = 0$ can be chosen.

Performance Characteristics. There are six geometric, structural, and operational parameters relevant to a foil journal bearing. These are β, α, (L/D), Λ, ϕ_L, and number of pads. There is also the eccentricity ratio and the attitude angle ϕ, the latter tied to the load angle ϕ_L. A set of standard conditions consisting of

$$(L/D) = \Lambda = \alpha = 1;\quad \epsilon = 0.6$$

is used, and any parametric variation commences from this set of reference values.

The Full Bearing. Table 9.15 gives a detailed listing of the performance of a vertically loaded ($\phi_L = 0$) full 360° bearing as a function of (L/D), α, and ϵ. Note should be taken of the fact that the start of the bearing, that is θ_s is so chosen as to avoid idle ($p = p_s$) regions at the upstream portion of the bearing. In effect, this requires that $\theta_s = \phi$. The case of nonvertically loaded foil bearings $\phi_L \neq 0$, is given in Table 19.15. Some of the noteworthy points emerging from these tabulations are:

- *Effect of α.* While in terms of ϵ there is a drastic drop in load capacity with a more compliant bearing, in terms of h_N there is actually an increase in load capacity.

  At large values of α, $\alpha > 10$, the load the bearing can support is low, due to the fact that the flexible foil deflects sufficiently to maintain high film thicknesses even at large eccentricities. Thus from a design standpoint, it may be advisable to use high compliance bearings at low loads; high loads, however, can be supported only with bearings of low values of α. In highly compliant bearings (particularly at high L/D ratios), an increase in eccentricity may produce an increase in h_N, a phenomenon opposite to rigid bearings where h_{min} is the inverse of ϵ.

TABLE 19.15 Performance of Λ 360° Foil Journal Bearing[9]

$\Lambda = 1; \quad \phi_L = C$

ϵ	α	$\phi = \theta_1$	$\theta_2 - \gamma$	$\overline{P}_{max}$	$\overline{h}_N$	$\overline{W} \cdot 10^2$	$\overline{T} \cdot 10$
0.3	0	63.5	81.5	$L/D = 0.5$	0.70	4.4	9.34
	1	59.0	87.0	1.046	0.72	4.2	9.15
	5	48.5	100.5	1.043	0.80	3.2	8.53
	10	60.0	114.0	1.037	0.85	2.9	8.09
	20	34.0	114.0	1.025	0.94	2.1	7.54
				1.017			
0.6	0	40.0	62.0	1.25	0.40	17.9	15.85
	1	36.0	76.0	1.144	0.51	13.7	13.93
	5	32.0	97.0	1.073	0.68	8.3	8.19
	10	30.0	105.0	1.05	0.795	6.1	7.33
	20	27.0	114.0	1.033	0.90	4.2	6.54
0.9	0	12.0	52.0	3.73	0.10	157.3	26.1
	1	19.0	71.0	1.33	0.41	34.7	13.8
	5	21.0	91.0	1.12	0.64	14.8	9.8
	10	21.0	99.0	1.077	0.76	9.8	8.5
	20	21.0	108.0	1.048	0.91	6.3	7.3
			$L/D = 1.0$				
0.3	0	37.0	97.0	1.173	0.70	27.9	22.7
	1	49.0	104.0	1.107	0.77	23.7	21.2
	5	36.0	117.0	1.061	0.94	14.8	18.6
	10	28.0	120.0	1.041	1.04	10.3	17.5
	20	20.0	132.0	1.025	1.14	6.37	16.5
0.6	0	36.0	77.0	1.539	0.40	94.9	31.1
	1	33.0	95.0	1.253	0.62	56.8	24.6
	5	28.5	112.0	1.114	0.90	28.8	19.1
	10	25.0	117.0	1.074	1.055	19.4	16.9
	20	20.0	120.0	1.046	1.22	12.2	15.0
0.9	0	13.0	59.0	4.850	0.10	504.5	58.2
	1	21.0	86.0	1.434	0.52	102.8	28.1
	5	22.0	94.0	1.154	0.86	42.9	19.5
	10	21.5	108.5	1.103	1.05	27.8	16.7
	20	19.0	127.0	1.063	1.26	17.2	14.4

TABLE 19.15 Performance of Λ 360° Foil Journal Bearing[9] (*Continued*)

$\Lambda = 1; \quad \phi_L = C$

ϵ	α	$\phi = \theta_1$	$\theta_2 - \gamma$	$\overline{P}_{max}$	$\overline{h}_N$	$\overline{W} \cdot 10^2$	$\overline{T} \cdot 10$
			$L/D = 1.5$				
0.3	0	52.0	103.0	1.218	0.70	70.0	33.4
	1	43.0	113.0	1.152	0.82	53.2	30.4
	5	29.0	119.0	1.076	1.03	28.5	26.4
	10	21.0	141.0	1.048	1.13	18.3	24.8
0.6	0	35.0	88.0	1.731	0.40	208.9	45.6
	1	32.0	104.0	1.311	0.68	112.0	34.3
	5	26.0	120.0	1.135	1.00	52.0	26.3
	10	22.0	137.0	1.084	1.18	34.1	23.1
0.9	0	14.0	68.0	5.300	0.10	298.9	85.1
	1	23.0	95.0	1.485	0.56	179.7	39.2
	5	23.0	112.0	1.184	0.96	74.2	26.7

- *Effect of* Λ. The performance of a foil bearing as a function of Λ conforms to the familiar pattern of compressible lubrication. After an initial rise in $\hat{W}$ with an increase in Λ, the load capacity, both in terms of an increase in $\hat{W}$ as well as a rise in $\tilde{h}_N$, tends to flatten off and approach an asymptotic value. The torque, however, rises almost as a linear function of the increase in Λ. The more compliant bearing shows lower power losses due to the prevailing higher film thickness.

The Multipad Bearing. The 3-pad design consists of three 120° arcs; the 5-pad design has five 72° arcs. In each case, the vertical line of symmetry bisects the bottom pad, so that $\phi_L = 0$ represents a load passing through the midpoint of the bottom pad. Tables 19.16 and 19.17 give a spectrum of solutions for the performance of the 3-pad bearing and these results show the following:

- *Variation with load angle.* Because of the cyclic nature of this bearing (symmetry for each 120°) there is much less variation in either W or T with a shift in load angle. In particular, there is no acute loss of load capacity when the line of centers passes between pads. The optimum load angle for $\alpha = 1$ is $\phi_L = -10°$; for $\alpha = 5$ it is $\phi_L = -14°$. The improvement in load capacity over that of central loading ($\phi_L = 0$) is of the order of 10 to 15%.
- *Variation with number of pads.* Figure 19.59 shows the variation of 1-, 2-, and 3-pad bearings as a function of load angle. The plot shows clearly a drop in load capacity with the number of pads, i.e., with a drop in extent of bearing arc β. As seen, the optimum for the 360° bearing occurs at $\phi_L = 0$, at which point the

TABLE 19.16 Performance of a Three-pad Bearing $(L/D) = \Lambda = 1, \beta = 20°$[9]

ϵ	θ_1	θ_L	ϕ	$\overline{P}_{max}$	$\overline{W} \times 10^2$	$T \times 10$
0.3	40	−140.0	79.2	1.073	12.1	21.4
	210	7.2	37.2	1.072	12.7	21.8
	217	−2.3	27.0	1.075	13.7	21.9
	220	−2.3	27.7	1.079	14.1	22.0
	225	−5.2	29.8	1.082	14.6	22.0
	245	−10.0	55.0	1.088	15.0	21.5
	273	−17.4	77.6	1.077	12.6	21.4
0.6	29	−145.0	69.2	1.188	24.0	27.8
	180	44.1	44.1	1,133	25.2	18.6
	200	−2.6	30.6	1.197	37.2	27.7
	2202	−10.6	29.4	1.215	29.8	28.8
	245	−17.0	48.0	1.284	34.9	29.0
	270	−25.5	64.5	1.185	28.7	27.4
0.9	15	−145.0	50.9	1.497	59.3	62.9
	194	0.0	16.3	1.340	69.5	34.7
	210	−10.8	19.2	1.375	76.9	45.7
	230	−14.1	333.9	1.572	74.3	71.8
	245	−14.7	50.5	1.412	52.4	77.3
	260	−26.2	53.8	1.463	55.3	56.6
			$\alpha = 5$			
0.3	38	143.0	74.9	1.049	0.01	20.7
	210	5.1	35.1	1.038	7.52	21.1
	214	0.0	34.3	1.040	7.87	21.1
	220	−5.3	34.7	1.043	8.31	21.2
	225	−8.6	36.4	1.045	8.60	21.2
	245	−13.3	51.7	1.053	9.07	20.8
	275	−11.1	73.9	1.050	8.18	20.7
0.6	12	−145.0	67.4	1.104	15.7	25.4
	180	55.0	55.5	1.048	12.6	16.5
	205	3.6	28.6	1.077	16.1	25.3
	220	−9.7	30.3	1.103	18.3	26.7
	245	−16.3	48.74	1.121	18.4	27.2
	270	−23.1	66.9	1.106	15.9	25.4
0.9	25	164.0	61.0	1.178	24.4	46.5
	200	3.5	23.5	1.122	25.3	34.4
	203	0.0	23.0	1.119	26.3	36.2
	210	−4.3	23.7	1.158	28.1	41.6
	230	−16.3	33.7	1.198	29.7	67.1
	245	−18.9	46.1	1.202	18.2	72.1
	260	−21.5	58.5	1.186	25.2	53.7

TABLE 19.17 Mode of Load of 3-Pad Bearing[9]
$(L/D) = \Lambda = 1; \beta = 120°$ each
Central Loading $\phi_L = 0$; Optimum Loading

α	ε	φ	$\overline{W}$	ϕ_L	φ	$\overline{W}$
1	0.3	37.5	13.8	−10	55.0	15.0
1	0.6	28.5	36.0	−10	29.0	39.8
1	0.9	16.0	68.0	−10	18.5	78.0
5	0.3	35.0	7.8	−14	53.0	9.0
5	0.6	30.0	17.0	−14	41.0	18.6
5	0.9	23.5	26.2	−14	30.0	29.8

torque also reaches it minimum value. The 3-pad bearing, as said previously, reaches an optimum at $\phi_L = -10°$; whereas the 5-pad bearing reaches an optimum at $\phi_L = -15°$.

Stiffness. Table 19.18 gives the values of the four spring coefficients for two values of compliance, the limiting case of $\alpha = 0$, and $\alpha = 1$. The $\alpha = 0$ case differs from a rigid gas bearing in that the subambient pressures are eliminated from the pressure profile. A comparative evaluation of the stability characteristics of the 1- and 3-pad bearings is, of course, best done in a study of a rotordynamic system, particularly when the cross coupling components vary not only in magnitude but also in sign. However, the following items can be deduced from the tabulated Λ data:

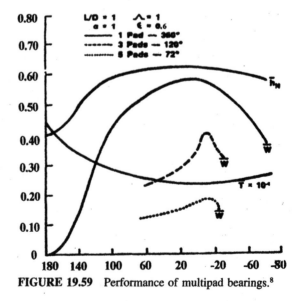

FIGURE 19.59 Performance of multipad bearings.[8]

TABLE 19.18 Values of Spring Coefficients[8]
$(L/D) = \Lambda; \phi_L = 0$

ϵ	α	ϕ	$\overline{W}$	$\overline{K}_{xx}$	$\overline{K}_{xy}$	$\overline{K}_{yx}$	$\overline{K}_{xx}$
				$\epsilon = 360°$			
0.6	0	35.7	0.951	1.920	−0.125	−2.345	3.237
0.75	0	24.1	1.894	3.416	−1.166	−3.989	8.981
0.9	0	12.8	5.055	7.202	−6.024	10.151	44.593
0.6	1	32.1	0.568	1.129	0.174	−0.693	1.130
0.75	1	26.3	0.7833	1.231	0.0254	−0.686	1.378
0.9	1	21.4	1.028	1.268	−0.098	−0.627	1.602
				3-pad-120° each			
0.6	0	26.0	0.635	1.123	−0.092	−2.05	2.635
0.75	0	17.4	1.321	2.102	−0.752	−3.710	7.432
0.9	0	8.6	3.695	4.728	−3.344	−8.768	37.103
0.6	1	25.5	0.359	0.5702	0.0451	−0.758	0.801
0.75	1	20.5	0.511	0.673	−0.017	−0.821	1.051
0.9	1	16.3	0.689	0.759	−0.057	−0.855	1.274

- When plotted against W the K_{yy}'s are about the same for both designs, whereas the K_{xx}'s are lower for the 3-pad configuration.
- With the more compliant case, the K's tend to level off with a rise in eccentricity, the values of the coefficients approaching the structural stiffness of the system.

In general, the advantage of compliant bearings in the area of stability is that levels of stiffness can be selected by the designer via a proper combination of structural and hydrodynamic stiffnesses. Thus, instead of making the bearing's inertia and support suit the inherent stiffnesses of purely hydrodynamic bearings, the designer may try to tailor and adjust bearing stiffness to the demands of his rotor-dynamic system.

Foil Thrust Bearings. Figure 19.60 shows the configuration of the thrust bearing considered next, which resembles that of a conventional tapered land design. All the postulates stated in connection with the journal bearing apply here as well, except, of course, that the film thickness is different. This is now given by

$$h = h_2 + g(r, \theta) + C(p - p_a)$$

where geometric function:

$$g(r, \theta) = (h_1 - h_2)[1 - \theta/b\beta] \quad \text{for } 0 \leq \theta \leq b\beta$$

$$g(r, \theta) = 0 \quad \text{for } b\beta \leq \theta \leq \beta$$

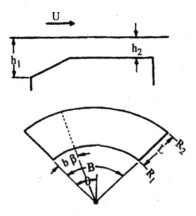

a. Nomenclature of the Thrust Bearing

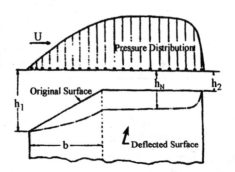

B. The Elastohydrodynamics of a Compliant Foil Bearing

FIGURE 19.60 The geometry of compliant surface thrust bearing.

Geometric Optimization. As shown in Table 19.19, the maximum pressure changes vary little with an increase or decrease in β or the value of h_1. The table also shows that when the total number of pads possible for a given value of β is accounted for in the calculation of the total load capacity, there is little difference in the choice of a particular arc length. The effect is shown in Fig. 19.61 for both a relatively stiff and soft bearing. In both cases, the nominal film thickness does not vary with β. The total load capacity in a full 2π thrust bearing shows a maximum somewhere between 45° and 50°. The effect of the proportion of ramp is shown in Fig. 19.62. For a relatively stiff bearing, the optimum is about 70%, which is close to the case of rigid bearings. For the more compliant case, the optimum is 50%. For high values of Λ, the optimum value of b recedes to values closer to 40%. It can be seen that near the highest values of W, the nominal film

TABLE 19.19 Effect of β on Bearing Performance[9]
$(L/R_2) = 0.5$; $b = 0.5$

$\bar{h}_1$	α^*	Λ^*	β	n	$\bar{W} \times 10^2$	$\bar{T} \times 10^3$	$\bar{p}_{max}$	$\bar{W}_{TOT} \times 10^2$
2.5	4	1.2	30	12	0.259	4.19	1.0288	3.12
			45	8	0.430	6.09	1.0320	3.44
			60	6	0.573	8.02	1.0329	3.44
			75	~5	0.686	9.98	1.0331	3.30
3.0	3-1/3	1.0	30	12	0.257	4.13	1.0293	3.08
			45	8	0.435	5.97	1.0334	3.48
			60	6	0.588	7.81	1.0349	3.53
			90	4	0.808	11.58	1.0349	3.23
4.0	2.5	0.75	30	12	0.238	4.01	1.0285	2.86
			45	8	0.415	5.82	1.0336	3.32
			60	6	0.574	7.56	0.0359	3.44
			90	4	0.814	11.08	1.0372	3.26

thickness is also at its highest, thus reinforcing a general load capacity optimum for compliant bearings at about $b = 0.5$.

The effect of varying $\tilde{h}_1$ is given in Table 19.20. It shows that while the value of $\tilde{W}$ changes little with a variation of $\tilde{h}_1$, the nominal film thickness goes up appreciably, doubling in value for a doubling of h_1. Thus, in terms of customary load capacity criteria, the highest values of $\tilde{h}_t$ seem desirable.

To summarize, the optimum geometry for a bearing with the common *OD* to *ID* ratio of 2 is:

$$\beta = 45°, \quad b = 0.5, \quad \tilde{h}_1 > 10$$

Performance Characteristics. The performance of a compliant foil bearing for the optimized parameters of $\beta = 45°$ and $b = 0.5$ are given in Table 19.21 for a wide range of parameters $\tilde{h}_1$, α^* and Λ^*. In terms of h_2 normalization the range of Λ extends to nearly 1000 and that of α to over 60. Figure 19.64 is a performance plot in terms of the basic variables involved in the bearing. The drop of load capacity with an increase in film thickness and with a decrease in the value of Λ are trends known from other studies of gas bearings. What is particularly noteworthy in Fig. 19.63 is the effect of α on load capacity. While at moderate Λ's high values of α yield the highest load capacity, at high Λ the optimum α is some intermediate value, in our case $\alpha^* = 1$. Note should be taken that all quantities, i.e., Λ^*, α^*, and h_N^* are all normalized by the geometric ramp height δ; and that the Fig. 19.63 plots contain implicitly various values of h_1. The relation between $\tilde{h}_1$ and $\tilde{h}_N$ is given in Fig. 19.64. This graph can be used to determine various ramp heights for different points of Fig. 19.63. The $\tilde{h}_1 - \tilde{h}_N$ graphs support the conclusion

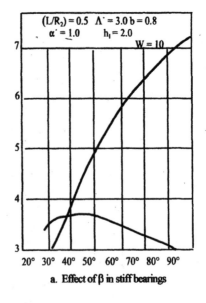

FIGURE 19.61 Effect of β on pad and total load.[8]

of the previous section as to the desirability of using high values of $\tilde{h}_1$ since they yield high nominal film thicknesses. It also shows that the higher $\alpha°$ and $\Lambda°$ are, the higher the film thickness.

Finally, Fig. 19.65 gives a plot of the spring constant for the bearing. The stiffness of the bearing is, of course, a function of both the structural stiffness as represented by K_B and of the hydrodynamic film stiffness. Since they are in parallel,

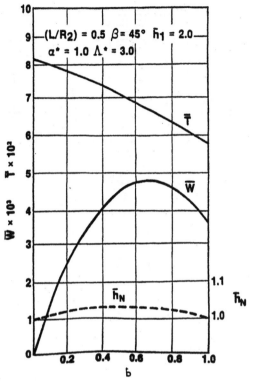

FIGURE 19.62 Effect of extent of ramp on performance of CS thrust bearing.[8]

high loads would tend to flatten the values of K for the softer bearings, leaving essentially the structural stiffness K_B as the dominating spring constant.

Advanced CSB Designs. The construction of CSB's lends itself to a number of modifications that can enhance a particular performance characteristic in accordance with operational requirements. Due to their analytical complexity and space

TABLE 19.20 Effect of $\bar{h}_1$ [8]
$\beta = 45°$ $L/R_2 = 0.5$, $b = 0.5$ $\alpha^* = 1.0$, $\Lambda^* = 10$

$\bar{h}_1$	$\overline{W} \times 10^2$	$\overline{T} \times 10$	$\bar{h}_N$	$\overline{K}^* \times 10^3$
10	8.169	44.5	5.106	114
15	8.626	72.2	7.675	116
17	8.730	83.3	8.730	116
19	8.811	94.5	9.730	117
20	8.845	100	10.265	122

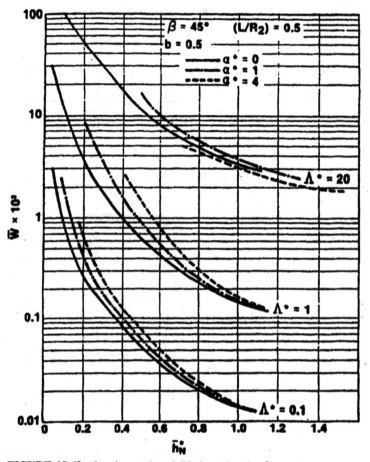

FIGURE 19.63 Load capacity of CS thrust bearing.[8]

limitations, they cannot be discussed to any extent but a mere listing of some of them will give an idea of the range of possibilities latent in this group of bearings.

CSB's with Variable Direction Stiffness. The bump foil design can be manipulated to provide a wide range of desirable dynamic properties. One such arrangement to vary the stiffness in both radial and circumferential directions is shown in Fig. 19.66. The stiffness gradient permits the formation of a variable hydrodynamic wedge in accordance with variation in load or speed. As speed increases, the particular arrangement can be made to increase film convergence which enhances stability precisely when it is needed. In one such application, an advanced air-lubricated journal bearing reached speeds of 135,000 rpm carrying a unit load of close to 100 psi. These advance bearings are capable of operation above rotors' bending critical speeds (super-critical operation), which is beyond any hydrodynamic bearing capabilities.[35]

Much of the above applies also to thrust bearings. A proper thrust bearing geometry is one that has a taper in the circumferential direction followed by a flat

TABLE 19.21 Compliant Foil Thrust Bearing Performance
$\beta = 45°$, $B = 0.5$, $(L/R_2) = 0.5$

		$\alpha^* = 0$				$\alpha^* = 1$				$\alpha^* = 4$				$\alpha = 20$		
Λ^*	$\bar{W} \times 10^2$	$\bar{h}_N^*$	$\bar{T} \times 10$	$\bar{K}^* \times 10^3$	$\bar{W} \times 10^2$	$\bar{h}_N^*$	$\bar{T} \times 10$	$\bar{K}^* \times 10^3$	$\bar{W} \times 10^2$	$\bar{h}_N^*$	$\bar{T} \times 10$	$\bar{K} \times 10^3$	$\bar{W} \times 10^2$	$\bar{h}_N^*$	$\bar{T} \times 10$	$\bar{K}^* \times 10^3$
							$\bar{h}_1 = 2$									
0.1	0.014	1.0	0.024	0.314	0.104	1.001	0.024	0.314	0.014	1.004	0.024	0.32				
1.0	0.144	1.0	0.241	3.24	0.145	1.011	0.24	3.20	0.147	1.046	0.23	3.35				
10.0	0.632	1.0	2.40	38.6	1.52	1.114	2.27	31.5	1.214	1.326	2.04	17.5				
20.0	3.33	1.0	4.79	76.4	2.736	1.197	4.34	45.5	1.828	1.461	3.78	18.8				
40.0	6.246	1.0	9.50	125.0	4.238	1.276	8.21	51.5	2.406	1.574	7.05	16.4				
							$\bar{h}_1 = 5$									
0.3	0.192	0.25	0.35	12.2	0.204	0.268	0.341	14.0	0.206	0.316	0.31	2.3	0.152	0.452	0.24	5.24
1.0	2.343	0.25	3.50	171.0	1.833	0.378	2.72	77.7	1.164	0.538	2.07	29.8	0.513	0.834	1.43	6.92
10.0	22.2	0.25	33.5	1230.0	6.633	0.635	17.3	104.0	2.697	0.905	12.9	26.1	0.854	1.15	10.2	3.64
20.0	34.66	0.25	65.2	1730.0	8.135	0.71	30.8	95.7	3.362	0.978	28.6	19.7				
40.0	46.4	0.25	127.0	2060.0	9.231	0.758	57.0	78.0	3.612	1.015	44.9	14.7	0.915	1.18	38.8	1.16
							$\bar{h}_1 = 10$									
0.1	0.66	0.111	1.63	91	0.661	0.161	1.30	61.1	0.571	0.192	1.15	40.5	0.464	0.234	0.98	25.0
1.0	8.818	0.111	16.1	1,120	2.170	0.307	7.81	113.0	2.309	0.38	6.50	64.8	1.619	0.485	5.38	35.6
10.0	60.38	0.111	150.0	6,120	8.109	0.566	44.5	114.0	5.345	0.695	37.5	56.7	3.323	1.825	32.3	25.0
20.0	86.84	0.111	292.0	8,240	9.457	0.630	79.5	95.6	6.020	0.785	68.0	45.0	3.627	0.88	59.8	18.5
							$\bar{h}_1 = 20$									
0.1	2.084	0.0526	6.60	575	1.103	0.16	3.54	90.0	0.843	0.162	2.91	52.2	0.623	0.209	2.39	29.7
1.0	22.35	0.0526	64.6	4,720	3.876	0.282	18.3	128.0	2.705	0.364	14.8	70.6	1.83	0.404	12.1	37.8
10.0	134.8	0.0526	605.0	26,600	8.845	0.54	100.0	117.0	5.674	0.666	83.6	56.2	3.467	0.790	71.7	24.3
20.0	185.5	0.0526	1179.0	34,200	10.24	0.596	179.0	97.5	6.280	0.722	152.0	44.0	3.733	0.835	113.0	17.7

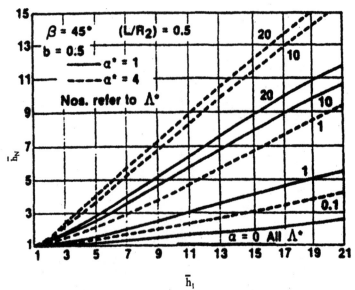

FIGURE 19.64 Relation between $\bar{h}_1$ and $\bar{h}_N$.[8]

portion. A CSB can be constructed with a foil possessing stiffening elements at the trailing edge, as shown in Fig. 19.66b. The stiffening elements placed between the top and bump foils provide a variable stiffness gradient from the leading to trailing edge yielding the desired converging shape.

CSB's with Controlled Coulomb Damping. A foil bearing can be constructed to improve internal damping and thus enhance its stability characteristics. This can be done by affecting the Coulomb damping due to the relative motion between the top and bump foil surfaces, as well as between the bump foil and the housing (see Fig. 19.67). This relative motion occurs, of course, when the bearing is loaded and the foils are radially deflected. To improve the friction characteristics of this relative motion, the rubbing surfaces are sputter coated with copper, silver, or some other high friction material. Sometimes the surfaces of the journal and mating top foil are coated with dry lubricants to minimize friction on start-up and shutdown. To further enhance stability, the bump foil can be circumferentially split along axial lines to improve alignment and axial compliance. A single pad design of this variety has carried unit loadings up to 100 psi.

CSB's with Cantilevered Leaves. A bending dominated foil bearing is the cantilevered leaf type design. A journal bearing of this type is shown in Fig. 19.68a. Here each leaf is preformed with a specific radius to induce a desirable film profile. The thrust bearing variety is shown in part b. It is essentially a thin plate with springs mounted below it which is separated by a backing plate from a foil assembly plate to which the leaves are attached. The plates are in the form of annular rings and the leaves in the form of annular sectors. The leaves should be thin

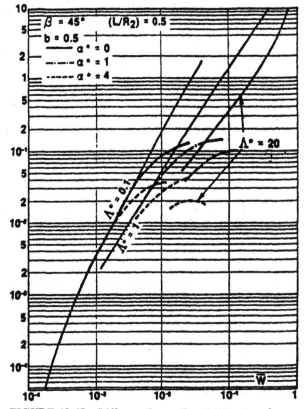

FIGURE 19.65 Stiffness of compliant foil bearings.[8]

enough to permit considerable flexibility but still thicker than the hydrodynamic film.

19.6.3 Magnetic Bearings

To illustrate the potential of magnetic bearings, a specific design example will be followed through. It will provide orientation and a guide for general cases where their use is contemplated, particularly with regard to controlling instability and resonances at high speeds.

General Principles. Practical AMBs are mostly the attractive type. Radial AMBs generally adopt an 8-pole stator configuration as shown in Fig. 19.69. Both the stator and journal consist of laminations of ferromagnetic material. The journal is shrunk on a shaft without windings. The use of laminations reduces eddy current, which not only causes power loss, but degrades the dynamic performance of the bearing. The stator poles are separated into four quadrants. In each quadrant, the

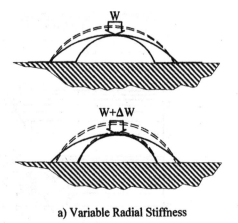

a) Variable Radial Stiffness

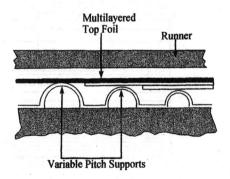

b) Variable Longitudinal Stiffness

FIGURE 19.66 Variable support stiffness in compliant bearings.

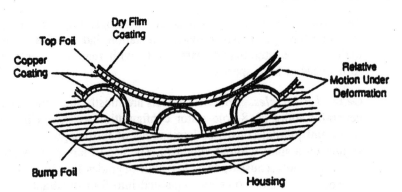

FIGURE 19.67 Mechanism for Coulomb damping in foil journal bearing.

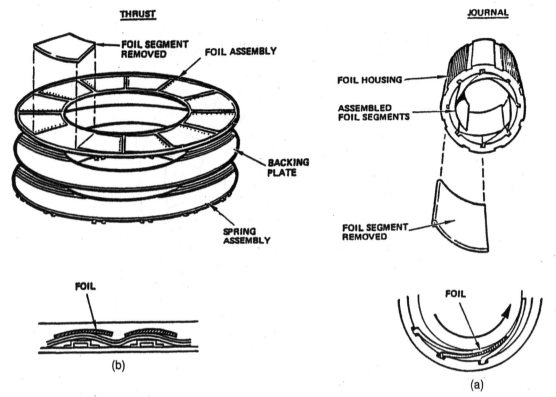

FIGURE 19.68 Garrett self-acting compliant foil bearing.

electromagnetic windings are wound in such a way that the magnetic flux circulates mainly inside the quadrants so that each quadrant of poles can be controlled independently.

Magnetic force is proportional to the ratio current to air-gap squared (Fig. 19.70). To support a load in a controlled axis (Fig. 19.71), unequal steady state or bias currents are induced in opposite pairs of poles, such that

$$W = f(I_1^2 - I_3^2)/C^2 \tag{19.48}$$

where $I_1 > I_3$

f = a magnetic pole constant for a given number of windings.

The bias current produces an I^2R loss, which is a major power loss in an AMB. However, the total resistance in the current path is not large; the AMB loss is in general insignificant compared to fluid film bearings. The journal floating in the magnetic field due to its bias currents alone is unstable. Linearized feedback control is achieved by making the air gap large relative to the journal's vibrations. For stable operation, the journal motion must be sensed and corrected instantaneously and continuously by superimposing a small control current to each bias current.

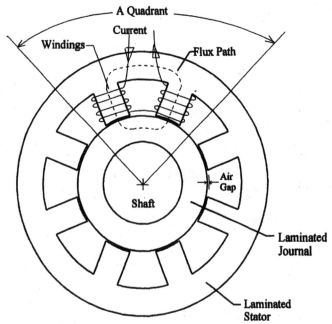

FIGURE 19.69 An 8-pole configuration of an active magnetic bearings.

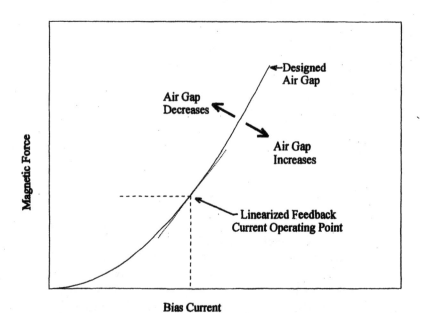

FIGURE 19.70 Nonlinearity of magnetic force.[2]

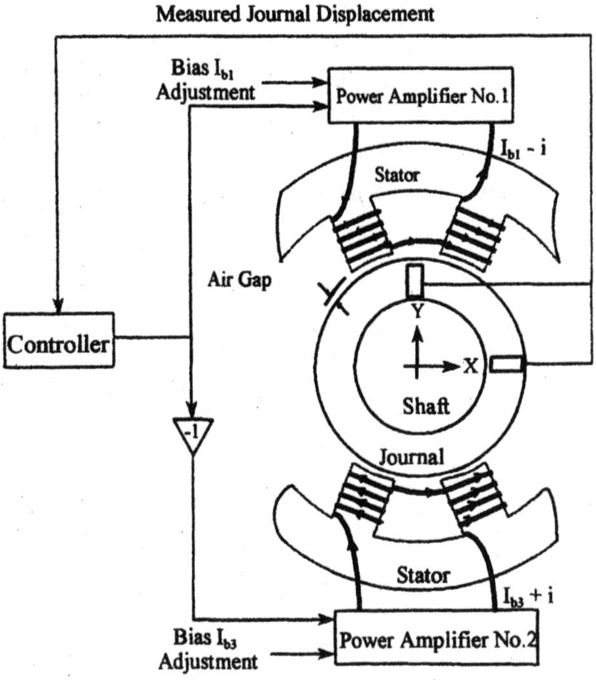

FIGURE 19.71 An independently controlled axis.

For example, as shown in Fig. 19.71, when the journal moves up by a small displacement, the current in the top quadrant will be reduced to a small amount, i, and the bottom quadrant increased by i. The control currents produce a net downward force, F, which pulls the journal back to the center. From sensing Y to producing F, a series of AMB components are involved, namely sensor, controller, power amplifier and electromagnets.

AMB Components

Electromagnets. For a given maximum load including static and dynamic loads, the AMB physical size is determined by the saturation flux density of the lamination material. For the 8-pole configuration,

$$F_{\max} = 5.75 \times 10^5 A_p B_s^2 \qquad (19.49)$$

where $F_{\max}$ = maximum load, N
A_p = surface area per pole, m^2
B_s = saturation flux density, Weber/m^2

The maximum value of the flux density in the linear range which is about 90% of the actual B_s value should be used in applying Eq. (19.49). Choosing the axial length L_p, the circumferential pole width is A_p/L_p. The radial dimensions can be determined from a given shaft diameter at the AMB. For sizing, the following guidelines should be followed:

- The cross-sectional area at any point of the flux path is not less than A_p.
- Adequate wiring space is provided.
- The axial length is no greater than the journal OD.
- As a rule, the air gap should be 10 times the expected journal vibration.

The pole surfaces are the most effective areas for heat dissipation via convection. The ampere-turns per pole is fixed for a given ferromagnetic material; the optimal choice of winding turns, M_t, is a trade-off of total current and inductance load, L, to the power amplifiers. The latter is proportional to $N_t^2 A_p C$ which is a crucial parameter causing control delay and bearing instability.

More than 8 poles can be designed for the stator, such as 16 or 24 evenly spaced. A large number of poles saves radial space because it localizes flux circulation. The coil pairs can be in series or parallel to a power amplifier with the same trade-off.

Power Amplifiers. Converting a low power control voltage signal to a high power control current and actuating the electromagnets requires power amplifiers. Two types are commercially available the linear and the pulse-width-modulation (PWM) type. The linear amplifier applies the control signal to a power transistor in an "active mode." The transistor continuously regulates the current through the windings from a DC source, V_s, with the current directly proportional to the control signal. The PWM type applies the control signal to generate high voltage pulses at a fixed frequency above audible range. The on-time period of each pulse is proportional to the input signal. The voltage pulse train produces current to the windings. The PWM type is electrically noisy and needs its own filters. The power transistors operate in a "saturation mode" with much less power loss.

There are three requirements for the power amplifiers. First, the control current, i, cannot be larger than the bias current. Second, the inductance of the electromagnets causes the control current to diminish and delay above a certain frequency (cut-off frequency). The PWM amplifiers usually apply their own current feedback to increase this frequency. Third, the value of V_s/L, called the current slew rate limit, is the maximum amperes per second that the amplifier can provide.

Sensors. Three displacement sensors prove to be practical: the capacitance probe, inductance, and eddy current probe. Each has its advantages and disadvantages, but all relate the small distance between the stationary sensor and the rotating shaft to an output electrical signal in volts. A low-pass filter is usually included in the sensor conditioning device to eliminate high frequency noise, including its own FM carrier. This filter, similar to the power amplifier cut-off characteristics, may cause a significant time delay in the frequency range of interest. A phase-lead circuit implanted in series in the feedback loop can reduce the delay.

Using displacement probes, journal velocity sensors are not needed in feedback control. Different analog circuits, such as a differentiator with a low-pass filter, and phase-lead circuits have been used to produce a pseudo velocity from the displacement measurement.

An analog surrogate called a *velocity observer,* instead of differentiating displacement, integrates journal force (equivalent to acceleration) to obtain velocity. The output of any pseudo velocity circuit is a combination of displacement and velocity signals. Thus, its feedback not only produces damping, but also contributes to the stiffness.

AMB Stiffness and Damping. From the previous discussion, a practical single axis control can be represented by the block diagram of Fig. 19.72. A radial AMB needs two independently controlled axes like this, while a thrust AMB needs only one.

A second-order, low-pass filter was assumed to be part of the sensor though it could have been a fourth-order or other type of filter. G_p is the sensor sensitivity; e.g., 1000 V/in. (40 V/mm). A phase-lead circuit is applied in series here for compensating the time delay caused by the inductance loads to the power amplifiers. One may set the phase-lead parameter "a" to be equal to the amplifier cutoff frequency, ω_a. Thus a system "zero" cancels a system "pole" in the 3-plane. This does not improve the current slew rate of the amplifier, but does increase the damping-to-stiffness ratio around ω_a. The other phase-lead parameter "b" is set in the range of $5a \leq b \leq 10a$.

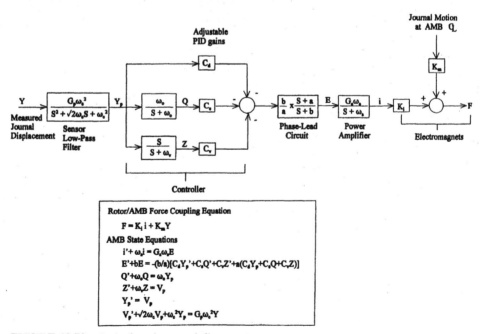

FIGURE 19.72 A single-axis control diagram.

The AMB stiffness and damping of this controlled axis can be calculated by using the equations below with S equal to $j\omega$.

$$-F/Y = K + j\omega b = -K_i(i/Y) - K_m \qquad (19.50)$$

$$i/Y = (Ts)(Tc)(Tp)(Ta) \qquad (19.51)$$

where

ω = excitation frequency, rad/s
$Ts = G_a \omega_n^2/(S^2 + \sqrt{2}\, \omega_n S + \omega_n^2)$
$Tc = -C_d - C_e \omega_o/(S + \omega_o) - C_v S/(S + \omega_v)$
$TP = (b/a)(S + a)(S + b)$
$Ta = G_a \omega_n/(S + \omega_n)$

The equations indicate that both K and B are functions of excitation frequency ω, not rotational speed. Numerical results are plotted in Fig. 19.73 using the AMB data. The frequency axis in this plot is normalized with respect to 50 Hz which is the average of two rigid-body critical speeds of a rotor. The amplitude is normalized with respect to $(G_p C_e K_i C_d - K_m)$.

At the low frequency range where the integral control dominates, the plot shows negative damping values. This should not cause alarm, however, since mechanical system resonances seldom exist in that low range. At the high frequency range, especially where the first two bending criticals exist, negative damping can cause resonances. This is discussed below.

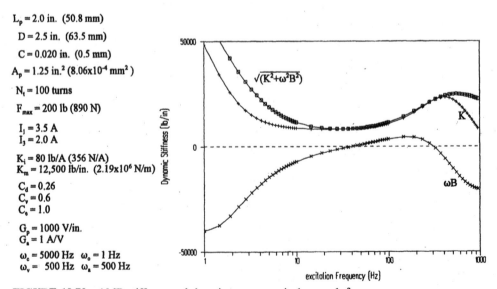

FIGURE 19.73 AMB stiffness and damping—a numerical example.[2]

Rotor-AMB System Dynamics

System Design Guidelines. AMBs are generally less stiff than rolling element or hydrodynamic oil-film bearings. Therefore, the first two system criticals have relatively rigid mode shapes, and their vibrations are easily controlled. The third and fourth criticals with bending mode shapes must be given careful consideration in high speed turbomachines.

Taking the rotor model in Fig. 19.74 as an example, its critical speed map shows that the rotor operates between the third and the fourth criticals. Two identical 8-pole AMBs are chosen to support the rotor with dimensions in Fig. 19.73. The first issue is finding the best method for determining the stiffnesses. In this case, the stiffness per bearing can be made 1000 lb/in. or 10,000 lb/in. (1.75×10^5 N/m or 1.75×10^6 N/m). The answer depends on rotor shock load. To take 1 g shock, this rotor of approximately 100 lb (45 kg) moves radially 50 mils and 5 mils (1.25 mm and .125 mm) respectively, for the lower and higher stiffnesses. The catcher bearing is set at 10 mil (0.25 mm) away from the rotor for a designed air gap of 20 mils (.5 mm). To avoid pounding the catcher bearing when shocked, the higher stiffness is chosen.

Reviewing the mode shapes at the chosen stiffness, Fig. 19.75 reveals that there are sufficient relative displacements at the bearings for control of the first and second modes. The third and the fourth modes are lacking the displacement at one bearing. To help control the third mode, the displacements sensor is mounted at the outboard side of each AMB where the sensor sees more than only the center motion.

The second design issue is to determine how many bending modes should be controlled. To keep the control electronics relatively simple, the frequency range with acceptable control response is limited by two factors, the inductance load and the filtering delay. It is imperative to have an adequately damped bending mode below the operating speed (the third mode in this case), because of the unbalance excitation during traversing the critical. The bending mode immediately above the operating speed (the fourth mode in this case) should be 15 to 20% away in frequency. However, it still can be excited by harmonics as the rotor is going up in speed, or by a shock load. But less damping is required for controlling this mode.

The higher bending modes normally are less likely to be excited. The rotor material damping is a source to resist the minor, or occasional excitation. Oil-film bearings always provide positive damping but there is no guarantee of this for AMBs. The control current at the high critical frequency may lay behind the displacement measurement, or the probe may be at the wrong side of the AMB. The AMB may become a small exciter for that mode. When it happens, a band-reject filter for the excitable mode can be inserted in series in the feedback loop to block the control at that modal frequency.

For the example herein, the power amplifier and the sensor low-pass filter are assumed to have the cut-off frequencies at 500 and 5000 Hz, respectively. Applying the normalized stiffness and damping of Fig. 19.73, the normalized frequency of 1.0 to 50 Hz, which is between the first and the second criticals as read from the

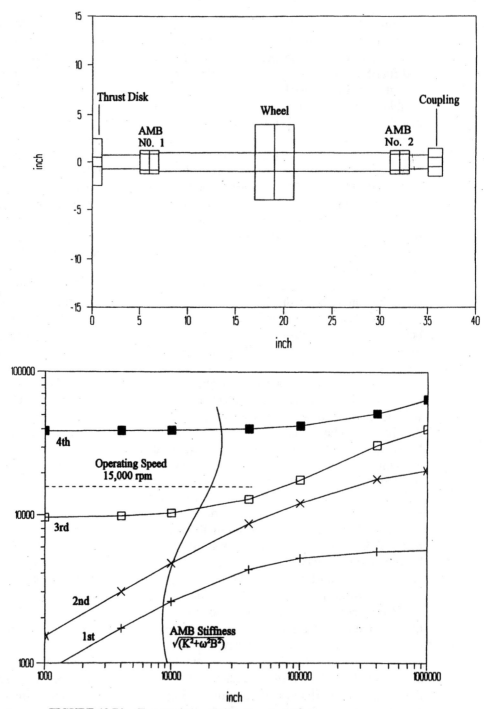

FIGURE 19.74 Characteristics of the rotor model.[2]

map. The values of $\omega B/K$ for the lower four modes range from 0.2 to 0.6, which is adequate for properly designed vibration modes. More damping can be achieved by increasing the value of C_v.

Rigorous Dynamic Analysis. After component sizing and cursory analysis, rigorous system analysis is needed to prove the expected rotor vibration behavior. Considering the fact that the AMB stiffness and damping are functions of excitation frequency, the state vector of the conventional rotor model is extended to include the state variables of the AMBs. The mathematical rotor model is the same as the conventional model including sections of shaft with specified *ID*, *OD*, length, and concentrated masses and inertias. The model for each bearing would be the bearing station number, the measurement station number, and the key parameters of Fig. 19.73.

Using this electromechanical model, the lower four damping frequencies of the rotor running at 15,000 rpm were computed and are presented in Table 19.22. The first three modes are adequately damped because the associated log decrement values are all significantly above 0.4. The latter is a damping value generally accepted for a rotor system supported in oil-film bearings. The fourth mode has a log decrement value of 0.06 without considering the rotor material damping. It should be acceptable since it is much higher in frequency that the third mode and the operating speed.

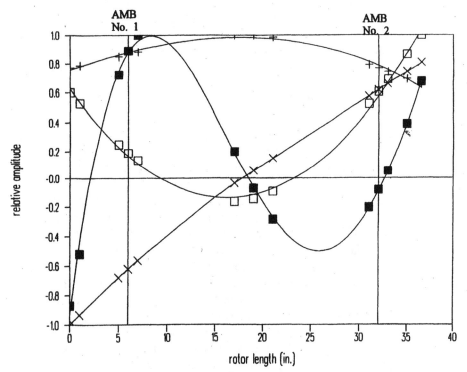

FIGURE 19.75 Rotor critical mode shapes.[2]

TABLE 19.22 Damped Natural Frequencies of Forward Modes[2] Rotor Speed = 15,000 rpm; Cross Coupling Stiffness (K_{xy}) at Wheel

Mode	$K_{xy} = 0$		$K_{xy} = -4000$ lb/in (-7×10^5 N/m)	
	Frequency (cpm)	Log decrement	Frequency (cpm)	Log decrement
1st	2,318	1.05	2,374	−0.00
2nd	5,263	2.20	5,623	2.19
3rd	11,154	0.89	11,144	0.88
4th	34,156	0.06	34,156	0.06

In Table 19.22, the cross-coupling stiffness produces a destabilizing seal effect at the wheel, but it only affects the first mode damping. The reason for this is that only the first mode shape has significant lateral displacement at the wheel. The rotor/AMB system can sustain a value of 4000 lb/in. (7×10^5 N/m) before becoming unstable.

Figure 19.76 presents an unbalance response at the third critical speed using the same electromechanical model. It indicates that the response peak is well damped and far away from the operating speed of 15,000 rpm. The peak dynamic current was calculated to be 0.45 (0-peak) at 11,500 rpm. It specifies that a current slew rate no less than 500 A/s must be provided by the power amplifier design.

19.7 CRYOGENIC APPLICATIONS

Due to the nature of the fluid, machinery employing cryogenic fluids such as liquid oxygen or liquid hydrogen, use predominantly rolling element bearings. These machines run at extreme speeds, are heavily loaded and do not permit the use of conventional lubricants. In all of these applications, wear of the bearing is the limiting factor so that the bearings have a very short life. Since REB's suffer from DN limits, they often are unable to run at the high speeds demanded for given rotor diameters. An additional problem with REB's is their high stiffness and near zero damping. The low viscosity of the process fluid makes squeeze film dampers ineffective and the low temperatures prevent the use of elastomeric damping. Introducing some of the novel type bearings discussed earlier offers a possibility of eliminating or ameliorating this critical situation in cryogenic turbomachinery.

The performance will be studied under the conditions of using liquid oxygen as the process fluid as well as the lubricant. In actual applications, such as turbopumps, these fluids are usually at very high pressures ranging from several hundred psi to perhaps 10,000 psi and this will constitute the ambient pressure for the bearings

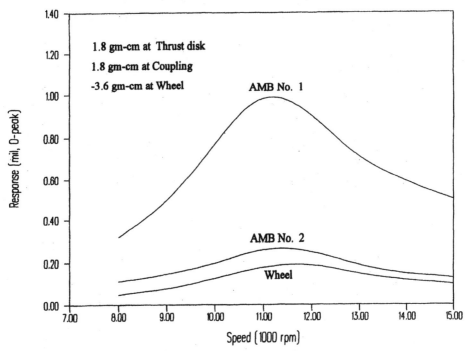

FIGURE 19.76 Unbalance response at 3rd critical speed.[2]

in use. Generally the lubricant should be treated as a compressible fluid. However, a parametric study of journal bearings shows that load capacity is a strong function of ambient conditions up to about a $p_a = 2,000$ psi; thereafter there is very little dependence on p_a. This is in conformity with the previously stated fact that a $\Lambda \rightarrow 0$, a compressible fluid acts as a liquid. Thus in some cases the use of the incompressible equations, which are much simpler, is justified.

19.7.1 Compliant Surface Bearings

Journal Bearings. Two designs will be considered, a full (360°) and a 3-pad (120°) bump foil bearing.

The Full 360° Bearing. Studies conducted for three different foil bearings indicate the following:

- W increases with bump stiffness.
- h_N decreases with stiffness.
- At an intermediate stiffness ($41 \cdot 10^3$ lb/in.), the foil deflection is of the same order as the H_n.
- Power loss rises slightly with stiffness.

- The vertical stiffness K_{yy} approaches that of K_{xx} at bump stiffnesses below 50,000 lb/in./in. of the fluid film.

Based on the above, a stiffness of 40×10^3 which corresponds to a foil thickness of 6 mils seems to be an optimum yielding a reasonable W at low deflections. Under these conditions, the structural stiffness dominates the overall bearing stiffness.

The 3-Pad CSB. One of the major parameters to be looked into here is the angular extent of the pad, or the number of pads. In the present 3-pad 120° configuration, actually only two of the pads would be loaded. The other parameter to be investigated is the load angle. This is particularly relevant when there is a rotating load in the bearing.

- **Steady Load.** Figure 19.77 shows the configuration of the 3-pad bearing with the load in the present case directed midway of the bottom pad. Bearing performance at operating speed is shown in Table 19.23 and Fig. 19.78. Load capacity increases exponentially with a decrease in h_N but the power loss is insensitive to it. The spring coefficients seem to have an inverse linear dependence on h_N which contributes much to bearing stability.

- **Variable Load Angle.** This parameter was studied under the fixed condition of $\epsilon = 0.9$ and a speed of 29,830 rpm. The results, shown in Table 19.25 and in the additional plots, reveal the following:

- Maximum load capacity is obtained for $\phi_L = 195°$.

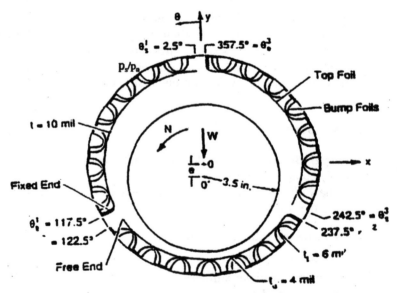

FIGURE 19.77 Three-pad foil journal bearing for SSME (space shuttle engine rotor) $L \times D \times C = 2.25 \times 3.5 \times 0.0015$ in., XKB = 40,960 lb/in./in.
$\tau = 10$ mil, $\tau_4 = 4$, $\tau_l = 6$

TABLE 19.23 Three-Pad Foil Journal Bearing Performance[12]

(n = 29,830 rpm)

LOX at 191°R and 2200 psia, Load at $\theta_L = 180°$

$L \times D \times C = 2.25 \times 3.5 \times 0.0015$ in; Bump Foil Thickness: $t_{lower} = 6$ mil; $t_{upper} = 4$ mil

ϵ	AA (°)	W (lb)	h_N (mil)	P_{max} (psig)	Power loss (hp)	Dynamic bearing coefficients (lb/in × 10⁶)			
						K_{xx}	K_{yx}	K_{xy}	K_{yy}
"	212.0	620.0	1.33	192.0	11.40	1.040	−1.300	0.45	0.500
0.6000	210.1	834.8	1.26	216.8	11.44	1.000	−1.202	0.40	0.620
"			1.22			0.980			
0.8000	207.8	1151.9	1.17	286.9	11.69	0.950	−1.062	0.28	0.750
0.9000	206.7	1307.5	1.12	320.8	11.94	0.940	−1.064	0.26	0.790
"			1.11			0.920			
0.9990	204.8	1463.6	1.06	350.0	12.41	0.910	−1.104	0.20	0.850
0.9995	204.8	1463.6	1.06	350.0	12.41	0.900	−1.105	0.21	0.801
"	203.0	1637.0	1.03	391.0		0.855	−1.130	0.18	0.890
"			1.02		13.00				
"	200.8	2158.0	0.95	487.0		0.850	−1.200	0.135	0.955
"	198.0	2720.0	0.89	589.0		0.850	−1.300	0.085	0.970
"		3000.0	0.87						
"			0.83	700.0				0.00	0.985
"	196.0		0.81			0.850			
"			0.79						1.000

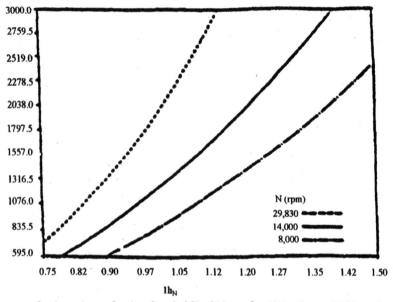

a. Load capacity as a function of nominal film thickness, $\theta_L = 180°$. Three-pad foil journal bearing with LOX; L x D x C = 2.25 x 3.5 x 0.0015 in., $t_u = 4$ and $t_\ell = 6$ mil.

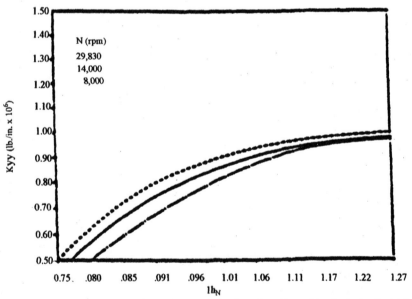

Three-pad foil bearing stiffness K_{yy} as a function of $1/h_N$; $\theta_L = 180°$, with LOX; bearing Lx D x C = 2.25 cx 3.5 x 0.0015 in.; $t_u = 4$ mil, $t_\ell = 6$ mil.

FIGURE 19.78 Performance of cryogenic 3-pad foil bearing.[12]

- h_N is smallest when $\phi_L = 168°$.
- The power loss here is 12 hp which compares with 11 for the steady load.

Total variation in W over the entire 360° span circumference is some 500 lb and of h_N one mil. Since a change in film thickness of 0.5 mils corresponds to a change in load capacity of 2,400 lb, a dynamic load from zero to a peak value of 2,400 lb at a fixed load angle would produce a journal orbit of 1/2 mils radius. This is quite an acceptable disturbance.

Thrust Bearings. From past studies, a compound thrust bearing emerges as the most promising design. Its geometry will be $\beta = 45°$ with an $(R_2 - R_1) = 2$ and with a $K_B = 40,960$ lb/in./in. Varying in this design, the value of h_N, from 3 to 16 the load capacity shows a maximum at $h_N = 14$. This is three and a half times the value in rigid bearings. Next, a variation in the extent of taper was conducted under different stiffnesses. Table 19.25 summarizes the performance of the bearing for the optimized variables which covers a range of operation from 1,000 to 30,000 rpm and ambient pressures from 30 to 2,200 psi. Satisfactory performance under all conditions is indicated by the rather healthy values of h_N which never falls below 0.3 mils and in most cases stay above half a mil.

TABLE 19.24 Three-Pad Foil Journal Bearing Performance vs. Load Angle ϕ_L[12]
LOX at 191°R and $P_a = 2200$ psia; $N = 29,830$, $\epsilon = 0.9$

θ_L (deg)	AA (deg)	W (lb)	h_N (mil)	$P_{ma} \times$ (psig)	Power loss (hp)	Dynamic bearing coefficients (lb/in × 10⁶)			
						K_{xx}	K_{yx}	K_{xy}	K_{yy}
120	234	903	1.90	278	12.00	1.15	−0.43	1.15	0.44
*132	227	914	1.94	258	13.25	1.27	−0.66	0.98	0.38
*144	218	967	1.98	234	14.25	1.40	−0.84	0.78	0.33
150	216	975	2.03	222	15.00	1.43	−0.95	0.68	0.30
*156	212	1040	1.59	234	13.25	1.4	−1.05	0.59	0.35
165	208	1105	0.92	255	10.00	1.51	−1.13	0.43	0.52
*168	207	1180	1.00	267	10.50	1.43	−1.0	0.38	0.57
180	206	1307	1.12	320	11.00	0.98	−1.06	0.27	0.78
*192	217	1422	1.32	395	10.60	0.32	−0.6	0.05	0.87
195	219	1451	1.40	406	10.70	0.15	−0.55	0.02	0.92
*204	239	1180	1.50	374	10.75	0.32	−0.48	0.50	0.82
210	247	1050	1.74	347	10.80	0.47	−0.40	0.77	0.69
*216	248	1026	1.72	338	11.10	0.61	−0.41	0.84	0.68
*228	241	967	1.81	305	11.90	0.90	−0.42	0.98	0.54
240	235	906	1.90	279	12.80	1.15	−0.43	1.15	0.44

*Extrapolated values.

TABLE 19.25 Performance of Thrust CFB Under Lox Conditions[12]
(8 pads at 45° each, $R_1 \times R_2 \times h_2 = 1.25$ in. $\times$ 2.5 in. $\times$ 0.0002 in.; $\bar{h}_1 = 14$, $K_{BR} = 40,950$ lb/in^2,
$t_L = 4$ mil; $t_U = 6$ mil, $\bar{K}_s = 0.5$, $\bar{K}_g = 1.5$)

Case no.	N (rpm)	P_s (psia)	W psi	W lb	h_N (mil)	ζ (mil/in)	H/Pad	K_{BS}	K_{BE}	Re no.	Remarks
1	29,830	2,200	260	3,835	1.23	1.9	2.1	20,480	61,440	17,685	The Re No. is computed for a 4-mil entrance gap
2	25,000	1,900	236	3,472	1.14	1.74	1.5	20,480	61,440	15,588	
3	20,000	1,400	211	3,104	1.07	1.57	1.01	20,480	61,440	12,618	
4	14,000	1,012	187	2,752	0.96	1.4	0.8	20,480	61,440	7,978	
5	11,000	900	168	2,472	0.92	1.3	0.4	20,480	61,440	—	
6	8,000	800	149	2,200	0.83	1.14	0.24	20,480	61,440	4,287	
7	4,000	500	108	1,560	0.65	0.84	0.08	20,480	61,440	2,144	Turbulent (inlet film)
8	4,000	500	73	1,080	0.52	0.6	0.045	20,480	61,440	1,000	Super Laminar (h_N)
9	2,000	370	38	536	0.37	0.3	0.01	20,480	61,440	800	Laminar
10	1,000	370	21.7	320	0.3	0.2	0.003	20,480	61,440	400	Laminar
11	1,000	370	12.5	184	0.53	0.6	0.002	3,530	1,059	400	Upper bump foil is considered (i.e. low stiffness)

Hybrid Bearings. If the purely hydrodynamic CSB does not meet the operational requirements, then a combination hydrodynamic-hydrostatic CSB can be used. Near zero speeds, the high pressure fluid would have to be taken from a reservoir until the unit starts generating its own high pressure supply.

Journal Bearings. The essence of a hydrostatic CSB fed through the rotating shaft is shown in Fig. 19.79. An LOX solution is here obtained for a 4-pad bearing with dimensions $D \times L \times C = 3.8$ in. $\times 2$ in. $\times 0.002$ in. with an orifice dimension of 0.056 in. This unit provides a 50% drop across the orifice, yielding a value of p_o of 250 psi and a load capacity of 2,000 lb. If a 2,000 psi supply pressure is available, such a bearing would carry a load of 5,100 lb at a value of h_N of about one mil.

Thrust Bearings. The thrust bearing design analyzed is one with $R_2 \times R_1 = 3.5$ in. $\times 2.5$ in. shown in Fig. 19.79. Using 8 orifices gives more or less square bearing pads which is an optimum arrangement. Three parameters determine the performance, namely

$$(R_2/R_1); \quad \bar{p}_s = (p_s/p_a); \quad \Lambda_s = 6\mu n a^2 (pv)^{1/2}/p_s h_N^3$$

For the present $(R_2/R_1) = 1.4$, the load capacities for several values of p_s and Λ_s are listed in Table 19.25. As seen, in order to meet a 6,000 thrust load over the range of 0 to 12,000 rpm, a supply pressure of some 3,000 psi is needed. For such a value of p_s and a 6,000 lbs load the other performance items would be

   Flow: $Q = 1.7$ gpm
   Axial stiffness: $K_z = 2.4 \times 10^6$ lb/in.
   Angular stiffness: $K_a = 0.96 \times 10^6$ lb-in./rad

To sum up the geometry and operating conditions of this bearing would be as follows:

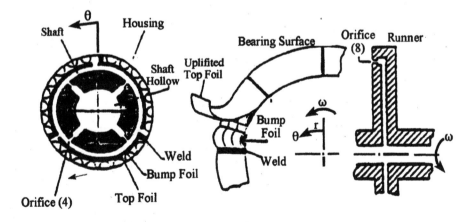

   a. Journal Bearings   b. Thrust Bearings
    (3 or 4 Pads)   8 Pads)

FIGURE 19.79 Hybrid CSB's for cryogenic applications.

$R_2 \times R_1 \times h_{\min} = 3.5 \times 2.5 \times 0.0005$ in.
8 orifices 0.05 in. in diameter
$(p_s/p_a) = 12$ or $p_s = 3{,}000$ psia
$\mu = 0.25 \times 10^{-7}$ reynes
$\rho = 10^{-4}$ lb-s^2/in.^{4}

19.7.2 Hydrostatic Bearings

For this section, the following special symbols will be used:

A = area of feeding recess = ld, in.^{2}
a = radius of orifice hole, in.
a_c = radius of orifice feeding hole chamber, in.
a_F = radius of orifice feeding hole, in.
C = radial clearance, in.
C_D = orifice discharge coefficient
D = diameter, in.
d = circumferential length of feeding recess, in.
e = eccentricity of journal, in.
G_θ, G_z, G_R = turbulent viscosity correction factors
$\overline{G}$ = dimensionless flow = $\dot{m}v_2/C^3(P_s - P_a)$
$\overline{G}'$ = dimensionless flow = $\dot{m}v_2 L(1 - \overline{Y})/C^3(P_s - P_a)DP_R G_z$
GRADPS = turbulence parameter or fluidic Reynolds number = $(P_s - P_a)C^3/R\mu_s v_s$
GRADP = modified turbulence parameter = (GRADPS) $P_R D/L \times (1 - Y)$
g_c = acceleration due to gravity = 386 in./s^2
h = local bearing film thickness, in.
h_R = depth of feeding recess, in.
$\overline{h}$ = dimensionless thickness = h/C
L = bearing length
l = axial length of feeding recess, in.
l_F = depth of orifice feeding hole, in.

TABLE 19.26 Thrust Bearing Forces as a Function of Supply Pressure[12]
$(R_1/R_2) = 1.4$; $pa - 250$ psia

p_a (psia)	p_s/p_a	Λ_s	W (lb)
1000	4	0.74	1650
2000	8	0.57	4620
300	12	0.49	6040

$\dot{m}$ = mass flow rate, lb-s²/in.
n = number of recesses
N = rotational speed, rpm
P = pressure, lb/in.²
$\tilde{P}$ = dimensionless pressure = $(P - P_a)/(P_s - P_a)$
R = radius, in.
r = coordinate in radial direction from feeding holes, in.
Ren = Reynolds number = UC/v_s
t = time, s
U = surface velocity of journal, ips
$\mathring{u}$ = velocity in circumferential direction, ips
$\bar{u}$ = mean total circumferential velocity, ips
$\bar{\bar{u}}$ = dimensionless circumferential velocity = $\bar{u}(\rho_s/P_s - P_a)^{1/2}$
u_p = mean circumferential velocity due to pressure gradient, ips
V = fluid volume, in.³
v = velocity normal to bearing surface, ips
W = load, lb
$\overline{W}$ = dimensionless load, = $\overline{W}/(P_s - P_a)LD$
w = velocity in axial direction, ips
$\bar{w}$ = mean axial velocity, ips
$\bar{\bar{w}}$ = dimensionless axial velocity = $\bar{w}(P_s - P_a)^{1/2}$
$\overline{X}$ = circumferential recess length ratio = $nd/\pi D$
$\overline{Y}$ = axial recess length ratio = $1/L$
z = coordinate in axial direction, in.
ϵ = eccentricity ratio = e/C
ζ = dimensionless axial coordinate = z/R
θ = coordinate in circumferential direction, rad
ψ = inertia parameter = $C(GRADP)^2 G_s^2/R(GRADPS)P_R$
ω = rotational speed of journal, rad/s
μ = absolute viscosity, lb-s/in.²
ν = kinematic viscosity; in.²/s
$\bar{\nu}$ = dimensionless kinematic viscosity = ν/ν_s
ρ = fluid mass density, lb-s²
$\bar{\rho}$ = dimensionless density = ρ/ρ_s

Subscripts

a = ambient conditions
F = film condition
R = recess condition
r = radial direction from feeding hole
s = supply conditions
z = axial direction
θ = angular direction (circumferential)

Since the cryogenic process fluid is usually available at very high pressures, a naturally attractive solution is to use fully hydrostatic bearings. The design chosen is a full 360 orifice compensated journal bearing. The continuous circumference provides a higher load capacity than a design with a number of discrete pads; a schematic of the bearing is shown Fig. 19.80. Due to the high speeds, the flow is assumed to be turbulent and a function of both the rotation and fluid velocity in the film. The density and viscosity of the fluid will vary here also with pressure according to

$$\bar{u} = \frac{h^2 G_\theta}{\mu} \frac{\partial P}{R \partial \theta} = U/2 \qquad (19.52a)$$

$$\bar{w} = \frac{h^2 G_s}{\mu} \frac{\partial P}{\partial z} \qquad (19.52b)$$

The turbulence level in a bearing can be due either to high rotation or to high film velocities induced by the pressure gradients, or both. In the last two cases, an iterative procedure is required in obtaining a solution since the pressure gradients are not known *a priori*. Concerning the variation of density and kinematic viscosity with pressure, this is approximated by the following isothermal relationships

$$\rho = \rho_s + (\rho_s - \rho_a) \frac{P - P_a}{P_s - P_a} \qquad (19.53)$$

$$\nu = \nu_s + (\nu_s - \nu_a) \frac{P - P_a}{P_s - P_a} \qquad (19.54)$$

where the subscripts s and a indicate conditions at supply pressure and ambient pressure, respectively. Since for LH_2 and LO_2 bearings, ρ and ν vary at most, only on the order of 20% from supply pressure to ambient pressure, this variation can be approximated quite well by the linear relations given in the foregoing.

Inertia forces result in a decrease in the static pressure when fluid accelerates and a corresponding increase in static pressure when fluid decelerates. In the LH_2

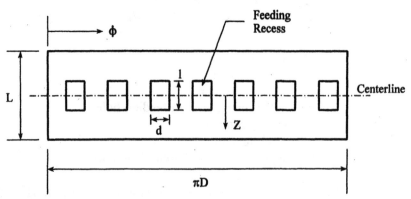

FIGURE 19.80 Full hydrostatic journal bearing.

and LO_2 bearings, the principal effect of inertia forces is that of a sudden decrease in static pressure at the edge of each bearing recess caused by the acceleration of fluid from relative stagnant conditions in each recess to a condition of high velocity in the thin bearing film. In the present analysis, this decrease in static pressure at the axial edge of each recess is calculated by means of the following relationship.

$$P \text{ (edge of recess)} = P_R - \rho \bar{w}^2/2 \qquad (19.55)$$

where P_R = recess pressure
$\bar{w}$ = mean local velocity in axial direction in bearing film at edge of recess

The relationship used to calculated static pressure at the circumferential edge of each recess is

$$P \text{ (edge of recess)} = P_R - \rho \bar{u}^2/2 \qquad (19.56)$$

where u is the mean velocity in the circumferential direction in the bearing film at the edge of the recess. In hydrostatic bearings, attainable load and stiffness are independent of lubricant viscosity. Hence, it was anticipated for the LH_2 and LO_2 bearings that once turbulence was fully established, load and stiffness would be essentially independent of the level of turbulence.

Although load capacity is relatively insensitive to turbulence, bearing mass flow rate depends strongly on the level of turbulence. In order to generalize the effects of turbulence and other operating parameters on flow rate, the following formulation for dimensionless flow was developed.

$$\text{Dimensionless flow,} \qquad \bar{G}' = \frac{\dot{m} v_x L (1 - \bar{Y})}{C^3 (P_s - P_a) D \bar{P}_R G_z}$$

The parameter

$$GRADP = (p_s - p_a) \cdot p_R C^3 \mu_s \nu_s (1 - \bar{Y})$$

provides an index of the axial flow Reynolds number. A plot of G_z versus $GRADP$ is given in Fig. 19.81. The dimensionless flow parameter $\bar{G}'$ is independent of P_R and ν and of the variables L, D and $\bar{Y}$ over the range of $0.5 < L/D < 1.5$ and $0.2 < \bar{Y} < 0.6$. Also, inertia effects are ignored here since they were found to have little effect on load capacity and flow rate.

In hydrostatic bearings designed for maximum stiffness at $\epsilon = 0$, load capacity increases linearly with ϵ out to values of $\epsilon \cong 0.7$. Consequently, bearing load data are presented here in the form of W/ϵ which is roughly independent of ϵ for values of $\epsilon \leq 0.7$.

Bearing flow rate tends to decrease slightly as ϵ increases, the decrease amounting to about 10% at $\epsilon = 0.7$. Bearing flow design data presented are for $\epsilon = 0$.

Effects of journal rotation on bearing flow and load capacity are neglected for the design data, i.e., the design charts pertain to purely hydrostatic bearings. For all anticipated operating conditions of the LH_2 and LO_2 bearing, the mode of operation was predominately hydrostatic. Also, the level of turbulence was always dominated by the pressure flow rather than by rotational shear.

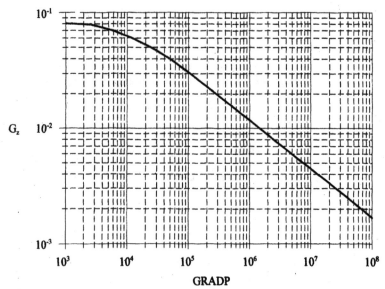

FIGURE 19.81 Turbulent viscosity relationship.[32]

Compressibility of liquid hydrogen and liquid oxygen was taken into account in the bearing analysis. However, this compressibility is so slight as to have a negligible effect on the steady-state performance of the bearings.

An approximately optimum design configuration was chosen for these bearings considering both load capacity and flow rate. The design chosen was

$L/D = 0.65$

$n = 6$

$\overline{X} = 0.5$

$\overline{Y} = 0.4$

$\tilde{P}_R = 0.5$

Design Charts

a. Effect of L/D Ratio. The effect of L/D on dimensionless stiffness $\overline{W}/\epsilon$ is shown in Fig. 19.82a. The decrease in $\overline{W}$ with L/D is typical of all hydrostatic bearings. Corrected dimensionless flow $\overline{G}'$ is independent of L/D, the effect of length to diameter ratio being taken account of by the inclusion of the ratio L/D in the expression $\overline{G}'$.

The reason that $\overline{W}/\epsilon$ decreases with L/D in hydrostatic bearings is that as L/D increases, circumferential flow in the bearing film becomes more significant. This tends to reduce load capacity because it tends to equalize the pressures in the various recesses.

The dimensionless variables $\overline{W}/\epsilon$ and $\overline{G}'$ somewhat obscure the effect that a change in bearing length has on the real load W and the real flow, m. In fact, if

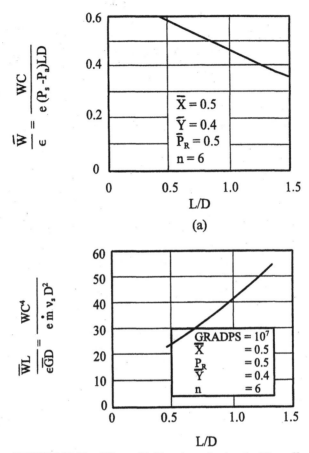

FIGURE 19.82 Effect of L/D ratio on load and stiffness.[32]

the variables C, ν, D, and e are kept fixed, the ratio of real load to real flow, $W/\dot{m}$ will increase quite steeply as L/D increases. To show this effect, $\overline{W}L/\epsilon D$ divided by the reduced dimensionless flow $\overline{G}$ is plotted where

$$\overline{G} = \frac{\dot{m}v_s}{C^2(P_s - P_a)} \tag{19.57}$$

The resulting curve is shown in Fig. 19.82. Note that one of the more effective ways of designing a hydrostatic bearing to have a large load capacity with low flow rates is to make L/D large.

b. Effect of Recess Pressure Ratio $\overline{P}_R$. The most significant design parameter with respect to optimization of load and stiffness is the recess pressure ratio $\overline{P}_R$. The influence of $\overline{P}_R$ on dimensionless stiffness is shown in Fig. 19.83a. As can be seen, maximum stiffness (load) occurs at $\overline{P}_R \approx 0.55$. It should be mentioned that this curve of $\overline{W}/\epsilon$ versus $\overline{P}_R$ is characteristic for a bearing with turbulent flow ($GRADP > 3000$) and with an orifice restrictor. For the case of laminar flow in

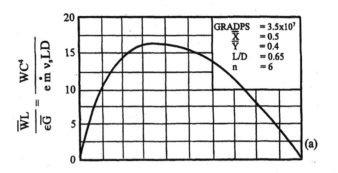

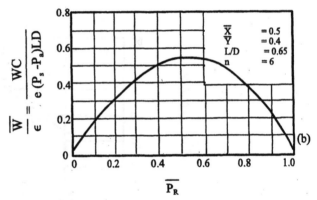

FIGURE 19.83 Effect of recess pressure ratio on dimensionless load-to-flow ratio and dimensionless stiffness.[32]

the bearing film, maximum load would occur at $\overline{P}_R \approx 0.586$, which is still very close to $\overline{P}_R = 0.55$.

Corrected dimensionless flow, $\overline{G}'$, is independent of $\overline{P}_R$, the effect of pressure ratio being taken into account in the parameter $\overline{G}'$, itself.

Actual flow, $\dot{m}$, will increase as $\overline{P}_R$ increases. Because of this effect, the maximum value of the ratio of stiffness to flow does not occur at $\overline{P}_R = 0.55$, but rather at a lower pressure ratio ($\overline{P}_R \approx 0.375$). This is illustrated by the curve shown in Fig. 19.83a.

Recommended design practice for LH_2 and LO_2 hydrostatic bearings would be to select $\overline{P}_R$ between 0.375 and 0.55. However, certain special requirements, such as limitations on flow rate may dictate the use of pressure ratios lower than $\overline{P}_R = 0.375$.

c. Effect of Axial Recess Length Ratio $\overline{Y} = 1/L$. The influence of the parameter Y on dimensionless stiffness is shown in Fig. 19.84b. As would be expected, $\overline{W}/\epsilon$ increases with $\overline{Y}$.

Corrected dimensionless flow $\overline{G}'$ is essentially independent of $\overline{Y}$ over the range $0.2 \le \overline{Y} \le 0.6$ due to the inclusion of the factor $(1 - \overline{Y})$ in the expression $\overline{G}'$.

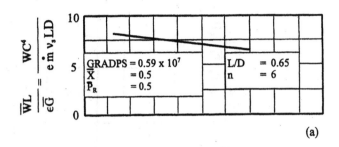

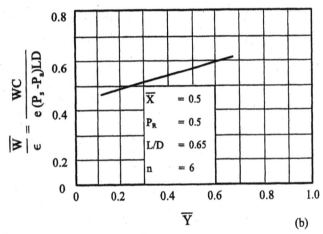

FIGURE 19.84 Effect of (axial recess/length) ratio on load and stiffness.[32]

Since actual flow increases with $\overline{Y}$ as does the actual load, W, it is of interest to see how the ratio of load to flow rate varies with $\overline{Y}$. This is shown by the curve in Fig. 19.84a. Note that the ratio of actual load to actual flow improves as the axial length of the feeding recess is reduced. Choice of an optimum recess length would depend on how much flow one was willing to sacrifice to obtain an increase in load capacity. It will be shown that bearing stability considerations also enter into the selection of optimum recess dimensions. This will be discussed in a subsequent section.

d. Effect of Circumferential Recess Length Ratio, $\overline{X} = nd/\pi D$. Dimensionless stiffness increases as the circumferential recess length is reduced, at least for values of $\overline{X} > 0.30$, Fig. 19.85b. The physical reason for this is that as the land width between pockets decreases ($\overline{X}$ increases), flow can more readily pass from one recess to the next, thereby tending to equalize the pressure in the different recesses.

The value $\overline{X}$ is one of the few parameters which influence the value of the dimensionless flow $\overline{G}'$, another parameter being n, the number of recesses. The variation of $\overline{G}'$ with $\overline{X}$ is shown in Fig. 19.85a.

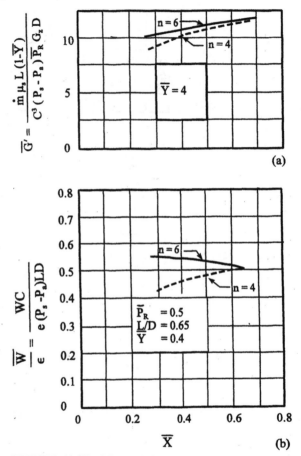

FIGURE 19.85 Effect of circumferential recess ratio on load and stiffness.[32]

e. Effect of Number of Feeding Recesses, n. The dimensionless flow $\overline{G}'$ increases very slightly with number of recesses n in the range from 4 to 8 recesses, Fig. 19.86a. Dimensionless stiffness increases somewhat in going from 4 to 6 recesses, but increases very little going from 6 to 8 recesses (Fig. 19.86b). The question of how many recesses to have in a bearing will depend primarily on whether the gain in load with increase in n is worth the added manufacturing complexity.

One point in favor of using a greater number of recesses is that it minimizes the effect that different directions of loading can have on load capacity. Other investigators report that there can be as much as a 25% difference in bearing load capacity depending on whether the load line passes through the center of a recess or between recesses. However, this study indicates only a slight change (<5%) in load with load direction in a bearing with four or more pockets.

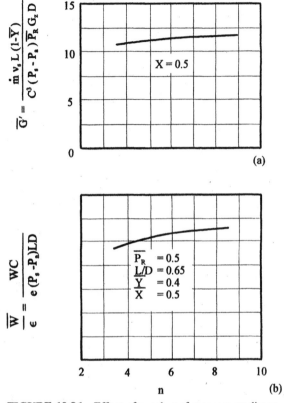

FIGURE 19.86 Effect of number of recesses on dimensionless flow and dimensionless stiffness.[32]

19.8 LUBRICANTS AND MATERIALS

This section is not meant to enter into a discussion of the very extensive field of bearing lubricants and materials, but merely to provide the basic data required in the use of the equations and tables given in previous sections.

19.8.1 Lubricants

The major functions of a lubricant in a bearing is to provide a fluid film at the interface; to convect the heat due to viscous dissipation; and to wet the surfaces during stops and starts when there is no full film at the interface. In a hydrodynamic bearing, the most important property in fulfilling the above functions is the viscosity. The level of generated hydrodynamic pressures and hence the load capacity, as well as all other performance characteristics such as temperatures, flow, power loss, etc., all depend very strongly on lubricant viscosity. Since viscosities are listed in a great variety of units, Table 19.27 is provided to facilitate conversion from one system to any other.

The oils used on most compressors are similar to those in turbines and gear sets. These are shown in Fig. 19.87 along with other common SAE grade oils. Some of the more pertinent characteristics of these oils are as follows:

Thermal conductivity:	0.08 BTU/(hr ft^2 °F)/ft [0.14 Joules/(sec-m-°C)]
Specific heat:	0.4 BTU/(lb °F) [2.52 K Joules/(Kg-°C)]
Heat of vaporization:	80 BTU/lb (187 K Joules/Kg)
Specific gravity:	0.88 at 60°F (16°C)
Flash point:	410° to 470°F (210° to 243°C)
Pour point:	−10°F (−23°C)

If the operating temperatures exceed 300°F or fall below −10°F, a shift to synthetic lubricants may be required and some such candidates are listed in Table 19.27. To increase the resistance and longevity of petroleum oils during prolonged usage, additives are often desirable and a list of such additives for particular contingencies are given in Table 19.29.

19.8.2 Bearing Materials

Selection of bearing materials for specific applications involves a scrutiny of the following characteristics: (1) compatibility; (2) embeddability and conformability; (3) corrosion resistance; and (4) compressive and fatigue strength. In hydrodynamic bearings, the most relevant items are the allowable maximum pressures before the material begins to deform or flow, and the value of T_{max} it can endure. For compressive strength, an alloy with intermediate strength is desirable; an alloy too low in strength is prone to extrude under load, while too strong a metal, being brittle, may crumble under impact loading. Fatigue strength is particularly important in applications with dynamic loading in order to prevent the formation of cracks or surface pits. The use of a thin soft layer bonded to a hard backing metal often gives the desired combination of fatigue and compressive strength; in such cases, however, the fatigue strength of the bond itself requires attention. When a material has low corrosion resistance, difficulties can be minimized by using oils with good oxidation inhibitors and by maintaining low bearing temperatures.

Babbitts. The most common bearing materials are babbitts, either tin-based or lead-based. The detailed properties of babbitts are given in Table 19.30. Babbitts can operate under conditions of boundary lubrication or dirty operation. They have excellent compatibility and non-scoring characteristics and are outstanding in tolerating errors in construction and operation. Their deficiencies with regard to fatigue strength can be improved by using an intermediate layer of high-strength material between a steel backing and a thin babbitt layer. Many of these, known under the name of trimetal bearings, use the following construction: (1) a low-

TABLE 19.27 List of Synthetic Lubricants[36]

Type	Kinematic viscosity, cs			Flash point, °F	Pour point, °F	Approximate cost per gallon	Typical uses
	210°F	100°F	−65°F				
Diester:							
Turbo Oil 15	3.6	14.2	12,600	430	−90	$10.00	MIL-L-7808 high load capacity, high temperature jet engine oil.
MIL-O-6085	3.5	13.5	10,000	450	−90	10.00	Low volatility aircraft hydraulic and instrument oil.
MIL-O-6387	4.6	15.8	5,000	410	<−80	—	Aircraft hydraulic fluid for alternator drives.
Phosphate:							
Tricresyl phosphate	3.8	30.7	—	465	—	3.60	Low flammability hydraulic fluid for diecasting machines.
Skydrole	3.85	15.5	>20,000	355	70	12.00	Nonflammable aircraft hydraulic oil. Nonflammable hydraulic oil for diecasting machines, punch pressures, etc.
Pydraul F-9c	5.8	54	—	430	+5	3.75	Air compressors.
Silicone:							
SF-96 (40)	16	40	850	600	<−100	30.0	Low-torque aircraft oil bearings, air craft hydraulic and damping fluid.
SF-96 (300)	122	300	7,000	605	<−55	30.00	Heat transfer, hydraulic, and damping applications.
SF-96 (1,000)	401	1,000	20,000	605	<−55	30.00	Heat transfer, hydraulic, and damping applications.
DC-710	40	275	—	575	−10	40.00	Heat transfer, high-temperature trolley bearings.
Silcate:							
OS-45	3.95	12.4	2,400	—	<−85	20.00	Wide-temperature-range aircraft hydraulic fluid.
Orsil BF-1	2.4	6.8	1,400	395	<−100		

TABLE 19.27 List of Synthetic Lubricants[36] (*Continued*).

Type	Kinematic viscosity, cs			Flash point, °F	Pour point, °F	Approximate cost per gallon	Typical uses
	210°F	100°F	−65°F				
Polyglycol:							
LB-140X	5.7	29.8	—	345	−50	2.40	Water-insoluble oils used for internal-combustion engines. (Prestone Motor Oil), high temp. bearings in ovens and furnaces and gears.
LB-300X	11.0	65.0	—	490	−40	2.40	
LB-650X	21.9	141.0	—	490	−20	2.40	
50-HB-55	2.4	8.9	—	260	−85	2.40	Water-soluble oils used in wire drawing, metal forming, and some machine tools.
50-HB-280X	11.5	60.6	—	500	−35	2.40	
50-HB-2000	72	433	—	545	−25	3.00	
Hydrolube 300N	—	666.3	—	None	−55	2.50	Water-polyglycol mixture used as non-flammable hydraulic fluid in die-casting and machine tool work.
Chlorinated aromatics:							
Aroclor 1248	3.1	48	—	380	20	2.30	Die-casting machines and high-pressure compressors.
Aroclor 1254	6.1	470	—	None	50	2.30	
Polybutenes:							
No. 8	7.9	72	—	310	−40	—	Electrical oils, hydraulic and shock absorbing fluids, kilns, and ovens, refrigerator compressors.
No. 20	106	3,600	—	410	10	1.05	
No. 128	4,000	—	—	450	70	1.40	High pressure compressors.
Fluorolubes:							
Fluorolubes FS	1.10	3.52	—	None	—	300.00	Equipment handling liquid oxygen, concentrated hydrogen perioxide, etc. Density of approximately 1.8 grams/cc.
Fluorolubes S	4.6	24.1	—	None	—	300.00	Process and natural gas compressors.

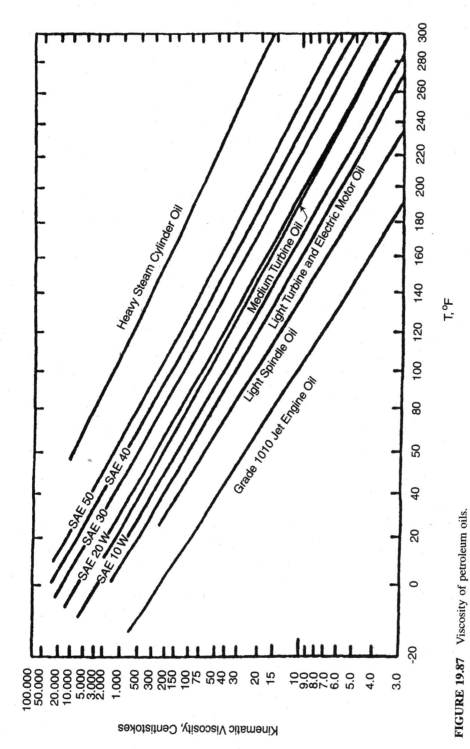

FIGURE 19.87 Viscosity of petroleum oils.

TABLE 19.28 Viscosity Conversion Factors

Multiply	By	To obtain
Stokes (cm^2/sec)	Density (g/cm^3)	Poises (gm/cm-sec)
Poises	100	Centipoises
Centistokes	Density (g/cm^3)	Centipoises
Centipoises	1.45×10^{-7}	Reynes (lb force-sec/in^2)
Centipoises	2.42×10^{-9}	(lb force-min/in^2)
Centipoises	5.6×10^{-5}	(lb mass in-sec)
Reyns (lb forc-sec/m$_2$)	6.895×10^3	Pascal-sec (N-sec/m^2)
Centipoises	10^{-3}	Pascal-sec (N-sec/m^2)

carbon-steel back; (2) an intermediate layer of copper or bronze; and (3) an overlay of lead-base babbitt from 0.001 to 0.020 in. thick. The intermediate layers increase the mechanical strength of the babbitt bearing and also provide reasonably good bearing surfaces in cases the thin babbitt surface layer is destroyed in operation.

Non-Babbitt Bearing Materials. Other common bearing materials used, whenever babbitt cannot be employed are:

- Bronze. Bearing bronzes may be grouped into lead bronzes, tin bronzes, and high-strength bronzes. The strength and high-temperature properties generally improve as one proceeds from the high-lead to high-tin to various high-strength bronzes. However, there is a loss in the compatibility properties as the amount of lead decreases. For this reason, it is generally advisable to use the highest lead

TABLE 19.29 General Types of Additives with Typical Chemical Compositions[36]

Function	Typical chemical type
Oxidation inhibitor	Phenolics
	Dithiophosphate
Detergent	Calcium petroleum sulfonate
Rust inhibitor	Organic acids
	Sodium petroleum sulfonate
Wear preventive	Trieresyl phosphate
Boundary lubrication	Chlorinated naphthalene
	Sulfurized hydrocarbon
Viscosity index improver	Polyisobutylene
Pour-point depressant	Polymethnerylate
Defoaming agent	Silicone oil

TABLE 19.30 Composition and Physical Properties of Babbitts[30]

Tin-base babbitts

Alloy	Specific gravity	Composition, %				Yield point* psi		Ultimate strength* psi		Brinell hardness		Melting point	Complete liquefaction
		Cu	Sn	Sb	Pb	66°F	212°F	66°F	212°F	68°F	212°F	°F	°F
1	7.34	4.56	90.9	4.52	None	4400	2680	12,850	6050	17.0	8.0	433	700
2**	7.39	3.1	39.2	7.6	0.03	6100	3000	14,900	8700	24.5	12.0	466	669
3**	7.46	8.3	83.4	8.3	0.03	6800	3100	17,600	9900	27.0	14.5	464	792
4	7.52	3.0	75.0	11.6	10.2	5550	2150	18,150	8900	34.5	12.0	363	583
5	7.75	2.0	65.5	14.1	18.3	2150	2150	18,060	8750	22.5	10.0	358	565

Lead-base babbitts

Alloy	Specific gravity	Composition, %				As (max)	Yield point* psi		Ultimate strength* psi		Brinell hardness		Melting point	Complete liquefaction
		Cu	Sn	Sb	Pb		66°F	212°F	66°F	212°F	68°F	212°F	°F	°F
6(e)	9.33	1.5	20	15	63.5	0.15	3800	2050	14,550	8060	21.0	10.6	358	581
7(f)	9.73	0.50	10	15	75	0.60	3550	1600	15,650	6150	22.5	10.5	464	514
8	10.04	0.50	5	15	80	0.20	3400	1760	15,600	6150	20.5	9.5	459	522
10	10.07	0.50	5	15	83	0.60	3550	1850	15,450	5450	17.6	9.0	468	507
11	10.28	0.50	—	15	85	0.25	3050	1400	12,800	5100	15.0	7.0	471	504
12	10.67	0.50	—	10	90	0.25	2800	1250	12,900	5100	14.5	6.5	473	498
15(g)	10.05	0.5	1	15	82	1.40					21.0	13.0	479	538
16(f)	9.88	0.5	10	12.5	77	0.20					27.5	13.6	471	495
19	10.50	0.50	5	9	95	0.20			15,600	6100	17.7	8.0	462	495

**In composites.
***Babbitts predominantly used by electric utilities (ASTM alloy B23).

content and the softest bronzes while still retaining the necessary strength and load-carrying capacity.

- Silver. Silver bearings normally consist of electro-deposited silver on steel backings with an overlay of 0.001 to 0.005 in. of lead. Indium is usually flashed on top of the lead overlay for corrosion protection. Silver bearings have outstanding metallurgical uniformity, excellent fatigue resistance and thermal conductivity, can carry very high loads, and can operated at high temperatures. Although the lead coating helps to relieve problem of poor embeddability and conformability, silver bearings are not recommended for applications where misalignment and dirt are present.

- Aluminum. Aluminum bearing alloys offer excellent resistance to corrosion by acidic oils, good load-carrying capacity, superior fatigue resistance, and good thermal conductivity. A smooth machine finish of the running surface is recommended along with a clean lubricant, a shaft hardness of 300 Brinell or higher, and a large enough clearance to allow for the high thermal expansion of the aluminum. Sometime the aluminum is overlaid with a thin coating of lead babbitt. This overlay assists in making up for the otherwise poor embeddability and conformability characteristics of the aluminum.

The range of temperatures that these various bearing materials, as well as some other materials, can endure is given in Table 19.31.

19.9 DESIGN CONSIDERATIONS

In practice, a designer must obtain quantitative data to ascertain on the one hand whether the bearing will meet his operational requirements, and on the other hand find out what the power losses, flows, temperatures, etc. will be to properly plan the layout of the facility. In Sections 19.3 to 19.7, the graphs and tables offer values for the performance of various bearing designs. These, however, do not exhaust the information required for rational design. What is needed is some orientation how the various geometrical and operational parameters affect bearing operation and how to go about improving or even optimizing a given bearing design. The following paragraphs should offer some guidance as to how to go about approaching this task.

19.9.1 Performance Parameters

The expressions required for calculating the more important items of bearing performance are the following:

- Film thickness. For an aligned journal, the film thickness is given by

$$\bar{h} = (h/C = 1 + \epsilon \cos(\theta - \phi)) \qquad (19.58)$$

TABLE 19.31 Approximate Temperature Limitations of Various Bearing Materials[36]

Material	Temperature range
Babbits	
Lead Base	~0 to ~150°C
Tin Base	~ -20 to ~150°C
Sintered Metals	
Bronze	~0 to ~150°C
Composites	~ -20 to ~200°C
Iron	~ -20 to ~230°C
Aluminum Alloys	~ -20 to ~180°C
Copper-lead	~ -20 to ~180°C
Bronzes	
Leaded	~ -20 to ~230°C
Tin	~ -20 to ~280°C
Aluminum	~ -20 to ~280°C
Cast Iron	~ -20 to ~300°C
Hardened Steels	~ -20 to ~300°C
Tool Steels	~ -20 to ~300°C
Carbongraphites (Untreated)	~ -40 to ~400°C
Carbongraphites (Treated)	~ -40 to ~540°C
Stellites	~ -40 to ~540°C
Nickel-Based Superalloys	~ -40 to ~650°C
Metal-Bonded Carbides	~ -40 to ~760°C
Metal-Bonded Oxides	~ -40 to ~760°C
Ceramics	~ -40 to >760°C

The attitude angle ϕ is defined as the angle between the line centers—a line passing the centers of bearing and journal—and the load vector. When the treatment is restricted to vertical loads, ϕ denotes the angle between location of h_{min} and the vertical and therefore the importance of ϕ lies in that it determines the location of h_{min}.

- Sommerfeld number (load parameter). The Sommerfeld number, given by

$$S = \frac{\mu N}{P}\left(\frac{R}{C}\right)^2 \qquad (19.59a)$$

has traditionally been the most important parameter. However, a more convenient quantity is the inverse of S, here called the load parameter, given by

$$\overline{W} = \frac{P}{\mu N}\left(\frac{C}{R}\right)^2 = \frac{W}{LD\mu N}\left(\frac{C}{R}\right)^2 \qquad (19.59b)$$

where $P = (W/LD)$ is the unit loading. What this parameter says is that any combination of P, μ, N, C, and R such as to leave the value of $\overline{W}$ unchanged, would result in the same bearing eccentricity ratio, ϵ, and attitude angle, ϕ.

- Minimum film thickness. The is the smallest distance between the journal and bearing surfaces and it is given by:

$$\overline{h}_{min} = \frac{h_{min}}{C} = (1 - \epsilon) \qquad (19.60)$$

What is normally referred to as load capacity relates to the load, W, which this h_{min} can support.

- Friction coefficient. This is the ratio between the frictional force and bearing load. It is normally expressed in the form of:

$$\left(\frac{R}{C}\right) f = \frac{(R/C)F_\tau}{W}$$

The general shape of f as a function s is given in Fig. 19.88. The region of sudden rise in f denotes the limit of hydrodynamic lubrication, followed by a regime of "boundary lubrication" characterized by partial contact between the mating surfaces.

- Power loss. This, of course, can be obtained from the value of F_τ, namely

$$H = F_\tau \cdot R \cdot \omega = f \cdot W \cdot R \cdot \omega$$

$$\overline{H} = \left(\frac{H}{H_0}\right) = \frac{H}{[\pi^3 \mu N^2 LD^3/C]} \qquad (19.61)$$

The quantity by which H is normalized, represents the power loss in an unloaded concentric journal bearing, i.e., one in which $\epsilon = 0$. It is known as the Petroff equation.

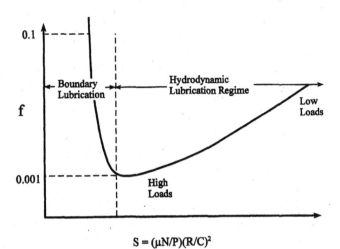

FIGURE 19.88 Behavior of friction coefficient in fluid film bearings.

- **Flow.** An amount of lubricant, Q_1, enters the bearing at the leading edge; an amount, Q_s leaks out the two sides of the bearing (one-half Q_s at each side), and an amount Q_2 leaves the trailing end of the pad. In most cases, since a journal bearing extends over circumference (2π), Q_2 is not discharged outside but reenters the next oil groove, so that the net amount of lubricant to be made up from an outside source is Q_s. The latter is referred to as side leakage. Clearly we must always have

$$Q_1 = Q_s + Q_2 \qquad (19.62)$$

All of these flows are given in dimensionless form as:

$$\overline{Q} = \frac{Q}{\frac{\pi}{2} NDLC} \qquad (19.63)$$

the denominator representing the flow in an unloaded, concentric bearing, i.e., at $\epsilon = 0$ (for which case $Q_s = 0$ and $Q_1 = Q_2$).

The above flows, Q_1, Q_s, and Q_2 are what may be called hydrodynamic flows induced by the shearing action and pressure gradients of the fluid film. Q_s is the minimum amount of oil to be delivered to the bearing to maintain a full fluid film with all its potentialities. In practice, designers supply more than this required minimum, using a supply pressure $p_s > p_a$. The effect of the supply pressure, usually of the order of 10 to 30 psig, can be ignored as far as bearing hydrodynamics are concerned.

- **Temperature rise.** A bulk temperature rise can be estimated from the values of power loss and side leakage, namely

$$\Delta T = (T_{av} - T_1) = \frac{H}{c_p w Q_s} \tag{19.64}$$

- *Dynamic coefficients.* The dimensionless stiffness is given by

$$\overline{K} = (K/2\mu NL)(C/R)^2$$

while the damping coefficient reads

$$\overline{B} = (\pi B/\mu L)(C/R)^3$$

from which the dimensional values of K and B can be obtained.

The coefficients ϵ, f, $\overline{H}$, $\overline{Q}_1$, $\overline{Q}_2$, $\overline{K}$ and $\overline{B}$ which serve to evaluate bearing performance are obtained from solutions of the Reynolds equation for the specific geometries and operating conditions of the various bearing designs. Many such solutions were given in Sections 19.3 through 19.7.

19.9.2 Bearing Configuration

The behavior of a bearing is naturally a function of its geometry. However, even for a given design there are a number of variables that will affect its performance. Among the more known parameters are the L/D and C/R ratios and the degree of preload. Of the less familiar ones one can cite load orientation, the geometry of the oil grooves or the relative proportions of a bearing's geometrical elements.

Journal Bearings. Although one often hears about the use of full, that is, 360° arc bearings, it is very rarely that such sleeves are employed in machinery. Most journal bearings consist of two or more pads separated by horizontal oil grooves making them in fact partial bearings, used either singly or in tandem. The number and distribution of these angular pads on bearing performance is one of the more important considerations in bearing design.

Partial Bearings. Whenever a single pad of an angular extent $\beta < 2\pi$ is used, it is called a partial bearing. When β is very small, its load capacity is low, as illustrated in Figs. 19.89 and 19.90. However, soon a limit is reached at about $\beta = 140°$ beyond which no further gains are registered. The reason for this asymptotic behavior is due to oil cavitation at the trailing end of the pad where the pressures decrease close to or even below ambient pressure. Thus, if a partial bearing is used there is no need to go beyond a 140° arc. The effect of temperature in partial bearings is a combination of two phenomena. The higher the arc the longer the dissipation path and the higher the temperatures; however, a longer arc produces thicker films and thus less heating. Consequently, as shown in Fig. 19.91, a crossover point occurs; at high loads low values of β are preferred, if low ΔT's are desired; at low loads a longer arc is preferred.

Grooved Bearings. Partial bearings are not used extensively. The most common designs are grooved bearings which consist of a number of pads arranged in tandem by cutting axial oil grooves around the 360° circumference. There is a great variety

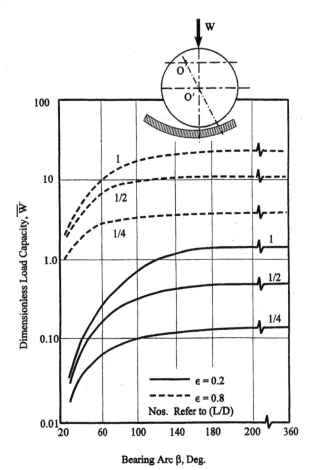

FIGURE 19.89 Effect of bearing arc on load capacity.[21]

of such designs, the most common being a 2-pad bearing with two grooves at the horizontal split. Others may have 3, 4 or 6 grooves forming the same number of individual pads. The more grooves the lower the load capacity, as shown in Figs. 19.92 and 19.93. Thus, if load capacity is the primary objective, a 2-groove bearing is best; however, those with a larger number of grooves are somewhat more stable. Related to the above is the fact that any hole or disruption in the bearing surface will reduce the load capacity. Figure 19.94 shows the effects on the pressure profile of cutting a slit or circular hole in the loaded part of a bearing. The larger the incursion, the more drastic the reduction in the hydrodynamic pressures which translates directly into reduced load capacity.

Tilting Pad Bearings. The primary characteristic of this family of bearings is that the individual pads are not fixed but are pivot-supported so that during operation not only does the journal move but so do the pads and each in a different fashion. A general picture of a tilting 3-pad bearing is shown in Fig. 19.95. The structural and analytical complexities of these bearings are more than compensated

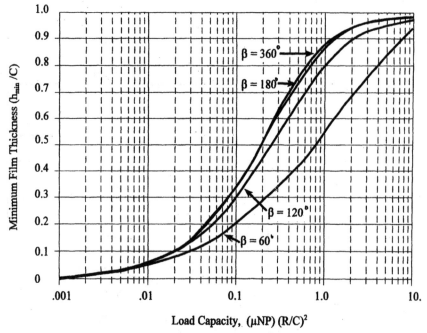

FIGURE 19.90 Effect of bearing arc on load capacity.[21]

by their great reliability and the fact that they have no rival in their stability characteristics.

The number of possible design parameters and operating modes in a tilting pad bearing is very large. Some of them are discussed below.

a. **Number of pads.** Table 19.32 gives a comparison of a 3-pad versus a 5-pad centrally pivoted bearing having zero preload. When the load is in line with the pivot, the 3-pad design has a higher load capacity but the reverse is true when the load direction is between the pads. For loads of engineering interest the 5-pad design consumes less power.

b. **Pivot location.** In order to assure two-directional rotation and for ease of assembly, most tilting pad bearings are centrally pivoted. However, a 10% or 15% displacement of the pivot in either direction would not significantly alter the general performance, a slight preference being a downward shift.

c. **Preload.** From many standpoints a high preload is desirable. Its effect on preventing the scraping of the top pads has been discussed previously and from this standpoint an m of at least 0.5 is required. High preloads also yield higher stiffness and damping. However, the penalty is that the film thickness over the pivot and often also the absolute h_{min} is reduced. Likewise, the power losses and temperatures rise with an increase in preload.

d. **Mode of loading.** In general, the shaft eccentricity will be lower when loaded over the pivot. It is characteristic of tilting pad bearings that, regardless of

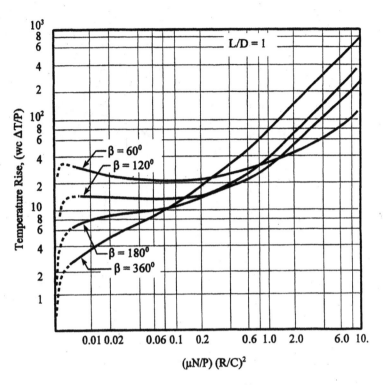

FIGURE 19.91 Effect of bearing arc on value of h_{min}.[31]

whether the load vector is over the pivot or between the pads, the locus of shaft center is along a vertical line which has a direct beneficial effect on stability. Results for the two modes of loading on stiffness and damping are given in Fig. 19.96 for a bearing of zero preload. As seen, both the spring and damping coefficients are lower for the between-pads mode of loading.

Oil-Ring Bearings. As pointed out previously, oil ring bearings operate under starved conditions. It is thus the main task of the designer to find ways to increase as much as possible the amount of oil delivered to the bearing surface. Some of the important parameters that play a role in accomplishing it are geometry shape of contact surface, weight, the material and size of the ring relative to the shaft. In an experimental study, a series of rings portrayed in Table 19.33 was tested with the purpose of both increasing the flow of lubricant and of extending the regime of stable ring operation. The conclusions reached were as follows:

a. An optimum ring shape is one with a quasi-trapezoidal cross-section and a series of straight teeth at the contact surface shown in Table 19.33 as Ring No. 2.

b. The best ring material is bronze with a weight of 0.135 lb per inch of ring circumference.

c. For bearing diameters in excess of 6 in., dual rings are recommended.

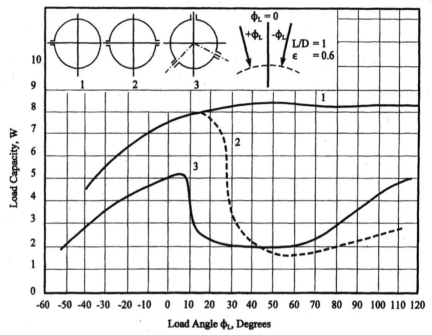

FIGURE 19.92 Load capacity of grooved bearings.[27]

d. An anchored spring leaf inserted between the ring and journal raises the amount of oil delivery and extends the ring's region of stable operation. One such stabilizer is shown in Fig. 19.97.

Load Angle. Bearing loads are usually directed midway of a pad or between grooves. However, improved performance can be obtained by shifting the load vector toward the trailing edge of the bearing pad. A comprehensive mapping of the effects of shifting the load vector around the circumference of a 2-groove bearing is shown in Fig. 19.98. Normally the load would be straight down, that is along $\phi_L = 0$. However, as seen in the figure by moving the load toward the trailing edge, improved performance is obtained for the entire range of bearing operation. At low loads an optimum occurs at a load of $\phi_L = 10°$; at high loads the value of ϕ_L is some 30°. The lowest load capacity would occur at a load angle of 60° from the midway point. Supplementary data is given in Table 19.34, where it is seen that the worst angular position results in a load capacity reduction of 70 to 80%. Similar data for a 3-groove bearing has been given in Fig. 19.92. Achievement of an optimum bearing position requires no special effort. It is sufficient to rotate the bearing in the housing the required 10° to 30° to obtain this. Attention should only be given to the oil delivery path since now the oil grooves would no longer be at the horizontal split. This can be taken care of by cutting a short oil supply channel on the outside of the bearing shell.

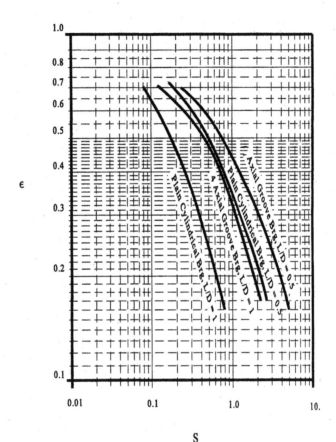

FIGURE 19.93 Comparisons of 2- and 4-axial groove bearings.

Misalignment. It was pointed out in an earlier section that an overhung impeller will cause bearing misalignment. As shown in Fig. 19.99, in severe misalignment the journal at one end may find itself in the upper half of the bearing even though the load is downward. As a consequence, a fluid film and hydrodynamic pressures may develop in both the lower and upper portions of the bearing. Stretching from the end where the hydrodynamic film is at the bottom, this film will wrap itself in helical fashion around the entire bearing circumference. In all cases the load capacity, that is the value of h for the imposed load, will be drastically reduced.

Thrust Bearings. Unit loads in thrust bearings are higher than in journal bearings and consequently their h_{min} will be smaller. But it should also be realized that, except for a bearing with a flat at the end, h_{min} in thrust bearings occurs not along a line as in journal bearings but at a point, namely the outer downstream edge of the pad. This point is also where T_{max} will occur and again it will be higher than in journal bearings. This is due to the low value of h_{min} but also to the higher linear velocities of the runner at the outer radius of the pad.

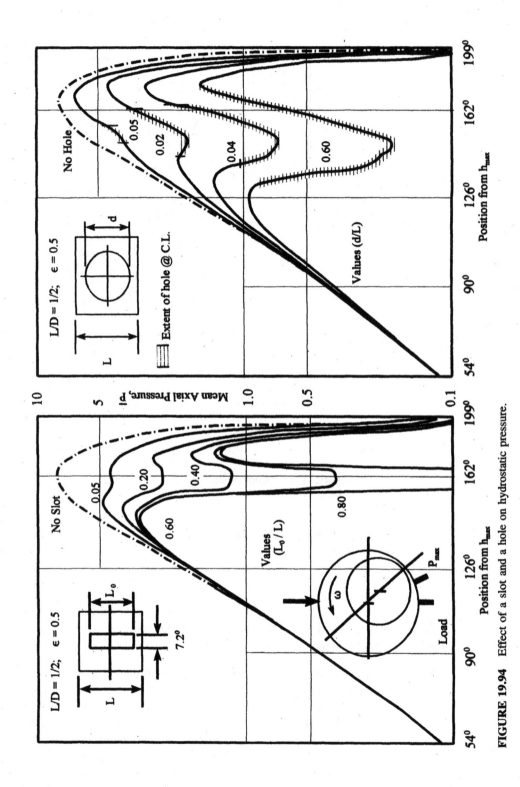

FIGURE 19.94 Effect of a slot and a hole on hydrostatic pressure.

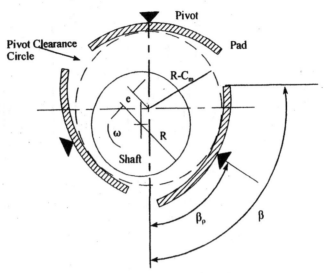

FIGURE 19.95 A tilting 3-pad journal bearing.

Tapered Land Bearings. A conventional tapered land bearing was shown in Fig. 19.35. There are three parameters here; the taper $(h_1 - h_2)$, the pad arc β and the (L/R_2) ratio. The angular extent also determines the number of pads in a thrust bearing. Table 19.35 shows the results of an optimization study giving the values of $(h_1 - h_2)$ and β for the entire range of (L/R_2) ratios. From this an optimum set of design parameters can be obtained for a particular application. It is worth noting that in general the optimum configuration is that which yields nearly square bearing pads.

An improved version of a plain tapered land bearing is one with a flat surface at the trailing end, as shown in Fig. 19.36. The additional merit of this design is that upon starting and stopping, the runner rides on a flat surface reducing wear.

TABLE 19.32 Relative Load Capacity 3- and 5-Pad Bearings[29]

$\overline{W}$	On-pivot load		Load between pivots	
	3 pads	5 pads	3 pads	5 pads
20	0.72	0.82	1.42	0.97
40	0.79	0.87	1.58	1.05
60	0.80	0.90	1.60	1.09
80	0.82	0.92	1.61	1.10
100	0.83	0.93	1.62	1.11

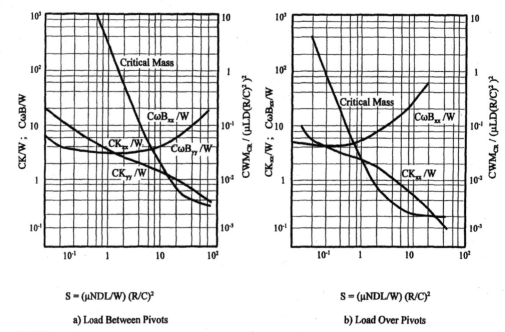

a) Load Between Pivots
b) Load Over Pivots

FIGURE 19.96 Effect of mode of loading on bearing stability in a 4-pad tilting pad bearing.

Here a new parameter is the ratio of the tapered to the flat portion. The plot in Fig. 19.100 shows such a variation from 60% to 100% taper, the latter being the tapered land bearing discussed previously. The load capacity peaks at a taper value of about 80% of the pad arc, that is the tapered portion should be four times that of the flat. Interestingly the value of (power loss/load capacity) achieves a minimum at the same point.

Misalignment. In properly operating thrust bearings, the load carried by each pad is the same. When the shaft and consequently the runner is misaligned, this is no longer true and some of the pads are much more heavily loaded than the other. A pictorial representation of this situation is given in Fig. 19.101. As seen, the loads carried by the heavily loaded pads as well as their maximum temperatures can be 10 times as high as the ones located opposite them where the runner is furthest from the pads. The values of h in the two sets of pads will be of the same ratio. The span of severity of bearing operation goes up with the number of pads used in the misaligned bearing. Thus, if misalignment is expected, one should not use more than 4 to 6 pads.

Hydrostatic Bearings. In a conventional hydrostatic bearing portrayed in Fig. 19.102, the load capacity is given by

$$W = \frac{\pi R_2^2 (p_o - p_a)[1 - (R_1/R_2)^2]}{2 \ln(R_2/R_1)}$$

There are therefore two parameters that determine the level of W; $(p_o - p_a)$ and

TABLE 19.33 Oil Ring Configuration[15]

Ring No.	Description	Cross Section (mm)	Unit Weight (N/m)
1	Split, T-Section Brass Ring Made from Rolled Stock Fastened Together	T → .133, .460, 133, A → .783	0.042
2	Split, Trapezoidal Section Machined From Bronze SAE 650 $\alpha = 30°, \beta^{**} = 0°$	.126 .562, .051, .094, 1.000	.135
3	Split Trapezoidal Section Machined From Bronze SAE 660 $\alpha = 30°, \beta^{**} = 90°$	.067, .118, β	0.142
4	Split Trapezoidal Section Machined From Bronze SAE 660 $\alpha = 30°, \beta^{**} = 45°$ or $135°$	.067, .118, β	0.139
5	Split and Relieved, Trapezoidal Section Machined from Bronze SAE 660 $\alpha = 30°, \beta^{**} = 0°$	.059, B, .205, .709	.131

* Unit weight equals ring weight/circumferential length.
** Angle β is measured from the direction of ring rotation.

(R_2/R_1). The variation of load with these quantities is shown in Fig. 19.102. As seen no optimum for load capacity occurs; it rises with Δp and drops with a rise in (R_1/R_2). A minimum occurs in the power loss, but power loss in a hydrostatic bearing is not of great concern and, when it is, it is due not to bearing geometry but to the onset of turbulence in the fluid.

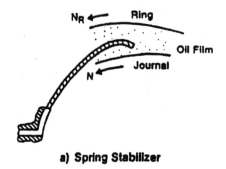

a) Spring Stabilizer

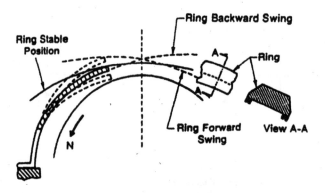

b) Stabilizer and Ring Positions during Oscillations

FIGURE 19.97 Configuration of oil ring stabilizer.[15]

19.9.3 Qualitative Guidelines

In selecting design parameters, it must be kept in mind that the choice often depends on the size of the bearing. Small bearings, less than 2 in. in diameter, can tolerate relatively lower values of h_{min}, higher unit loads, P; and operate close to isothermal conditions, whereas larger bearings require larger values of h_{min}, lower values of P, and tend to run close to adiabatic conditions. On the other hand, (C/R) ratios must be higher for small bearings. With this as an introduction, Table 19.36 gives some typical design practices in the field of journal bearings. The performance characteristics of one's design should fall somewhere within the range of values listed in the table.

Nearly all the bearing data given here are for bearings operating under laminar conditions. Should turbulence set in, the operating characteristics will change. One may expect turbulence when the bearing Reynolds number reaches a value between 750 and 1500. The higher the Reynolds number, the more intense will be the effect of turbulence. Table 19.37 shows what will be the impact of the turbulent regime on the major items of bearing operation.

PRINCIPLES OF BEARING DESIGN 19.143

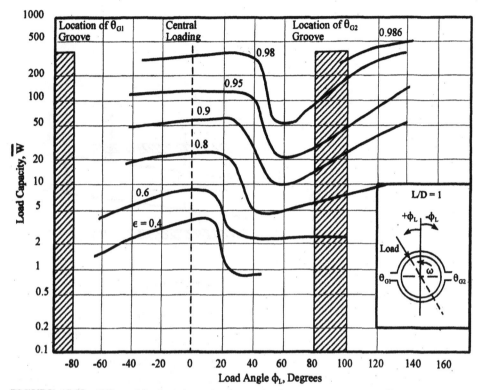

FIGURE 19.98 Effect of load angle on load capacity in two-groove bearing.[37]

TABLE 19.34 Effect of Load Angle on Load Capacity in Conventional Two-Groove Bearing[13]

L/D	ϵ	$\overline{W}$ at $\phi_L = 0$	Worst condition ϕ_L	$\overline{W}$	$\left[\dfrac{\overline{W} \text{ at Worst } \phi_L}{\overline{W} \text{ at } \phi_L = 0}\right]$
0.5	0.6	3.1	50	1.05	0.32
	0.95	83	55	14.5	0.175
1.0	0.6	8	52	2	0.25
	0.95	115	60	20	0.17

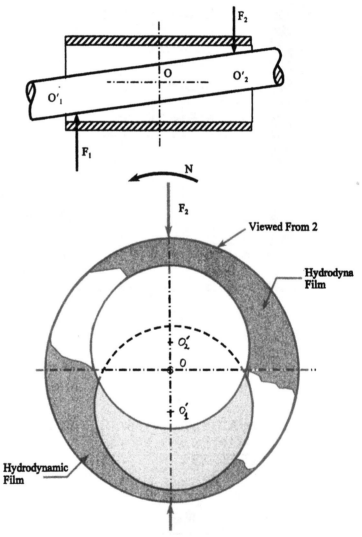

FIGURE 19.99 Hydrostatic forces and films under misalignment.

TABLE 19.35 Optimum Pad Arrangement[25]

L/R_2	h_1/δ_θ	β, deg	Number of pads
1/3	1	<30	>10
	1/2	<30	>10
	1/4	35	9
	1/8	40	8
1/2	1	40	8
	1/2	45	7
	1/4	50	6
	1/8	60	5
2/3	1	50	6
	1/2	60	5
	1/4	80	4
	1/3	>80	4

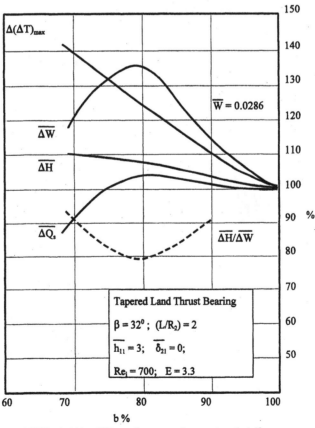

FIGURE 19.100 Effect of extent of taper (or flat).[13]

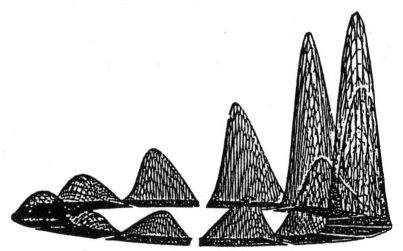
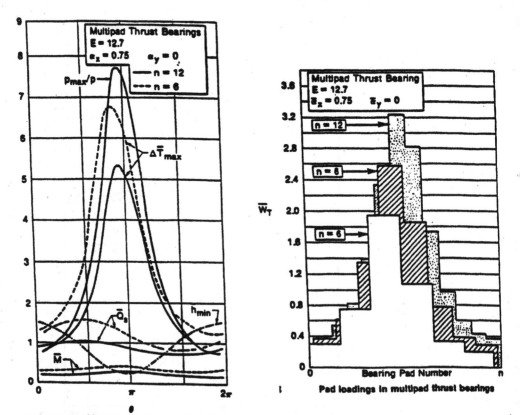

FIGURE 19.101 Effects of misalignment in thrust bearings.[28]

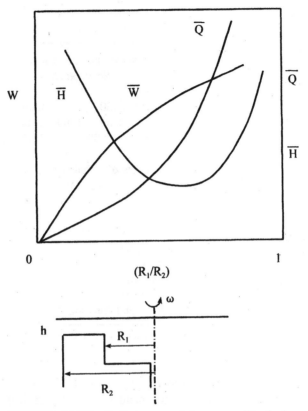

FIGURE 19.102 Performance of incompressible hydrostatic bearing.

TABLE 19.36 Typical Design Limits for Journal Bearings

Minimum film thickness	0.001–0.01 in. (0.0025 to 0.25 mm)
Temperature rise	Up to 80°F (27°C) (on babbitt)
Maximum temperature	Up to 300°F (150°C) (on babbitt)
Loads	500 psi (3.4 MPa)
L/D ratio	0.25 to 1.0
C/R ratio	0.001 to 0.002
Preload, m	0.25 to 0.75
Bearing arcs	150° to 60° for fixed pad
	80° to 30° for tilting pad
Inlet oil temperature, T	80°F to 130°F (27° to 55°C)

TABLE 19.37 Effect of Turbulence on Bearing Performance

Reynolds number a parameter
$$Re = \rho R\omega h/u$$

Regimes
- $Re < 750$ Laminar
- $750 < Re < 1500$ Transition
- $Re > 15000$ Turbulence

Effects of turbulence on bearing performance

Item	Effect
Load capacity	⊕
Oil flow	⊖
Power loss	⊕
Temperatures	⊕
Stiffness and damping	⊕ or ⊖

⊕-Increase; ⊖-Decrease.

In a more comprehensive way, Table 19.38 provides a guide in which direction design modifications should head in order to ameliorate unsatisfactory results in a chosen design. Finally, Table 19.39 offers a cursory look at the relative advantages and disadvantages in choosing journal bearings of different designs.

TABLE 19.38 Effect of Design Parameters on Journal Bearing Performance

Code: (+) means increase magnitude of parameter to achieve effect in left-hand column.
(−) means decrease magnitude of parameter to achieve effect in left-hand column.

Example: To decrease temperature rise, one or more of the following can be done: decrease L/D ratio, increase C/R, decrease oil viscosity, etc.

Objective	L/D	C/R	Geometry	Viscosity	Preload, m	Arc, β	Supply oil pressure, p_s
To decrease temperature rise	(−)	(+)	Elliptical	(−)	(−)	(−)	(+)[b]
To reduce power loss	(−)	(+) or (−)	Circular	(−)	(−)	(−)	(−)
To reduce T_{max}	(+)	(+)	Elliptical	(−)	(−)	(+) or (−)	No effect
To increase oil flow	(+)	(+)	Elliptical	(−)	(−)	(+)	(+)
To improve stability	(−)	(−)	See (a) below	(+) or (−)	(+)	(−)	No effect
To increase load capacity	(+) or (−)	(−)	Circular	(+)	(−)	(−)	No effect
To avoid turbulence	(+) or (−)	(−)	3-lobe	(+)	(+)	(−)	(−)
Stability	(+)	(−)	Tilting Pad	(+) or (−)	(+)	(+)	No effect

a) The stability of a journal bearing increases in the following order: circular, pressure, elliptical, 3-lobe, tilting pad.
b) Apparent effect only.

TABLE 19.39 Characteristics of Various Journal Bearings Journal Bearing Summary Table

Bearing type	Advantages	Disadvantages	Comments
Axial Groove	1. Easy to make 2. Low cost	1. Subject to oil whirl	Round bearings are nearly always "crushed" to make elliptical or multi-lobe
Elliptical	1. Easy to make 2. Low cost 3. Good damping at critical speeds	1. Subject to oil whirl at high speeds 2. Load direction must be known	Probably most widely used bearing at low or moderate speeds
Three and Four Lobe (Tapered Land, etc.)	1. Good suppression of whirl 2. Overall good performance 3. Moderate cost	1. Some types can be expensive to make properly 2. Subject to whirl at high speeds	Currently used by some manufacturers as standard bearing design
Pressure Dam (Single Dam)	1. Good suppression of whirl 2. Low cost 3. Good damping at critical speeds 4. Easy to make	1. Goes unstable with little warning 2. Dam may be subject to wear or build up over time 3. Load direction must be known	Very popular with petrochemical industry. Easy to convert elliptical over to pressure dam.
Hydrostatic	1. Good suppression of oil whirl 2. Wide range of design parameters 3. Moderate cost	1. Poor damping at critical speeds 2. Requires careful design 3. Requires high pressure lubricant supply	Generally high stiffness properties used for high precision rotors
Tilting Pad	1. Will not cause whirl (no cross coupling) 2. Wide range of design parameters	1. High cost 2. Requires careful design 3. Poor damping at critical speeds 4. Hard to determine actual clearances 5. High horsepower loss	Widely used bearing to stabilize machines with subsynchronous non-bearing excitations

19.10 REFERENCES

The selection of the following references was made with the intent of providing sources from which additional data could be culled for bearing design purposes.

1. Allaire, P. E., D. F. Li, and K. C. Choy, "Transient Unbalance Response of Four Multilobe Journal Bearings," *Journal of Lubr. Technology,* Trans. ASME, July 1980.
2. Chen, H. M., "Active Magnetic Bearing Technology: A Conventional Rotordynamic Approach," 15th Leeds-Lyon Symposium on Tribology, September 1988.
3. Chen, H. M., "Magnetic Bearings and Flexible Rotor Dynamics," STLE Annual Meeting at Cleveland, Ohio, May 9–12, 1988.
4. Chen, H. M., et al., "Stability Analysis for Rotors Supported by Active Magnetic Bearings," 2nd International Symposium on Magnet Bearings, July 12–14, 1990, Tokyo, Japan, pp. 325–328.
5. Chen, H. M., "Design and Analysis of a Sensorless Magnetic Damper," presented at ASME Turbo Expo, June 5–8, 1995, Houston, Texas, 95GT180.
6. Compressor Handbook, Gulf Publishing Co., Book Division.
7. Gross, W. A., "Gas Film Lubrication," John Wiley, 1962.
8. Heshmat, H., J. A. Walowit, and O. Pinkus, "Analysis of Gas-Lubricated Compliant Thrust Bearings," ASME Paper 82-LUB-39, 1982.
9. Heshmat, H., J. A. Walowit, and O. Pinkus, "Analysis of Gas-Lubricated Foil Journal Bearings," ASME Paper 82-LUB-40, 1982.
10. Heshmat, H., and J. Dill, "Fundamental Issue in Cryogenic Hydrodynamic Lubrication," Proc. AFOSR/ML Fundamentals of Tribology Workshop (February 1987).
11. Heshmat, H., "Analysis of Compliant Foil Bearings with Spatially Variable Stiffness" presented at AIAA/SAE/ASME/ASEE 27th Joint Propulsion Conference, June 24–26, 1991, Sacramento, CA, Paper No. AIAA-91-2101.
12. Heshmat, H., "A Feasibility Study on the Use of Foil Bearings in Cryogenic Turbopumps," presented at AIAA/SAE/ASME/ASEE 27th Joint Propulsion Conference, June 24–26, 1991, Sacramento, CA, Paper No. AIAA-91-2103.
13. Heshmat, H., and P. Hermel, "Compliant Foil Bearing Technology and Their Application to High Speed Turbomachinery," 19th Leeds-Lyon Symposium on Thin Film in Tribology—From Micro Meters to Nano Meters, Leeds, U.K., Sept. 1993, D. Dowson, et al. (eds) (Elsevier Science Publishers B.V., 1993), pp. 559–575.
14. Heshmat, H., and O. Pinkus, "Performance of Starved Journal Bearings with Oil Ring Lubrication," *Journal of Tribology,* Trans. ASME 107, no. 1 (1985): 23–32.
15. Heshmat, H., and O. Pinkus, "Experimental Study of Stable High-Speed Oil Rings," *Journal of Tribology,* ASME 107, no. 1 (1985): 14–22.
16. Heshmat, H., and O. Pinkus, "Performance of Oil Ring Bearing," International Science Conf. on Friction, Wear, Lubr., Tashkent, U.S.S.R., May 1985.
17. Hustek, J. F., and O. J. Peer, "Design Considerations for Compressors with Magnetic Bearings," Proc. 3rd Int. Symposium on Magnetic Bearings, July 1993, Alexandria, VA.
18. Jones, G. J., and F. A. Martin, "Geometry Effects in Tilting-Pad Journal Bearings," ASLE Paper No. 78-AM-@A-2, 1978.

19. Ku, C.-P. R., and H. Heshmat, "Compliant Foil Bearing Structural Stiffness Analysis: Part I—Theoretical Model Including Strip and Variable Bump Foil Geometry," *Journal of Tribology,* Trans. ASME, vol. 114, no. 2 (1992): 394–400.

20. Pinckney, F. D., and J. M. Keesee, "Magnetic Bearing Design and Control Optimization for a Four-Stage Centrifugal Compressor," Proceedings of Mag. '92, pp. 218–227.

21. Pinkus, O., and B. Sternlicht, "Theory of Hydrodynamic Lubrication" (New York: McGraw-Hill, 1961).

22. Pinkus, O., and D. F. Wilcock, "Low Power Loss Bearings for Electric Utilities: Volume II: Conceptual Design and Optimization of High Stability Journal Bearings; Volume III: Performance Tables and Design Guidelines for Thrust and Journal Bearings," MTI Report Nos. 82TR42, 82TR43, April 1982.

23. Pinkus, O., "Analysis of Elliptical Bearings," Trans. ASME, vol. 78, 1956, pp. 965–973.

24. Pinkus, O., "Analysis and Characteristics of the Three-Lobe Bearing," Trans. ASME, Ser.D., vol. 81, March 1959.

25. Pinkus, O., "Solution of the Tapered-Land Sector Thrust Bearing," Trans. ASME, vol. 80, Oct. 1958.

26. Pinkus, O., "Analysis of Non-circular Gas Journal Bearings," *Journal of Lubr. Technology,* Trans. ASME, Oct. 1975.

27. Pinkus, O., "Solution of Reynolds Equation for Arbitrarily Loaded Journal Bearings," Trans. ASME, Series D, vol. 83, no. 2, June 1961.

28. Pinkus, O., "Misalignment in Thrust Bearings Including Temperature and Cavitation Effects," *Journal of Tribology,* Oct. 1986.

29. Pinkus, O., "Optimization of Tilting Pad Journal Bearings Including Turbulence and Thermal Effects," *Israel Journal of Technology,* vol. 22, nos. 2–3, 1984/85.

30. Pinkus, O., "Manual of Bearing Failure and Repair in Power Plant Rotating Equipment," EPRI, July 1991.

31. Raimondi, A. A., and J. Boyd, "A Solution for the Finite Journal Bearing and Its Application to Analysis and Design—III," Trans. ASLE, vol. 1, no. 1, 1959.

32. Reddickoff, J. M., and J. H. Vohr, "Hydrostatic Bearings for Cryogenic Rocket Engine Turbopumps," *Journal Lubr. Technology,* July 1969.

33. Schmied, J. L. and J. C. Predetto, "Rotor Dynamic Behaviour of a High-Speed Oil-Free Motor Compressor with a Rigid Coupling Supported on Four Radial Magnetic Bearings," Proceedings of 4th International Symposium on Magnetic Bearings, August 23–26, 1994, ETH Zurich, Switzerland, pp. 441–447.

34. Vohr, J. H., "The Design of Hydrostatic Bearings," Columbia University, NY.

35. Walton, J. F., and H. Heshmat, "Compliant Foil Bearings for Use in Cryogenic Turbopumps," Proceedings of Advanced Earth-to-Orbit Propulsion Technology Conference Held at NASA/MSFC May 17–19, 1994, NASA CP3282, vol. 1, Sept. 19, 1994, pp. 372–381.

36. Wilcock, D. F., and Booser, E. R., "Bearing Design and Application" (New York: McGraw Hill, 1957).

CHAPTER 20
COMPRESSOR VALVES

Walter J. Tuymer
Hoerbiger Corporation of America, Inc.

Dr. Erich H. Machu
Consulting Mechanical Engineer
Hoerbiger Corporation of America, Inc.

20.1 PURPOSE

Compressor valves are check valves that control the flow into and out of a compressor cylinder. There has to be at least one suction valve and one discharge valve in every compression chamber.

20.2 HISTORY

The modern era of compressor valves started in 1897 when the Austrian engineer, Hanns Hoerbiger, designed and patented a steel plate valve intended for a low pressure air blower in a steel mill application. Interestingly, this design employed a frictionless guided valve plate useful for non-lubricated compressors, a feature, which until the introduction of non metallic materials to compressor valves was poorly achieved by most other valve designs.

In 1910, Hans Mayer patented a different valve design that became known as the "Feather valve." This design uses several flexible steel strips as sealing elements and became the standard for the Worthington Company for many years.

In 1931, Ingersoll Rand patented yet another valve design that became known as the "Channel valve." The channel valve employed several strips with a cross section like a "U" (therefore the name "channel"), each supported by a leaf spring. It was the standard valve for the Ingersoll Rand Company and was probably produced in higher numbers than any other compressor valve.

A valve design older than all the others was rediscovered in the late 1950's with the event of pipeline compressor applications and the availability of high strength plastics, in this case nylon. Thompson Industries developed this poppet valve for the Clark Brothers, a compressor manufacturer in Olean, New York. This valve proved to be extremely efficient in low compression ratio applications.

20.3 SURVEY OF VALVE DESIGN

All automatic compressor valves have several basic components in common:

- Valve seat
- Sealing element(s)
- Lift constraint (guard)
- Spring(s)

The flow passages in a valve can be arranged in various different ways:

- Circular rows of ports
- Parallel series of ports
- Irregular arranged number of holes

20.3.1 Valve Designs Used in Air and Gas Compressors

Ring Valve. Ring valves are probably the most commonly used valves in air as well as gas compressors. A properly manufactured valve ring is a very simple element with a perfectly uniform stress distribution. It therefore has a high tolerance for impacts. From a valve designer's point of view, rings can easily be guided and the utilization of the available area is good (for manufacturing reasons most valves are round). The disadvantage of individual rings is the need for each ring to be separately spring loaded. Since the specific spring load for each ring cannot be perfectly identical and because the flow distribution across a valve is not uniform, it is difficult if not impossible to make all rings move uniformly. On the other hand, due to the possibility of individual motion of each ring, this design is more tolerant to liquids than other designs.

Several ring valve designs which are used are discussed below.

Simple Ring Valve. This design uses plain rings guided in the valve guard. The valve springs can either be small individual coil springs, separate for each ring, or slightly larger springs supporting two rings. Some older valve designs use one large coil per ring. Wafer or lentoid springs are used in smaller valves. The advantage of these springs is their low space requirement, allowing for extremely thin valve guards and therefore low clearance volume. The disadvantage of these springs is the limited possibility of adaptation to different operating conditions. For this reason, they are mostly used in air compressors.

Damped Ring Valve. The only damping system used in ring valves is *gas damping*. This design uses very thick valve rings guided on the full diameter in a closely fit groove in the valve guard. In theory, when the valve opens, the gas beneath the

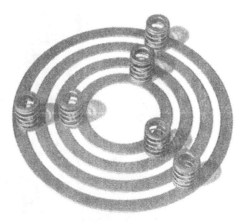

FIGURE 20.1 Ring and spring set for ring valve with one spring covering 2 rings.

valve ring is trapped in the guard groove and has to be squeezed out through the narrow passages on either side of the valve ring. The effect of damping largely depends on production tolerances, state of wear, and/or presence of liquid or solid contamination in the gas. On valves with non-metallic valve rings and steel guards, the different coefficient of thermal expansion makes this an almost impossible task for the designer. It follows that the application for this damping system is limited to metallic sealing elements.

Contoured Ring Valves. Some valve designs utilize a contoured valve ring together with a heavily chamfered or contoured valve seat groove. This is done to achieve a lower flow resistance in the lift area where, with conventional designs, the gas has to make two 90° turns. On the other hand, for given valve lift and given total length of seat lands, the geometric valve port area is reduced in the proportion of sine of the angle of flow deviation.

All these valves use non-metallic valve ring materials. Since these materials have a different thermal expansion coefficient than the valve seat material, there may be a leakage problem either in the cold or hot condition of the valve. This problem is reduced by a certain flexibility of the valve ring which allows it to "roll into" the valve seat groove.

FIGURE 20.2 Ring valve with wafer springs.

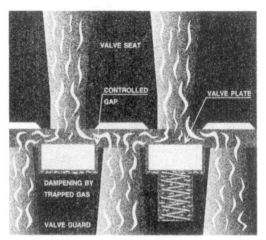

FIGURE 20.3 Gas dampened ring valve.

Ported Plate Valve. There are basically two types of ported plate valves in existance:

Simple Plate Valve. Plate valves, for the most part, are ring valves with their individual rings connected by bridges. The advantage of a plate valve over an individual ring valve is that there is only one sealing element to be controlled. The simultaneous opening or closing of all ports is automatically given. This advantage however comes at a price. From a stress point of view, the simple, very uniform valve ring has been transformed into a much more complicated element. When a valve plate impacts at one point it starts to bend, causing a rather complex, non uniform stress distribution.

Double or Mass Damped Plate Valve. The designation *double* damped valves is actually a misnomer, since there is only one real damping feature used. The valve springs themselves are not to be considered a damping feature.

Plate valves allow for a very effective, mechanical damping system. So called damping plates are positioned between the valve plate and the valve guard and sometimes are spring loaded separately. During the valve opening event, the valve

FIGURE 20.4 Contoured ring valve (*courtesy Cook Manly*).

FIGURE 20.5 Double damped valve.

plate travels the first portion of the lift alone and then collides with the damping plates. During this impact, linear momentum is conserved, but kinetic energy is destroyed. In other words, the energy required to accelerate the mass of the damping plates results in a reduction of the velocity of the valve plate itself and consequently reduces the final impact velocity against the valve guard. As an extra bonus, the valve plate tends to be leveled when contacting the damping plates.

Channel Valve. By numbers, this valve design is by far the most widely used compressor valve in the Western hemisphere. It uses a number of straight sealing elements with the U shaped cross section, therefore the name *channel valve*. Each individual channel is spring loaded by a leaf spring and guided on both ends of

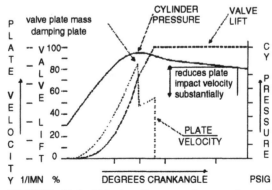

FIGURE 20.6 Velocity diagram of mass damped valve.

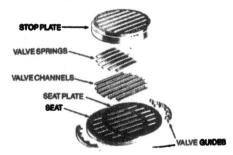

FIGURE 20.7 Channel valve (*courtesy of Dresser Rand*).

the channel by a comb like guide. This valve design is very efficient in low to medium pressure, low to medium speed compressors.

Feather Valve. This valve design is intriguing for its apparent simplicity. It uses a leaf type sealing element which is allowed to bow into a machined recess in the guard. The leaf is therefore also its own spring. Feather valves are no longer used in new compressors and their application was limited to low and medium pressure compressors with clean gas service.

Poppet Valve. Poppet valves use rather large (approx. 7/8 inch diameter) holes in the valve seat and for each of these holes there is a mushroom shaped sealing element, called a poppet. Each poppet has its own valve spring. The original poppets were made of bronze, which due to its high mass, was practically useless. The advent of Nylon made this design the valve of choice for low compression ratio, low speed compressors. Valve lifts of up to 3/8 inch were commonly used and in pipeline service this valve's efficiency was unequaled.

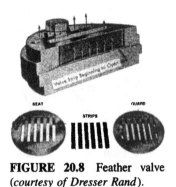

FIGURE 20.8 Feather valve (*courtesy of Dresser Rand*).

FIGURE 20.9 Poppet valve (*courtesy of Dresser Rand*).

When PEEK (Polyetheretherketone,[1] a polymer) became available, poppet valves were also successfully used in higher compression ratios. With reduced valve lifts necessary for these applications, the efficiency of the standard poppet valve should be checked against other valve designs.

For high speed compressors, a variation of the common poppet valve was developed. This design utilizes a much smaller poppet—called *Minipoppet*—and therefore a smaller valve seat hole. Lower valve lifts without reducing flow area are possible with this design.

Double Deck Valve. In conventional cylinder designs, there is limited space available for compressor valves. If the valve area achieved is insufficient and added clearance volume can be tolerated, double deck valves can be used. Double deck valves can be of different basic valve designs such as ring-, plate-, poppet- or feather valves. In all but the feather valve design, a second valve is positioned upside down above the first valve. The two valves are separated by a spacer and held together by means of a sleeve. The gas flow to the bottom deck is straight from the cylinder; the flow to the top deck is through the sleeve and then reversed by means of a cover. The combined valve area of a double deck valve is approximately 40% larger than that of the same size single deck valve.

Deck-and-One-Half Valve. Deck-and-one-half valves are valves where a full deck valve is positioned above a ring type bottom valve and the flow through both valves is in the same direction. These valves provide an even larger valve area than double deck valves, but require a very deep valve pocket since the cylinder porting has to be above the top deck valve. A cone type spacer is normally used to separate the two valve decks and provides flow access to the top deck. The total valve area of a deck-and-one-half valve can be 60 to 65% larger then the same size single deck valve.

---

[1] See e.g. Purdue proceedings 1990, page 701.

FIGURE 20.10 Double deck poppet valve.

Concentric Valves. Concentric valves are valves where a suction and a discharge valve are arranged concentrically either within one body, or as two individual valves. Either the suction or the discharge valve can be arranged in the center, the other valve is then positioned around it. These valves are normally used on single acting cylinders and are positioned on the face of the cylinder. The original objective was to make maximum use of available cover area with minimum clearance volume. This is needed in compressors with very high compression ratios as seen in refrigeration compressors and small air compressors. Small displacement high pressure cylinders also use this valve design.

Reed Valve. The domain of the reed valve encompasses small refrigeration and air compressors. A reed valve normally consists of a single seat plate with valve reeds positioned on both sides of the seat plate. The reeds themselves are made of thin strip steel and can have almost any shape. The reed is held on one end and covers a port with the other end. The valve opens when the differential pressure starts to bend the reed away from the seat plate. This elastic deformation of the reed also acts as the valve spring and no other springs are used.

There are reed valve designs using two seat plates with a spacer (normally a gasket) and the valve reeds positioned between the seat plates. The advantage of this design is a positive stop for the valve reed when the valve is open; the disadvantage is its high clearance volume and a higher cost compared to a single seat plate design. Due to the low mass of the moving element—the valve reed—this design can be used at very high compressor speeds. Reed valves are commonly used in compressors up to 3600 rpm and have been successfully tested at speeds up to 7000 rpm.

20.4 THEORY

20.4.1 Principle of operation

The essential part of every compressor valve is the sealing element. It is usually spring loaded and allowed to move between two stops, the valve seat and the valve guard. When resting against the valve seat the valve is closed, when against the valve guard, it is considered to be fully opened. The distance the sealing element can travel between these two stops is called the *valve lift*.

The sealing element is made to move by the action of a gas force and a spring force rather than mechanically. The forces of gravity or inertia forces due to machine vibration are usually small and can be neglected. Other forces are undesired like the viscous sticking force due to the presence of liquids, but sometimes have to be taken into account[9].

When the valve is opened, the flowing gas will exert a dynamic gas drag force on the valve plate which is of the order of magnitude of the pressure drop of the gas flow across the valve multiplied by valve plate area. When closed, the valve has to be able to support the full static pressure difference between suction and discharge pressures, which can be very high. The flexural strength required to resist this pressure difference is supplied by the valve seat.

Special attention has to be given to the motion of the sealing element during the opening and closing events. In fact, in most cases, it is the impact of the sealing element against one of its stops that causes the valve to fail. Small impact velocities are a condition necessary to assure long valve life. Other reasons for premature valve failure may be wear, erosion and clogging due to solid or liquid contaminants, poor maintenance or simply mechanical overload.

20.4.2 Relationship between Valve Design and Compressor Efficiencies and Work

The compressor efficiencies of interest are

- The energy efficiency or isentropic efficiency η_{is}, comparing work requirements of the real machine and ideal machine operating under identical conditions of nominal suction pressure and temperature, nominal discharge pressure and gas composition
- The volumetric efficiency η_{vol} or better efficiency of delivery λ, comparing stroke volume with the volume of gas at nominal suction conditions delivered per cycle.

The amount of work required by a compression cycle is the amount of energy required by one working chamber (called *head end* or *crank end* depending on the location with respect to the crankshaft) to take a mass of gas M from the suction line, compressing it inside the cylinder from suction to discharge pressure, then delivering M to the discharge line, and re-expanding the gas remaining in the cylinder to suction pressure.

- The isentropic part W_{is} which is theoretically required by the ideal machine of same size working without losses under the same external conditions, i.e. with identical gas composition, suction and discharge pressures, suction temperature. In the real machine however, the gas inside the cylinder at the beginning of the compression stroke will be at a temperature $T_{1,s}$ generally higher (for reasons explained below and given in more detail in[12]) than T_1, consequently already the isentropic work W_{is} to be expended per unit mass of gas will be increased in the proportion of $T_{1,s}/T_1$.

- And an additional ΔW covering the losses occurring in the real machine, corresponding to the work losses due to pressure drops in valves and valve pockets, repeated expenditure of work for recompressing gas that has leaked back during a previous cycle, heat exchange "in the wrong moment" causing the process to be polytropic, mechanical friction between moving parts like piston rings and packings, etc. The term *work loss* is in fact a misnomer and in apparent conflict with the principle of conservation of energy or the first law of thermodynamics. What really happens is that energy of a high value (mechanical or electrical energy coming from the driver) is converted into useless energy, useless in the sense of the second law of thermodynamics, i.e. heat energy at a low temperature level whose presence is even a nuisance.

Since the days of James Watt some 200 years ago, it is common practice to measure total work exchange between a piston and the gas inside a cylinder as the area of the *PV*-diagram or indicator diagram. The ratio between the work required by the ideal machine and the work required by the real machine is called *energy efficiency* or *isentropic efficiency*

$$\eta_{is} = \frac{100 \times W_{is}}{W_{is} \dfrac{T_{1,s}}{T_1} + \Delta W} \ [\%]$$

Volumetric efficiency η_{vol} is defined geometrically and can be taken from an indicator diagram.

$$\eta_{vol} = 100 - s \left[\frac{Z_1}{Z_2} \left(\frac{P_2}{P_1} \right)^{1/\kappa} - 1 \right] \ [\%]$$

where

- s total cylinder clearance [%]
- Z_1, Z_2 real gas compressibility at nominal suction and discharge conditions
- P_1, P_2 nominal suction and discharge pressures
- κ exponent of an isentropic temperature change in the range between P_1 and P_2, with an ideal gas, $\kappa\ C_p/C_v$

The efficiency of delivery λ is equal to volumetric efficiency η_{vol} minus a sum of volume losses

$\lambda \quad \eta_{vol} - \Sigma$ (volume losses) [%]

While the influence of the valve on η_{vol} is limited to its contribution to clearance volume s, the volume losses responsible for the difference between η_{vol} and the efficiency of delivery λ depend to a great part on design and state of wear of the valves, in particular:

- Temperature corrections to account for the fact that the gas inside the cylinder at the beginning of the compression stroke will be at a higher temperature than T_1 and hence lighter. This heating results from the conversion of the part of ΔW corresponding to the work loss in suction valves and pockets into heat dissipated in the gas during the intake cycle. In addition, there can be heat exchange between the gas and the walls of the suction plenum, of the suction valve and of the cylinder.
- Volume losses due to leakage in valves, piston rings and packings. With the exception of packings, the impact of leakage is twofold: it reduces the amount of gas and increases the temperature.
- Volume losses due to excessive valve throttling
- Volume losses due to late valve closure
- Pulsations can cause compression to start from a lower pressure than p_1 thus producing a volume loss.

An analytical treatment of this problem area based on differential equations is given in Ref. 11, based on simple algebraic equations in Refs. 12 and 14. It follows that both, isentropic efficiency volume losses due to leakage in valves, piston rings and packings. Except for leaking packings, the impact of leakage and efficiency of delivery are affected by valve losses, which therefore have to be minimized. On the other hand, valve life should be as long as possible.

Unfortunately, both postulates are contradictory: For a given valve design, good efficiencies mean small ΔW and therefore small pressure drops in the valves, hence large flow areas and a *high valve lift*. On the other hand, a long lasting valve means small impact velocities and hence a *small lift*. It is up to the skill and the experience of the valve engineer to find the best possible compromise by properly selecting valve design, lift, springs and materials.

20.4.3 Motion of Sealing Element

Under ideal conditions the opening motion of the valve sealing element is initiated by the pressure differential across the valve overcoming the pre-load of the valve springs. The acceleration of the element is a function of the gas drag force and an elastic force provided by the valve springs.

During the opening period of the valve its instantaneously available valve area starts at zero and gradually increases to its full flow area. At this time however the piston moves with a certain velocity. Depending on which changes faster, the volume displaced by the piston or the gas flow rate through the valve, the differential pressure across the valve increases or decreases. This condition causes the "opening hump" seen on most *pv*-diagrams (Fig. 20.11). Once the valve is open, the volume flow through the valve should more or less equal the piston displacement, such that the pressure changes inside the cylinder are small. During this phase, the instantaneous gas velocity in the valve lift area is approximately equal to the instantaneous piston velocity multiplied by the ratio of the piston area and the valve lift area. As the piston approaches the dead center position, its velocity approaches zero and the gas drag force decreases. The valve springs should be sized to overcome this diminishing gas drag force and start the closing motion of the sealing element in order to make sure that the valve is closed at the moment the piston reaches dead center.

If the media to be compressed contains foreign elements which have an adhesive effect (water, oil, etc.), the valve motion is adversely affected. Depending on the contact area between the sealing element and the valve seat or valve guard, the adhesive force (commonly called valve "sticktion") can be sizable. The differential pressure required to open the valve is equal to the pre-load of the valve springs plus the adhesive force. The valve plate motion will be delayed while the pressure differential continues to increase and to reach high values. Once the valve plate has left the seat, the sticking force suddenly drops to zero, the higher gas drag force is left without any counteracting force, thus producing a much higher acceleration than would normally be the case. Consequently, the opening impact of the sealing element against the valve guard is very high. In addition, the "opening

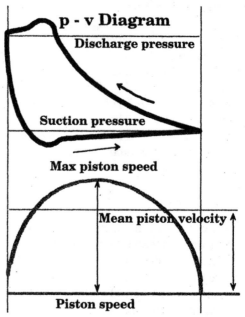

FIGURE 20.11 *P-V* diagram with piston velocity.

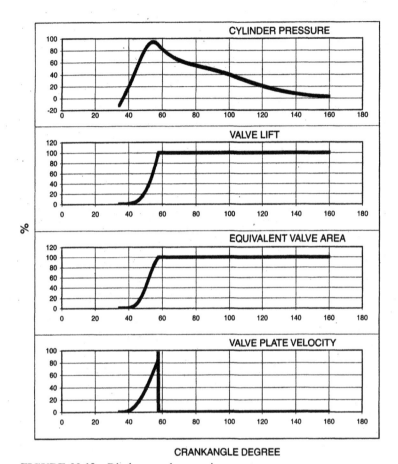

FIGURE 20.12 Discharge valve opening event.

hump" on the *pv*-diagram is larger than normal. When closing the valve, the spring force has to overcome the gas drag force and in addition the adhesive force of the sealing element against the guard. This can cause the valve to close well after the piston reaches dead center, resulting in gas flowing back through the valve again causing a high impact of the sealing element against the valve seat.

20.4.4 Valve lift

There are only two technical items of interest to a valve user: the valve losses or efficiency, and the valve life. Valves presently used are pressure activated which implies a pressure differential and therefore losses. However a minimum loss is required in order to control the valve motion properly. The valve life is normally a function of the valve lift. Assuming clean gas and no liquids in the gas-stream, valves fail due to fatigue caused by impacts. A valve sealing element normally increases its velocity throughout the whole lift, which means the higher the valve

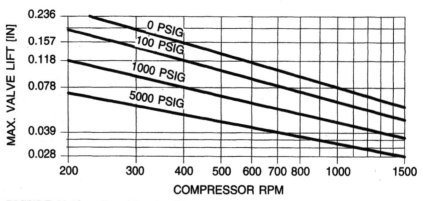

FIGURE 20.13 Allowable valve lifts for steel.

lift, the higher the impact velocity and the shorter the valve life. Impact velocities of valve sealing elements can only be calculated approximately when performing a complete valve motion study. There are however guidelines for maximum recommended valve lifts depending on the speed of the compressor and the operating pressure. One guideline that has been used for decades and which closely reflects curves published by a major valve manufacturer is given by the following empirical formula for steel valve plates

$$h_{steel} = \frac{13}{RPM^{2/3} p^{1/6}}$$

where

- h allowable lift [in]
- p line pressure [psia]
- RPM speed [rpm]

For non metallic plates, add .040 in to h_{steel}, for poppets simply double this value.

20.4.5 Valve Areas

As the gas flows through a compressor valve, it passes through areas which are defined more or less arbitrarily, but physically meaningful, in the following way.

Seat Area. The seat area is normally defined as the area of the seat grooves minus the area of all radial webs in the vertical projection, in other words the area in which, if a valve seat is held against a light source, light becomes visible. As already stated, the valve seat has to provide flexural strength against the high pressure load when the valve is closed. This is achieved by giving it a certain height and providing enough radial webs thus reducing flow area. The best solution can only be found by sophisticated calculations of mechanical strength, usually by using finite element methods.

Lift Area. Lift area equals the total length of seat lands multiplied by the valve lift. The seat lands area is where the sealing element seals against the valve seat. With increasing total length of the seat lands, the risk of valve leakage also increases.

Guard Area. Its definition is analogous to seat area.

Required Geometric Valve Lift Area. In the past, valves have been sized for an application primarily by using the valve velocity as the main criterion. Guidelines for acceptable velocities have been established by a number of companies. The valve velocity certainly has an influence in sizing a valve, but in itself is not sufficient. In terms of losses the important criteria are the pressure drop across the valve and the assurance of full pressure recovery inside the cylinder at piston dead center, i.e. allow the cylinder pressure to recover fully to the line pressure at the end of the stroke. An indicator—the q—value was introduced[11] representing the adequacy of a valve.

$$v_m = \frac{\frac{bore^2 \times \pi}{4} \times 2 \times stroke \times RPM}{n \times f_e \times 12}$$

where

v_m	valve velocity	ft/min
bore	cylinder bore	inches
stroke	piston stroke	inches
n	number of suction (discharge) valves per cylinder end	
f_e	lift area for one suction (discharge) valve in sq. inches	
π	3.14159	

$$q_{sv} = \frac{\pi^2}{8} \times \frac{\rho_s}{p_s} \times \left(v_m \times \frac{f_e}{\Phi}\right)^2 \times \frac{100}{144 \times 3600 \times g}$$

where

- ρ_s suction gas density lb/cu.ft
- p_s suction pressure psia
- Φ equivalent area sq.inches
- q_{sv} relative pressure drop % of suction line pressure
- g acceleration due to gravity = 32 ft/s²

q_{sv} is a dimensionless parameter used in the differential equations describing pressure and temperature evolution inside the cylinder.[11] For normal flow through a suction valve and small pressure drops, q_{SV} can be interpreted as the maximum pressure drop in % of line pressure. Discharge valves, providing they do not leak and close in time, have almost no influence on compressor capacity, but high q_{DV} signify excessive pressure drops and work losses. With a good valve design, q-values should be within the limits 2% and 12% for both, suction and discharge valves. Too small q-values mean too small pressure drops, hence too small gas drag forces and too small spring forces since these have to be sized to match pressure drop. The valve cannot be actuated properly; it will flutter and total work loss will be higher than necessary. Too high q-values mean high pressure drops and losses of work and capacity. In the suction valve, a q_{SV} value in excess of approximately 16% indicates no pressure recovery in the cylinder at dead center. This will cause a late closing of the valve and capacity losses.

Equivalent Area. The equivalent area is the effective orifice area of a valve when the valve is completely open. It is *the geometric valve area* affected by its coefficient of flow contraction and is therefore the area which should be used to calculate valve losses accurately. Most reputable valve manufacturers flow test their valves in order to establish this value. The equivalent area is influenced by the basic valve design, the valve lift and also by the manufacturing quality. It is for the last item that the exact value of the equivalent area can vary from one valve to another of the same design.

Given that the valve is designed using acceptable industry standards, the equivalent area can be estimated for plate or ring valves by formula developed by Hunt Davis.[2]

$$\Phi = \sqrt{\frac{1}{\dfrac{1}{k_l^2 \times A_l^2} + \dfrac{1}{k_{be}^2 + A_{be}^2} + \dfrac{1}{k_s^2 \times A_s^2}}}$$

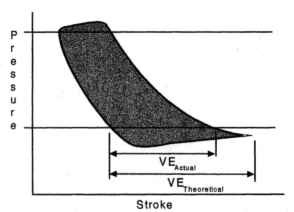

FIGURE 20.14 *P-V* diagram for cylinder with insufficient suction valve area (q-value > 16).

where

A_1	lift area	$k_1 = .85$
A_{be}	area between elements	$k_{be} = .66$
A_s	seat area	$k_s = 1$

From the suction flange of a cylinder to the compression chamber, the gas passes not only through the valve but also through internal cylinder passages, the valve cage and from the valve through the valve port. Every one of these passages potentially contributes to the total losses incurred which are normally called total ventilation losses. The ratio of total ventilation losses over valve losses can be quantified using a pocket loss factor.[8]

20.4.6 Valve Springs

The function of valve springs is to close the valve in time. The objective is that all valves should be closed when the compressor piston reaches dead center. Since the valve sealing element has to travel the whole valve lift from the fully opened to the fully closed position, the closing motion has to start before dead center and therefore against the gas flow. If the valve springs are too weak, the valve will close late (after the piston passes the dead center position) and gas is allowed to flow back through the valve causing the valve sealing element to slam closed. On the other hand if the valve springs are too heavy, the valve will close early. Since the piston continues to travel toward dead center, the pressure differential between cylinder and line will increase and the valve will reopen. An extreme case of too heavy springs leads to valve flutter. Tendency to valve flutter increases if the spring load considerably exceeds the values given in the two formulas below. Small pressure drops together with high valve lifts require springs with low spring rates and thin wires lacking sturdiness. In addition, these springs have to be long and have a tendency to buckle.

$$p_f = .4 \times q_{SV} \times p_s$$

for a suction valve and

$$p_f = .4 \times q_{DV} \times p_s \times \left(\frac{p_d}{p_s}\right)^\kappa$$

for a discharge valve, provided q_{DV} is calculated as if it was a suction valve, i.e. using v_m, f_e and Φ of the discharge valve, but ρ_s and p_s.

Both above formulas are approximate and valid for pressures up to about 500 [psia]. For other applications, it is indicated to verify spring selection by means of motion study. Properly selected valve springs have practically no effect on the valve opening.

20.4.7 Theoretical Valve Motion

All the formulas shown above have been found empirically and are valid only for a typical range of parameters. They can be used to conventionally size valves, lifts and springs for one and only one given operating condition. However if parameters are outside typical ranges or if operating conditions are variable, these methods fail. They cannot tell us whether a valve sized and tuned for one condition can safely be used for another condition as well. To answer such a question, the interaction between valve and compressor has to be modeled. The result is a system of at least three ordinary differential equations: one for the changes of valve lift; one for cylinder pressure; and one for cylinder temperature. These equations are too complex to be integrated analytically and have to be solved numerically in steps by one of the well known algorithms (Runge-Kutta, predictor-corrector and similar), necessitating a considerable amount of computation.

In its simplest form, this approach was published almost half a century ago.[1] Since then, particularly since the availability of powerful computers, the models have been extended. For lack of space only some typical improvements shall be mentioned here, for example including gas pressure pulsations,[5] considering gas inertia in the valve ports,[4,7,10] the so called gas spring effect,[10] gas leakages,[11] heat exchange,[6,11,12] higher modes of a valve reed,[3] more degrees of freedom of the valve plate's motion to predict the higher impact velocities resulting from cocking of the valve plate,[13] etc. All this is to show that the problem is in fact rather complex.

When interpreting computed valve motions one should therefore always bear in mind that

- The models currently in use are far from being physically comprehensive, and real motions may look more or less different.
- The coefficients of flow resistance and drag force and their variation with valve lift as used in the model may be different from those in the real valve.
- External data like pressures, temperatures, gas analysis, pulsation patterns may vary.
- The gas may be contaminated, liquid may be present.
- Numerical algorithms are always subject to local loss of precision, particularly in the vicinity of piston dead center positions and when valve pressure drops are small (usually with q-values $< 2\%$).

The results of these motion studies depend therefore to a large degree on their interpretation. As general rules the following can be used:

- A perfect looking valve motion is usually not as good as it looks, since the absence of rebounds on the seat is often due to late closure and therefore dangerous.
- A normal valve opening with no or little rebound on the guard is desirable. Although computed rebounds do not necessarily represent reality, they largely

depend on the way the impact is modeled and the arbitrary choice of coefficients of restitution. Usually they are inoffensive.
- The valve should stay open without flutter during most of the piston stroke.
- An early valve closing (before dead center) is preferred.
- Ignore rebounds which start shortly before the piston reaches dead center at the end of valve closure.
- The model computes impact velocities. A sealing element fails due to excessive local stresses in the material. One of the problems, therefore, is to convert impact velocities into stresses, and set allowable limits for them. These limits depend on the valve design, material choice, quality of production and also on the computational model used and the experience gained with it so far.

Some hints may be taken from the theory of one-dimensional wave propagation in an elastic medium. With this assumption, the pressure amplitude Δp (= change in pressure = stress σ) is related to Δu (= velocity amplitude = impact velocity) by $\Delta p = \rho \times c \times \Delta u$, where ρ is the density and c the velocity of sound in the sealing element material, given by $c = \sqrt{E/\rho}$. Thus, knowledge of very basic material properties such as density ρ, modulus of elasticity E and allowable stress ρ is sufficient to estimate allowable impact velocities Δu. This method finds about 1100 ft/min for valve plate steel, and 2200 to 3900 ft/min for PEEK depending on temperature. With all the imperfections of theoretical valve motion studies explained above and doubts as to the applicability of linear wave propagation theory, it seems indicated not to exceed one-third of these values ("factor of safety = factor of ignorance"). A more sophisticated approach, still using the theory of wave propagation in elastic media, but considering the properties of the seat as well, can be found in Ref. 16, laboratory results in Ref. 17.

The proper interpretation of a theoretical valve motion study is a matter of experience. People with little or no experience should consult with an expert before drawing wrong conclusions.

20.5 VALVE MATERIALS

For most applications, normal carbon or alloy steels are used for the valve body. The determining factor is the necessary valve seat strength to withstand the differential pressure the valve is subjected to. For corrosive gases, martensitic or austenitic stainless steels are used and some highly corrosive applications require Monel, Hastelloy or similar materials. For aggressive gases such as hydrogen sulfide, separate specifications[18] have been established and should be followed.

20.5.1 Valve Sealing Elements

In the past, most valve sealing elements have been manufactured from heat treated martensitic stainless steel like AISI 410 or 420. For more corrosive gases, precipitation hardening steels (17-7 PH) or Inconel X750 were used. These materials are still in use for high temperature and high pressure applications or when finger unloaders are used and the valve plate deflection in unloaded condition is high.

Today's materials of choice are injection molded composite materials with Nylon 6/6 (30% glass-filled) being the best choice providing the operating temperature is within the limit of this material. The maximum allowable operating temperature for Nylon 6/6 varies from manufacturer to manufacturer, but should be approximately 250°F or less, depending on the operating pressure. For higher temperature, PEEK (Polyetheretherketone) is used. The temperature limit for this material again varies depending on the manufacturer and is approximately 450°F. Constantly, new materials are developed especially by the aerospace industry, which are then used for compressor valve parts.

20.5.2 Valve Springs

The most widely used compressor valve spring material is a precipitation hardening stainless steel of type 17-7 PH. For highly corrosive application nickel alloys such as Inconel X750 or Nimonic 90 are used, with some special applications using a multiphase material type MP 35 N or Elgiloy.

20.6 VALVE LIFE

In ideal conditions, a properly designed valve has almost infinite life. Unfortunately, this condition is never found. Contamination in the gas in the form of foreign particles or liquids, including oil, corrosion as well as poorly designed or applied valves with inferior dynamic behavior result in finite life of compressor valves.

- Foreign Particles: Depending on the materials used for the valve components and the size and substance of these particles, they do more or less damage. The important thing to recognize is that they do damage. Very small particles normally pass through any valve design, but they cause abrasion on the valve seat and the sealing element, eventually leading to valve failure.

    Somewhat larger but still small particles can cause breakage of the sealing element when metallic sealing elements are used but are normally embedded into the material on non-metallic sealing elements.

    No valve will survive if large, hard parts are passed through it.

- Liquid in the gas is also detrimental to valve life. Valve designs with individual sealing elements possibly made of non-metallic materials tolerate liquids better than valves with one solid sealing element. Liquids drops impact on the sealing element causing a deformation on a single element valve which can result in instant breakage. In single element valves, the one element struck by the liquid drop can move independently not resulting in any damage.

    The effect of "oil sticktion" was explained above in "Motion of Sealing Element."

- Poor valve design or application almost always results in valve flutter or late valve closing. In either case, there are either multiple impacts per compression cycle or too high impacts resulting in premature valve failure.

20.7 METHODS TO VARY THE CAPACITY OF A COMPRESSOR

A compressor running at constant speed, with constant suction pressure and temperature, discharge pressure and gas composition will always deliver the same amount of gas per unit time. If capacity is to be varied, special methods have to be applied. Some of these are mentioned hereafter.

20.7.1 Clearance Volume Regulation

The disadvantage of this type of control is its initial cost and problems of space when applied to the crank end of a cylinder.

Varying the effective clearance volume in a compressor cylinder affects the intake volume of a compression chamber (see the definition of η_{vol}). The amount of compressed gas remaining in the clearance volume has to expand to the intake pressure before new gas can be drawn into the cylinder through the suction valve. Depending on the size of the clearance volume, the minimum capacity of a compression chamber can be as low as 0%. This method recovers all of the energy used to compress the gas with the exception of mechanical and fluid dynamic friction and heat losses. Several different methods are known to accomplish this.

Fixed Clearance Volume. A fixed volume is attached to the compressor cylinder and connected to it by means of some shutoff mechanisms. This shutoff device can be a normal compressor valve equipped with an unloading device or a plug unloader activated by a mechanical or pneumatic actuator. Fixed clearance volumes can be attached to a valve port if either single or double deck valves with a center hole are used. On the head end of a cylinder clearance volume, bottles can easily be added. Compressor capacity can be varied in steps only: Opening the pocket valve means minimum capacity, and closing it gives full load.

Variable Clearance Volume. Variable clearance volumes are added almost exclusively to the head end of a compressor cylinder. They consist of a volume bottle with a movable piston and normally a manually operated spindle to move the piston. Since the position of the clearance volume piston can be varied continuously, the same holds true for compressor capacity. The problems with these designs are leakage through the packing of the clearance volume piston as well as the high actuating force required when adjusting the volume size while the compressor is in operation. When the gas is not very clean and the clearance volume piston has not been moved for some time, it may stick and cause problems when trying to change its position.

20.7.2 Fixed Volume, Variable Pressure Clearance Volume

The amount of gas stored in a fixed volume can be varied if the pressure in that space is varied. This can be achieved when the clearance volume is separated from the compression chamber by means of an additional valve equipped with an unloader and actuator, opening and closing once per crankshaft revolution at instants of crank angle that can be selected by varying the unloader force. To do this, it is sufficient to vary the unloader force: Zero force gives full load, maximum force gives minimum load. Intermediate values allow for a stepless adjustment of compressor capacity.[15]

20.7.3 Reverse Flow Capacity Control

The reverse flow capacity control uses an unloader and an actuator on the suction valves in order to hold the suction valves open during a portion of the discharge stroke.[15] The delayed closing of the suction valves causes the gas in the compression chamber to be pushed back into the suction line rather then being compressed for as long as the suction valves are kept open. As soon as the suction valves are closed, or allowed to be closed, the remaining gas in the compression chamber is compressed and discharged.

Air pressure is applied to the diaphragm of the actuator that causes the control spring to be compressed and the spring pushes the unloader against the valve plate, opening the valve. During the compression stroke, the gas is pushed back through the suction valve into the suction line with a velocity approximately proportional to the piston speed. The gas velocity therefore increases steadily over the first half of the piston stroke* and the drag force on the valve plate increases with the gas velocity. As soon as the drag force on the valve plate overcomes the force the control spring exerts on the unloader, the valve will close. By varying the air

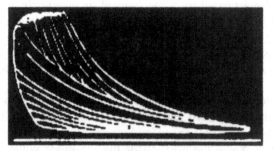

FIGURE 20.15 *P-V* diagram at various loads.

---

*This is an approximation and technically incorrect. On the head end of a compressor cylinder, the maximum piston speed is always reached after mid-stroke, on the crank end before mid-stroke. The exact position depends on the ratio of crank throw to connecting rod length.

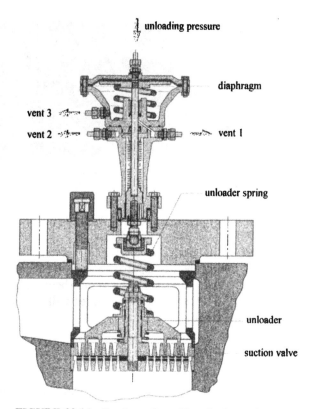

FIGURE 20.16 Suction valve with unloader and actuator for reverse flow capacity control.

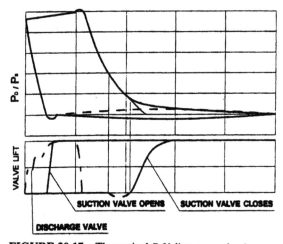

FIGURE 20.17 Theoretical P-V diagram and valve motion at partial load.

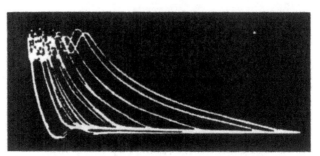

FIGURE 20.18 Actual *P-V* diagram at various loads.

pressure on the actuator, the control spring force is changed and therefore the time when the suction valves start to close.

The range of the capacity variation is limited to approximately 60% (i.e. from 100% or full load down to approximately 40%). Once the maximum of reverse flow pressure drop in the suction valve has been reached without having overcome the control spring force, the suction valves will not close at all and the compressor is then fully unloaded.

20.7.4 Variable Speed Regulation

This method of capacity control is normally used on compressors with engine drives. Compressors driven by electric motors normally have fixed speeds for cost reasons. The speed range where this system is economically effective depends primarily on the driver, but limits are set on the compressor side as well.

When lowering speed, the fluctuations of rotational speed will increase with the square of speed reduction. For example, if at full speed the coefficient of fluctuation equals 1%, it will go up to 4% at half speed. The resulting vibrations may be detrimental to foundations, buildings and the compressor itself. In addition, lower speeds may cause malfunction of main bearing lubrication and rod load reversal (API 618, point 2.4.4) which is necessary for lubricating the cross-head pin bearing. Finally, valve layout may also have to be revised if speed is to be variable. Practical experience shows that for all these reasons speed reduction is rarely lower than 40% of nominal.

20.7.5 Bypass Control

Of all available capacity control methods, this is the one with the worst energy efficiency. The basic principle is to spill back any excess gas from the discharge side of a compressor to the suction side via an intercooler and a control valve. The energy used to compress this excess gas is completely wasted. This method is normally used when for some reason other more energy efficient methods cannot be applied.

20.7.6 Plug Unloaders

Plug unloaders are used in place of finger unloaders to unload the cylinder end where they are positioned. Though mechanically reliable, the application of plug unloaders has to be governed by the resulting losses in loaded as well as unloaded condition.

The plug unloader is basically an opening in the center of the suction valve or a separate port, which connects the compressor cylinder with the suction line when the port is open. A "plug" is used to close this port during loaded operation, hence the name *Plug Unloader*. When the plug is positioned in the suction valve or in a separate valve port, it occupies area otherwise used for valve ports and therefore reduces the total valve area. This may be meaningless in some applications—especially when light gases are compressed—but may result in high valve losses in other applications. The effective plug area also has to be checked for its adequacy in order to insure total unloading of the cylinder side. When a cylinder is unloaded using plug unloaders, the suction valve is operating normally which means the plug area and the valve area are effective during the intake stroke, but only the plug area is available during the discharge stroke.

A thorough loss calculation is recommended before deciding on the application of this type of unloading.

20.7.7 Valve Lifters

Valve lifters were used in the past to lift the whole suction valve from its valve pocket thereby unloading a cylinder end. From an efficiency point of view, this is a very effective method to unload a cylinder since not only the valve area but the

FIGURE 20.19 Plug unloader (*courtesy of Dresser Rand*).

whole valve pocket area is available for flow. API 618 does not recommend this method due to seating problems of the valve and therefore it is not used very often.

20.7.8 Valve Depressors (Unloaders)

Valve depressors, also called *finger unloaders,* are mechanical devices to hold the valve sealing element open during the intake and the compression stroke. The gas taken into the cylinder during the intake stroke is therefore pushed back into the suction line during the discharge stroke. No productive compression work is performed when the unloaders are depressed. There is some compression work performed—and transferred into heat—due to the throttling when the gas passes through the valve. The equivalent area of a suction valve with unloader is somewhat affected by the unloader since the unloader is positioned directly in front of the valve. On reasonable designs, this should not result in more than 5 to 10% difference and is for all practical purposes the same for the flow in either direction.

On multiple valve cylinders, all suction valves should be unloaded in order to assure proper unloading and low heat build up.

Unloading Force. When unloading suction valves, it is necessary that the valve sealing elements are depressed rapidly and also that the unloading actuator is vented adequately so there are no unnecessary impacts between the unloader and the sealing element(s). More important however is a proper unloading force which has to be high enough to keep the suction valve open during the compression stroke without being so high that the unloader, or suction valve can be destroyed.

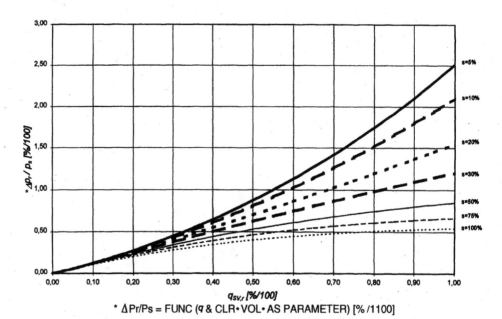

$$F_u = \Delta p_r \times A_0 \times C_w + F_{cs}$$

$$\Delta p_r \cong 1.2 \times \frac{q_{SV,r}}{100} \times p_s$$

where

- F_u required force to unload the valve [lbs]
- Δp_r pressure build up inside the cylinder on reverse flow through the suction valves [psi]
- A_o valve plate area
- C_w pressure drag coefficient [–], where $0.6 \leq C_w \leq 0.9$ with usual valve designs
- f_{es} force of closing springs when valve is open [lbs]
- $q_{SV,r}$ q-value for reverse flow through suction valve(s) [%]
- p_s suction pressure [psia]

As can be seen from the above diagram, the approximate formula $\Delta p_r \cong 1.2 \times q_{SV,r}/100 \times p_s$ given above is valid for $q_{SV,r}$—values up to about 20% and all clearance volumes. For example, with $q_{SV,r} = 20[\%]$, the above diagram gives $\Delta p_r / p_s \cong 0.25$ even for clearance volumes as small as 5%. For higher values of $q_{SV,r}$ and smaller clearance volumes, $\Delta p_r / p_s$ will be higher: With $q_{SV,r} = 50\,[\%]$ and clearance volume 5%, the above diagram gives $\Delta p_r / p_s \cong 0.9$. Care should therefore be taken not to underestimate the unloading forces when q-values are high (will be the case when not all suction valves are equipped with unloaders) and clearance volumes are small.

20.7.9 Active Valves

With the advancement of electronics, valves can be controlled in their opening as well as closing through mechanical devices. This means the differential pressure across the valve is no longer the factor when a valve opens and when it closes. Some external mechanisms are determining the timing and the differential pressure may assist in the actual motion. In very simple terms, this consists of a normal suction valve with an unloader and a special electro-hydraulic actuator. A solenoid valve—electronically activated—allows very high hydraulic pressure to act on a small cylinder, depressing the unloader and opening the suction valve. When the hydraulic pressure on the cylinder is vented, the gas pressure in the cylinder acting on the suction valve plate closes the valve. For this system to function, it is extremely important that the solenoid valve has a response time of less then one millisecond and also that the amount of hydraulic fluid moved is extremely small. Opening impacts can be controlled by properly selecting the instant of valve opening. Closing impacts are moderate and controlled by a simple patented hydraulic damping system. The opening time of the suction valve is not extremely important, but it should be slightly before the cylinder pressure reaches the line pressure. The closing time can be any time during the compression stroke, thereby allowing a

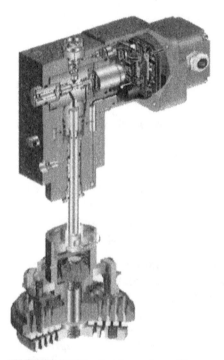

FIGURE 20.20 Suction valve with unloader and actuator for active valve control.

capacity variation from 100 to 0%. At this time this system is only available on suction valves. Discharge valves require even higher unloading forces and also provide a safety problem in case of malfunctioning of the unloading system.

The advantage of such a system is the perfect valve action even in varying operating conditions, larger valve areas due to higher allowable valve lifts compared to conventional values, and infinite capacity regulation. The disadvantages are high initial cost (this may change in due time) and a relatively complex electronic-hydraulic system as well as the limited unloading force available.

20.8 REFERENCES

1. Costagliola, M., *Dynamics of a Reed Type Valve,* PhD thesis, Massachusetts Institute of Technology, 1949.
2. Davis, Hunt, *Gas Compressor Valve,* Air and Gas Engineering, January 1968.
3. Soedel, W., *Introduction to Computer Simulation of Positive Displacement Type Compressors,* Ray W. Herrick Laboratories, Purdue University, 1972.

4. Trella, T. J. and W. Soedel, *Effect of Valve Port Gas Inertia on Valve Dynamics, Part I: Simulation of a Poppet Valve, Part II: Flow Retardation on Valve Opening* Purdue Compressor Technology Conference 1974, proceedings.
5. Soedel, W., *Gas Pulsations in Compressor and Engine Manifolds*, Ray W. Herrick Laboratories, Purdue University, 1978.
6. Brok, S. W., S. Touber, J. S. van der Meer, *Modeling of Cylinder Heat Transfer, Large Effort, Little Effect?*, Purdue Compressor Technology Conference 1980, proceedings.
7. Boeswirth, L., *A Model for Valve Flow Taking Non Steady Flow into Account*, 1984 International Compressor Engineering Conference at Purdue, proceedings.
8. Bauer, Dr. F., *Valve Losses in Reciprocating Compressors*, 1988 Compressor Engineering Conference At Purdue, proceedings and Hoerbiger Engineering Report 52.
9. Bauer, Dr. F., *The Influence of Liquids on Compressor Valves*, 1990 Compressor Engineering Conference At Purdue, proceedings, and Hoerbiger Engineering Report 53.
10. Boeswirth, L., *Non Steady Flow in Valves*, 1990 International Compressor Engineering Conference at Purdue, proceedings.
11. Machu, Dr. E. H., *How Leakages in Valves Can Influence the Volumetric and Isentropic Efficiencies of Reciprocating Compressors*, 1990 Compressor Engineering Conference At Purdue, proceedings, and Hoerbiger Engineering Report 54.
12. Machu, E. H., *Valve Throttling, its Influence on Compressor Efficiency and Gas Temperatures, Part I: Full Load Operation, Part II: Zero Load and Half Load Operation*, 1992 International Compressor Engineering Conference at Purdue, proceedings.
13. Machu, E. H., *The Two-Dimensional Motion of the Valve Plate of a Reciprocating Compressor Valve*, 1994 International Compressor Engineering Conference at Purdue, proceedings.
14. Machu, E. H., *Reciprocating Compressor Valve Selection, Some Points to be Considered for Optimizing Life Time and Compressor Efficiency*, European Conference on Developments in Industrial Compressors and their Systems, London, proceedings C477/020 IMechE 1994.
15. Machu, E. H., *New Developments in the Stepless Reverse Flow Capacity Control System for Reciprocating Compressors*, European Conference on Developments in Industrial Compressors and their Systems, London, proceedings C477/019 IMechE 1994.
16. Soedel, W., *On Dynamic Stresses in Compressor Valve Reeds or Plates During Colinear Impact in Seats*, Purdue Compressor Technology Conference 1974, proceedings.
17. Svenzon, M., *Impact Fatigue of Valve Steel*, Purdue Compressor Technology Conference 1976, proceedings.
18. NACE International, NACE Standard MRO175-95, Item No. 21302, Standard Material Requirements Sulfide Stress Cracking Resistant Metallic Materials for Oilfield Equipment.

CHAPTER 21
COMPRESSOR CONTROL SYSTEMS

Robert J. Lowe
T. F. Hudgins, Inc.

21.1 CONTROLS—DEFINITIONS

"Compressor controls" have different definitions among users, designers, and suppliers. These can be, but are not limited to:

1. Machinery protection
   a. Simple alarm system for several parameters
   b. Simple shutdown system for several parameters
2. Safety shutdown system monitoring multiple parameters for machinery protection and safety of personnel
3. Same as 2, with loading/capacity control
4. Same as 3 with communications and remote monitoring. Computer monitoring provides enhanced automation and energy management in addition to the capability to store and manipulate data for maintenance and diagnostic purposes.
5. Environmental regulations have made the addition of emissions controls on compressors and engine drivers necessary for certain geographical areas.

21.2 RECIPROCATING COMPRESSOR MONITORING

Monitoring systems have progressed to measuring many parameters beyond those required for basic safety. These include capacity, pressure, volume, time, vibration, calculations of gas polytropic exponents, rod load, power, fuel consumption and efficiency.

The industry is also rapidly moving to a "predictive maintenance" mode as compared to "preventive maintenance." Monitoring systems must be more complex to accomplish predictive maintenance.

Modern technology provides the means to track progressive machine conditions which, when compared to machine history, can provide the user with sufficient data to predict when a repair or maintenance must be completed as compared to routine scheduled maintenance.

21.2.1 Centrifugal/Axial Compressor Monitoring

Centrifugal and axial compressors are less forgiving than slower speed reciprocating machines because of the operating speeds.

The controls for centrifugal machines must have very quick response times in order to prevent catastrophic failures and are much more complex than the systems required for reciprocating compressors.

Real time on-line performance monitoring and trending is now commonplace because of available high speed computer systems.

Surge control on axial and centrifugal machines is essential to prevent damage in the event of rapidly changing flow conditions. Bearing vibration monitoring in three planes and axial trip sensors, along with bearing trip monitors, are common practice on centrifugal machinery.

Refer to API Standard 670.

21.3 SYSTEM CONSIDERATIONS

The control system must be selected to suit the machine, the operator, and the application. These considerations include, but are not limited to:

1. Is this machine part of a critical process?
2. What is the cost/benefit ratio to be provided by controls?
3. What are the personnel hazards?
4. Is this machine in a hazardous area?
5. What codes such as NEC, API, NFPA, OSHA, EPA or other regulatory agencies apply?
6. What degree of complexity is required for safety shutdown, capacity control, loading control, remote start-stop capability, and communications?
7. Should the safety devices be monitored by the information/operations/capacity controls system or should safety monitoring be in a separate stand alone system?
8. Will the system be maintained by:
   a. Mechanics and/or operators
   b. Instrument technicians
   c. Electronics technicians
   d. Communications technicians
   e. Users or contract personnel
9. Will the system require expansion at a later date?

21.4 SYSTEM SELECTION—DEFINE THE SCOPE

Answers to the preceding questions will generate additional questions and provide direction for selection of the proper system.

The systems available at present, range from simple pneumatic and electric systems to highly complex electronic systems that operate, not only as safety systems, but provide control and operation of compressors while storing and trending data. These systems are also capable of predictive maintenance and repair management when properly structured.

Instrumentation and control technology developments have provided many improvements, allowing expansion of monitoring and control systems. Fiber optics, Ethernet, and modern technology provide massive information flow over great distances and allow centralized control of multiple machine locations.

Prior to computer systems, machinery managers did manual calculations to determine fuel efficiency and gas measurement. The calculations are so complex that results were obtainable once per hour, at best.

There are over 100 parameters to be considered in the American Gas Association (AGA3) and NX19 Fuel Flow Calculations and Super Compressibility Calculations.

Computers can provide these results every 60 to 90 seconds, enabling operators to manage fuel efficiency, engine emissions, while monitoring torque, compressor rod load, horsepower, capacity, and other conditions.

The goal is to improve efficiency and increase run time by managing operating costs and maintenance.

It should be noted that specialties have developed in certain applications such as speed control, vibration monitoring, emissions monitoring, and centrifugal compressor surge control. These can be separate "stand alone" systems interfaced with the overall controls system. Each requires careful analysis and system selection using the compression manufacturer's specific data.

21.5 HUMAN FACTORS

Selection of controls must take into consideration the equipment and the personnel operating and maintaining the system. The control system designer must be thoroughly familiar with the operational and mechanical aspects of the machines to be monitored. The designer must make a careful evaluation of material compatibility in the environment and all parts of the process. The mechanics and operators must be involved early in the design discussions is essential to the system success. Training requirements and the "user friendly" features of the system are necessary for reliability and acceptance by the operators.

The system goals, sequencing, interlocks, peripherals, transient conditions, time delays, response times, accuracy, reset differentials, passwords, and security keys must be defined early in the process of system design.

Component access for calibration and maintenance is necessary on all systems and a major factor determining system layout and assembly. Inspections of similar installations and interviews with existing users will save time and expense, minimizing errors.

Documentation, detailed drawings, schematics, component instructions, and parts list are as necessary as the hardware. Instructions for calibration, testing, and maintenance must be provided with the system in order to have satisfactory operation.

21.6 ELECTRICAL AND ELECTRONIC CONTROLS

21.6.1 Applicable Codes

Specific codes govern the use of electrical equipment in hazardous as well as nonhazardous areas. Some guidelines on these ratings are as follows.

Hazardous area electrical classifications for USA and Canada—Class, Divisions and Groups.

Class I	Combustible gasses and vapor.
Class II	Combustible dust
Class III	Fibers and lint
Division 1	Combustible present or probably present in normal operation.
Division 2	Combustible not present in normal operation.
Group A:	Acetylene
Group B:	Hydrogen

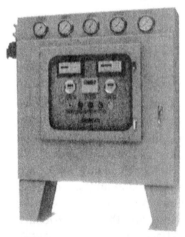

FIGURE 21.1 Nitrogen compressor control panel (Class 1, Group C & D, Division II Area) front panel view. (*Courtesy of Autocator Products Division T. F. Hudgins, Incorporated*).

FIGURE 21.2 Nitrogen Compressor Control Panel (*Class 1, Group C & D, Division II Area*) *inside panel view.* (*Courtesy of Autocator Products Division T. F. Hudgins, Incorporated*).

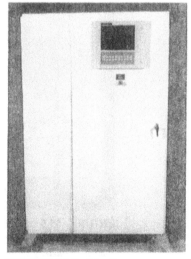

FIGURE 21.3 Pipeline compressor panel (programmable logic controller panel) front view.

FIGURE 21.4 Pipeline compressor panel (separate electrical compartment). (*Courtesy of Advanced Control Engineering Services*).

FIGURE 21.5 Pipeline compressor panel (separate pneumatic compartment). (*Courtesy of Advanced Control Engineering Services*).

FIGURE 21.6 Pipeline compressor panel (programmable logic controller and temperature scanner) front panel view. (*Courtesy of Advanced Control Engineering Services*).

Group C: Acetaldehyde, Ethylene, Methyl Ether
Group D: Acetone, Gasoline, Methanol, Propane
Group E: Metal Dust
Group F: Carbon Dust
Group G: Grain Dust

FIGURE 21.7 Pipeline compressor panel (programmable logic controller and temperature scanner) inside panel view. (*Courtesy of Advanced Control Engineering Services*).

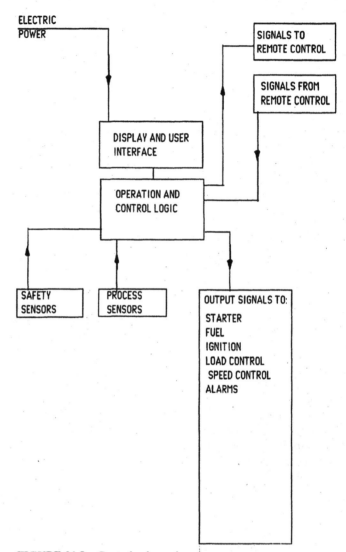

FIGURE 21.8 Control schematic.

Refer to NFPA 325M, 497M, CSA22.1, FM3610, UL913, NFPA70 and CSAC22.2, No. 157 ANSI/ISARP12.6. Also see National Elec. Code articles 500 to 504.

Intrinsically safe equipment is safe by nature of its operation and design. The following description is from the Instrument Society of America, Standard ISA-RP12.6:

"Intrinsically safe equipment and wiring is equipment which is incapable of releasing sufficient electrical or thermal energy under normal or abnormal conditions, to cause ignition of a specific hazardous atmosphere mixture in its most ignitable concentration.

Intrinsically safe electrical equipment and wiring may be installed in any hazardous location of any Group classification for which it is accepted without requiring explosion proof housings or other means of protection."

21.6.2 Electrical Controls

Electrical controls for compressors include:

1. Switch gage systems employing pressure and temperature gages with the gage pointer as an electrical contact for alarm or shutdown by activating a master control circuit. Annunciation is a pop-out or digital annunciator. Improvements in recent years provide micro-switch contacts on gage systems for Division II hazardous areas.
2. Digital annunciator systems using switch sensors. These annunciators display a number which is compared on a printed legend on the panel listing the alarm and shutdown sensors which are monitored. These can be certified for use in Class 1, Group C and D, Division I and II hazardous areas.
3. Relay logic systems require pilot lights or annunciators for indication. These are generally used on applications in non-hazardous areas and use standard switch sensors. Relays are also available for Class 1, Group C and D, Division II areas.
4. Electronic systems which are factory programmed with logic sequences and displays are available for general purpose and hazardous areas. The number of input and output relays are fixed by the manufacturer.
5. Electronic systems which are pre-programmed micro processor based, and field configurable with compressor logic sequences were introduced in the mid 1980's. These are applied in hazardous areas and are generally intrinsically safe for Class 1, Group C and D, Division I or II areas. Displays are alphanumeric. These systems use both discrete switch and analog sensors. The analog systems are scanners for monitoring multiple inputs and have configurable set points.

    The quantities of inputs and outputs are not infinitely expandable but can be field selected within limits. (Communications capability allows this system to operate as a "stand alone" system communicating to a host or process computer.)
6. Process controllers and distributed controls systems (DCS) can be used when available in lieu of a stand alone control system at the machine. This is important when capacity control from a central control system is essential for the process. There must be an evaluation as to the inclusion of the safety controls within the central control system or to have the safety system as a separate system.

Note: Most electrical compressor control systems interface with pneumatic actuators for valve actuation power.

7. **Programmable Logic Controllers** Programmable Logic Controllers are microprocessor based electronic systems which are modular and expandable to hundreds of inputs and outputs. Modules are available to accept all standard sensors. Input and output modules can receive and send the commonly used signals. These include variable 4 to 20 mA, 1 to 5 Volt DC, switch and communication signals. Trained programmers using hand held or other computers set up ladder logic, Proportionate-Integer-Derivative (PID) loops, timers and other control functions. PID controllers provide compensation to control or limit lead, lag and oscillation in the system response. Graphic and alphanumeric displays are readily available. First introduced in the 1960's, these controls have been industrially hardened and are now available for Class 1, Group C and D, Division II hazardous areas.

7a. Logic controllers built specifically for compressors have been designed with the programmability and flexibility of a small Programmable Logic Controllers (PLC) and using similar logic systems and programming. The alphanumeric display is incorporated into the single unit with logic module, along with input and output cards and communications. These small systems are modular, but do have limited input/output quantities.

Note 1: Electronic analog controls provide the distinct advantage of multiple alarm set points for a single sensor. Alarm can occur prior to reaching a dangerous shutdown condition.

Note 2: Electrical or electronic panel can be made explosion proof for Division I areas when used with a monitored air or inert gas purge system designed specifically for this purpose to purge the control cabinet with air. The purge system includes a pressure sensor to alarm or shutdown when purge pressure is low.

The Z and X purge air systems are used in hazardous areas. Type X purge air systems usually alarm but not shutdown in the event of pressure loss. Type Z purge air systems cause a shutdown in the event purge air pressure is lost.

21.7 PNEUMATIC CONTROLS

Pneumatic controls were developed for hazardous area safety control and operation of compressors early in the natural gas and process industries. They are inherently explosion proof and can be interfaced with electric or electronic systems using pressure switches and transmitters. Air or gas provides the power for valve actuation on all compressor systems, especially process or pipeline gas compressors. These valves include:

Fuel valve on engine driver
Starting air control valve
Suction valve
Discharge valve

FIGURE 21.9 Marine seismic compressor (pneumatic control panel). (*Courtesy of Eureka Energy Systems, Inc.*).

Bypass valve
Purge valve
Blowdown valve
Capacity control valves
Capacity control actuators

21.7.1 Indication

The systems include differential pressure operated three-way valves with a window in the side of the valve body and red/green stripes on the spool. This device acts as an indicating relay or "pneumatic pilot light" and sequencing valve.

Pneumatic sensors for temperature, pressure, level vibration, and speed monitoring are connected by pneumatic tubing to a pneumatic indicating relay valve. These can be tubed in parallel or series circuits, depending on the manufacturer's design to provide a shutdown or alarm indication.

Pneumatic indicating relays have been operating since the early 1950's and proven reliable under severe operating condition.

System circuitry can be designed to provide many complex sequences.

21.8 MANUAL CONTROLS

Manual start-stop and loading sequences are used but for safety reasons, usually in conjunction with a control/monitoring system. Loading-unloading and capacity control can be manually regulated but this procedure is labor intensive and more precisely handled automatically.

21.9 PRELUBE-POST LUBE SYSTEM

Prior to start:
Prelube the machine to distribute oil to metal wear parts for a time determined by the manufacturer, usually 90 to 300 seconds. This is accomplished using an electric or air motor driven auxiliary pump controlled by the control system.

A pressure switch or valve can initiate the start sequence upon completion of prelube or cause a "READY TO START" indication, once the prelube pressure and/or time is adequate.

Post lube:
Operation of the auxiliary oil pump for a timed period at the time shutdown provides cooling to remove residual heat from the compressor and lubricate the machine.

21.10 LOADING-UNLOADING

Automatic loading and unloading of a small air compressor is common. The compressor discharge is automatically opened to atmosphere as the compressor reaches the selected discharge pressure.

Loading and unloading of process or pipeline compressors pertains to operation of valves in the suction and discharge piping permitting or stopping flow through the machine. The sequence of operation will be clearly defined in the manufacturer's instruction manual.

21.11 CAPACITY CONTROL

Loading and unloading should not be confused with capacity control. Capacity control is used for the following reasons:

1. Process flow control
2. Fuel/power efficiency
3. Rod load regulation
4. Pressure regulation

Capacity can be changed in several ways.

1. Speed regulation
2. Control of supply gas to the machine
3. Bypassing the discharge flow back to the suction side of the machine
4. Finger unloaders—(manual or pneumatic) finger unloaders are rods projecting from an actuated hub located in the suction valve chamber. The rods press the

suction valve disc to the open position to unload the cylinder. These can be operated on multiple valves on a cylinder to step-unload or load the machine increasing or decreasing capacity according to the number of active valves.

5. Plug unloaders serve the same purpose of the finger unloaders but are constructed with an actuated plug positioned to open or close a hole in the center of the valve. Gas flows in and out of the center hole when the valve is deactivated by opening the plug. These unloaders can be manual or air cylinder operated.

6. Clearance pockets—cylinder volume can be changed by the addition of clearance (volume) pockets to valve ports or the head end of a compressor cylinder. Control of flow in and out of these clearance pockets can be by automatic air, hydraulic, or manual valve actuation.

Note: Operating and discharge temperatures must be monitored while using capacity control methods 3 through 6 listed above. Gas surging in and out of the cylinder can result in heat build up if improperly controlled.

21.12 LOADING AND UNLOADING

Gas compressors, depending on the service, will use either a three valve sequence, four valve sequence, or five valve sequence.

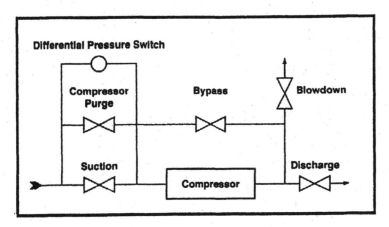

3 Valve Sequence	4 Valve Sequence	5 Valve Sequence
Suction	Suction	Suction
Discharge	Discharge	Discharge
Blowdown	Blowdown	Blowdown
	Bypass	Bypass
		Purge

21.12.1 Three Valve Loading Sequence

Start the driver with the compressor "unloaded."
Discharge valve closed
Suction valve closed
Blowdown valve open

To load—

1. Open discharge valve.
2. Open suction valve.
3. Close blowdown valve after a delay which is long enough to allow all air to be purged from the machine, by the gas, vented through the blowdown valve.

To unload—

1. Open blowdown valve.
2. Close suction valve.
3. Close discharge valve.

21.12.2 Four Valve Loading Sequence

Start the driver with the compressor "unloaded"
Discharge valve closed
Suction valve closed
*Blowdown valve open (or closed)
Bypass valve open

To load—

1. Open discharge valve.
2. Open suction valve.
3. Close blowdown valve if open.
4. Close bypass valve.

To unload—

1. Open blowdown valve.
2. Open bypass valve.
3. Close suction valve.
4. Close discharge valve.

21.12.3 Five Valve Loading Sequence Machine Not Running

  Blowdown valve open
  Bypass valve open
  Suction valve closed
  Discharge valve closed
  Purge valve closed

Prior to start—

1. Close bypass valve.
2. Open purge valve.
3. After time delay for pressurized purge, open bypass.

Start machine—

1. After warm-up or appropriate time, initiate **load** sequence.
2. Open purge valve (pressurize compressor and confirm pressure).
3. Close purge valve.
4. Open discharge valve.
5. Open suction valve.
6. Close bypass valve to begin flow through the compressor. (Bypass valve can be used to control flow through the machine.)

Five Valve Unloading Sequence

1. Open bybass valve.
2. Open blowdown valve (optional)
3. Close suction valve.
4. Close discharge valve.

  Emergency shutdown should cause blowdown valve to immediately open.

Note: Starting with the blowdown valve closed allows the compressor to remain charged with gas when stopped. Special packing seals are necessary to retain the gas when the machine is in the stopped position. These seals grip the rod at the packing case when the compressor stops and are interlocked with the control system to release seals prior to starting and to grip the rod only after the machine stops.

  The compressor loading sequences can be completely automated and should always include valve position sensors to confirm valve positions. Appropriate time delays between steps must be included. In addition, if valve positions are not confirmed within 30 to 45 seconds (adjust as necessary), the system should shut down for safety reasons.

Certain systems use either the suction valve or the bypass valve for capacity control.

Caution: All sequences should be in accordance with the manufacturer's instructions. Careful attention must be given to the gas dynamics when using clearance pockets and valve unloaders to avoid temperature build-up within the cylinders.

21.13 SENSOR CLASSIFICATION—(ALARM CLASSES)

Sensors can be classed as one of the three main groups for purposes of monitoring on reciprocating machines. There are subgroups, but the three main classes are:

Class A Sensors
Class A sensors are monitored continuously and must be healthy at all times. Typical would be liquid levels and discharge temperature.

Class B Sensors
Class B Sensors—Certain conditions which are not present when the machine is not running are not monitored during the start sequence and are locked out by use of a timer. Low oil pressure and water pressure are examples. Vibration can be time-latched out during the start sequence.

Class C Sensors
Class C Sensors—Process conditions, low suction, low discharge pressure and other sensors do not become healthy until the machine is running and loaded. Class C sensors become armed and monitored after the sensor becomes healthy and resets.

Recap:
In order to start, all Class A sensors must be healthy; Class B sensors are armed at the expiration of a Class B timer or as the sensor becomes healthy. Class C sensors become armed as the sensor or the complete group of C sensors reset in healthy condition. This is dependent upon the system logic.

21.14 SENSORS

Operating conditions on and around compressors include high temperatures, grease, oil, vibration and pulsation. Rugged sensors are necessary to operate reliably under these conditions.

Sensors are available as:

1. 4–20mA transmitters
2. 1–5 volt transmitter
3. Switches
4. Pneumatic valves

5. Thermocouples—the common alloys are (K) Chrome-Alumel and (J) Iron-Constantan.
6. (RTD) Resistance temperature detectors
7. Mechanical trip valves
8. Proximity sensors

Starting: (Sensors)

Starting can begin automatically by control system or manually upon completion of prelube time.

Engine drivers rotate on start command driven by the starting system—the automated system measures the speed and applies ignition and fuel according to manufacturer's instructions. The engine should complete warm-up cycle determined by a timer or a temperature sensor prior to being loaded.

Electric motor starting should also incorporate the compressor prelube cycle if necessary, and in certain applications, require a fan driven air purge prior to starting. The control system should incorporate a timer or a pressure switch to acknowledge completion of the prelube and/or air purge cycle. This occurs in a manner similar to compressor prelube and should be according to manufacturer's instructions.

21.14.1 Monitoring Compressor Parameters

The following list includes points to monitor on a typical large compressor system.

Ambient temperature
Compressor crosshead guide temperature
Compressor packing case temperature (high)
Compressor packing case vent (leak detection) (high)
Compressor rod bearing temperature
Compressor rod drop (rider band wear)
Compressor valve temperature
Coolant flow rates
Coolant pressure
Crankcase pressure (high)
Differential pressure—oil filter
Emergency shutdown
Engine fuel flow rates
Engine fuel temperature
Engine manifold pressure (low)
Engine power cylinder exhaust temperature
Engine turbocharger bearing coolant in and out
Engine turbocharger bearing temperature

Engine turbocharger lubricating oil pressure
Engine turbocharger lubricating oil temperature
Engine/compressor main bearing temperature
Fire detection system shutdown
Fuel level (low)
Lube oil pressure (low)
Lubrication flow (low)
Lubrication oil consumption
Lubricator flow (low)
Metal (in oil) particle detection
Motor bearing temperature
Motor bearing vibration
Motor overload
Motor power failure
Motor purge fan failure
Motor vibration
Motor winding failure
Motor winding temperature
Oil cooler temperature in and out (high)
Oil filter Differential (high)
Oil level–compressor (low)
Oil level—driver (low)
Oil level—lubricator (low)
Scrubber level (high)
Scrubber level (low)
Suction, discharge, interstage pressure (high)
Suction, discharge, interstage pressure (low)
Suction, discharge, interstage temperature
Vibration—compressor (high)
Vibration—cooler (high)
Vibration—driver (high)
Water cooler temperature in and out (high)
Water pump differential pressure (low)

21.15 SPECIAL COMPRESSOR CONTROLS

The compressor industry has created a demand for special control systems and sensors for problems unique to compressors. Among these are:

FIGURE 21.10 Pneumatic compressor packing case purge control panel to control packing leakage. (*Courtesy of Autocator Products Division, T. F. Hudgins, Incorporated*).

Packing case purge control
Rod drop measurement and alarms
Vibration monitoring
Vibration sensors
Mounting vibration sensors
Metal particle detection in lubrication oil
Lubrication flow sensors
Sensors for pulsating compression pressures
Eutectic temperature sensors
Energy management
Cylinder pressure measurement

21.15.1 Packing Case Purge Controls

Control systems used to maintain purge gas on packing cases designed to prevent leakage of process gas along the rod to atmosphere have proven effective. Measurement of vent gases is used as a diagnostic tool to determine the rod packing wear and replacement schedule.

21.15.2 Surge Control

Centrifugal and axial compressor surge control is a specialty area of compressor capacity control and safety. These systems are necessary to provide stable operation on high speed centrifugal compressors.

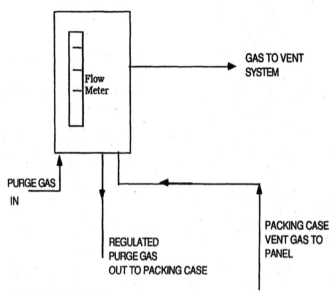

FIGURE 21.11 Flow meter connection for measuring purge and packing leakage flow.

21.15.3 Compressor Rod Drop

Measuring rod drop is an effective tool in determining when rider rings should be replaced and can minimize cylinder wear. Rod drop occurs in a horizontal compressor cylinder as piston or rider rings wear, allowing the piston to descend. The intent of a compressor rod drop measuring system is to alarm this event prior to the piston contacting the cylinder.

Three measurement systems are in use today. These are:

1. Pneumatic eutectic sensor is a bracket mounted sensor located beneath the rod. As the rod rubs on the sensor, the friction causes the eutectic solder to melt, allowing the pneumatic control system to vent and alarm or shut down the machine.
2. Proximity rod drop measurement without the crank angle monitoring system is an electronic equivalent of the pneumatic system alarming when the rod reaches a set point. The operator can also observe a read-out and trends from the system display.
3. Proximity rod drop sensor in conjunction with a crank angle detection system which measures rod position at exactly the same point within the rod stroke. The proximity probe is an eddy current sensor. The output varies in response to rod position. Measuring a specific crank angle allows the operator to be assured of the piston position at the time of measurement.

21.15.4 Vibration Monitoring

Vibration monitoring systems are necessary for determining the health of a machine. An experienced analyst can determine the exact cause of a noise or condition by analyzing the vibration patterns. The system complexity can vary with speed and critical nature of the machine.

If a machine is equipped with vibration alarms, these signals can alert the operator to call for an extensive vibration analysis by a specialist using portable diagnostic equipment to determine the exact cause of the vibration. The other option is to provide the extensive analysis diagnostic system as part of the installation and machine controls.

21.15.5 Vibration Sensors

1. Inertia sensors are available as pneumatic valves and electric switches. These are intended to respond to vibration on low speed equipment and in ranges from 0 to 3000 cpm to 0 to 12000 cpm, at vibration sensing ranges 0 to 5g's and 0 to 10 g's. These devices are sensitive to vibration parallel to the axis of the sensor mechanism.
2. Eddy current sensors measure machinery motion and are frequently used to monitor moving shafts because direct contact is not necessary. Position, dynamic motion, and wear are now measurable using these sensors.
3. Accelerometer based sensors sense impact events and are available with and without conditioning. Conditioning, either in the sensor or in the control system, converts the accelerometer signal to velocity or displacement signals proportional to vibration. The conditioned accelerometers convert the low powered signal from the sensor into a 4 to 20mA or 1 to 5 volt DC signal, and can be connected to electronic monitors that are not especially designed for monitoring vibration. Unconditioned sensors have low output signal and are intended to be connected to an electronic monitor that includes the conditioning and vibration monitoring system.
4. There is a wide selection of read-out devices and systems that provide indications and alarms, computerized systems data storage, and trending of the following data. This is a very specialized technology and there are manufacturers

FIGURE 21.12 Metal particle detector.

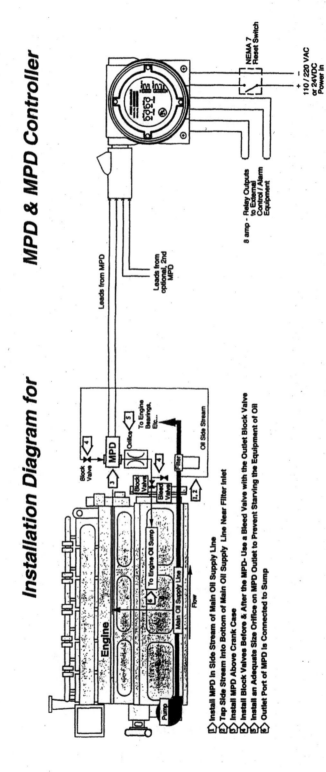

FIGURE 21.13 Metal particle detection diagram.

providing this service worldwide. Typical displays, either in velocity or displacement, would be:

Running speed vibration

Crank case deflection

Compressor crosshead vibration

Compressor cylinder head vibration

5. Compressor valve covers can be monitored temporarily by accelerometers as part of a portable compressor analysis diagnostic system to determine valve condition and operation.

Selection of vibration sensors requires careful analysis of machine speed, sensor location, and vibration amplitude based on experience or manufacturer's data. Vibration on low speed machines (120 to 1,500 rpm) is commonly measured in mils or mm displacement. High speed machine (360 to 90,000 rpm) vibration is measured in inches per second or mm/s velocity.

21.15.6 Mounting Vibration Sensor

"Bearings take the load during vibration" and are the desired vibration monitoring points. The restraint provided by the base can limit vertical movement, therefore, the sensor must be mounted where the most movement is predicted.

21.15.7 Metal Particle Detectors

Metal particles in oil are an indication of:

A. Contamination

B. Machinery wear

Metal particles from bearings, pump wear, cylinder wear, and other failures have been detected by these sensors prior to the build up of temperature or vibration. These systems provide a constant monitoring for metal particles in the lubricating fluid.

The application requires a small oil side stream from the dirty side of the filter to flow through a perforated printed circuit card within a suitable housing. As the particles complete the circuit, the alarm circuit is energized. These detectors respond to all conductive particles.

Sensors for ferrous only particles are also available.

21.15.8 Lubrication Low Flow Sensors

Compressor lubrication system failure can be detected by monitoring switches or valves which remain in a "healthy" condition with flow. As lubrication system flow diminishes, or stops, the sensor trips.

More sophisticated electronic systems actually measure the flow and transmit the data to computers. The alarm set point can be in the flow monitor or in the computer.

21.15.9 Pressure Sensors

The need to measure engine firing pressures and compression pressures has caused several very rugged sensor types to be developed to withstand the pulsation and high temperatures. The rugged construction is necessary because these sensors are mounted on the cylinder for measuring internal pulsating cylinder pressures.

The sensors have piezo-electric and strain gage elements and are proving valuable for controlling operating conditions by computer and evaluating the machine performance. These are available with and without the requirement for water cooling.

21.15.10 Cautions

Sensors must be installed in accordance with the ISA recommendations and ASME codes and Regulations Agency codes.

Sensors must be protected from corrosive fluids and excessively high pressures and temperatures.

Temperature sensors can be protected by suitable thermowells.

Static pressure sensors can be protected by suitable gage isolators where necessary. These are sealed diaphragm devices especially designed for this purpose.

21.16 TEMPERATURE CONTROL (OIL AND WATER)

Controlling oil and water temperatures is essential in order to maintain proper running clearances. In addition, proper temperatures minimize wear, maintain clearances and proper lubricant viscosity, and prevent condensation in the crankcase. Temperatures are controlled by either self-contained thermostatic valves or temperature control valves with controllers.

21.16.1 Self Contained Thermostatic Valves

Special self contained three way valves using internal copper impregnated expanding wax elements were developed for the compressor and engine industries. These valves are located in the oil and water piping.

During a cold start-up, the fluid is recirculated back to the heat source (bypass position). As the machine warms up, and approaches the element set point, within

5°F to 7°F, the element begins to close the bypass port and send fluid to the cooler. The valve will modulate flow to bypass and cooler in order to maintain the set temperature maintaining a constant flow volume through the machine.

Sizes are selected for a pressure drop of 2 to 7 psi in order to be certain of proper velocities and to maintain heat transfer to the element for proper response.

These valves are available with a variety of body materials, trim, and seals. Material compatibility and codes must be considered. As an example, API recommends that piping systems containing hydrocarbons be of steel construction.

Certain lubricants are not compatible with standard o-rings and the thermostatic valves must be specified accordingly. Commercially available sizes range from 1/2 inch to 8 inch flanges. Available materials are cast iron, steel, ductile iron, bronze, and stainless steel.

21.16.2 Cooling Water and Lube Oil

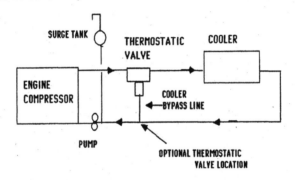

COOLING WATER SYSTEM DIAGRAM

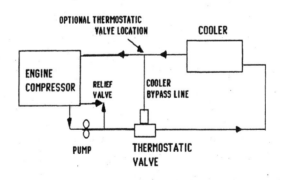

LUBE OIL SYSTEM DIAGRAM

FIGURE 21.14 Cooling water and lube oil systems.

21.17 ELECTRIC MOTOR AND PNEUMATICALLY OPERATED TEMPERATURE CONTROL VALVES

Three way temperature control valves have been designed for use with pneumatic cylinder and electric motor operators. The need to provide very low pressure differential, yet maintain control, were considerations. External actuators allow the use of remote sensing elements and controller to suit the application.

Pneumatic and electronic industrial controllers can be used with these valves and operators to regulate temperatures. Another option is to use the central plant process control system. The pneumatic controller allows the valve to be placed in a hazardous area such as a gas pipeline compressor station.

Proportional controls which vary the output according to the input, proportional, integral, derivative (PID) controls provide a more precise control regulation by eliminating the offset between the desired and actual temperatures. The derivative action attempts to anticipate temperature changes based on the cyclic rates of change experienced by the system. I to P converters are used to interface an electronic controller to a pneumatic operator. Commercially available in sizes range from 2 inch to 16 inch flanges. Industrial control valves can be used in this service, but do operate at higher differential pressures.

21.18 ENERGY MANAGEMENT SYSTEMS

Energy management systems provide dramatic cost savings where multiple machines are used in industries such as foundries, automotive or process plants, as well as pipeline and process industries.

Careful analysis and control of machine run time, along with capacity control, will optimize power consumption. An added benefit is reduced machine wear and maintenance.

21.19 SPECIFICATIONS, CODES, AND STANDARDS

Refer to the following publication for suggested specifications, codes, and standards.

Instrument Society of America—ISA

S5.1 Instrumentation Symbols and Identification

S5.4 Instrument Loop Diagrams

S5.5 Graphic Symbols for Process Displays

RP12.6 Installation of Intrinsically Safe Systems for Hazardous (Classified) Locations

Standards and Practices for Instrumentation
National Electrical Manufacturers Association—NEMA
ICS 3–1978 Industrial Systems
ICS 6–1978 Enclosures for Industrial Controls and Systems
American Petroleum Institute—API

API 670–Vibration, Axial-Position, and Bearing-Temperature Monitoring Systems
API 618–Recommended Practice for Compressor Emissions Monitoring
API RP–550 Manual on Installation of Refinery Instruments and Controls
API RP–520 Design and Installation of Refinery Instruments and Controls
API RP–521 Guide for Pressure Relief and Depressuring Systems
API RP–14F Recommended Practice for Design and Installation of Electrical Systems for Offshore Production Platforms
Underwriter's Laboratories, Inc.—UL
Standards for Safety
Factory Mutual System—FM
Approved Standards and Data Sheets
National Fire Protection Association—NFPA
National Fire Codes Volumes 1 through 16

CHAPTER 22
COMPRESSOR FOUNDATIONS

Robert L. Rowan, Jr.
Robert L. Rowan & Associates, Inc.

22.1 FOUNDATIONS

The key to rotating and reciprocating machinery reliability is the foundation. One of the main functions of foundations is to support the machines at a precise elevation, thus allowing the original precision alignment to be maintained over the life of the machine.

Besides the critical task of maintaining the alignment of the machine, the foundation must supply enough mass to absorb the unbalanced forces that the operating machine produces. Good engineering input from the manufacturer of the machine is essential to the designer of the foundation, but equally as important is a geotechnical analysis of the soil on which the foundation will rest.

22.1.1 Types of Foundations

Reciprocating and centrifugal compressors can be packaged or unitized on a fabricated skid (Fig. 22.1), block mounted (Fig. 22.2), or set on a pile cap foundation (Fig. 22.3). Large centrifugal machines are also sometimes set on "table top" foundations, which are shaped much like a kitchen table with multiple legs (Fig. 22.4). This style is popular for larger machines and allows the space underneath to be used for long radius, large diameter piping and auxiliary equipment.

The above types represent practices in the United States. No review, though, would be complete without mentioning a new option that is starting to be seen in the United States because of successful installations in Europe. This option in the type of compressor foundations, is the use of spring supports. The advantages of this option include good isolation of the dynamic forces, good definition of support properties and additional possibilities for future modifications or corrections. Spring support systems for compressor applications will typically have vertical natural frequencies in the range of 3 to 5 Hz. Horizontal frequencies usually are slightly less than the vertical frequency. As these values are less than comparable frequen-

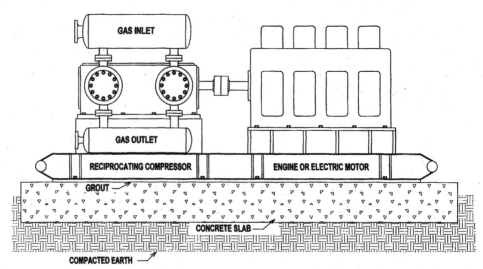

FIGURE 22.1 Skid mounted/packaged compressor. (Illustration courtesy of Robt. L. Rowan & Assoc., Inc.)

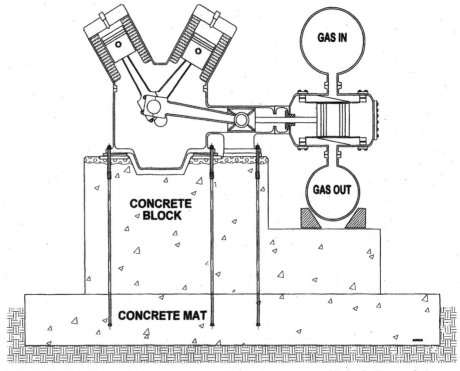

FIGURE 22.2 Block mounted compressor. (Illustration courtesy of Robt. L. Rowan & Assoc., Inc.)

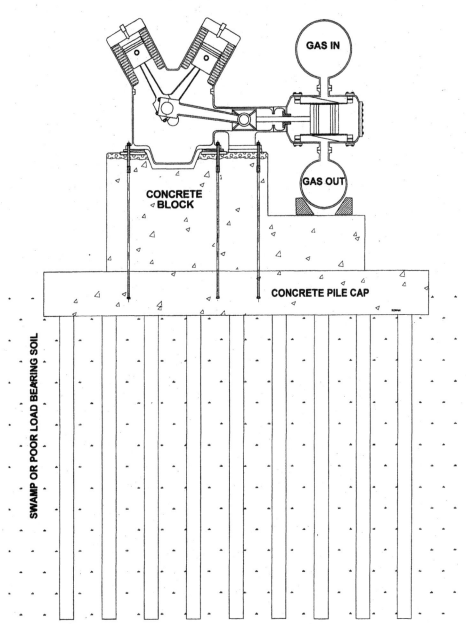

FIGURE 22.3 Pile cap foundation. (Illustration courtesy of Robt. L. Rowan & Assoc., Inc.)

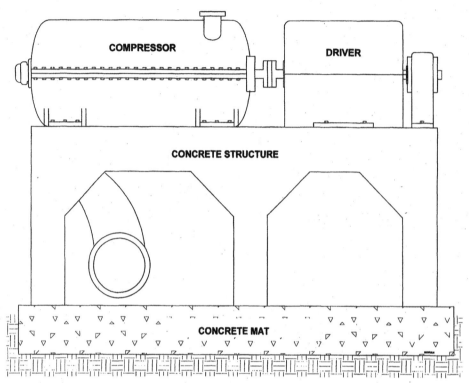

FIGURE 22.4 Table top foundation. (Illustration courtesy of Robt. L. Rowan & Assoc., Inc.)

cies for soil or pile supported systems, the spring system typically provides better isolation of the dynamic forces of the compressor. The springs themselves, usually steel coil designs, provide well defined stiffnesses both horizontally and vertically. This advantage simplifies the dynamic analysis of the foundation eliminating the need to incorporate a range of soil properties in this analysis. By including viscous dampers in the design, the complete dynamic system can be put together with great confidence. Finally, the discrete nature of the spring support system permits easy replacement of the elements if a change to the stiffness or damping characteristics becomes necessary. Similarly, misalignment from settlement and similar sources can be corrected at the spring support level.

22.1.2 Design

The detailed design of any of the above foundations is beyond the scope of this chapter. Unfortunately, there are no established building codes at this time (1996), but under the auspices of the American Concrete Institute, a committee is working to develop a report that could eventually become a foundation design guide document. Major engineering firms, operating companies, and equipment manufacturers that have their own in-house guidelines are represented on the committee. Additionally, under the sponsorship of the Pipeline Compressor Research Counsel, Southwest Research Institute, along with interested industry users of compressors, much needed data on both dynamic and thermal stresses in foundations is being developed. With such input data, along with the work of ACI, the design of foundations in the future can be more precise.

22.1.3 Soil Frequency and Vibration

While there have been many technical articles written on theories of foundation design and vibration, a very comprehensive reference is the work of Prakash and Puri, *Foundations for Machines: Analysis and Design*.[1] With a good background in geotechnical engineering, the authors tie together very well the interaction between the cyclic vibrations caused by the machine with the natural frequency of the soil-foundation system. Foundations must be designed to avoid the dreadful consequences of harmonic resonance, which occurs when the frequency of the vibrating machine matches the natural frequency of the foundation (block and soil). Prakash and Puri teach that by applying the principles of soil engineering and soil dynamics with theories of vibration, low tuned or high tuned foundations can be designed so as to avoid resonance. Their work leads the way to designing foundations for dynamic machines which will have acceptable levels of vibration. Good engineering at this stage will pay off with a smoother running machine, better maintenance of alignment, and lower maintenance costs for the replacement of wear parts (bearing, seals, etc.).

22.1.4 Collection of Data for the Design Step*

While readers of this handbook may not ever be called on to design a foundation, they may very well be asked to supply data to an engineering design firm working under its direction.

While the work of ACI committee is incomplete at this stage, probable recommended data collection steps will be as follows:

1. Data gathering
   a) Design goal
   b) Site factors
   c) Sub-soil data
   d) Machine data
2. Design criteria
   a) Static loads

---

*Based on preliminary draft of ACI 351-2 Sub-Committee

b) Dynamic loads
3. Concrete strength/stresses
   a) Compressive
   b) Flexural
   c) Tension
   d) Bearing
   e) Fatigue
4. Concrete deflection/deformation
5. Soil strength/stresses
6. Soil deformation/settlement
7. Vibration limits
8. Psychological factors

22.1.5 Materials of Construction

Portland Cement Concrete: Reinforced portland cement concrete is the usual material of construction for either the foundation proper, or for the mat under a fabricated steel skid. A mix design, based on locally available ingredients, can be developed that yields a compressive strength of 4,000 psi in 28 days. The amount of steel will depend on the tensile and bending loads, as well as thermal stresses. Many foundations designed over the past 30 years have been under-reinforced, as evidenced by cracking. Cracking can lead to deterioration of the alignment condition and even catastrophic failure. Extra steel, to increase flexural and tensile strength is very prudent. Steel, put in initially, does not cost very much, but a foundation repair later, because of an under-reinforced foundation, is very costly. Figures 22.5 and 22.6 show a modern design with extra rebar vs. a design done twenty years ago.

Polymer Modified Concrete: While reinforced portland cement concrete is almost universally used today, many older foundations have been repaired using a more technologically advanced material called "Polymer Modified Concrete." Substituting a polymer for the usual water in portland cement concrete, produces an improved concrete. The polymer, along with fiber reinforcing, produces a very dense product with *low* heat of hydration, stronger physical properties in tensile and flexure, and cures in 24 hours[2].

22.1.6 Anchor Bolts

Anchor bolts are a vital link between the compressor and the foundation. Unfortunately, designers often overlook important points concerning anchor bolts, such as how long and strong they should be and the amount of preload. Anchor bolts, as well as other parts of the support system, such as sole plates and chocks (to be

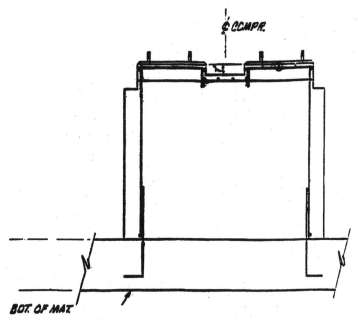

FIGURE 22.5 Typical perimeter steel reinforcing—era 1960s.

discussed in section 22.1.7), can be one of the principle points of failure on new construction projects. Failure usually occurs during the first year of operation.

While the number and size of the anchor bolts are set by the equipment manufacturer, their length, configuration, and material of construction are in the hands of the foundation designer. Figure 22.7 shows good and bad designs.

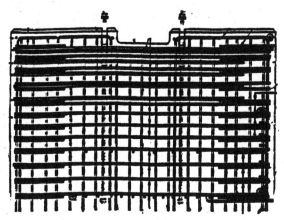

FIGURE 22.6 Dense steel reinforcing based on current design practices.

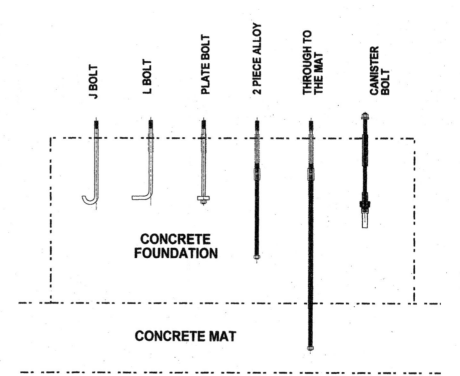

FIGURE 22.7 Evolution of anchorbolt designs. (Illustration courtesy of Robt. L. Rowan & Assoc., Inc.)

Length: Short anchor bolts have historically caused problems in compressor foundations. Horizontal cracks in the foundation often result. The best practice today is to make them as long as possible, terminating them in the concrete mat under the concrete foundation. In this manner, they do not contribute to horizontal cracking and have the added benefit of adding a post-tensioning effect.

Material: Anchor bolts for any dynamic machine cannot be too strong. Today, anchor bolts made from steel, conforming to ASTM A-193 with a yield strength of 105,000 psi, are not much more expensive than steel half as strong. As the need for high clamping forces for compressors is being recognized, alloy steel bolts to ASTM A-193 provide the necessary capacity without going to a larger anchor bolt.

Preload: While some compressor manufacturers will specify an initial torque value for the initial installation, often field experience will show a much higher (maybe two to three times) clamping force will be required to lower frame movement/vibration. Unless the anchor bolts put into the foundation to start with have extra capacity, the machine will not perform as it should, or a costly retrofit will have to be done.

22.1.7 Support Systems

Figure 22.8 shows a range of options on how to support a gas compressor from the older method of full bed grouting, to the latest technology of adjustable support

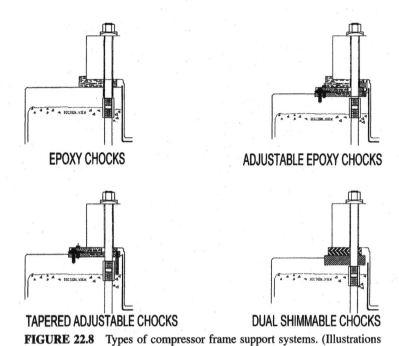

FIGURE 22.8 Types of compressor frame support systems. (Illustrations courtesy of Robt. L. Rowan & Assoc., Inc.)

systems. Adjustable supports are the system of choice today, because they eliminate a potential problem of poor initial alignment which happens from time to time with full bed grouting. Adjustable systems also allow the optimum hot running condition to be achieved as the frame can be re-aligned to correct for the alignment changes that occur as the machine heats up during its first 100 hours of operation.

22.1.8 Grout

Since the introduction of epoxy grouts for gas compressor grouting in 1957, the use of cementitious grouts mixed with water has virtually stopped. Epoxy grouts are stronger, resist oil and many chemicals, and perform well in dynamically loaded situations.

While grout need not be stronger in compressive strength than the concrete underneath, a good grout will be tough enough to take impact and cyclical loads from the dynamic machine it supports. For that reason, compressive strengths above 5,000 psi and tensile strength above 1,000 psi are all that are required. Higher compressive strengths are not necessarily better if the product is brittle and cracks excessively in service. Almost all good machinery grouts can crack, so expansion joints are required. The expansion joints should be strategically placed so cracks will not develop in the prime load transfer area adjacent to the anchor bolts.

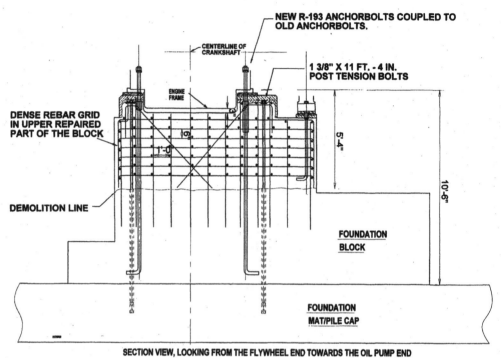

FIGURE 22.9 Section view, looking from the flywheel end towards the oil pump end. (Illustration courtesy of Robt. L. Rowan & Assoc., Inc.)

22.1.9 Repair of Foundations

Almost every foundation 20 twenty years old and designed with only minimal steel reinforcing is a candidate for replacement or repair. Common repair techniques include removing the top 24 inches to 30 inches of grout and concrete, cutting off and up-grading the anchor bolts, adding a heavy rebar layout in the excavated area and post-tensioning the repair to old remaining concrete.

What to use for the post-tensioned repair described above is extremely important. If the job schedule will allow 21 to 28 days, portland cement concrete is the best choice. If a 24-hour curing product is needed, then a polymer modified concrete should be used. Either product will have a modulus of elasticity of at least 4,000,000 psi, and will have negligible creep at typical compressor foundation temperatures. What should *not* be used as a deep pour repair material to replace the removed concrete is the epoxy grout material that is used as the final cap on top of the foundation. Epoxy grouts are just that—a grout designed to be used in 2 to 4 inch thicknesses. Epoxy grouts, as a class of material, have a modulus of elasticity ranging from under 1,000,000 psi up to 2,500,000 psi, with the lower range being the most prevalent. This means epoxy grout will compress under load, two to five times more than concrete. Additionally, some epoxy grouts creep enough at typical foundation temperatures to cause equipment misalignment. There have been catastrophic machine failures as a result of deep pours of epoxy grout. A new compressor foundation should not be designed with a 14-inch thick upper pour of epoxy grout nor should an older concrete foundation be repaired that way.

Besides up-grading the anchor bolts, an adjustable support system is also added to allow easier realignment. Figure 22.9 shows a typical foundation repair design.

22.2 REFERENCES

1. Prakask, Shamsher, and Vijay K. Puri, *Foundations for Machines: Analysis & Design*, Wiley Series in Geotechnical Engineering.
2. Rowan, Robert L. & Associates, Inc., *Re-Grouting Reciprocating Gas Compressors, 5 Year Repairs vs. 20 Year Reliability Criteria*, 1:12 Grouting Technology Newsletter.

CHAPTER 23
PACKAGING COMPRESSORS

Judith E. Vera
Project Engineer
Energy Industries, Inc.

Compressor packages can be used for a variety of applications including gas boosting, gas gathering, gas lifting, gas injection, gas turbine compression, vapor recovery, landfill, and digester gas compression, or propane/butane refrigeration compression. Two basic types of positive displacement compressor packages will be discussed in this section: reciprocating gas compressors, and rotary screw gas compressors.

Reciprocating gas compressors can handle from 60 horsepower to 7,200 horsepower (using a 6-throw compressor frame and an electric motor driver). Although reciprocating units are the most common due to their ability to handle large horsepower requirements, high pressure, and varying conditions, rotary screw units are advantageous when limited space or limited package weight specifications apply. Rotary screw units are ideal for large volume/low suction pressure applications and have fewer maintenance and vibration problems.

23.1 COMPRESSOR SIZING

Most packagers have a compressor sizing program available to choose a compressor frame and driver to compress the gas under the desired conditions. To properly size a compressor, the following must be provided: suction pressure, discharge pressure, gas analysis or gas specific gravity, elevation, ambient temperature, suction temperature, and desired capacity.

The driver can be a gas engine, gas turbine, or electric motor depending on the end user's specific requirements. Although the primary components of the packaged gas compressor unit are the compressor frame and its driver, numerous other parts are essential to the efficient operation of the unit.

23.2 BASE DESIGN

Gas compressor package bases are designed of sufficient mass to support the weight of the entire package. Bases can be designed as structural steel only when the package must be kept within certain weight limitations. However, the preferred base design is a steel frame filled with reinforced concrete. This design is portable and eliminates the need for field-poured foundations.

The base can be designed with or without the drain and vent piping inside the skid frame. An ecology rail should be designed to keep any oil or water leakage onto the skid contained into a drain system. This rail is usually made of 2 × 2 or 3 × 3 angle around the perimeter of the base and skid drains located at four or more locations at the skid edge.

Lifting eyes, designed with a 4:1 safety factor that can be added to the skid, or draw bars are located at the edge of the skid to aid in pulling the package up onto a truck.

Scrubber plates are preferred and are usually made of 1/2 inch steel plate and are located on the skid for scrubber placement.

The engine and compressor set on a pedestal which is preferably filled with concrete. It is important that this pedestal is one solid piece on reciprocating compressor packages. The natural vibrating forces in a reciprocating package on a pedestal that is two separate pieces can cause skid failure or misalignment of the engine and compressor during normal operation. A typical skid design prior to the concrete fill is shown in Fig. 23.1. The skid is then filled with concrete, as shown in Fig. 23.2.

23.3 SCRUBBER DESIGN

Scrubbers are designed to remove solids and liquids from the gas before they reach the compressor cylinder. This is necessary because the tolerances in a compressor are such that any foreign matter can damage the internal parts of the compressor. Scrubbers should be placed before each stage of compression. On a multi-stage unit, the interstage scrubbers are required to remove any liquids formed by condensation during the cooling process.

Mist pad scrubbers are the most common scrubber design. They are vertical vessels which have a wire mesh mist pad, typically 4 to 6 inches thick with 9 to 12 lb/cft bulk density. (Gas Processors Suppliers Association's Engineering Data Book, Volume I, 10th Edition[1] contains a method for sizing mist pad scrubbers.) Scrubbers are sized based on the critical or terminal gas velocity required for particles to drop or settle out of gas. Equation 23.1 defines the maximum gas velocity entering the mist pad scrubber as a function of the gas density and the liquid density.

$$Vt = K \left[\frac{pl - pg}{pg} \right]^{0.5} \tag{23.1}$$

FIGURE 23.1 Typical skid design prior to concrete fill.

where Vt = terminal gas velocity (ft/sec)
K = empirical constant for mist pad scrubber sizing (ft/sec), (Fig. 23.3)[1]
pl = liquid phase density, droplet or particle (lb/cft). This is assumed to be water and is equal to 62.4 lb/cft.
pg = gas phase density (lb/cft).

Equation (23.1) is an empirical equation based on Stoke's Law. The inside area, A (sq ft) of the mist pad scrubber can then be found by:

$$A = \frac{Qa}{Vt} \qquad (23.2)$$

where: Qa = actual gas flow rate (cft/sec).

From the inside area of the mist pad scrubber, we can then determine the minimum required inside diameter of the mist pad scrubber to ensure all liquids are removed from the gas of a maximum specified flow rate.

Figure 23.4 shows a typical design of a mist pad scrubber and the instrumentation normally used on scrubbers. The gas enters the scrubber and is directed downward toward an impingement or baffle plate. The liquids and solids collected on this plate are forced through holes that allow the liquids and solids to drain into

FIGURE 23.2 Typical skid design with concrete fill.

an accumulation chamber. The change in direction, decrease in velocity and gravitational force further help the dropout of the liquids and solids. The gas then passes through a mist extractor which allows the fine droplets of liquid to be collected until they grow heavy enough to fall to the impingement plate. Liquid collected in the accumulation chamber rises to a level sensed by a liquid level controller and then is dumped from the scrubber through a diaphragm-operated dump valve. If the liquid flow is greater than the dump valve can handle, a high liquid level shutdown switch is added to prevent liquid overflow into the cylinders. A liquid level sight gauge allows visual monitoring of the scrubber liquid level. A manual drain is supplied to completely drain the scrubber.

Vane scrubbers are much more efficient than mist pad scrubbers and therefore it is possible to use a smaller diameter vessel with a vane scrubber than is required by a mist pad scrubber. Vane scrubbers use a vane pack made up of several corrugated plates with liquid drainage traps. Manufacturers of vane scrubbers should be contacted on the details of the design of their vessels since vane sizing designs are proprietary.

	K Factor (ft/sec)	C Factor (ft/hr)
Horizontal (w/vert. pad)	0.40 to 0.50	1440 to 1800
Spherical	0.20 to 0.35	720 to 1260
Vert. or horz. (w/horz. pad)	0.18 to 0.35	648 to 1296
@ Atm. pressure	0.35	1296
@ 300 psig	0.33	1188
@ 600 psig	0.30	1080
@ 900 psig	0.27	972
@ 1500 psig	0.21	756
Wet steam	0.25	900
Most vapors under vacuum	0.20	720
Salt & caustic evaporators	0.15	540

Note: (1) K = 0.35 @ 100 psig - Subtract 0.01 for every 100 psi above 100 psig.
(2) For glycol and amine solutions, multiply K by 0.6 - 0.8.
(3) Typically use one-half of the above K or C values for approximate sizing of vertical separators without woven wire demisters.
(4) For compressor suction scrubbers and expander inlet separators, multiply K by 0.7 - 0.8.

FIGURE 23.3 Typical K & C factors for sizing woven wire demisters. (Reprinted with permission from *Gas Processor Suppliers Association Engineering Data Book, Volume I*, Section 7, Fig. 7-9, p. 7-7, GPSA, 1987).

23.4 LINE SIZING

The gas lines on a compressor package are typically sized for a gas velocity of less than or equal to 3500 ft/min. One method used to determine the gas flow rate is the Weymouth equation[2]:

$$Q = 433.5 \frac{T_b E}{P_b} \left[\frac{P_1^2 - P_2^2}{SL_m T_{avg} Z_{avg}} \right]^{0.5} (d)^{2.667} \qquad (23.3)$$

where: Q = gas flow rate at base conditions (cft/day).
d = pipe inside diameter (in)
L_m = length of pipe (mi)
T_b = temperature at base conditions (°R), (ANSI 2530 specification: T_b = 520°R)
T_{avg} = average temperature (°R), [T_{avg} = 1/2 ($T_{in} + T_{out}$)]

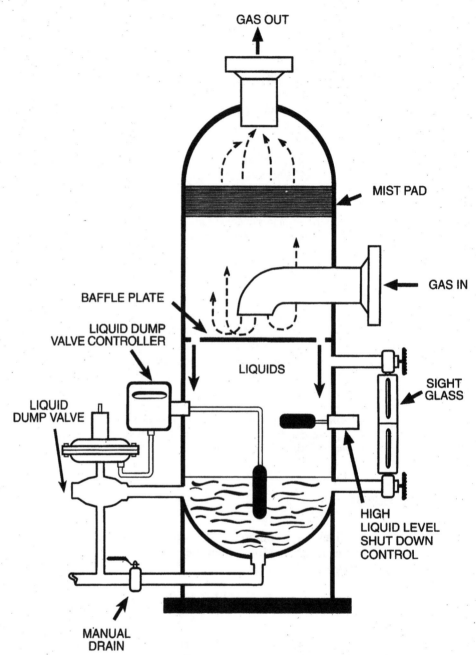

FIGURE 23.4 Typical scrubber dump instrumentation.

P_b = base absolute pressure (psia) (ANSI 2530 specification: P_b = 14.73 psia)
E = pipeline efficiency factor, (usually assumed to be 1.00)
P_1 = inlet pressure (psia)
P_2 = outlet pressure (psia)
Z_{avg} = average compressibility factor
S = specific gravity of the gas with respect to air (where $S = 1$)

Then velocity can be confirmed by using $V = Q/A$, where A = line flow area, making sure to keep the units uniform.

Gas piping on a compressor package includes, but is not limited to, process gas piping, a valved bypass line from the aftercooler outlet to the first stage scrubber for unloaded starting, a vent line or valve for purging the machine, scrubber dump piping, compressor cylinder packing and distance piece vents and drains, and engine gas start and fuel system piping. All lines should be hydrotested to the maximum allowable working pressure of the flange class. Flange ratings should be in accordance with ANSI B16.5[3]. Piping should also be x-rayed in accordance with ANSI B31.3[4].

23.5 PULSATION BOTTLE DESIGN

On reciprocating gas compressors, some form of vibration control is required. Without the pulsation bottle or some other device to change the frequency of gas pulsations caused by the movement of the piston, the harmonic rhythm of their frequency would cause a progressive increase in vibration and eventual damage to the unit.

On small horsepower units (less than 350 horsepower), pulsation bottles are not required to eliminate vibration. However, orifice plates are required in the suction and discharge lines on each stage of compression. These plates can be welded into the lines or placed between flanges to allow flexibility when pressure and flow conditions change.

On units greater than 350 horsepower, pulsation bottles are required. There are two different approaches to pulsation bottle design. One approach, adequate for many compressor applications, is to design a volume bottle only. This approach, however, does not guarantee elimination of harmful vibration. To design a volume bottle, the following empirical relation can be used.

Bottle volume = 10 × swept volume of cylinder
(Swept volume = cylinder area × stroke)

This gives the volume required to determine the diameter and seam-to-seam dimension of the bottle. Bottles should be mounted as close as possible to the cylinder and the nozzles should be located in the center of the bottle to reduce unbalanced forces. Nozzle flanges should be designed to meet the required pressure and temperature operating conditions (refer to ANSI 16.5[3] for flange ratings).

Another approach that does offer the guarantee of no harmful vibration is the volume-choke-volume method that involves choking the flow prior to the volume bottle or choking the flow inside the volume bottle. Analog and/or digital studies can be performed on a particular machine to further verify the guarantee of no harmful pulsations. Design of these pulsation bottles takes into account the rotational speed of the compressor, the gas composition, and the operating conditions. Chokes should be designed for about .5% pressure drop, but not more than 1%. See API Standard 618, Reciprocating Compressors for General Refinery Services[5], for a more detailed discussion on recommended design approaches for pulsation bottles. All pressure vessels, bottles and scrubbers, should be designed to meet Section VIII of ASME Code[6] and be of sufficient pressure rating to cover all operating conditions.

23.6 PRESSURE RELIEF VALVE SIZING

Pressure relief valves or pressure safety valves are required on the discharge bottles or lines of the compressor. These valves are designed to open when the operating pressure exceeds the set pressure of the relief valve. The set pressure of these valves should be determined by the lowest working pressure of the compressor package of the system the valve is designed to protect. Figure 23.5 is a typical Piping and Instrumentation Diagram. This diagram is a schematic that follows the flow of the compressor package. The table on the top of the diagram shows the maximum allowable working pressure (MAWP) of all of the vessels and coolers. This diagram also shows all the valves and instrumentation on the compressor package. By using this diagram, the relief valve set pressure can be determined.

There are two types of relief valves: spring-operated, and pilot-operated. Pilot-operated relief valves are useful when the operating pressure is closer than 15% of the relief valve set pressure. The spring-operated relief valves tend to pop frequently when the operating pressure is close to the set pressure. Pilot-operated valves are less sensitive and will not relieve the pressure until the exact set pressure is reached. The orifice sizes required in the relief valves can be sized by the manufacturer or by formulas found in the ASME Boiler and Pressure Vessel Code, Appendix 11, Section VIII, Division 1[6].

$$W = CKAP \, (M/T)^{.5} \quad (23.4)$$

where: W = weight flow of gas that valve is relieving (lb/hr)
K = flow coefficient for gas [see UG-131(d) and (e)][6]
A = actual discharge flow area of the safety valve or orifice (sq in)
P = (set pressure × 1.1 plus atmospheric pressure*) (psia)
(* Relief valves are typically rated at 10% over set pressure.)
M = molecular weight
T = absolute temperature at inlet (°F + 460)
C = constant for gas or vapor which is a function of ratio of specific heat, Cp/Cv

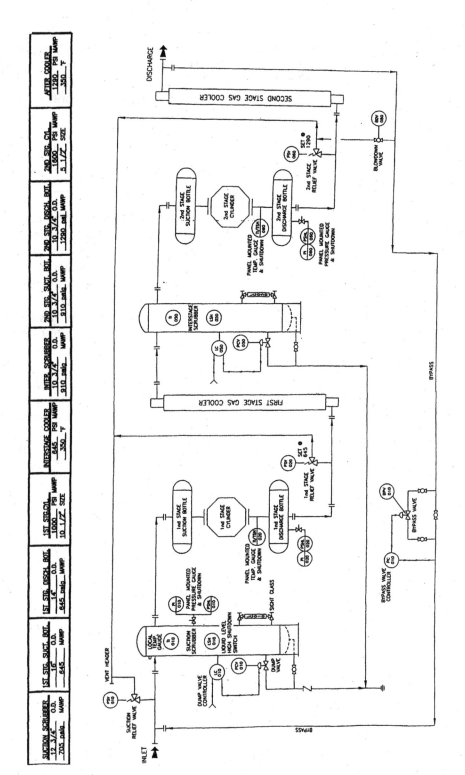

FIGURE 23.5 Typical piping and instrumentation diagram.

The K value is determined by a series of tests conducted at the National Board of Boiler and Pressure Vessels in Columbus, Ohio in accordance to the ASME Section VIII Code. Equation 23.4 should be used to solve for A, or the orifice size of the relief valve, and then pick a relief valve that has a standard orifice size larger than the A value calculated.

23.7 COOLER DESIGN

Governed by the basic gas law, the gas temperature increases in proportion to the increase in pressure so that each stage of compression requires cooling prior to entering the next stage of compression. The cooler also acts as a radiator to cool the engine and compressor frame. The cooler should be fitted with a surge tank to vent the system of all entrained air and also serve as a fill point for the cooling system. It is outside the scope of the packager to design the cooler, however, it is the responsibility of the packager to ensure the cooler manufacturer has the correct requirements in order to do a proper cooler design. The packager must furnish the cooler manufacturer with the performance run which gives the pressures, temperatures, flow conditions, elevation and ambient temperature. The packager must also provide any additional information required to correctly design the cooler, such as customer specifications. It is important that the packager ensure that the pass arrangement designed by the cooler manufacturer provides the most efficient piping arrangement available.

Packagers normally use air-cooled heat exchangers. These units use ambient air to cool fluids and gases. Typically, the design ambient air used is 100°F for summer conditions. Manual or automatic louvers are used over the gas sections to ensure the cooler does not overly cool in the winter months.

Heatload information provided by the engine manufacturer is used to determine the proper sizing of the engine jacket water sections. The gas coil sections are generally designed to take the discharge temperature from the cylinders and cool it down to either 120°F or 130°F on the interstage sections and 120°F on the aftercooler section.

The basic formula that cooler manufacturers use to determine the heatload required for the gas sections is[7]:

$$Q = mc_p(T_d - T_c) \qquad (23.5)$$

where: Q = heatload (Btu/hr)
m = mass flowrate (lbs/hr)
c_p = specific heat at average temperature (Btu/lb°F)
T_d = discharge temperature (°F)
T_c = cooled temperature (°F)

23.8 COMPRESSOR LUBRICATION

To keep size and weight down, packages are usually built around high speed compressors. This makes lubrication of the cylinders and piston and rod packing critical. The lubrication system is determined by the size of the cylinder and rod. A separate force-feed lubricator driven off the crankshaft supplies oil to the compressor cylinders and piston rod packing. Figure 23.6 shows a typical compressor lubrication system. Divider valve blocks which contain metering pistons are mounted on a base block which contains passageways and built-in check valves to ensure proper oil flow to the cylinders. A cycle indicator is used on one of the divider valve blocks to provide a positive indication of system operation. The indicator pin is an extension of the piston and cycles back and forth as the piston moves. The pints of oil per day and a maximum cycle time is determined by the compressor packager.

23.9 CONTROL PANEL & INSTRUMENTATION

The control panel is the brain of the compressor package. It operates in conjunction with numerous safety switches to monitor the compressor package and protect it

FIGURE 23.6 Typical compressor lubrication system.

DISPLAY		LOCKED OUT BY TIMER		POINTS NOT LOCKED OUT BY TIMER					
10/57	SHUTDOWN FUNCTIONS	10		20	VIBRATION	30	HIGH SUCT. SCRUBBER LEVEL	40	OVERSPEED	50	HIGH TEMPERATURE CYLINDER #1
00	RESET READY TO START	11	LOW ENGINE OIL PRESSURE	21		31	HIGH INTSTG. 1 SCRUBBER LEVEL	41	HIGH SUCTION PRESSURE	51	HIGH TEMPERATURE CYLINDER #2
80	BATTERY OK	12	LOW COMP. OIL PRESSURE	22		32	HIGH INTSTG. 2 SCRUBBER LEVEL	42	HIGH INTSTG. 1 PRESSURE	52	HIGH TEMPERATURE CYLINDER #3
00	RUNNING-TIMER ACTIVE	13	CYL. OIL NO FLOW	23		33	LOW JACKET WATER LEVEL	43	HIGH INTSTG. 2 PRESSURE	53	HIGH TEMPERATURE CYLINDER #4
01	NORMAL RUN	14	LOW SUCTION PRESSURE	24		34	LOW ENGINE OIL LEVEL	44	HIGH DISCHARGE PRESSURE	54	
89	POWER OK	15	LOW INTSTG. 1 PRESSURE	25		35	LOW COMPRESSOR OIL LEVEL	45	HIGH ENGINE WATER TEMP.	55	
09	TEST	16	LOW INTSTG. 2 PRESSURE	26		36		46		56	
60	STOP	17	LOW DISCHARGE PRESSURE	27		37		47		57	

FIGURE 23.7 Typical electric panel display. (Reprinted with permission from Altronic Controls, Inc.)

FIGURE 23.8 Typical electric panel.

from major failures. The panel may be electric or pneumatic. Although instrumentation will vary from package to package, some of the most common safety switches on the panel monitor low engine lube oil pressure/level, high engine jacket-water temperature, engine overspeed, low jacket-water level, low compressor crankcase lube oil pressure/level, lubricator no flow, cooler and compressor vibration, high/low inlet pressure, high/low interstage pressure, high/low discharge

FIGURE 23.9 Typical pneumatic panel.

pressure (each stage), high discharge temperature (each cylinder), high liquid level (inlet and interstage scrubbers), and automatic fuel shutoff valve.

An electric control panel uses power generated by the magneto to perform the safety shutdown functions of the unit. Under normal operating conditions, the electric current flows from the magneto to the ignition system. The magneto is also wired to the control panel which is wired to the safety switches. When a safety switch is tripped and goes to ground, the electric current is diverted from the

FIGURE 23.10 Trailer-mounted CAT G3304NA/C12 sullair unit.

ignition to the control panel. In the control panel, a switch is tripped causing the magneto to ground and shut off the engine. Figure 23.7 shows what's typically monitered and displayed on a panel. Figure 23.8 is a photograph of the same panel.

A pneumatic control panel uses gas pressure instead of electricity to trigger a shutdown. Tubing is run between the safety switches, control panel relays, and the engine fuel shut-off valve. The safety switches maintain a low pressure signal locked into the control panel relays as long as the unit is within safe operating limits. If the limits are exceeded, the safety switch vents gas from the indicator relay, thus triggering the shutdown. The vented gas also causes the relay valve to change positions, allowing additional gas to be vented from the fuel shut-off valve as well as from a pressure switch which causes the magneto to ground and shut off the engine. The unit will remain down until the problem is corrected and the panel is reset. The unit can then be restarted. Figure 23.9 shows a typical pneumatic panel.

23.10 ROTARY SCREW GAS COMPRESSORS

A rotary screw compressor consists of two intermeshing helical rotors inside a housing. The helical rotor grooves are filled with gas as they pass over the suction

FIGURE 23.11 Typical reciprocating unit, CAT G3516LEA/FE665-21, Energy Industries.

port. As the rotors turn, the grooves are closed by the housing, forming a compression chamber. Lubricant is injected into the screw to provide sealing, cooling, and lubrication. Because the lubricant is recirculated, the screw compressor uses less lubricant than a reciprocating compressor. The gas/lubricant mixture exits from the compressor and passes over the discharge port. As with the reciprocating units, a scrubber is required ahead of the screw compressor to keep solids and liquids out of the compressor. However, unlike the reciprocating units, the screw unit requires the use of a separator to separate the lubricant from the gas. The separator is usually constructed according to ASME code and consists of an upper chamber that houses a coalescing filter element to separate the lubricant from the gas. The bottom chamber is the impingement section and reservoir with drain. As with the reciprocating units, the screw package also includes a driver, either electric motor or gas engine, gas piping, control panel and instrumentation, and cooler. Pulsation bottles are not required since there are no cylinders and since there is little or no vibration caused by the screw compressor. The units are simple with few components, and it is possible to design a self-contained, trailer-mounted unit, such as the CAT G3304NA/Sullair C-12 unit shown in Fig. 23.10. A typical reciprocating unit is shown in Fig. 23.11.

23.11 REGULATORY COMPLIANCE & OFFSHORE CONSIDERATIONS

The regulatory documents most commonly referred to when designing compressor packages are the American Petroleum Institute (API) API 11P[8] and API 618[5].

When units are packaged for offshore use, several other factors must be considered. The Minerals Management Service requires that gas compressors used offshore must adhere to their regulations. Also the API RP14C, Section A8[9] requires specific safety devices.

23.12 TESTING

Units should be tested in the shop prior to shipment to the customer. Although it is usually not possible to test the unit under fully-loaded conditions, the unit can be run with air, checked for vibration problems, and examined to ensure all safety shutdowns operate properly.

23.13 REFERENCES

1. Gas Processors Suppliers Association, *Engineering Data Book, Volume I,* Section 7, Fig. 7-9, Typical K & C Factors For Sizing Woven Wire Demisters, p. 7-7. Reprinted with permission.
2. Gas Processors Suppliers Association, *Engineering Data Book, Volume II,* Section 17, Eq. 17-22, p. 17-6, 1987.
3. ASME/ANSI B16.5 "Pipe Flanges and Flanged Fittings." ASME, 1988.
4. ASME/ANSI B31.3 "Chemical Plant and Petroleum Refinery Piping." ASME, 1990.
5. API STANDARD 618, "Reciprocating Compressors for General Refinery Services." API, 3rd Ed., February 1986.
6. ASME "Boiler and Pressure Vessel Code." ASME, Section VIII, Division 1, 1995 Ed.
7. Lindberg, Peter, P.E., "Thermodynamics," *Professional Engineering Registration Program,* 6th Ed.
8. API SPECIFICATION 11p, "Specification for Packaged Reciprocating Compressors for Oil and Gas Production Services." API, 2nd Ed., November 1989.
9. API RP 14C, "API Recommended Practice for Analysis, Design, Installation and Testing of Basic Surface Safety Systems for Offshore Production Platforms." API, 3rd Ed., April 15, 1984.

APPENDIX

A.1 DEFINITIONS OF COMPRESSOR TERMS

A.2 CONVERSION FACTORS

A.3 TEMPERATURE CONVERSION

A.4 AREAS AND CIRCUMFERENCES OF CIRCLES

A.5 PROPERTIES OF SATURATED STEAM

A.6 PARTIAL PRESSURE OF WATER VAPOR IN AIR

A.7 ATMOSPHERIC PRESSURE AND BAROMETRIC READINGS AT DIFFERENT ALTITUDES

A.8 DISCHARGE OF AIR THROUGH ORIFICE

A.9 LOSS OF AIR PRESSURE DUE TO PIPE FRICTION

A.10 LOSS OF PRESSURE THROUGH SCREW PIPE FITTINGS

A.11 HORSEPOWER TO COMPRESS AIR

A.12 n VALUES AND PROPERTIES OF GASES

A.13 TEMPERATURE RISE VS. COMPRESSION RATIO

A.14 SINGLE-STAGE COMPRESSOR CALCULATIONS

A.15 TWO-STAGE COMPRESSOR CALCULATIONS

A.16 SINGLE-STAGE ADDED CYLINDER CLEARANCE

A.1 DEFINITIONS OF GAS COMPRESSOR ENGINEERING TERMS

Absolute pressure is total pressure measured from absolute zero, i.e., from an absolute vacuum. It equals the sum of gauge pressure and atmospheric pressure corresponding to the barometer (expressed in pounds per square inch).

Absolute temperature equals degrees Fahrenheit plus 459.6 or degrees centigrade plus 273. These values are referred to as degrees Rankine and degrees Kelvin, respectively.

Adiabatic or isentropic compression of a gas is effected when no heat is transferred to or from the gas during the compression process.* The characteristic equation relating pressure and volume during adiabatic compression is

$$pv^k = C$$

where k is the ratio of the specific heat at constant pressure to the specific heat at constant volume.

Polytropic compression is effected when heat is transferred to or from gas during the compression process at such a precise rate that the relation between pressure and volume can be expressed by the equation

$$pv^n = C$$

in which n is constant.

Where the actual compression path for a particular compressor is known, and where the heat transfer to or from the gas is at the proper rate, the value of n may be determined from the equation.

Isothermal compression is effected when interchange of heat between air or gas and surrounding bodies occurs at a rate precisely sufficient to maintain the air or gas at constant temperature during compression. It may be considered as a special case of polytropic compression.

The characteristic equation for isothermal compression is

$$pv = C$$

k value is the value of the exponent defined by the equation under *Adiabatic or Isentropic Compression* for any particular gas.

n value is the value of the exponent as defined by the equation under *Polytropic Compression* for any particular gas.

Compressibility factor is a factor expressing the deviation from the perfect-gas law.

Pressure ratio or compression ratio is the ratio of the absolute discharge pressure to the absolute inlet pressure.

---

*In compressor practice, this definition refers only to the reversible adiabatic or isentropic compression process.

Free air is defined as air at atmospheric conditions at any specific location. Because altitude, barometer, and temperature may vary at different localities and at different times, it follows that this term does not mean air under identical or standard conditions.

Free air as a measure of volume may be applied either to displacement or capacity, and in no way distinguishes between these two terms.

Standard air is defined as air at a temperature of 68°F, a pressure of 14.70 psia, and a relative humidity of 35% (.0750 density). This agrees with the definitions adopted by ASME, but in gas industries the temperature of "standard air" is usually given as 60°F.

Displacement of a compressor is the volume displaced per unit of time and is usually expressed in cubic feet per minute. In a reciprocating compressor it equals the net area of the compressor piston multiplied by the length of stroke and by the number of compression strokes per minute. The displacement rating of a multistage compressor is the displacement of the low-pressure cylinder only.

Capacity (*actual delivery*) of an air or gas compressor is the actual quantity of air or gas compressed and delivered, expressed in cubic feet per minute at conditions of total temperature, total pressure, and composition prevailing at the compressor inlet. Capacity is always expressed in terms of air or gas at intake conditions rather than in terms of standard air or gas.

Theoretical horsepower is defined as the horsepower required to compress adiabatically the air or gas delivered by the compressor through the specified range of pressures. For a multistage compressor with intercooling between stages *theoretical horsepower* assumes equal work in each stage and perfect cooling between stages.

Theoretical power (*polytropic*) is the mechanical power required to compress polytropically and to deliver, through the specified range of pressures, the gas delivered by the compressor.

Air indicated horsepower is the horsepower calculated from compressor-indicator diagrams. The term applies only to displacement-type compressors.

Brake horsepower or shaft horsepower is the measured horsepower input to the compressor.

It should be noted that horsepower, either indicated or brake, for any displacement compressor varies with compression ratio as well as absolute intake and discharge pressures. Performance guarantees are expressed in terms of horsepower per cubic foot capacity. In comparing test results with performance guarantees, corrections should be made for any deviation from specified values of absolute intake pressures and ratio of compression.

Intercooling is the removal of heat from the air or gas between stages or stage groups.

Degree of intercooling is the difference in air or gas temperatures between the inlet of the compressor and the outlet of the intercooler.

Perfect intercooling prevails when the air temperature leaving the intercoolers is equal to the temperature of the air at the compressor intake.

Volumetric efficiency is the ratio of the capacity of the compressor to displacement of the compressor. The term does not apply to centrifugal compressors.

Mechanical efficiency is the ratio of the horsepower imparted to the air or gas to brake horsepower. In the case of a displacement-type compressor it is the ratio of air or gas indicated horsepower to indicated horsepower of the power cylinders for a steam engine or internal-combustion engine-driven compressor or to the brake horsepower delivered to the shaft in the case of a power-driven compressor.

Compression efficiency (adiabatic) is the ratio of the theoretical horsepower to horsepower imparted to the air or gas actually delivered by the compressor. Power imparted to the air or gas is brake horsepower minus mechanical losses.

Efficiency of the compressor is the ratio of the theoretical horsepower to brake horsepower. It is equal to the product of compression efficiency times mechanical efficiency.

Compressor efficiency (polytropic), for which alternate terms are "hydraulic efficiency" and stage efficiency," is the ratio of theoretical power (polytropic) to shaft power.

Temperature-rise ratio is the ratio of computed isentropic temperature rise to measured total temperature rise during compression. For a perfect gas, this is equal to the ratio of isentropic enthalpy rise to actual enthalpy rise. Consequently, for gases which do not deviate seriously from the perfect-gas law, the temperature-rise ratio is sometimes referred to as *"temperature-rise efficiency."*

Inlet pressure is the absolute total pressure at the inlet flange of a compressor.

Discharge pressure is the absolute total pressure at the discharge flange of a compressor. It is commonly stated in terms of gauge pressure; unless the associated barometric pressure is included, this is an incomplete statement of discharge pressure.

Inlet temperature is the total temperature at the intake flange of the compressor.

Discharge temperature is the total temperature at the discharge flange of the compressor.

Gas specific weight is the weight of air or gas per unit volume. Unless otherwise specified, it refers to the weight per unit volume at conditions of total pressure, total temperature, and composition prevailing at the inlet of the compressor.

Specific gravity is the ratio of specific weight of air or gas to that of dry air at the same pressure and temperature.

Speed refers to the revolutions per minute of the compressor shaft.

Electrical input is measured at the motor terminals. For synchronous motors with separately driven exciters, the excitation input as measured at the slip rings is added to the input to the stator. For synchronous motors with direct-connected exciters, the exciter losses are deducted from the measured stator input.

Load factor is the ratio of the average compressor load during a given period of time to the maximum rated load of the compressor.

A.2 CONVERSION FACTORS (MULTIPLIERS)

MULTIPLY	BY	TO OBTAIN	MULTIPLY	BY	TO OBTAIN
Acres	160.	Square Rods	Meters	3.28	Feet
Acres	43,560.	Square Feet	Meters	39.37	Inches
Acres	0.001562	Square Miles	Meters	1.094	Yards
Acres	4,840.	Square Yards	Miles	5,280.	Feet
Acre-feet	43,560.	Cubic Feet	Miles	1.609	Kilometers
			Miles	1,760.	Yards
Barrels, Water	350.	Pounds	Miles per Hour	88.	Feet per Minute
Barrels, Water	31.5	Gallons	Miles per Hour	1.609	Kilometers per Hour
Barrels, Oil	42.0	Gallons	Miles per Hour	.8684	Knots per Hour
Barrels per Day	0.02917	Gallons per Minute	Milligrams	1,000.	Grams
B T U	778.3	Foot-Pounds	Milliliters	.001	Liters
B T U per Minute	0.02357	Horse-Power	Millimeters	.1	Centimeters
			Minutes	.01667	Hours
			Minutes (Angle)	.0002909	Radians
Centimeter	0.3937	Inches	Minutes (Angle)	60.	Seconds (Angle)
Circular Mills	0.7845	Square-Mils	Months	30.42	Days
Cubic Feet	1,728.	Cubic Inches	Months	730.	Hours
Cubic Feet	0.03704	Cubic Yards			
Cubic Feet	7.481	Gallons	Ounces	437.5	Grains
Cubic Feet	0.1781	Barrels (Oilfield)	Ounces	28.35	Grams
Cubic Meters	35.31	Cubic Feet	Ounces	.0625	Pounds
Cubic Yards	27.	Cubic Feet	Ounces (Fluid)	1.805	Cubic Inches
Days	24.	Hours	Parts per Million	0.05835	Grains per Gallon
Days	1,440.	Minutes	Pints	28.87	Cubic Inches
Degrees (Angle)	60.	Minutes (Angle)	Pints	.125	Gallons
Degrees (Angle)	0.01745	Radians	Pounds	7,000.	Grains
			Pounds	453.6	Grams
Fathoms	6.	Feet	Pounds of Water	.01602	Cubic Feet of Water
Feet	30.48	Centimeters	Pounds of Water	27.68	Cubic Inches of Water
Feet	12.	Inches	Pounds of Water	.1198	Gallons
Feet	.0001894	Miles	Pounds per Cubic Foot	.01602	Grams per Cubic Cm.
Feet	0.36	Varas	Pounds per Square Inch	2.307	Feet of Water
Feet of Water (depth)	.4335	Pounds per Square Inch	Pounds per Square Inch	2.036	Inches of Mercury
Feet per Minute	.01136	Miles per Hour			
Feet per 100 feet	1.	Per cent Grade	Quadrants (Angle)	90.	Degrees
Foot	0.3048	Meters	Quarts (Liquid)	57.75	Cubic Inches
Foot	0.0667	Rods	Quarts (Liquid)	946.4	Cubic Centimeters
Foot Pounds	.001285	B T U			
Foot Pounds per Minute	.0000303	Horsepower	Radians	57.30	Degrees
Furlongs	40.	Rods	Radians per Second	9.549	Revolutions per Minute
Furlong	660.	Feet	Reams	500.	Sheets
			Revolutions	360.	Degrees
Gallons (Imperial)	1.209	Gallons (U.S.)	Revolutions	6.283	Radians
Gallons	3,785.	Cubic Centimeters	Revolutions per Minute	.1047	Radians per Second
Gallons	.02381	Barrels, Oil	Revolutions per Minute	6.	Degrees per Second
Gallons	.1337	Cubic Feet	Rods	16.5	Feet
Gallons	3.785	Liters			
Gallons per Minute	.002228	Cubic Feet per Second	Square Centimeters	.1550	Square Inches
Gallons per Minute	34,286.	Barrels per Day	Square Feet	144.	Square Inches
Grams	.03527	Ounces	Square Feet	.00002296	Acres
			Square Feet	929.	Square Centimeters
Hectares	2.471	Acres	Square Inches	6.452	Square Centimeters
Horsepower	42.40	B T U's per Minute	Square Inches	.006944	Square Feet
Horsepower	33,000.	Foot pounds per Minute	Square Miles	640.	Acres
Horsepower	745.7	Watts	Square Miles	259.	Square Kilometers
Horsepower (boiler)	33.52	B T U per Hour	Square Kilometer	247.1	Acres
Hours	.04167	Days	Square Meters	10.76	Square Feet
			Square Meters	.0002471	Acres
			Square Yards	9.	Square Feet
Inches	.08333	Feet	Square Yards	.8361	Square Meters
Inches	2.54	Centimeters			
Inches of Mercury	.4912	Pounds per Square Inch	Temperature (Degrees Cent.)	1.8 (Add 32 Deg.)	Temp. (Degrees Fahr.)
Inches of Water	.03613	Pounds per Square Inch	Temperature (Degrees Fahr.)	5/9 or 0.5556 (Subtract 32 Deg.)	Temp. (Degrees Cent.)
Kilograms per Sq.Cent.	14.22	Pounds per Sq. In.			
kilometers	3,281.	Feet	Tons (Long)	2,240.	Pounds
Kilometers	.6214	Miles	Tons (Metric)	2,205.	Pounds
Kilometers per Hour	54.68	Feet per Minute	Tons (Metric)	7.454	Barrels (36 Degree Oil)
Kilometers per Hour	.5396	Knots per Hour	Tons (Short)	2,000.	Pounds
Kilometers per Hour	.6214	Miles per Hour	Tons (Short)	2.	1,000 Pounds
Kilowatts	1,000.	Watts			
Kilowatts	1.341	Horsepower	Watts	3.412	B T U per Hour
Kilowatts	44,270.	Foot Pounds per Minute	Watts	.001341	Horsepower
Knots	6,080.	Feet	Watts	.001	Kilowatts
Knots	1.152	Miles	Watt Hour	3.413	B T U
Knots per Hour	1.152	Miles per Hour	Watt Hour	2,656.	Foot Pounds
			Watt Hours	.001	Kilowatt Hours
League (Statute)	3.	Miles	Weeks	168.	Hours
Links (Surveyor's)	7.92	Inches			
Liters	1,000.	Cubic Centimeters	Yards	.9144	Meters
Liters	.03531	Cubic Feet	Yards	91.44	Centimeters
Liters	.2642	Gallons	Years	8,760.	Hours

A.3 TEMPERATURE CONVERSION CHART (CENTIGRADE—FAHRENHEIT)

NOTE: The numbers in boldface refer to the temperature in degrees, either Centigrade or Fahrenheit, which is desired to convert into the other scale. If converting from Fahrenheit to Centigrade degrees, the equivalent temperature will be found in the left column; while if converting from degrees Centigrade to degrees Fahrenheit, the answer will be found in the column on the right.

Centigrade		Fahrenheit	Centigrade		Fahrenheit	Centigrade		Fahrenheit	Centigrade		Fahrenheit
−73.3	−100	−148.0	2.8	37	98.6	33.3	92	197.6	293	560	1040
−67.8	−90	−130.0	3.3	38	100.4	33.9	93	199.4	299	570	1058
−62.2	−80	−112.0	3.9	39	102.2	34.4	94	201.2	304	580	1076
−59.4	−75	−103.0	4.4	40	104.0	35.0	95	203.0	310	590	1094
−56.7	−70	−94.0	5.0	41	105.8	35.6	96	204.8	316	600	1112
−53.9	−65	−85.0	5.6	42	107.6	36.1	97	206.6	321	610	1130
−51.1	−60	−76.0	6.1	43	109.4	36.7	98	208.4	327	620	1148
−48.3	−55	−67.0	6.7	44	111.2	37.2	99	210.2	332	630	1166
−45.6	−50	−58.0	7.2	45	113.0	37.8	100	212.0	338	640	1184
−42.8	−45	−49.0	7.8	46	114.8	43	110	230	343	650	1202
−40.0	−40	−40.0	8.3	47	116.6	49	120	248	349	660	1220
−37.2	−35	−31.0	8.9	48	118.4	54	130	266	354	670	1238
−34.4	−30	−22.0	9.4	49	120.2	60	140	284	360	680	1256
−31.7	−25	−13.0	10.0	50	122.0	66	150	302	366	690	1274
−28.9	−20	−4.0	10.6	51	123.8	71	160	320	371	700	1292
−26.1	−15	5.0	11.1	52	125.6	77	170	338	377	710	1310
−23.3	−10	14.0	11.7	53	127.4	82	180	356	382	720	1328
−20.6	−5	23.0	12.2	54	129.2	88	190	374	388	730	1346
−17.8	0	32.0	12.8	55	131.0	93	200	392	393	740	1364
−17.2	1	33.8	13.3	56	132.8	99	210	410	399	750	1382
−16.7	2	35.6	13.9	57	134.6	100	212	414	404	760	1400
−16.1	3	37.4	14.4	58	136.4	104	220	428	410	770	1418
−15.6	4	39.2	15.0	59	138.2	110	230	446	416	780	1436
−15.0	5	41.0	15.6	60	140.0	116	240	464	421	790	1454
−14.4	6	42.8	16.1	61	141.8	121	250	482	427	800	1472
−13.9	7	44.6	16.7	62	143.6	127	260	500	432	810	1490
−13.3	8	46.4	17.2	63	145.4	132	270	518	438	820	1508
−12.8	9	48.2	17.8	64	147.2	138	280	536	443	830	1526
−12.2	10	50.0	18.3	65	149.0	143	290	554	449	840	1544
−11.7	11	51.8	18.9	66	150.8	149	300	572	454	850	1562
−11.1	12	53.6	19.4	67	152.6	154	310	590	460	860	1580
−10.6	13	55.4	20.0	68	154.4	160	320	608	466	870	1598
−10.0	14	57.2	20.6	69	156.2	166	330	626	471	880	1616
−9.4	15	59.0	21.1	70	158.0	171	340	644	477	890	1634
−8.9	16	60.8	21.7	71	159.8	177	350	662	482	900	1652
−8.3	17	62.6	22.2	72	161.6	182	360	680	488	910	1670
−7.8	18	64.4	22.8	73	163.4	188	370	698	493	920	1688
−7.2	19	66.2	23.3	74	165.2	193	380	716	499	930	1706
−6.7	20	68.0	23.9	75	167.0	199	390	734	504	940	1724
−6.1	21	69.8	24.4	76	168.8	204	400	752	510	950	1742
−5.6	22	71.6	25.0	77	170.6	210	410	770	516	960	1760
−5.0	23	73.4	25.6	78	172.4	216	420	788	521	970	1778
−4.4	24	75.2	26.1	79	174.2	221	430	806	527	980	1796
−3.9	25	77.0	26.7	80	176.0	227	440	824	532	990	1814
−3.3	26	78.8	27.2	81	177.8	232	450	842	538	1000	1832
−2.8	27	80.6	27.8	82	179.6	238	460	860	566	1050	1922
−2.2	28	82.4	28.3	83	181.4	243	470	878	593	1100	2012
−1.7	29	84.2	28.9	84	183.2	249	480	896	621	1150	2102
−1.1	30	86.0	29.4	85	185.0	254	490	914	649	1200	2192
−0.6	31	87.8	30.0	86	186.8	260	500	932	677	1250	2282
0.0	32	89.6	30.6	87	108.6	266	510	950	704	1300	2372
0.6	33	91.4	31.1	88	190.4	271	520	968	732	1350	2462
1.1	34	93.2	31.7	89	192.2	277	530	986	760	1400	2552
1.7	35	95.0	32.2	90	194.0	282	540	1004	788	1450	2642
2.2	36	96.8	32.8	91	195.8	288	550	1022	816	1500	2732

The formulas at the right may also be used for converting Centigrade or Fahrenheit degrees into the other scale.

$$\text{Degrees Cent., } C° = \frac{5}{9}(F° + 40) - 40 \qquad \text{Degrees Fahr., } F° = \frac{9}{5}(C° + 40) - 40$$

APPENDIX A.7

A.4 AREAS AND CIRCUMFERENCES OF CIRCLES

Diam.	Circ.	Area	Diam.	Circ.	Area	Diam.	Circ.	Area	Diam.	Circ.	Area	Diam.	Circ.	Area	Diam.	Circ.	Area	Diam.	Circ.	Area
1/32	.0981	.00076	7	21.99	38.484	15	47.12	176.71	23	72.25	415.47	31	97.38	754.76	39	122.5	1194.5
1/16	.1963	.00306	1/8	22.38	39.871	1/8	47.51	179.67	1/8	72.64	420.00	1/8	97.78	760.86	1/8	122.9	1202.2
1/8	.3926	.01227	1/4	22.77	41.282	1/4	47.90	182.65	1/4	73.04	424.55	1/4	98.17	766.99	1/4	123.3	1209.9
3/16	.5890	.02761	3/8	23.16	42.718	3/8	48.30	185.66	3/8	73.43	429.13	3/8	98.56	773.14	3/8	123.7	1217.6
1/4	.7854	.04908	1/2	23.56	44.178	1/2	48.69	188.69	1/2	73.82	433.73	1/2	98.96	779.31	1/2	124.0	1225.4
5/16	.9817	.07669	5/8	23.95	45.663	5/8	49.08	191.74	5/8	74.21	438.36	5/8	99.35	785.51	5/8	124.4	1233.1
3/8	1.178	.1104	3/4	24.34	47.173	3/4	49.48	194.82	3/4	74.61	443.01	3/4	99.74	791.73	3/4	124.8	1240.9
7/16	1.374	.1503	7/8	24.74	48.707	7/8	49.87	197.93	7/8	75.00	447.69	7/8	100.1	797.97	7/8	125.2	1248.7
1/2	1.570	.1963	8	25.13	50.265	16	50.26	201.06	24	75.39	452.39	32	100.5	804.24	40	125.6	1256.6
9/16	1.767	.2485	1/8	25.52	51.848	1/8	50.65	204.21	1/8	75.79	457.11	1/8	100.9	810.54	1/8	126.0	1264.5
5/8	1.963	.3068	1/4	25.91	53.456	1/4	51.05	207.39	1/4	76.18	461.86	1/4	101.3	816.86	1/4	126.4	1272.3
11/16	2.159	.3712	3/8	26.31	55.088	3/8	51.44	210.59	3/8	76.57	466.63	3/8	101.7	823.21	3/8	126.8	1280.3
3/4	2.356	.4417	1/2	26.70	56.745	1/2	51.83	213.82	1/2	76.96	471.43	1/2	102.1	829.57	1/2	127.2	1288.2
13/16	2.552	.5184	5/8	27.09	58.426	5/8	52.22	217.07	5/8	77.36	476.25	5/8	102.4	835.97	5/8	127.6	1296.2
7/8	2.748	.6013	3/4	27.48	60.132	3/4	52.62	220.35	3/4	77.75	481.10	3/4	102.8	842.39	3/4	128.0	1304.2
15/16	2.945	.6902	7/8	27.88	61.862	7/8	53.01	223.65	7/8	78.14	485.97	7/8	103.2	848.83	7/8	128.4	1312.2
1	3.141	.7854	9	28.27	63.617	17	53.40	226.98	25	78.54	490.87	33	103.6	855.30	41	128.8	1320.2
1/8	3.534	.9940	1/8	28.66	65.396	1/8	53.79	230.33	1/8	78.93	495.79	1/8	104.0	861.79	1/8	129.1	1328.3
1/4	3.927	1.227	1/4	29.06	67.200	1/4	54.19	233.70	1/4	79.32	500.74	1/4	104.4	868.30	1/4	129.5	1336.4
3/8	4.319	1.484	3/8	29.45	69.029	3/8	54.58	237.10	3/8	79.71	505.71	3/8	104.8	874.85	3/8	129.9	1344.5
1/2	4.712	1.767	1/2	29.84	70.882	1/2	54.97	240.52	1/2	80.10	510.70	1/2	105.2	881.41	1/2	130.3	1352.6
5/8	5.105	2.073	5/8	30.23	72.759	5/8	55.37	243.97	5/8	80.50	515.72	5/8	105.6	888.00	5/8	130.7	1360.8
3/4	5.497	2.405	3/4	30.63	74.662	3/4	55.76	247.45	3/4	80.89	520.77	3/4	106.0	894.61	3/4	131.1	1369.0
7/8	5.890	2.761	7/8	31.02	76.588	7/8	56.16	250.94	7/8	81.28	525.83	7/8	106.4	901.25	7/8	131.5	1377.2
2	6.283	3.141	10	31.41	78.539	18	56.54	254.46	26	81.68	530.93	34	106.8	907.92	42	131.9	1385.4
1/8	6.675	3.546	1/8	31.80	80.515	1/8	56.94	258.01	1/8	82.07	536.04	1/8	107.2	914.61	1/8	132.3	1393.7
1/4	7.068	3.976	1/4	32.20	82.516	1/4	57.33	261.58	1/4	82.46	541.18	1/4	107.6	921.32	1/4	132.7	1401.9
3/8	7.461	4.430	3/8	32.59	84.540	3/8	57.72	265.18	3/8	82.85	546.35	3/8	107.9	928.06	3/8	133.1	1410.2
1/2	7.854	4.908	1/2	32.98	86.590	1/2	58.11	268.80	1/2	83.25	551.54	1/2	108.3	934.82	1/2	133.5	1418.6
5/8	8.246	5.411	5/8	33.37	88.664	5/8	58.51	272.44	5/8	83.64	556.76	5/8	108.7	941.60	5/8	133.9	1426.9
3/4	8.639	5.939	3/4	33.77	90.762	3/4	58.90	276.11	3/4	84.03	562.00	3/4	109.1	948.41	3/4	134.3	1435.3
7/8	9.032	6.491	7/8	34.16	92.885	7/8	59.29	279.81	7/8	84.43	567.26	7/8	109.5	955.25	7/8	134.6	1443.7
3	9.424	7.068	11	34.55	95.033	19	59.69	283.52	27	84.82	572.55	35	109.9	962.11	43	135.0	1452.2
1/8	9.817	7.669	1/8	34.95	97.205	1/8	60.08	287.27	1/8	85.21	577.87	1/8	110.3	968.99	1/8	135.4	1460.6
1/4	10.21	8.295	1/4	35.34	99.402	1/4	60.47	291.03	1/4	85.60	583.20	1/4	110.7	975.90	1/4	135.8	1469.1
3/8	10.60	8.946	3/8	35.73	101.62	3/8	60.86	294.83	3/8	86.00	588.57	3/8	111.1	982.84	3/8	136.2	1477.6
1/2	10.99	9.621	1/2	36.12	103.86	1/2	61.26	298.64	1/2	86.39	593.95	1/2	111.5	989.80	1/2	136.6	1486.1
5/8	11.38	10.320	5/8	36.52	106.13	5/8	61.65	302.48	5/8	86.78	599.37	5/8	111.9	996.78	5/8	137.0	1494.7
3/4	11.78	11.044	3/4	36.91	108.43	3/4	62.04	306.35	3/4	87.17	604.80	3/4	112.3	1003.7	3/4	137.4	1503.3
7/8	12.17	11.793	7/8	37.30	110.75	7/8	62.43	310.24	7/8	87.57	610.26	7/8	112.7	1010.8	7/8	137.8	1511.9
4	12.56	12.566	12	37.69	113.10	20	62.83	314.16	28	87.96	615.75	36	113.0	1017.8	44	138.2	1520.5
1/8	12.95	13.364	1/8	38.09	115.46	1/8	63.22	318.09	1/8	88.35	621.26	1/8	113.4	1024.9	1/8	138.6	1529.1
1/4	13.35	14.186	1/4	38.48	117.85	1/4	63.61	322.06	1/4	88.75	626.79	1/4	113.8	1032.0	1/4	139.0	1537.8
3/8	13.74	15.033	3/8	38.87	120.27	3/8	64.01	326.05	3/8	89.14	632.35	3/8	114.2	1039.1	3/8	139.4	1546.5
1/2	14.13	15.904	1/2	39.27	122.71	1/2	64.40	330.06	1/2	89.53	637.94	1/2	114.6	1046.3	1/2	139.8	1555.2
5/8	14.52	16.800	5/8	39.66	125.18	5/8	64.79	334.10	5/8	89.92	643.54	5/8	115.0	1053.5	5/8	140.1	1564.0
3/4	14.92	17.720	3/4	40.05	127.67	3/4	65.18	338.16	3/4	90.32	649.18	3/4	115.4	1060.7	3/4	140.5	1572.8
7/8	15.31	18.665	7/8	40.44	130.19	7/8	65.58	342.25	7/8	90.71	654.83	7/8	115.8	1067.9	7/8	140.9	1581.6
5	15.70	19.635	13	40.84	132.73	21	65.97	346.36	29	91.10	660.52	37	116.2	1075.2	45	141.3	1590.4
1/8	16.10	20.629	1/8	41.23	135.29	1/8	66.36	350.49	1/8	91.49	666.22	1/8	116.6	1082.4	1/8	141.7	1599.2
1/4	16.49	21.647	1/4	41.62	137.88	1/4	66.75	354.65	1/4	91.89	671.95	1/4	117.0	1089.7	1/4	142.1	1608.1
3/8	16.88	22.690	3/8	42.01	140.50	3/8	67.15	358.84	3/8	92.28	677.71	3/8	117.4	1097.1	3/8	142.5	1617.0
1/2	17.27	23.758	1/2	42.41	143.13	1/2	67.54	363.05	1/2	92.67	683.49	1/2	117.8	1104.4	1/2	142.9	1625.9
5/8	17.67	24.850	5/8	42.80	145.80	5/8	67.93	367.28	5/8	93.06	689.29	5/8	118.2	1111.8	5/8	143.3	1634.9
3/4	18.06	25.967	3/4	43.19	148.48	3/4	68.32	371.54	3/4	93.46	695.12	3/4	118.6	1119.2	3/4	143.7	1643.8
7/8	18.45	27.108	7/8	43.58	151.20	7/8	68.72	375.82	7/8	93.85	700.98	7/8	118.9	1126.6	7/8	144.1	1652.8
6	18.85	28.274	14	43.98	153.93	22	69.11	380.13	30	94.24	706.86	38	119.3	1134.1	46	144.5	1661.9
1/8	19.24	29.464	1/8	44.37	156.69	1/8	69.50	384.46	1/8	94.64	712.76	1/8	119.7	1141.5	1/8	144.9	1670.9
1/4	19.63	30.679	1/4	44.76	159.48	1/4	69.90	388.82	1/4	95.03	718.69	1/4	120.1	1149.0	1/4	145.2	1680.0
3/8	20.02	31.919	3/8	45.16	162.29	3/8	70.29	393.20	3/8	95.42	724.64	3/8	120.5	1156.0	3/8	145.6	1689.1
1/2	20.42	33.183	1/2	45.55	165.13	1/2	70.68	397.60	1/2	95.81	730.61	1/2	120.9	1164.1	1/2	146.0	1698.2
5/8	20.81	34.471	5/8	45.94	167.98	5/8	71.07	402.03	5/8	96.21	736.61	5/8	121.3	1171.7	5/8	146.4	1707.3
3/4	21.20	35.784	3/4	46.33	170.87	3/4	71.47	406.49	3/4	96.60	742.64	3/4	121.7	1179.3	3/4	146.8	1716.5
7/8	21.59	37.122	7/8	46.73	173.78	7/8	71.86	410.97	7/8	96.99	748.69	7/8	122.1	1186.9	7/8	147.2	1725.7

A.8 APPENDIX

A.5 PROPERTIES OF SATURATED STEAM

Vacuum Inches of Mercury	Temp. Fahrenheit	Total Heat Above 32°F. in Water Ht. Unit	Total Heat Above 32°F. in Steam Ht. Unit	Volume Cu. Ft. in 1 Lb. of Steam	Gauge Pressure Lbs. Per Sq.In.	Temp. Fahrenheit	Total Heat Above 32°F. in Water Ht. Unit	Total Heat Above 32°F. in Steam Ht. Unit	Volume Cu. Ft. in 1 Lb. of Steam
29.74	32	0.00	1073.4	3294	52.3	300.0	269.6	1179.0	6.47
29.67	40	8.05	1076.9	2438	54.3	302.0	271.6	1179.6	6.29
29.56	50	18.08	1081.4	1702	56.3	303.9	273.6	1180.1	6.12
29.40	60	28.08	1085.9	1208	58.3	305.8	275.5	1180.6	5.96
29.18	70	38.06	1090.3	871	60.3	307.6	277.4	1181.1	5.81
28.89	80	48.03	1094.8	636.8	62.3	309.4	279.3	1181.6	5.67
28.50	90	58.00	1099.2	469.3	64.3	311.2	281.1	1182.1	5.54
28.00	100	67.97	1103.6	350.8	66.3	312.9	282.9	1182.5	5.41
27.88	101.83	69.8	1104.4	333.0	68.3	314.6	284.6	1183.0	5.28
25.85	126.15	94.0	1115.0	173.5	70.3	316.3	286.3	1183.4	5.16
23.81	141.52	109.4	1121.6	118.5	72.3	317.9	288.0	1183.8	5.05
21.78	153.01	120.9	1126.5	90.5	74.3	319.5	289.7	1184.2	4.94
19.74	162.28	130.5	1130.5	73.33	76.3	321.1	291.3	1184.6	4.84
17.70	170.06	137.9	1133.7	61.89	78.3	322.6	292.9	1185.0	4.74
15.67	176.85	144.7	1136.5	53.56	80.3	324.1	294.5	1185.4	4.65
13.63	182.86	150.8	1139.0	47.27	82.3	325.6	296.1	1185.8	4.56
11.60	188.27	156.2	1141.1	42.36	84.3	327.1	297.6	1186.2	4.47
9.56	193.22	161.1	1143.1	38.38	87.3	329.3	299.8	1186.7	4.347
7.52	197.75	165.7	1144.9	35.10	91.3	332.0	302.7	1187.4	4.192
5.49	201.96	169.9	1146.5	32.36	95.3	334.8	305.5	1188.0	4.047
3.45	205.87	173.8	1148.0	30.03	99.3	337.4	308.3	1188.7	3.912
1.42	209.55	177.5	1149.4	28.02	103.3	340.0	311.0	1189.3	3.786
Lbs.Gauge					107.3	342.5	313.6	1189.8	3.668
0.0	212	180.0	1150.4	26.79	111.3	345.0	316.2	1190.4	3.556
0.3	213.0	181.0	1150.7	26.27	115.3	347.4	318.6	1191.0	3.452
1.3	216.3	184.4	1152.0	24.79	119.3	349.7	321.1	1191.5	3.354
2.3	219.4	187.5	1153.1	23.38	123.3	352.0	323.4	1192.0	3.263
3.3	222.4	190.5	1154.2	22.16	127.3	354.2	325.8	1192.5	3.175
4.3	225.2	193.4	1155.2	21.07	131.3	356.3	328.0	1192.9	3.092
5.3	228.0	196.1	1156.2	20.08	135.3	358.5	330.2	1193.4	3.012
6.3	230.6	198.8	1157.1	19.18	139.3	360.5	332.4	1193.8	2.938
7.3	233.1	201.3	1158.0	18.37	143.3	362.6	334.6	1194.3	2.868
8.3	235.5	203.8	1158.8	17.62	147.3	364.6	336.7	1194.7	2.801
9.3	237.8	206.1	1159.6	16.93	151.3	366.5	338.7	1195.1	2.737
10.3	240.1	208.4	1160.4	16.30	155.3	368.5	340.7	1195.4	2.675
11.3	242.2	210.6	1161.2	15.72	159.3	370.4	342.7	1195.8	2.616
12.3	244.4	212.7	1161.9	15.18	163.3	372.2	344.7	1196.2	2.560
13.3	246.4	214.8	1162.6	14.67	167.3	374.0	346.6	1196.6	2.507
14.3	248.4	216.8	1163.2	14.19	171.3	375.8	348.5	1196.9	2.455
15.3	250.3	218.8	1163.9	13.74	175.3	377.6	350.4	1197.3	2.406
16.3	252.2	220.7	1164.5	13.32	179.3	379.3	352.2	1197.6	2.358
17.3	254.1	222.6	1165.1	12.93	183.3	381.0	354.0	1197.9	2.312
18.3	255.8	224.4	1165.7	12.57	185.3	381.9	354.9	1198.1	2.290
19.3	257.6	226.2	1166.3	12.22	190.3	384.0	357.1	1198.5	2.237
20.3	259.3	227.9	1166.8	11.89	200.3	388.0	361.4	1199.2	2.138
22.3	262.6	231.3	1167.8	11.29	210.3	391.9	365.5	1199.9	2.046
24.3	265.8	234.5	1168.9	10.74	220.3	395.6	369.4	1200.6	1.964
26.3	268.7	237.6	1169.8	10.25	230.3	399.3	373.3	1201.2	1.887
28.3	271.7	240.5	1170.7	9.80	245.3	404.5	378.9	1202.1	1.782
30.3	274.5	243.4	1171.6	9.39	265.3	411.2	386.0	1203.1	1.658
32.3	277.2	246.1	1172.4	9.02	285.3	417.5	392.7	1204.1	1.551
34.3	279.8	248.8	1173.2	8.67	305.3	423.4	399.1	1204.9	1.456
36.3	282.3	251.4	1174.0	8.35	325.3	429.1	405.3	1205.7	1.372
38.3	284.7	253.9	1174.7	8.05	345.3	434.6	411.2	1206.4	1.298
40.3	287.1	256.3	1175.4	7.78	365.3	439.8	416.8	1207.1	1.231
42.3	289.4	258.7	1176.0	7.52	385.3	444.8	422	1208	1.17
44.3	291.6	261.0	1176.7	7.28	435.3	456.5	435	1209	1.04
46.3	293.8	263.2	1177.3	7.06	485.3	467.3	448	1210	0.93
48.3	295.9	265.4	1177.9	6.85	535.3	477.3	459	1210	0.83
50.3	298.0	267.5	1178.5	6.65	585.3	486.6	469	1210	0.76

A.6 PARTIAL PRESSURE OF WATER VAPOR IN SATURATED AIR 32° TO 212°F

Temp., °F.	Pressure, p.s.i.a.	Temp., °F.	Pressure, p.s.i.a.	Temp., °F.	Pressure, p.s.i.a.
		75	.4295	120	1.692
32	.0887	76	.4440	121	1.739
33	.0923	77	.4590	122	1.788
34	.0961	78	.4744	123	1.838
		79	.4903	124	1.889
35	.1000	80	.5067	125	1.941
36	.1041	81	.5236	126	1.995
37	.1083	82	.5409	127	2.049
38	.1126	83	.5588	128	2.105
39	.1171	84	.5772	129	2.163
40	.1217	85	.5960	130	2.221
41	.1265	86	.6153	131	2.281
42	.1315	87	.6352	132	2.343
43	.1367	88	.6555	133	2.406
44	.1420	89	.6765	134	2.470
45	.1475	90	.6980	135	2.536
46	.1532	91	.7201	136	2.603
47	.1591	92	.7429	137	2.672
48	.1652	93	.7662	138	2.742
49	.1715	94	.7902	139	2.814
50	.1780	95	.8149	140	2.887
51	.1848	96	.8403	141	2.962
52	.1918	97	.8663	142	3.039
53	.1989	98	.8930	143	3.118
54	.2063	99	.9205	144	3.198
55	.2140	100	.9487	145	3.280
56	.2219	101	.9776	146	3.363
57	.2300	102	1.0072	147	3.449
58	.2384	103	1.0377	148	3.536
59	.2471	104	1.0689	149	3.625
60	.2561	105	1.1009	150	3.716
61	.2654	106	1.1338	155	4.201
62	.2749	107	1.1675	160	4.739
63	.2848	108	1.2020	165	5.334
64	.2949	109	1.2375	170	5.990
65	.3054	110	1.274	175	6.716
66	.3162	111	1.311	180	7.510
67	.3273	112	1.350	185	8.382
68	.3388	113	1.389	190	9.336
69	.3506	114	1.429	195	10.385
70	.3628	115	1.470	200	11.525
71	.3754	116	1.512	205	12.769
72	.3883	117	1.555	210	14.123
73	.4016	118	1.600	212	14.696
74	.4153	119	1.645		

In many process applications, it is necessary to determine the compressor capacity required to deliver a given quantity of dry gas, i.e., after all moisture is removed. Since water vapor carried in incoming gas stream will be eliminated in interstage coolers, aftercoolers and driers, capacity must be increased to compensate for vapor which, though it enters the compressor as volume, has no value to the process. Thus, if process requires 1000 cfm of dry air at 100°F and intake is 14.7 p.s.i.a., 100°F and saturated, required capacity at compressor intake is:

$$1000 \times \frac{14.7}{14.7 - .949} = 1068 \text{ c.f.m.}$$

Thus, capacity has been corrected by multiplying by the suction pressure and dividing by the suction pressure minus the partial pressure of the water vapor at suction temperature.

If, in the example above, the air was only 70% saturated (70% relative humidity), the partial pressure is multiplied by the percent saturation. Thus:

$$\text{Capacity at intake} = 1000 \times \frac{14.7}{14.7 - (.7)(.949)} = 1047 \text{ c.f.m.}$$

A.7 ATMOSPHERIC PRESSURE AND BAROMETRIC READINGS AT DIFFERENT ALTITUDES

Altitude above sea level, ft.	Atmospheric pressure psi.	Barometer reading, in. Hg.	Altitude above sea level, ft.	Atmospheric pressure, psi.	Barometer reading, in. Hg.
0	14.69	29.92	7,500	11.12	22.65
500	14.42	29.38	8,000	10.91	22.22
1,000	14.16	28.86	8,500	10.70	21.80
1,500	13.91	28.33	9,000	10.50	21.38
2,000	13.66	27.82	9,500	10.30	20.98
2,500	13.41	27.31	10,000	10.10	20.58
3,000	13.16	26.81	10,500	9.90	20.18
3,500	12.92	26.32	11,000	9.71	19.75
4,000	12.68	25.84	11,500	9.52	19.40
4,500	12.45	25.36	12,000	9.34	19.03
5,000	12.22	24.89	12,500	9.15	18.65
5,500	11.99	24.43	13,000	8.97	18.29
6,000	11.77	23.98	13,500	8.80	17.93
6,500	11.55	23.53	14,000	8.62	17.57
7,000	11.33	23.09	14,500	8.45	17.22
			15,000	8.28	16.88

APPENDIX   A.11

A.8 DISCHARGE OF AIR THROUGH AN ORIFICE

In cu. ft. of Free Air per Min. at Atmospheric Pressure of 14.7 psia, and 70°F

Gauge pressure before orifice, psi.	Diameter of orifice, in.										
	1/64	1/32	1/16	1/8	1/4	3/8	1/2	5/8	3/4	7/8	1
	Discharge, cu. ft. free air per min.										
1	.028	0.112	0.450	1.80	7.18	16.2	28.7	45.0	64.7	88.1	115
2	.040	0.158	0.633	2.53	10.1	22.8	40.5	63.3	91.2	124	162
3	.048	0.194	0.775	3.10	12.4	27.8	49.5	77.5	111	152	198
4	.056	0.223	0.892	3.56	14.3	32.1	57.0	89.2	128	175	228
5	.062	0.248	0.993	3.97	15.9	35.7	63.5	99.3	143	195	254
6	.068	0.272	1.09	4.34	17.4	39.1	69.5	109	156	213	278
7	.073	0.293	1.17	4.68	18.7	42.2	75.0	117	168	230	300
9	.083	0.331	1.32	5.30	21.1	47.7	84.7	132	191	260	339
12	.095	0.379	1.52	6.07	24.3	54.6	97.0	152	218	297	388
15	.105	0.420	1.68	6.72	26.9	60.5	108	168	242	329	430
20	.123	0.491	1.96	7.86	31.4	70.7	126	196	283	385	503
25	.140	0.562	2.25	8.98	35.9	80.9	144	225	323	440	575
30	.158	0.633	2.53	10.1	40.5	91.1	162	253	365	496	648
35	.176	0.703	2.81	11.3	45.0	101	180	281	405	551	720
40	.194	0.774	3.10	12.4	49.6	112	198	310	446	607	793
45	.211	0.845	3.38	13.5	54.1	122	216	338	487	662	865
50	.229	0.916	3.66	14.7	58.6	132	235	366	528	718	938
60	.264	1.06	4.23	16.9	67.6	152	271	423	609	828	1,082
70	.300	1.20	4.79	19.2	76.7	173	307	479	690	939	1,227
80	.335	1.34	5.36	21.4	85.7	193	343	536	771	1,050	1,371
90	.370	1.48	5.92	23.7	94.8	213	379	592	853	1,161	1,516
100	.406	1.62	6.49	26.0	104	234	415	649	934	1,272	1,661
110	.441	1.76	7.05	28.2	113	254	452	705	1,016	1,383	1,806
120	.476	1.91	7.62	30.5	122	274	488	762	1,097	1,494	1,951
125	.494	1.98	7.90	31.6	126	284	506	790	1,138	1,549	2,023

Based on 100% coefficient of flow. For well-rounded entrance multiply values by 0.97. For sharp-edged orifices a multiplier of 0.65 may be used.

This table will give approximate results only. For accurate measurements see ASME Power Test Code, Velocity Volume flow Measurement.

Values for pressures from 1 to 15 psig, calculated by standard adiabatic formula.

Values for pressures above 15 psig, calculated by approximate formula proposed by S. A. Moss:

$$W = \frac{0.5303\, a C P_1}{\sqrt{T}}$$

W = discharge in lb. per sec.
a = area of orifice in sq. in.
C = coefficient of flow
P = upstream total pressure in psia.
T = upstream temperature in deg. F. abs.

Values used in calculating above table were C = 1.0, P = gauge pressure + 14.7 psi., T = 530 F. abs.

Weights (W) were converted to volumes using density factor of 0.07494 lb. per cu. ft. This is correct for dry air at 14.7 psia, and 70 F.

Formula cannot be used where P is less than two times the barometric pressure.

A.9 LOSS OF AIR PRESSURE DUE TO PIPE FRICTION

To determine the pressure drop in psi, the factor listed in the table for a given capacity and pipe diameter should be divided by the ratio of compression (from free air) at entrance of pipe, multiplied by the actual length of the pipe in feet, and divided by 1,000.

Cu.ft. free air per min.	Nominal diameter (inches)															
	1/2	3/4	1	1-1/4	1-1/2	1-3/4	2	2-1/2	3	3-1/2	4	4-1/2	5	6	8	10	12
5	12.7	1.2	0.5														
10	50.7	7.8	2.2	0.5													
15	114.1	17.6	4.9	1.1													
20	202	30.4	8.7	2.0	0.9												
25	316	50.0	13.6	3.2	1.4	0.7											
30	456	70.4	19.6	4.5	2.0	1.1											
35	621	95.9	26.6	6.2	2.7	1.4											
40	811	125.3	34.8	8.1	3.6	1.9											
45	.....	159	44.0	10.2	4.5	2.4	1.2										
50	.....	196	54.4	12.6	5.6	2.9	1.4										
60	.....	282	78.3	18.2	8.0	4.2	2.2										
70	.....	385	106.6	24.7	10.9	5.7	2.9	1.1									
80	.....	503	139.2	32.3	14.3	7.5	3.8	1.5									
90	.....	646	176.2	40.9	18.1	9.5	4.8	1.9									
100	.....	785	217.4	50.5	22.3	11.7	6.0	2.3									
110	.....	950	263	61.1	27.0	14.1	7.2	2.8									
120	.....	.....	318	72.7	32.2	16.8	8.6	3.3									
130	.....	.....	369	85.3	37.8	19.7	10.1	3.9	1.2								
140	.....	.....	426	98.9	43.8	22.9	11.7	4.6	1.4								
150	.....	.....	490	113.6	50.3	26.3	13.4	5.2	1.6								
160	.....	.....	570	129.3	57.2	29.9	15.3	5.9	1.9								
170	.....	.....	628	145.8	64.6	33.7	17.6	6.7	2.1								
180	.....	.....	705	163.3	72.6	37.9	19.4	7.5	2.4								
190	.....	.....	785	177	80.7	42.2	21.5	8.4	2.6								
200	.....	.....	870	202	89.4	46.7	23.9	9.3	2.9								
220	.....	.....	.....	244	108.2	56.5	28.9	11.3	3.5								
240	.....	.....	.....	291	128.7	67.3	34.4	13.4	4.2								
260	.....	.....	.....	341	151	79.0	40.3	15.7	4.9								
280	.....	.....	.....	395	175	91.6	46.8	18.2	5.7								
300	.....	.....	.....	454	201	105.1	53.7	20.9	6.6								
320	.....	.....	.....	.....	.....	.....	61.1	23.8	7.5	3.5							
340	.....	.....	.....	.....	.....	.....	69.0	26.8	8.4	3.9	2.0						
360	.....	.....	.....	.....	.....	.....	77.3	30.1	9.5	4.4	2.2						
380	.....	.....	.....	.....	.....	.....	86.1	33.5	10.5	4.9	2.5						
400	.....	.....	.....	.....	.....	.....	94.7	37.1	11.7	5.4	2.7						
420	.....	.....	.....	.....	.....	.....	105.2	40.9	12.9	6.0	3.1						
440	.....	.....	.....	.....	.....	.....	115.5	44.9	14.1	6.6	3.4						
460	.....	.....	.....	.....	.....	.....	125.6	48.8	15.4	7.1	3.7	2.0					
480	.....	.....	.....	.....	.....	.....	137.6	53.4	16.8	7.8	4.0	2.2					
500	.....	.....	.....	.....	.....	.....	150.0	58.0	18.3	8.5	4.3	2.4					
525	.....	.....	.....	.....	.....	.....	165.0	64.2	20.2	9.4	4.8	2.6					
550	.....	.....	.....	.....	.....	.....	181.5	70.2	22.1	10.2	5.2	2.9					
575	.....	.....	.....	.....	.....	.....	197	76.7	24.2	11.2	5.7	3.1					
600	.....	.....	.....	.....	.....	.....	215	83.5	26.3	12.2	6.2	3.4					
625	.....	.....	.....	.....	.....	.....	233	92.7	28.5	13.2	6.8	3.7					
650	.....	.....	.....	.....	.....	.....	253	98.0	30.9	14.3	7.3	4.0	2.2				
675	.....	.....	.....	.....	.....	.....	272	105.7	33.3	15.4	7.9	4.3	2.4				
700	.....	.....	.....	.....	.....	.....	294	113.7	35.8	16.6	8.5	4.6	2.6				
750	.....	.....	.....	.....	.....	.....	337	130.5	41.1	19.0	9.7	5.3	2.9				
800	.....	.....	.....	.....	.....	.....	382	148.4	46.7	21.7	11.1	6.1	3.3				
850	.....	.....	.....	.....	.....	.....	433	168	52.8	24.4	12.5	6.8	3.8				
900	.....	.....	.....	.....	.....	.....	468	188	59.1	27.4	14.0	7.7	4.2				
950	.....	.....	.....	.....	.....	.....	541	209.4	65.9	30.5	15.7	8.6	4.7				
1,000	.....	.....	.....	.....	.....	.....	600	232	73.0	33.8	17.3	9.5	5.2	1.9			
1,050	.....	.....	.....	.....	.....	.....	658	256	80.5	37.3	19.1	10.4	5.8	2.1			
1,100	.....	.....	.....	.....	.....	.....	723	280.6	88.4	40.9	21.0	11.5	6.3	2.4			
1,150	.....	.....	.....	.....	.....	.....	790	306.8	96.6	44.7	22.9	12.5	6.9	2.6			
1,200	.....	.....	.....	.....	.....	.....	850	344.0	105.2	48.8	25.0	13.7	7.5	2.8			
1,300	.....	.....	.....	.....	.....	.....	.....	392.0	123.4	57.2	29.3	16.0	8.8	3.3			
1,400	.....	.....	.....	.....	.....	.....	.....	.....	.....	66.3	33.9	18.6	10.2	3.8			

APPENDIX A.13

A.9 LOSS OF AIR PRESSURE DUE TO PIPE FRICTION (CONTINUED)

To determine the pressure drop in psi, the factor listed in the table for a given capacity and pipe diameter should be divided by the ratio of compression (from free air) at entrance of pipe, multiplied by the actual length of the pipe in feet, and divided by 1,000.

Cu.ft. free air per min.	Nominal diameter (inches)																
	1/2	3/4	1	1-1/4	1-1/2	1-3/4	2	2-1/2	3	3-1/2	4	4-1/2	5	6	8	10	12
1,500	....	....	....	....	....	....	....	....	....	76.1	39.0	21.3	11.8	4.4			
1,600	....	....	....	....	....	....	....	....	....	86.6	44.3	24.2	13.4	5.1			
1,700	....	....	....	....	....	....	....	....	....	97.8	50.1	27.4	15.1	5.7			
1,800	....	....	....	....	....	....	....	....	....	110	56.1	30.7	16.9	6.4			
1,900	....	....	....	....	....	....	....	....	....	122	62.7	34.2	18.9	7.1	1.6		
2,000	....	....	....	....	....	....	....	....	....	135	69.3	37.9	20.9	7.8	1.8		
2,100	....	....	....	....	....	....	....	....	....	149	76.4	40.8	23.0	8.7	2.0		
2,200	....	....	....	....	....	....	....	....	....	166	83.6	45.8	25.3	9.5	2.2		
2,300	....	....	....	....	....	....	....	....	....	179	91.6	50.1	27.6	10.4	2.4		
2,400	....	....	....	....	....	....	....	....	....	195	99.8	54.6	30.1	11.3	2.6		
2,500	....	....	....	....	....	....	....	....	....	212	108.2	59.2	32.6	12.3	2.9		
2,600	....	....	....	....	....	....	....	....	....	229	117.2	64.0	35.3	13.3	3.1		
2,700	....	....	....	....	....	....	....	....	....	247	126	69.1	38.1	14.3	3.3		
2,800	....	....	....	....	....	....	....	....	....	265	136	74.3	41.0	15.4	3.6		
2,900	....	....	....	....	....	....	....	....	....	285	146	79.8	43.9	16.5	3.9		
3,000	....	....	....	....	....	....	....	....	....	305	156	85.2	47.0	17.7	4.1		
3,200	....	....	....	....	....	....	....	....	....	347	177	97.1	53.5	20.1	4.7		
3,400	....	....	....	....	....	....	....	....	....	391	200	109.5	60.4	22.7	5.3		
3,600	....	....	....	....	....	....	....	....	....	438	224	122.8	67.7	25.4	5.6	1.8	
3,800	....	....	....	....	....	....	....	....	....	488	250	137	75.5	28.4	6.6	2.0	
4,000	....	....	....	....	....	....	....	....	....	542	277	151	83.6	31.4	7.3	2.2	
4,200	....	....	....	....	....	....	....	....	....	....	305	168	92.1	34.6	8.1	2.4	
4,400	....	....	....	....	....	....	....	....	....	....	335	183	101.2	38.1	8.9	2.7	
4,600	....	....	....	....	....	....	....	....	....	....	366	200	110.5	41.5	9.7	2.9	
4,800	....	....	....	....	....	....	....	....	....	....	399	218	120.4	45.2	10.5	3.2	
5,000	....	....	....	....	....	....	....	....	....	....	433	236	131	49.1	11.5	3.4	
5,250	....	....	....	....	....	....	....	....	....	....	477	260	144	54.1	12.6	3.8	
5,500	....	....	....	....	....	....	....	....	....	....	524	286	158	59.4	13.9	4.2	1.6
5,750	....	....	....	....	....	....	....	....	....	....	....	313	173	64.9	15.2	4.6	1.8
6,000	....	....	....	....	....	....	....	....	....	....	....	341	188	70.7	16.5	5.0	1.9
6,500	....	....	....	....	....	....	....	....	....	....	....	402	222	82.9	19.8	5.9	2.3
7,000	....	....	....	....	....	....	....	....	....	....	....	464	256	96.2	22.5	6.8	2.6
7,500	....	....	....	....	....	....	....	....	....	....	....	532	294	110.5	25.8	7.8	3.0
8,000	....	....	....	....	....	....	....	....	....	....	....	....	335	125.7	29.4	8.8	3.6
9,000	....	....	....	....	....	....	....	....	....	....	....	....	423	159	37.2	11.2	4.4
10,000	....	....	....	....	....	....	....	....	....	....	....	....	523	196	45.9	13.8	5.4
11,000	....	....	....	....	....	....	....	....	....	....	....	....	....	237	55.5	16.7	6.5

A.10 LOSS OF PRESSURE THROUGH SCREW PIPE FITTINGS

(Given in equivalent lengths (feet) of straight pipe.)

Nominal pipe size, (inches)	Actual inside diameter, (inches)	Gate valve	Long radius ell or on run of standard tee	Standard ell or on run of tee reduced in size 50 per cent	Angle valve	Close return bend	Tee through side outlet	Globe valve
1/2	0.622	0.36	0.62	1.55	8.65	3.47	3.10	17.3
3/4	0.824	0.48	0.82	2.06	11.4	4.60	4.12	22.9
1	1.049	0.61	1.05	2.62	14.6	5.82	5.24	29.1
1-1/4	1.380	0.81	1.38	3.45	19.1	7.66	6.90	38.3
1-1/2	1.610	0.94	1.61	4.02	22.4	8.95	8.04	44.7
2	2.067	1.21	2.07	5.17	28.7	11.5	10.3	57.4
2-1/2	2.469	1.44	2.47	6.16	34.3	13.7	12.3	68.5
3	3.068	1.79	3.07	6.16	42.6	17.1	15.3	85.2
4	4.026	2.35	4.03	7.67	56.0	22.4	20.2	112
5	5.047	2.94	5.05	10.1	70.0	28.0	25.2	140
6	6.065	3.54	6.07	15.2	84.1	33.8	30.4	168
8	7.981	4.65	7.98	20.0	111	44.6	40.0	222
10	10.020	5.85	10.00	25.0	139	55.7	50.0	278
12	11.940	6.96	11.0	29.8	166	66.3	59.6	332

APPENDIX A.15

A.11 HORSEPOWER (THEORETICAL) REQUIRED TO COMPRESS AIR FROM ATMOSPHERIC PRESSURE TO VARIOUS PRESSURES—MEAN EFFECTIVE PRESSURES

Discharge Pressure			Isothermal Compression Single or Multi-Stage		Adiabatic Compression*									Per Cent of Power Saved by Two-Stage Over Single Stage and Three Stage over Two Stage Adiabatic Compression
					Single Stage		Two-Stage				Three-Stage				
Lbs. Per Sq.In. Gage	Lbs. Per Sq.In. Absolute	Atmospheres Absolute	M.E.P.	Theoretical H.P.per 100 Cu.Ft.	M.E.P.	Theoretical H.P.per 100 Cu.Ft.	Theoretical Intercooler Gage Pressure	M.E.P. Lbs. per Sq. In. referred to Low-Press. Air Cyl.	Theoretical H.P.per 100 Cu.Ft.	Theoretical Intercooler Gage Pressure 1st / 2nd		M.E.P. Lbs. per Sq. In. referred to Low-Press. Air Cyl.	Theoretical H.P.per 100 Cu.Ft.	
5	19.7	1.34	4.13	1.8	4.48	1.96	....	....	....	....	....	....	....	....
10	24.7	1.68	7.57	3.3	8.21	3.58	....	....	....	....	....	....	....	....
15	29.7	2.02	10.31	4.5	11.4	5.0	....	....	....	....	....	....	....	....
20	34.7	2.36	12.62	5.5	14.3	6.2	....	....	....	....	....	....	....	....
25	39.7	2.70	14.68	6.4	16.9	7.4	....	....	....	....	....	....	....	....
30	44.7	3.04	16.30	7.1	19.2	8.4	....	....	....	....	....	....	....	....
35	49.7	3.38	17.90	7.8	21.4	9.3	....	....	....	....	....	....	....	....
40	54.7	3.72	19.28	8.4	23.4	10.2	....	....	....	....	....	....	....	....
45	59.7	4.06	20.65	9.0	25.2	11.0	....	....	....	....	....	....	....	....
50	64.7	4.40	21.80	9.5	27.0	11.8	....	....	....	....	....	....	....	....
55	69.7	4.74	22.95	10.0	28.7	12.6	....	....	....	....	....	....	....	....
60	74.7	5.08	23.90	10.4	30.3	13.3	....	....	....	....	....	....	....	....
65	79.7	5.42	24.80	10.8	31.9	13.9	....	....	....	....	....	....	....	....
70	84.7	5.76	25.70	11.2	33.3	14.6	20.6	29.2	12.8	....	....	....	....	12.3
75	89.7	6.10	26.62	11.6	34.7	15.2	21.6	30.3	13.3	....	....	....	....	12.5
80	94.7	6.44	27.52	12.0	36.0	15.7	22.7	31.3	13.7	....	....	....	....	12.7
85	99.7	6.78	28.21	12.3	37.3	16.3	23.6	32.3	14.1	....	....	....	....	13.5
90	104.7	7.12	28.93	12.6	38.6	16.9	24.5	33.2	14.5	....	....	....	....	14.2
95	109.7	7.46	29.60	12.9	39.8	17.4	25.5	34.2	14.9	....	....	....	....	14.4
100	114.7	7.80	30.30	13.2	40.9	17.9	26.3	35.0	15.3	....	....	....	....	14.5
110	124.7	8.48	31.42	13.7	43.2	18.9	28.1	36.7	16.1	....	....	....	....	14.8
120	134.7	9.16	32.60	14.2	45.2	19.8	29.8	38.3	16.8	....	....	....	....	15.1
130	144.7	9.84	33.75	14.7	47.2	20.7	31.5	39.6	17.3	....	....	....	....	16.4
140	154.7	10.52	34.67	15.1	49.2	21.5	32.9	40.8	17.9	....	....	....	....	16.7
150	164.7	11.20	35.59	15.5	51.0	22.3	34.5	42.3	18.5	....	....	....	....	17.1
160	174.7	11.88	36.30	15.8	....	....	36.1	43.6	19.0	....	....	....	....	....
170	184.7	12.56	37.20	16.2	....	....	37.3	44.7	19.5	....	....	....	....	....
180	194.7	13.24	38.10	16.6	....	....	38.8	45.8	20.0	....	....	....	....	....
190	204.7	13.92	38.80	16.9	....	....	40.1	46.8	20.4	....	....	....	....	....
200	214.7	14.60	39.50	17.2	....	....	41.4	47.8	20.9	....	....	....	....	....
250	264.7	18.00	42.70	18.6	....	....	47.6	52.5	22.7	23.8	86.3	48.95	21.3	6.2
300	314.7	21.40	45.30	19.7	....	....	53.4	56.5	24.5	26.1	98.6	52.34	22.8	6.9
350	364.7	24.81	47.30	20.6	....	....	58.5	59.6	26.1	28.2	130.4	55.27	24.1	7.7
400	414.7	28.21	49.20	21.4	....	....	63.3	62.7	27.4	30.1	131.6	57.85	25.2	8.0
450	464.7	31.61	51.20	22.3	....	....	67.8	65.3	28.6	31.8	132.2	60.17	26.2	8.4
500	514.7	35.01	52.70	22.9	....	....	72.1	67.8	29.7	33.4	142.6	62.28	27.2	8.5
550	564.7	38.41	53.75	23.4	....	....	76.3	70.0	30.7	34.9	152.6	64.20	28.0	8.8
600	614.7	41.81	54.85	23.9	....	....	80.5	72.3	31.9	36.3	162.4	65.97	28.8	9.1
650	664.7	45.23	56.02	24.4	....	....	....	....	....	37.7	172.0	67.64	29.6	....
700	714.7	48.63	57.08	24.9	....	....	....	....	....	38.9	181.1	69.18	30.2	....
750	764.7	52.03	58.08	25.3	....	....	....	....	....	40.2	191.1	70.61	30.8	....
800	814.7	55.44	59.01	25.7	....	....	....	....	....	41.3	199.0	71.98	31.4	....
850	864.7	58.84	59.88	26.1	....	....	....	....	....	42.5	207.6	73.27	32.0	....
900	914.7	62.21	60.71	26.5	....	....	....	....	....	43.5	216.1	74.49	32.5	....
950	964.7	65.64	61.49	26.8	....	....	....	....	....	44.6	224.5	75.67	33.0	....
1000	1014.7	69.05	62.23	27.2	....	....	....	....	....	45.6	232.6	76.78	33.5	....
1100	1114.7	75.85	63.62	27.8	....	....	....	....	....	47.5	248.6	78.86	34.4	....
1200	1214.7	82.66	64.88	28.3	....	....	....	....	....	49.3	264.2	80.78	35.2	....
1300	1314.7	89.46	66.04	28.8	....	....	....	....	....	51.0	279.2	82.57	36.0	....
1400	1414.7	96.23	67.12	29.3	....	....	....	....	....	52.7	294.0	84.23	36.7	....
1500	1514.7	103.07	68.12	29.7	....	....	....	....	....	54.2	308.4	85.79	37.4	....

*Based on a value for n of 1.3947.

A.12 n VALUE AND PROPERTIES OF VARIOUS GASES AT 60°F. and 14.7 P.S.I.A.

Gas	Symbol	C_p/C_v = n	Sp. gr. Air = 1.00	Mol. wt.	Lb. per cu. ft.	Cu. ft. per lb.	Boiling point at atmos. press., deg. F	Crit. temp., deg. F	Crit. pressure, lb. abs.
Acetylene	C_2H_2	1.30	0.91	26.02	0.07	14.53	—118	96	910
Air		1.41	1.00	28.98	0.08	13.06	—317	—221	546
Ammonia	NH_3	1.32	0.60	17.03	0.05	22.18	— 28	270	1,638
Argon	A	1.67	1.38	39.94	0.11	0.46	—302	—187	705
Benzene	C_6H_6	1.08	2.69	78.05	0.21	4.85	176	551	700
Butane	C_4H_{10}	1.11	2.07	58.08	0.15	6.51	31	307	528
Butylene	C_4H_8	1.11	1.94	56.06	0.15	6.75	21	291	621
Carbon dioxide	CO_2	1.30	1.53	44.00	0.12	8.59	—109	88	1,072
Carbon disulfide	CS_2	1.20	2.63	76.12	0.20	4.97	115	523	1,116
Carbon monoxide	CO	1.40	0.97	28.00	0.07	13.50	—313	—218	514
Carbon tetrachloride	CCl_4	1.18	5.33	153.83	0.41	2.46	170	541	661
Carbureted water gas		1.35	0.41						
Chlorine	Cl_2	1.33	2.49	70.91	0.19	5.33	— 30	291	118
Dichloromethane	CH_2Cl_2	1.18	3.01	84.93	0.22	4.46	105	421	1,490
Ethane	C_2H_6	1.22	1.05	30.05	0.08	12.59	—127	90	717
Ethyl chloride	C_2H_5Cl	1.13	2.37	64.50	0.17	5.87	54	370	764
Ethylene	C_2H_4	1.22	0.97	28.03	0.07	13.50	—155	50	747
Flue gas		1.40							
Freon (F-12)	CCl_2F_2	1.13	4.52	120.91	0.32	3.13	— 21	233	580
Helium	He	1.66	0.14	4.00	0.01	94.51	—452	—450	33
Hexane	C_6H_{14}	1.08	2.74	86.11	0.23	4.39	156	454	433
Hexylene	C_6H_{12}		2.92	84.09	0.22	4.50			
Hydrogen	H_2	1.41	0.07	2.02	0.01	188.62	—423	—400	188
Hydrogen chloride	HCl	1.41	1.27	36.46	0.10	10.37	—121	124	1,198
Hydrogen sulfide	H_2S	1.30	1.19	34.08	0.09	11.10	— 75	212	1,306
Isobutane	C_4H_{10}	1.11	2.02	58.08	0.15	6.51	14	273	543
Isopentane	C_5H_{12}	1.07	2.50	72.09	0.19	5.25	82	370	483
Methane	CH_4	1.32	0.55	16.03	0.04	23.63	—258	—116	672
Methyl chloride	CH_3Cl	1.20	1.79	50.48	0.13	7.49	— 11	289	966
Naphthalene	$C_{10}H_8$		4.42	128.06	0.34	2.95			
Natural gas† (app. avg.)		1.27	0.67	19.46	0.05	19.45		— 80	670
Neon	Ne	1.64	0.70	20.18	0.05	18.78	—410	—380	389
Nitric oxide	NO	1.40	1.04	30.01	0.08	12.61	—240	—137	954
Nitrogen	N_2	1.41	0.97	28.02	0.07	13.46	—320	—232	492
Nitrous oxide	N_2O	1.31	1.53	44.02	0.12	8.60	—129	98	1,053
Oxygen	O_2	1.40	1.11	32.00	0.08	11.82	—297	—182	730
Pentane	C_5H_{12}	1.06	2.47	72.09	0.19	5.25	97	387	485
Phenol	C_6H_5OH		3.27	94.05	0.25	4.02	360	786	889
Propane	C_3H_8	1.15	1.56	44.06	0.12	8.59	— 48	204	632
Propylene	C_3H_6	1.15	1.45	42.05	0.11	9.00	— 52	198	661
Refinery gas† (app. avg.)		1.20							
Sulfur dioxide	SO_2	1.26	2.26	64.06	0.17	5.90	14	315	1,141
Water vapor (steam)	H_2O	1.33‡	0.62	18.02	0.05	21.00	212	706	3,206

†To obtain exact characteristics of natural gas and refinery gas, the exact constituents must be known.
‡This n value is given at 212°F. All others are at 60°F. Authorities differ slightly; hence all data are average results.

A.13 TEMPERATURE RISE FACTORS VS. COMPRESSION RATIO

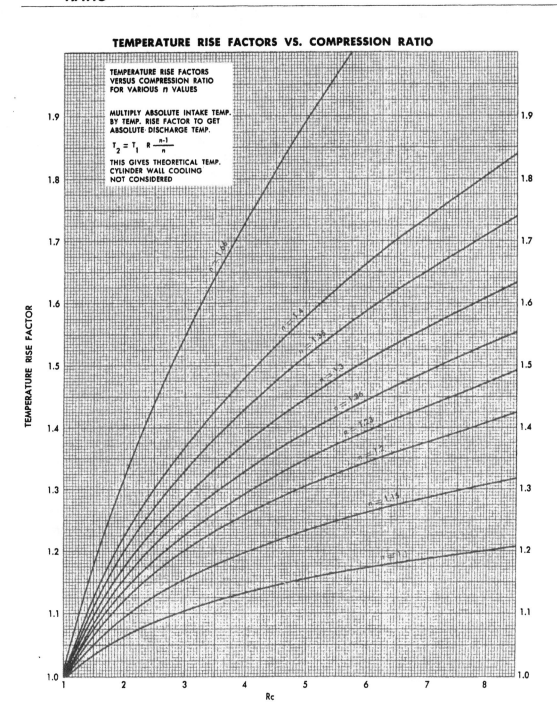

A.14 PROCEDURE FOR DETERMINING SIZE AND PERFORMANCE OF HORIZONTAL DOUBLE-ACTING SINGLE-STAGE COMPRESSORS FOR GAS OR AIR

Step 1. OBTAIN OPERATING CONDITIONS OF COMPRESSOR
(a) Suction pressure
(b) Discharge pressure
(c) Capacity required
(d) n value of gas to be compressed

Step 2. DETERMINE COMPRESSION RATIO
Divide absolute discharge pressure by absolute suction pressure. Must be less than 6.0. If greater than 6.0, two-stage compressor must be used. (Absolute pressure equals gauge pressure plus *atmospheric* pressure.)

Step 3. DETERMINE BRAKE HORSEPOWER REQUIRED PER MILLION CUBIC FEET OF GAS PER DAY
This is found by reference to Charts A.26, A.27, A.28, A.29, using n-value from Step 1 and compression ratio from Step 2.
NOTE—For conditions which require continuous operation at compression ratios less than (2), refer inquiry to the compressor manufacturer for calculation of exact horsepower requirements.

Step 4. CALCULATE TOTAL COMPRESSOR BRAKE HORSEPOWER REQUIRED
Multiply required capacity (expressed in million cubic feet per day) by brake horsepower required per million cubic feet (determined in Step 3).

Step 5. CHOOSE APPROPRIATE FRAME SIZE FOR COMPRESSOR
Refer to data sheets showing maximum allowable frame horsepower at full speed. The total brake horsepower found in Step 4 must be less than the rating of frame selected.

Step 6. ESTIMATE PROBABLE DISPLACEMENT REQUIRED
Divide capacity required (from Step 1) by suction pressure (expressed in number of atmospheres)* and by an average volumetric efficiency (usually chosen as 0.8).
*Number of atmospheres equals absolute suction pressure divided by base pressure, at which gas volume is to be measured.

Step 7. CHOOSE PROBABLE CYLINDER SIZE
Refer to cylinder sheet for frame size chosen in Step 5. Choose cylinder with displacement just larger than estimated displacement required (Step 6). Check maximum allowable pressures of cylinder chosen against operating pressures (Step 1). Do *not* exceed maximum allowable pressures.

Step 8. DETERMINE EXACT VOLUMETRIC EFFICIENCY FOR GAS BEING COMPRESSED
Refer to volumetric efficiency chart, for appropriate volumetric efficiency curve for cylinder chosen (Step 7). From volumetric efficiency curve, read exact value of volumetric efficiency at compression ratio (Step 2) and appropriate curve (above).

Step 9. DETERMINE "n" VALUE CORRECTION FACTOR*
Refer to curve for n value correction. Read correction factor at n value of gas (Step 1) and compression ratio (Step 2).

*This correction factor is used only when gas other than air is being compressed. No correction is necessary for air.

Step 10. DETERMINE THE CAPACITY OF THE COMPRESSOR
Multiply the displacement of chosen cylinder times the volumetric efficiency (Step 8) times the n value correction if needed (Step 9) times the number of atmospheres (Step 6).

Step 11. CALCULATE TOTAL COMPRESSOR BRAKE HORSEPOWER REQUIRED
Multiply the capacity (expressed in million cubic feet per day), determined in Step 10, times the brake horsepower per million cubic feet per day determined in Step 3. This brake horsepower must not exceed the maximum frame horsepower rating chosen in Step 5. Should the total brake horsepower required exceed the maximum frame horsepower, the next larger frame size will be used. New calculations must then be made starting with Step 5.

Step 12. CORRECTION FACTORS
The calculation procedure as outlined above is based on the assumption of standard conditions of the air or gas being compressed, that is, 14.7 psia base pressure for measurement of gas volume and 60°F gas inlet temperature. No corrections are necessary when using these standard conditions.

Should a base pressure other than 14.7 psia be used in determining number of atmospheres (Step 6), a correction must be applied to the total compressor brake horsepower required (Step 11). A ratio of the base pressure being used to the standard base pressure (14.7 psia), will provide the proper correction.

Example: Base pressure = 15.025 psia

Multiple total compressor brake horsepower required (Step 11) by 15.025/14.7.

Should an inlet temperature other than 60°F exist, proper correction of the compressor capacity (Step 10) must be made if volumes are to be expressed at standard conditions. A ratio of the absolute base temperature to the absolute inlet temperature, when multiplied by the compressor capacity, will convert the gas volume to standard temperature (60°F).

Absolute temperature = Temp. °F + 460.

Example: Gas inlet temperature = 80°F

Multiply capacity of compressor (Step 10) by (60 + 460)/(80 + 460) or 520/540.

NOTE: Gas inlet temperature does not affect total compressor brake horsepower required, therefore, this correction should be made *after* all other calculations and corrections have been made.

FINAL RESULTS OF CALCULATIONS

The complete operating performance of the compressor has not been determined. It is usually expressed as follows:

Suction pressure − PSIG...(Step 1)
Discharge pressure − PSIG..(Step 1)
Capacity − CFD or CFM ....(Step 10 or 12)
Base Conditions − PSIA___°F___ ....(Step 12)
Compressor BHP ....(Step 11)
Compressor speed − RPM...(Step 5)
n value of gas..(Step 1)
Suction temperature °F ....(Step 12)
Atmospheric pressure..(Step 12)

EXAMPLE:

Step 1. Suction pressure − 50 PSIG
Discharge pressure − 150 PSIG
Capacity − 500,000 CFD
n value − 1.26

Step 2. $CR = \dfrac{164.7}{64.7} = 2.54$

Step 3. Brake horsepower per million cubic feet = 56.3

Step 4. Total compressor brake horsepower = (.50)(56.3) = 28.15

Step 5. Use for example a frame rated at 35 brake horsepower maximum at 450 RPM

Step 6. Approximate displacement $= \dfrac{500{,}000}{(.8)(4.4)} = 142{,}000$ CFD

Number of atmospheres $= \dfrac{64.7}{14.7} = 4.4$

Step 7. Use for example 6″ cylinder – Displacement = 144,000 CFD
Maximum pressures:
Suction 125 PSIG
Discharge – 200 PSIG

Step 8. Volumetric efficiency = .878

Step 9. n value correction factor = .974

Step 10. Capacity = 144,000(.878)(.974)(4.4) = 541,000 CFD

Step 11. Total compressor brake horsepower = (.541)(56.3) = 30.5

FINAL RESULTS:

Suction pressure – 50 PSIG
Discharge pressure – 150 PSIG
Capacity – 541,000 CFD
Compressor BHP – 30.5
Compressor RPM – 450
n value – 1.26
Suction temperature – 60°F
Atmospheric pressure – 14.7 PSIA

A.15 PROCEDURE FOR DETERMINING SIZE AND PERFORMANCE OF HORIZONTAL DOUBLE-ACTING TWO-STAGE COMPRESSORS FOR GAS OR AIR

Step 1. OBTAIN OPERATING CONDITIONS OF COMPRESSOR
(a) Suction pressure
(b) Discharge pressure
(c) Capacity required
(d) n value of gas to be compressed

Step 2. DETERMINE COMPRESSION RATIO
Divide absolute discharge pressure by absolute suction pressure. (Absolute pressure equals gauge pressure plus atmospheric pressure.)

Step 3. DETERMINE COMPRESSION RATIO FOR EACH STAGE OF COMPRESSION (CALLED IDEAL COMPRESSION RATIO)
This is found by taking the square root of the compression ratio found in Step 2.

Step 4. **DETERMINE BRAKE HORSEPOWER REQUIRED PER MILLION CUBIC FEET OF GAS PER DAY**
This is found by reference to the Charts A.26, A.27, A.28 and A.29 using n value from Step 1 and compression ratio per stage from Step 3.

Step 5. **CALCULATE TOTAL BRAKE HORSEPOWER REQUIRED FOR EACH STAGE**
Multiply required capacity (expressed in million cubic feet per day) by brake horsepower required per million cubic feet (from Step 4).

Step 6. **CHOOSE APPROPRIATE FRAME SIZE FOR COMPRESSOR**
Refer to sheets showing maximum allowable frame horsepower at full speed. The total brake horsepower found in Step 5 must be less than rating of the frame selected.

Step 7. **ESTIMATE PROBABLE DISPLACEMENT REQUIRED BY LOW PRESSURE CYLINDER (FIRST STAGE)**
Divide capacity required (from Step 1) by suction pressure (expressed in number of atmospheres)* and by an average volumetric efficiency (usually chosen as 0.8)
*Number of atmospheres equals absolute suction pressure divided by base pressure at which gas volume is to be measured.

Step 8. **CHOOSE PROBABLE LOW PRESSURE CYLINDER SIZE**
Refer to cylinder sheet for frame size chosen in Step 6. Choose the low pressure cylinder with displacement just larger than estimated displacement required (Step 7) check suction pressure (Step 1) against maximum allowable suction pressure of cylinder chosen. Do *not* exceed maximum allowable pressures.

Step 9. **DETERMINE PROBABLE DISPLACEMENT OF HIGH PRESSURE CYLINDER (SECOND STAGE)**
Divide actual displacement of low pressure cylinder chosen (Step 8) by compression ratio per stage (Step 3).

Step 10. **CHOOSE HIGH PRESSURE CYLINDER SIZE**
Refer to cylinder sheet for frame size chosen in Step 6. Choose the high pressure cylinder with displacement just larger than probable displacement required (Step 9) check discharge pressure (Step 1) against maximum allowable discharge pressure of cylinder chosen. Do *not* exceed maximum allowable pressures.

Step 11. **DETERMINE MECHANICAL COMPRESSION RATIO BETWEEN HIGH AND LOW PRESSURE CYLINDERS**
Divide actual displacement of low pressure cylinder by the actual displacement of high pressure cylinder.

Step 12. **DETERMINE BRAKE HORSEPOWER PER MILLION CUBIC FEET PER DAY OF GAS**

From Charts A.26, A.27, A.28 and A.29, using mechanical compression ratio (Step 11) and N value (Step 1).

Step 13. **DETERMINE EXACT VOLUMETRIC EFFICIENCY FOR GAS BEING COMPRESSED.**
Refer to volumetric efficiency chart, for appropriate volumetric efficiency reference curve for low pressure cylinder chosen for low pressure cylinder chosen (Step 8). From volumetric efficiency curve, read exact value of volumetric efficiency at M.C.R. (Step 11) and appropriate curve (above).

Step 14. **DETERMINE n-VALUE CORRECTION FACTOR***
Refer to curve n value correction. Read correction factor at n value of gas (step 1) and M.C.R. (step 11).
*This correction factor is used only when gas other than air is being compressed. No correction is necessary for air.

Step 15. **DETERMINE THE CAPACITY OF THE COMPRESSOR**
Multiply the displacement of the low pressure cylinder (Step 8) times the volumetric efficiency (Step 12) times the n value correction, if needed, (Step 13) times the number of atmospheres (Step 7).

Step 16. **CALCULATE BRAKE HORSEPOWER REQUIRED BY LOW PRESSURE COMPRESSOR**
Multiply the capacity (expressed in million cubic feet per day) (Step 15) times the brake horsepower per million cubic feet per day (Step 12). This brake horsepower required should not exceed the maximum frame horsepower rating chosen in Step 6.

Step 17. **CALCULATE INTERSTAGE PRESSURE**
Multiply absolute suction pressure (Step 1) by the M.C.R. (Step 11) and subtract atmospheric pressure. This is interstage pressure in gauge pressure. This pressure must not exceed the maximum allowable discharge pressure of the L.P. cylinder (Step 8), of the maximum allowable suction pressure of the H.P. cylinder (Step 10).

Step 18. **DETERMINE COMPRESSION RATIO OF HIGH PRESSURE CYLINDER**
Divide absolute discharge pressure (Step 1) by absolute interstage pressure (Step 17).

Step 19. **DETERMINE BRAKE HORSEPOWER PER MILLION CUBIC FEET PER DAY FOR HIGH PRESSURE CYLINDER**
This is found by reference to Charts A.26, A.27, A.28 and A.29 using n value (Step 1) and C.R. for high pressure cylinder (Step 18).

Step 20. **CALCULATE BRAKE HORSEPOWER REQUIRED BY HIGH PRESSURE COMPRESSOR**
Multiply capacity (Step 15) in million cubic feet per day times brake horsepower per million cubic feet per day for high pressure cylinder

(Step 19). This brake horsepower must not exceed maximum allowable horsepower for frame chosen (Step 6).

Step 21. **TOTAL COMPRESSOR BRAKE HORSEPOWER**
This is the sum of low pressure brake horsepower required and high pressure brake horsepower required (Step 16 and 20).

RESULTS:

Suction pressure	Step 1
Discharge pressure	Step 1
Compressor BHP	Step 21
Capacity	Step 15
Compressor speed	Step 6
n value	1.26
Suction temperature	60°F
Atmospheric pressure	14.7 PSIA

Corrections for changes in atmospheric pressure and suction temperature are same as in single-stage calculations.

EXAMPLE:

Step 1. Suction pressure − 20 PSIG
Discharge pressure − 650 PSIG
Capacity − 330,000 CFD
n value − 1.26

Step 2. $CF = \dfrac{664.7}{34.7} = 19.15 \quad \sqrt{19.15}$

Step 3. CR per stage = 4.38 (ICR)

Step 4. Brake horsepower per million cubic feet − 91.4

Step 5. Total compressor BHP per stage = (.33)(91.4) = 30.1

Step 6. Use for example a frame rated at
35 BHP maximum at 450 RPM

Step 7. Approximate displacement (L.P.) = $\dfrac{330,000}{(.8)(2.36)}$ = 175,000 CFD

Number of atmospheres = $\dfrac{34.7}{14.7}$ = 2.36

Step 8. Use 7″ cylinder − low pressure
Displacement − 197,000 CFD
Maximum pressure:
  Suction − 150 PSIG
  Discharge − 150 PSIG

Step 9. Approximate displacement (H.P.) = $\dfrac{197,000}{4.38}$ = 45,000 CFD

Step 10. Use 3-½″ cylinder − high pressure
Displacement − 46,560 CFD
Maximum pressure:
  Suction − 600 PSIG
  Discharge − 800 PSIG

Step 11. MCR − $\dfrac{197,000}{46,560}$ − 4.23

Step 12. Brake horsepower per million cubic feet = 88.9

Step 13. Volumetric efficiency = .793

Step 14. n value correction factor = .954

Step 15. Capacity = (197,000)(.793)(.954)(2.36) = 352,000 CFD

Step 16. Brake horsepower (low pressure) = (.352)(88.9) = 31.3

Step 17. Interstage pressure = 34.7(4.23) = 146.8 PSIA
  146.8 − 14.7 = 132.1 PSIG

Step 18. Compression ratio (high pressure) = $\dfrac{664.7}{146.8}$ = 4.53

Step 19. Brake horsepower per million cubic feet = 94.0

Step 20. Brake horsepower (high pressure) = (.352)(94.0) = 33.1

Step 21. Total compressor brake horsepower = 31.3 + 33.1 = 64.4

FINAL RESULTS:

Suction pressure − 20 PSIG
Discharge pressure − 640 PSIG
Capacity − 352,000 CFD
Compressor brake horsepower − 64.4
Compressor RPM − 450
n value − 1.26
Suction temperature − 60°F
Atmospheric pressure − 14.7 PSIA

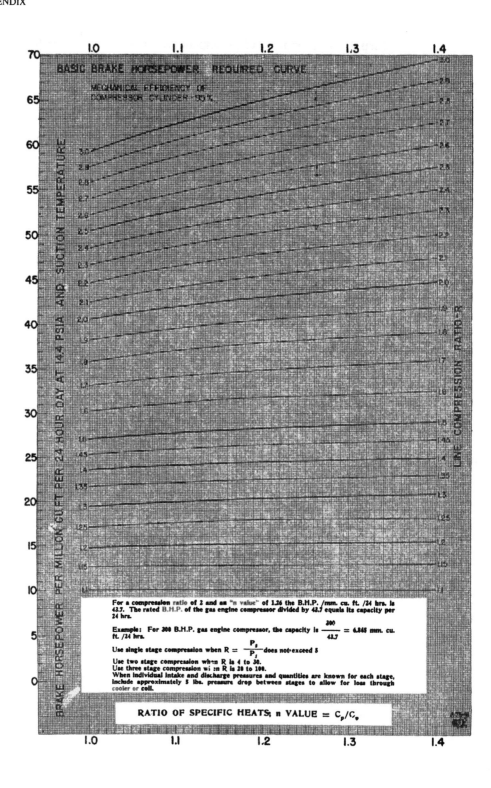

APPENDIX **A.27**

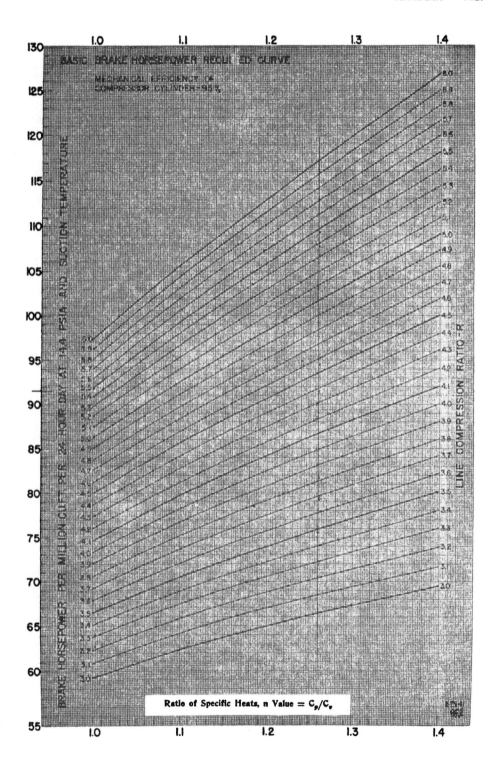

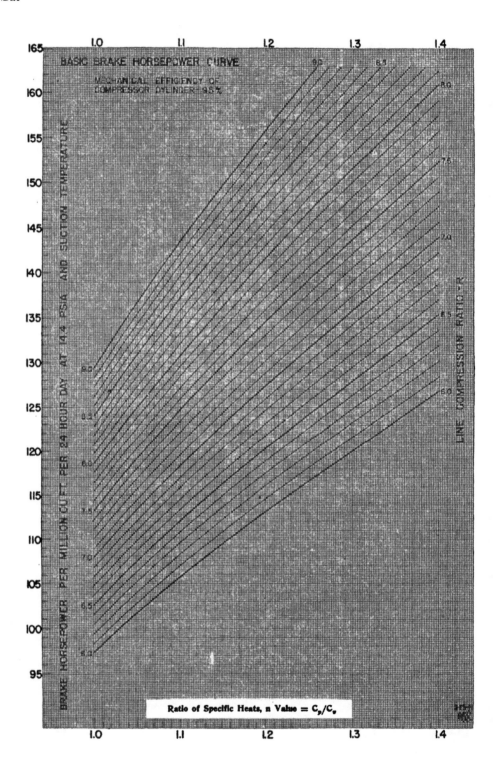

APPENDIX A.29

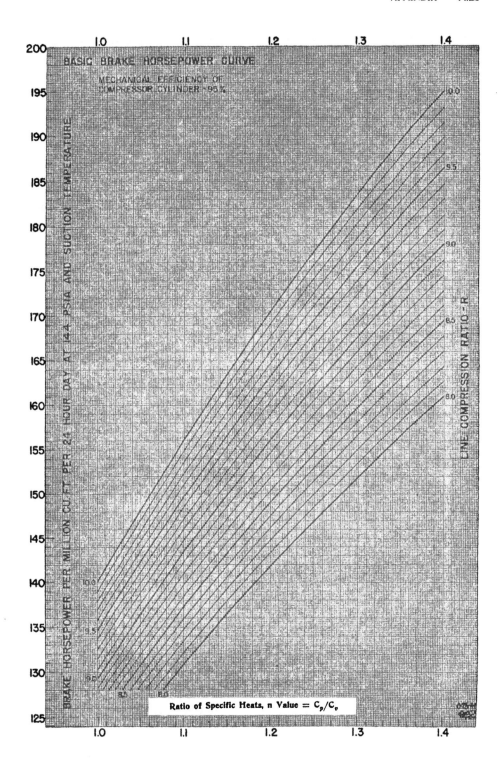

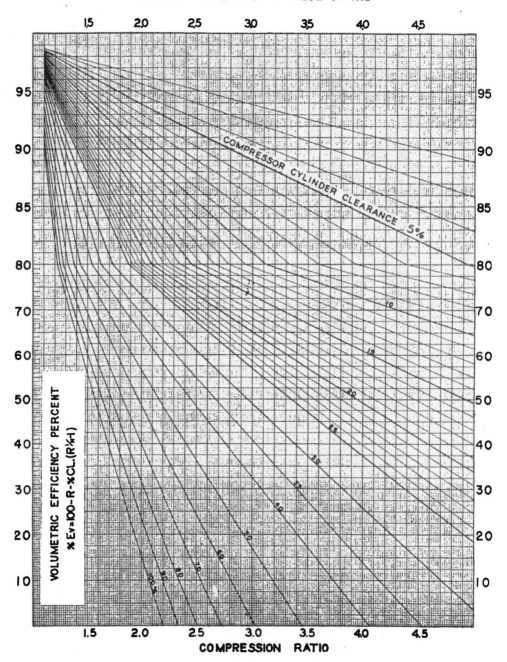

COMPRESSOR VOLUMETRIC EFFICIENCY CURVES
FOR GAS WITH K OR N VALUE OF 1.15

APPENDIX A.31

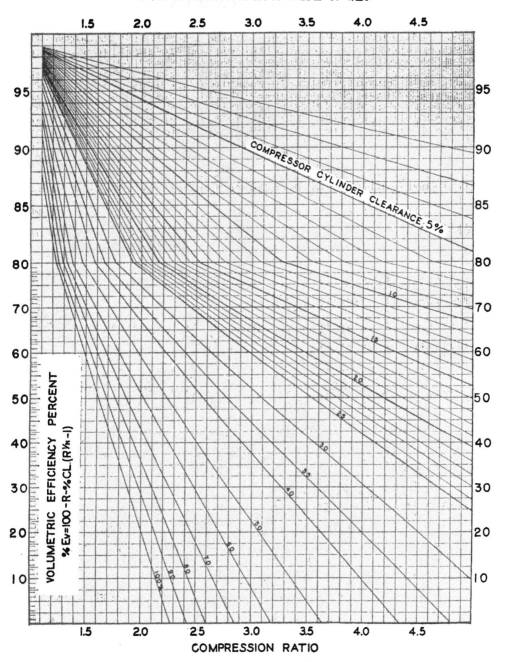

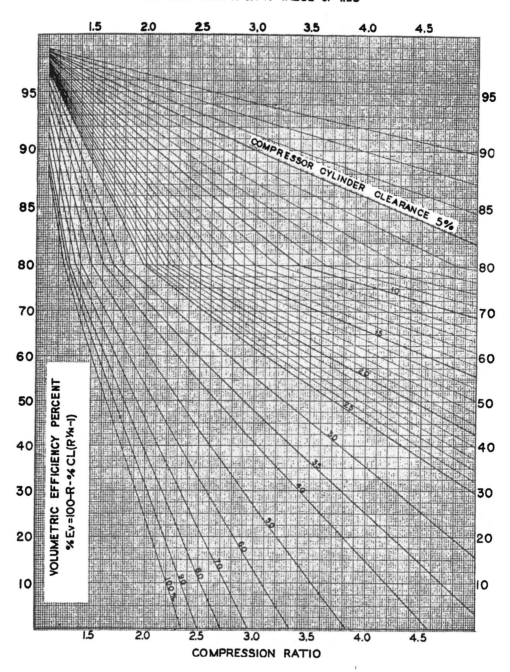

APPENDIX **A.33**

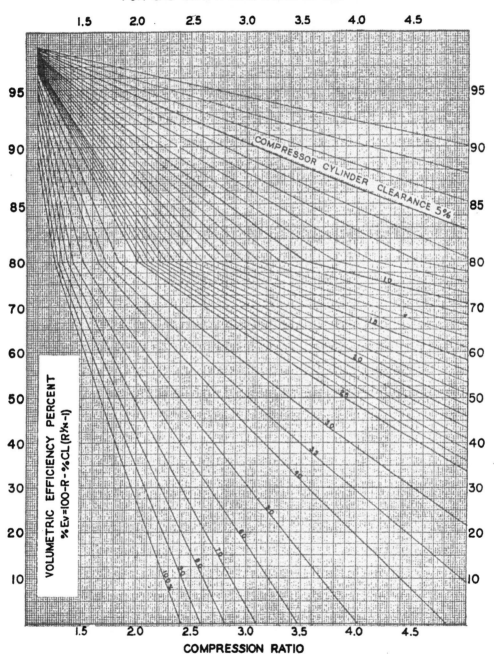

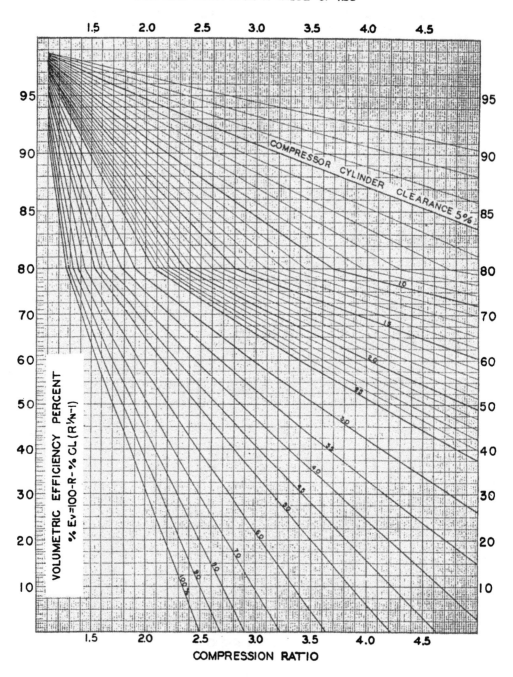

APPENDIX **A.35**

COMPRESSOR VOLUMETRIC EFFICIENCY CURVES
FOR GAS WITH K OR N VALUE OF 1.40

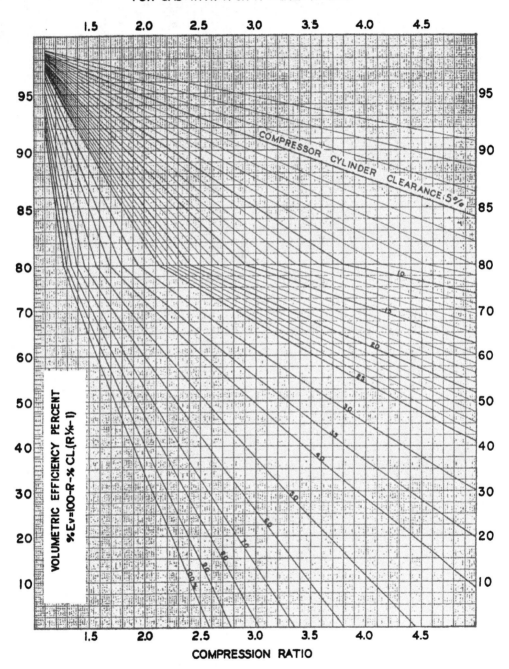

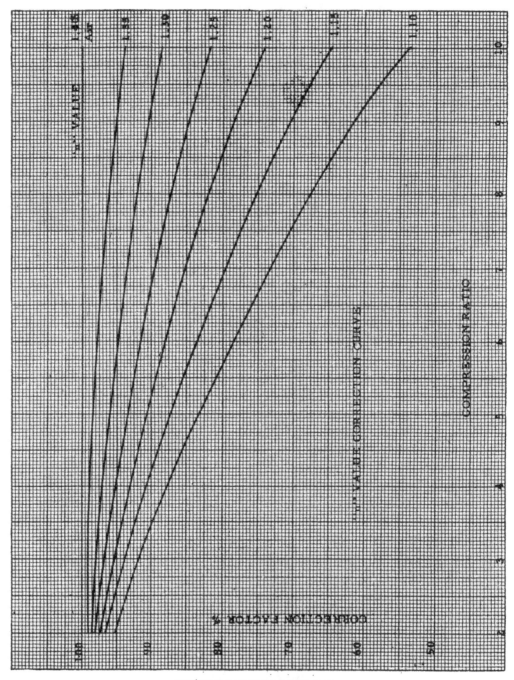

N VALUE CORRECTION CURVES

A.16 SINGLE-STAGE ADDED CYLINDER CLEARANCE

Discussion and Theory

In calculating air or gas compressor performance, certain conditions of compression ratio and cylinder size may cause a particular compressor to be overloaded; that is, the compressor brake horsepower will exceed the maximum frame horsepower allowable for the machine. In this case, the capacity of the compressor must be reduced in order to lower the required brake horsepower to within the frame rating. This is accomplished by adding clearance bottles to the cylinder, thereby reducing the volumetric efficiency of the compressor. As the volumetric efficiency is reduced, the capacity of the compressor and brake horsepower required are also lowered.

Volumetric efficiency may be defined as the ratio of the volume of gas actually discharged from the cylinder per unit of time to the total volume swept by the piston per unit of time. A reduction in the volume of gas discharged, therefore, will lower the volumetric efficiency of the compressor.

As the piston moves forward in the cylinder, the gas present is compressed and discharged from the cylinder. With clearance bottles present, a part of the gas is displaced into the bottles rather than being discharged. Then, as the piston begins its return stroke, the compressed gas expands from the clearance bottles into the cylinder. The force of the gas under pressure expanding against the piston reduces the amount of work required from the power unit to compress the gas on the opposite side of the piston on the return stroke.

In addition, the presence in the cylinder of the gas from the clearance bottles reduces the amount of gas that can be taken in from the suction line. The effect reduces the actual capacity of the compressor unit.

Calculations

(1c) Determine the maximum capacity of the compressor, not exceeding maximum horsepower allowable.

Multiply the total compressor capacity (Step 10 from single-stage calculations) by the maximum horsepower allowable for the compressor (cylinder date sheets) and divide by the total compressor brake horsepower required (Step 11 from single-stage calculations).

(2c) Calculate new volumetric efficiency for compressor that will reduce capacity and horsepower to frame rating.

Divide maximum capacity at frame rating (Step 1c) by cylinder displacement and by suction pressure (expressed in number of atmospheres).

(3c) Determine total clearance required.

This is found by reference to compressor volumetric efficiency curves. Use proper curve based on n value of gas being compressed (Step 1). At intersection of compression ratio (Step 2) and new volumetric efficiency (Step 2c), read total cylinder clearance required.

(4c) Determine amount of clearance to be added.

From total clearance required (Step 3c) subtract the inherent cylinder clearance. The average inherent clearance in a cylinder is 5%.

With the amount of clearance found in Step 4c added to the cylinder, the new capacity of the compressor will be that found in Step 1c, and the brake horsepower required will be equal to the maximum allowable frame horsepower.

Example Problem

Typical performance data for a 5 × 11 compressor:

Suction pressure − PSIG	300
Discharge pressure − PSIG	700
Capacity − CFD	2,050,000
Compressor BHP required	103
Compressor speed − RPM	327
n value of gas	1.26

Because the maximum horsepower allowable on this 11″ compressor is 100, clearance must be added in order to reduce brake horsepower required to stay within the frame rating.

Procedure is as follows:

(1c) Maximum capacity @ 100 HP

$$= \frac{\text{compressor capacity} \times \text{maximum HP allowable}}{\text{compressor BHP required}}$$

$$= \frac{2,050,000(100)}{103}$$

$$= 1,990,000 \text{ CFD}$$

(2c) New volumetric efficiency = $\dfrac{\text{maximum capacity @ 100HP}}{\text{cylinder displacement} \times \text{number of atmospheres}}$

$= \dfrac{1{,}990{,}000}{108{,}000\ (21.4)}$

$= .86$

(3c) Total clearance required = 12.7%

Compression ratio = $\dfrac{714.7}{314.7} = 2.27$

(4c) Amount of clearance to add =   12.7%
                                  − 5.0%
                                  ──────
                                    7.7%

INDEX

Absolute pressure, **A.**2
Absolute temperature, **A.**2
Adiabatic compression, **A.**2
Air padding, **9.**2
Aluminakyle, **7.**3
Ammonia synthesis, **3.**73, **7.**1
Antisurge protection, **3.**82, **4.**43
Areas and circumferences of circles, **A.**7
Asphalt production, **3.**73
Atmospheric pressure and barometric readings at altitudes, **A.**10
Autofrettage, **7.**11, **7.**15, **7.**16, **7.**19, **7.**23, **7.**35, **7.**37

Barrel compressor (*see* Vertically split compressor)
Bearing:
  circular, **19.**31, **19.**63
  compliant surface, **19.**16, **19.**75, **19.**94
  cryogenic, **19.**104
  damper, **3.**55
  elastohydrodynamic, **19.**9
  elliptical, **19.**33, **19.**66
  film thickness, **19.**28, **19.**78, **19.**130
  flow, **19.**3
  fluid film, **19.**9
  gas, **19.**62
  hydrostatic, **19.**15, **19.**67, **19.**112, **19.**140
  journal, **3.**70, **19.**28, **19.**74, **19.**132
  liner, **4.**43
  materials, **19.**122
  magnetic, **4.**43, **19.**9, **19.**21, **19.**93
  multipad, **19.**82
  rolling element, **19.**18
  tapered land, **19.**46, **19.**139
  temperature, **19.**30
  three-lobe, **19.**37, **19.**66
  thrust, **3.**70, **19.**45, **19.**67, **19.**85, **19.**109, **19.**137
  tilt pad, **4.**43, **19.**42, **19.**54, **19.**133

Benedict-Webb-Rubin-Starling model, **3.**29

Capacity, **1.**10, **2.**12, **5.**31, **A.**3
Capacity control:
  bypass, **2.**6, **20.**24, **21.**12
  clearance pocket, **2.**7, **20.**21, **A.**37
  finger unloader, **2.**7, **20.**26, **21.**12
  port unloader, **2.**7, **20.**25
  screw compressor slide, **2.**6
  variable speed, **2.**6, **20.**24, **21.**12
Centrifugal compressor:
  balance piston, **3.**2, **3.**5, **3.**66, **4.**11
  casing, **3.**2, **3.**60, **4.**1
  coupling, **3.**67
  diaphragm, **3.**2, **4.**3
  diffuser, **3.**40, **4.**10, **4.**23
  discharge nozzle, **4.**10
  discharge volute, **3.**6, **4.**12
  electrical system, **8.**8
  foundations, **22.**1
  guide vane, **4.**10
  impeller, **3.**2, **3.**58, **3.**62, **4.**3
  intercooling, **4.**4
  inlet nozzle, **3.**3, **4.**2
  inlet volute, **3.**3
  multistage, **3.**29, **19.**7
  off design operation, **4.**25
  oil system, **3.**7
  performance, **3.**28, **4.**16, **4.**51, **4.**63
  slope, **4.**16
  splitter vane, **4.**10
  thrust bearing, **3.**3

Choking, **3.**26, **4.**16, **4.**21
Clearance volume, **1.**9, **1.**11, **2.**1
CNG codes and standards, **8.**19
CNG station, **8.**8, **8.**14

CNG pressure vessel, **8**.12, **8**.18
CNG compressor:
   blow down gas recovery, **8**.5
   crankcase, **8**.4
   design, **8**.1, **8**.7
   lubrication, **8**.6
   sealing, **8**.3
CNG dispenser, **8**.10
CNG fill system, **8**.14
Coatings:
   fluorocarbon, **4**.6
   nickel, **4**.6

Compressibility, **1**.3, **1**.10
Compressibility factor, **A**.2
Compression ratio, **1**.7, **5**.30
Conversion factors, **A**.5
Critical speed, **3**.47, **4**.28
Cracking, **3**.73
Crosshead:
   auxiliary, **7**.5, **7**.6, **7**.7
   connection, **7**.13
   guide, **7**.6
   pin load reversal, **2**.3, **5**.15
Cylinder:
   autofrettage, **7**.12
   construction, **7**.35
   elastic simulation, **7**.27
   fatigue tests, **7**.19
   finish, **7**.35
   heavy walled, **7**.13
   hypercompressor, **7**.8
   materials, **7**.16, **7**.35
   stress distribution, **7**.14
   tie rods, **7**.8, **7**.27

Damped systems, **4**.31, **4**.39
Dewhirl vanes, **4**.41
Diaphragm compressor:
   accessories, **15**.10
   applications, **15**.11
   cleaning and testing, **15**.11
   design, **15**.4
   head assembly, **15**.6
   limitations, **15**.12
   materials, **15**.9
   operation, **15**.1
   pressures, **15**.3
Discharge of air through an orifice, **A**.11

Displacement, **A**.3
Double flow compressor, **4**.72

Effective head, **3**.15
Efficiency:
   compression (adiabatic), **A**.4
   compressor (polytropic, hydraulic, stage), **3**.16, **3**.19, **A**.4
   delivery, **20**.10
   isentropic, **1**.7, **1**.15, **3**.16, **20**.10
   mechanical, **A**.4
   volumetric, **1**.11, **1**.14, **5**.31, **6**.9, **10**.11, **20**.10, **A**.4, **A**.30
Emissions control, **16**.2
Energy equation, **1**.6
Entering sleeve, **17**.10
Euler equation, **3**.17

Finite element method, **7**.25, **7**.32
Fixed clearance, **1**.9
Flow:
   coefficient, **1**.6, **3**.18, **3**.22, **3**.30
   subsonic, **1**.7
Foundations:
   anchor bolts, **22**.6
   grout, **22**.9
   materials, **22**.6
   pile cap, **22**.3
   reinforcing, **22**.7
   repair, **22**.10
   skid mounted/packaged, **22**.2, **23**.2
   soil frequency/vibration, **22**.5
   table top, **22**.4
   types, **22**.1

Frame load, **2**.2
Free air, **A**.3
Friction coefficient, **17**.33, **19**.30, **19**.130

Gas sampling, **4**.49
Gas booster:
   applications, **11**.1
   construction, **11**.2
   cooling, **11**.6
   drive, **11**.3
   flow chart, **11**.14
   pressure ratio, **11**.9
   storage tank, **11**.15
   valves, **11**.4

Gas laws, **1.**2
Gas reinjection, **3.**78
Gas sampling, **4.**49
Gas transportation, **3.**79

Hans Hoerbiger, **20.**1
Hans Mayer, **20.**1
Heat transfer, **2.**17, **2.**21, **2.**24, **4.**55, **17.**33
Horizontally split compressor, **3.**7, **3.**58
Horsepower:
   air indicated, **A.**3
   brake (shaft), **A.**3, **A.**26
   theoretical (polytropic), **1.**9, **A.**3, **A.**15
Hypercompressor, **7.**1
Hyper packing, **17.**23

Impeller:
   backward leaning, **3.**23, **4.**19
   discharge section, **3.**38
   forward leaning, **3.**23, **4.**11
   inlet section, **3.**37
   manufacturing, **3.**62
   overhung, **4.**3, **19.**6
   radial, **3.**23, **4.**19
   thrust, **3.**3, **4.**14, **4.**17
Incidence, **4.**23
Indicator card, **1.**9
Intercooling, **A.**3
Isentropic compression (*see* Adiabatic compression)
Isentropic head, **3.**15
Isentropic temperature exponent, **1.**4
Isentropic volume exponent, **1.**4, **1.**10
Isothermal compression, **A.**2

Liquids:
   effect in cylinders, **5.**2
   in gas stream, **4.**48, **4.**66
Liquefaction, **3.**79
Load factor, **A.**4
Loss of air pressure due to pipe friction, **A.**12
Loss of pressure through pipe fittings, **A.**14
Lubricant production, **3.**73
Lubrication:
   additives, **18.**5, **19.**126
   feed rate, **18.**8, **7.**37
   gas absorption, **18.**5
   hydrocarbon dilution, **18.**5
   low flow sensor, **21.**23
   lubricators, **18.**11

   oil ring, **19.**57, **19.**135, **19.**141
   pre-post lube, **21.**12
   removal, **18.**7
   synthetic, **18.**10, **19.**123
   viscosity, **18.**3, **19.**125

Mach number, **3.**18
Methanol synthesis, **3.**76
Modeling (*see* Simulation)
Mollier chart, **1.**5, **3.**14
Myhlestad-Prohl calculation, **3.**51

n Value and properties of gases, **A.**16, **A.**36
Non-lube compressor, **17.**8, **18.**8

Oxygen compression, **3.**80

Packaging compressors:
   base design, **23.**2
   cooler design, **23.**10
   line sizing, **23.**5
   pressure relief valve, **23.**6
   pulsation bottle design, **23.**7
   scrubber design, **23.**2
Packing:
   breaker rings, **7.**44, **17.**4
   cooling, **17.**30
   cup stress, **7.**20
   distance piece venting, **17.**18
   emissions control, **17.**20, **21.**19
   friction, **17.**29, **17.**33
   heat generation, **17.**32
   high pressure, **7.**43, **17.**22
   leakage, **2.**19, **2.**22, **17.**13, **17.**18, **18.**2
   lubrication, **8.**6, **17.**8, **18.**8
   partition, **17.**3
   purged, **17.**21
   materials, **17.**7, **17.**29, **18.**8
   nomenclature, **17.**2
   rod size effects, **17.**11
   seal rings, **7.**44, **17.**5
   static sealing, **17.**20
   thermal effects, **17.**10
   wiper, **17.**3, **17.**22
Partial pressure of water vapor in air, **A.**9
Piping:
   acoustics, **6.**4
   flow straightener, **4.**67
   velocity profile, **4.**67

Pipeline compressor, **3**.11
Piston ring:
   friction, **17**.29
   leakage, **2**.20, **2**.22, **5**.2, **5**.7, **5**.25, **17**.26, **17**.28
   instantaneous pressure, **17**.27
   materials, **17**.28, **18**.9
   rider rings, **17**.25
Polyethylene:
   compressors, **7**.4
   high density (HDPE), **7**.3
   low density (LDPE), **7**.3
   safety aspects, **7**.12
Polymer buildup, **4**.5
Polytropic compression, **A**.2
Polytropic head, **3**.15
Pressure distribution, **7**.44, **17**.4, **17**.27
Pressure drop:
   contour, **3**.26
   endwall, **3**.26
   friction, **3**.23
   impact, **3**.26
   incidence, **3**.24
   mixing, **3**.26
   overall, **3**.25
Pressure ratio, **3**.29, **A**.2
Pressure/Time (PT) patterns, **5**.11
Pressure-volume (PV) diagram, **2**.19, **5**.3, **6**.11, **20**.12
Procedure for determining compressor size, **A**.18
Pulsation, **2**.16, **2**.21, **2**.24, **5**.9, **6**.3, **13**.2, **23**.7
Purging:
   nitrogen, **7**.13

Redlich-Kwong equation of state, **1**.3
Reciprocating compressor:
   capacity control, **21**.12
   cooling, **21**.24
   crankcase lubrication, **18**.2
   cylinder lubrication, **18**.2, **21**.9, **21**.25, **23**.11
   electrical controls, **21**.4, **21**.9, **23**.11
   foundations, **22**.1
   loading/unloading, **21**.12, **21**.13
   manual controls, **21**.11
   monitoring, **21**.1, **21**.17
   pneumatic controls, **21**.10, **23**.14
   rod drop, **21**.20
   sensors, **21**.16, **21**.21, **21**.23
Reforming, **3**.72
Regasification, **3**.79

Reynolds number, **3**.19, **19**.13, **19**.115, **19**.148
Rod load, **2**.2, **5**.17, **5**.32
Rotor dynamics, **3**.50, **4**.27, **19**.103
Rotor balancing, **3**.58, **4**.28
Rotary screw compressor:
   adiabatic efficiency, **14**.8
   advantages, **14**.5
   applications, **14**.6
   helical rotors, **14**.3
   lubrication, **18**.1
   packaging, **23**.15
   sizing, **14**.7

Screw compressor:
   heat transfer, **2**.24
   leakage, **2**.25
   port losses, **2**.23
   pulsation, **2**.24
Scroll compressor:
   advantages, **12**.3
   construction, **12**.4
   lubricated, **12**.5
   oilless, **12**.7
   principal of operation, **12**.2
Seals:
   balance piston, **4**.11
   bellows, **16**.7
   bushing, **16**.4
   carbon ring, **16**.3
   circumferential, **16**.5
   double, **16**.12
   honey comb, **3**.56, **4**.41
   interstage, **4**.6
   labyrinth, **3**.3, **3**.56, **3**.68, **4**.6, **4**.40, **16**.3
   leakage, **4**.55
   mechanical, **3**.69
   oil, **3**.69
   performance, **16**.10
   rotary contacting, **16**.1
   rotary non contacting, **16**.2
   sleeve, **4**.43
   sliding, **7**.36
   tandem, **16**.9
   tip, **12**.5
Simulation:
   buried piping, **6**.15
   dynamic fluid transient systems, **6**.3
   pulsation, **6**.3

Simulation (*cont.*):
  static systems, **6.**2
  stress, **6.**3, **6.**9
  thermal flexibility, **6.**12
  vibration, **6.**3
Sommerfield number, **19.**29, **19.**130
Speed triangle (*see* Velocity vector diagram)
Specific heat ratio, **3.**19
Spring:
  valve, **5.**10, **20.**17, **20.**20
Stall, **4.**24
Standard air, **A.**3
Steam properties, **A.**8
Stonewall (*see* Choking)
Straight lobe compressor:
  bearings, **13.**7
  construction, **13.**3
  installation, **13.**8
  noise, **13.**2
  operating principle, **13.**1
  pulsation, **13.**2
  PV diagram, **13.**4
  seals, **13.**7
Strain gage method, **7.**32
Stress:
  axial, **7.**16, **7.**24
  circumferential, **7.**16, **7.**24
  factor of safety, **7.**17, **7.**19
  radial, **7.**16, **7.**24
  risers, **7.**17
  Von Mises, **7.**18
Supercompressibility, **1.**3, **21.**3
Surge, **3.**26, **4.**16, **4.**22, **4.**43, **21.**19
Swirl (*see* Dewhirl)

Tank car unloading:
  compressor, **9.**4
  pump, **9.**1
  vapor recovery, **9.**6
Temperature conversion chart, **A.**6
Temperature rise vs compression ratio, **A.**17
Temperature-rise ratio, **A.**4
Testing centrifugal compressors:
  economizer nozzles, **4.**55
  efficiency, **4.**60
  field data analysis, **4.**62
  instrumentation, **4.**50

  internal temperatures, **4.**57
  iso-cooled compressors, **4.**54
  power, **4.**59
  thermodynamic performance, **3.**4
Trouble shooting, **4.**63
Tungsten carbide, **7.**4, **7.**15
Turbocompressor, **3.**13
Turbomachinery, **3.**12
Turbulence, **19.**13, **19.**115, **19.**148

Urea production, **3.**77, **7.**1

Vacuum remelted steel, **7.**17
Valve:
  active, **20.**27
  area, **20.**14
  axial, **7.**9, **7.**30
  channel, **20.**1, **20.**5
  combined suction/discharge, **7.**12
  concentric, **20.**8
  deck-and-one-half, **20.**7
  double deck, **20.**7
  dynamics, **2.**12, **6.**10
  equivalent area, **1.**12, **20.**16
  feather, **20.**1, **20.**6
  leakage, **5.**4, **5.**14, **5.**25
  lift, **20.**11, **20.**13
  losses, **1.**11, **2.**15, **2.**18, **5.**10, **5.**30
  materials, **20.**19
  motion, **20.**18
  poppet, **7.**9, **7.**30, **20.**1, **20.**6
  ported plate, **20.**4
  radial, **7.**8
  reed, **20.**8
  ring, **20.**2
Vehicle fueling:
  appliance, **10.**1
  compressor, **10.**6, **10.**13
Vertically-split compressor, **3.**8, **3.**59
Velocity vector diagram, **3.**20, **3.**38, **3.**39, **4.**16
Vibration analysis, **3.**51, **5.**19
Viscosity, fluid, **19.**12, **19.**30

Water injection, **4.**6
Wear:
  detector, **21.**20, **21.**23
  distribution, **7.**46

ABOUT THE EDITOR

Paul C. Hanlon is manager of product design with C. Lee Cook in Louisville, Kentucky. A mechanical engineering graduate of the University of Cincinnati, he has worked for over 40 years in the design, application, and troubleshooting of seals for engines, compressors, and other major equipment used throughout the chemical, oil, and gas-processing industries. Mr. Hanlon is also the author of numerous articles for leading technical journals.

Printed in the USA
CPSIA information can be obtained
at www.ICGtesting.com
JSHW050758210824
68268JS00005B/128